574.192

LIPIDS
Chemistry, Biochemistry, and Nutrition

LIPIDS
Chemistry, Biochemistry, and Nutrition

James F. Mead
Emeritus Professor of Biological Chemistry and Nutrition
Schools of Medicine and Public Health
University of California
Los Angeles, California

Roslyn B. Alfin-Slater
Professor of Nutrition and Biological Chemistry
Schools of Public Health and Medicine
University of California
Los Angeles, California

David R. Howton
Professor of Allied Medical Sciences
John A. Burns School of Medicine
University of Hawaii at Manoa
Honolulu, Hawaii

and

George Popják
Emeritus Professor of Biological Chemistry
School of Medicine
University of California
Los Angeles, California

PLENUM PRESS • NEW YORK AND LONDON

Library of Congress Cataloging in Publication Data

Main entry under title:

Lipids.

Includes bibliographies and index.
1. Lipids. 2. Lipids—Metabolism. 3. Lipids in nutrition. I. Mead, James F. (James Franklyn) [DNLM: 1. Lipids. QU 85 L7646]
QP751.L556   1985   574.19′247   85-19304
ISBN 0-306-41990-4

© 1986 Plenum Press, New York
A Division of Plenum Publishing Corporation
233 Spring Street, New York, N.Y. 10013

All rights reserved

No part of this book may be reproduced, stored in a retrieval system, or transmitted in any form or by any means, electronic, mechanical, photocopying, microfilming, recording, or otherwise, without written permission from the Publisher

Printed in the United States of America

PREFACE

For a number of years, we have been teaching a course on the chemical, metabolic, and nutritional aspects of lipids to graduate and postdoctoral students in the health sciences at the University of California, Los Angeles. The course was taught in an "open-ended" manner—that is, after each subject was introduced, a discussion was continued for as long as necessary to explore thoroughly its current status. It became evident that no text was available that could supplement the lectures adequately. This book is an attempt to remedy the deficiency and to treat extensively—although not exhaustively—most aspects of lipids, ranging from their physical and chemical properties, through biochemistry and metabolic pathways, to their role in nutrition. It is intended, as were the lectures, for those who have a basic knowledge of biochemistry, but wish to increase their knowledge of lipids. These may include graduate or advanced students, investigators preparing to enter research in lipids, clinicians, and, indeed, all those who may profit by an increased awareness of the importance of lipids.

Although most of the subjects treated are given adequate references, it was not our intention that this book would be a complete reference source. Rather, those publications cited are deemed by the authors of each chapter to be important and to lead to further reading. Inevitably, many more references are given for those subjects in which rapid advances are now being made than for those that are somewhat more static. In the former case, only very recent research papers can give the more complete story; in the latter, good current reviews are usually available. In a field in which explosive developments are occurring, important discoveries may have been made in the unavoidable interval between writing and publishing, but it is hoped that the knowledge imparted in each chapter may form a sufficient base for understanding new advances. This is the major purpose of this book and is the reason for the way in which the subjects are presented. In each chapter, the personal interests of the author will unavoidably be evident.

We wish to thank our many associates who have contributed their time, knowledge, and good wishes without any formal recognition—in particular Lilla Aftergood and Hasel Popják. It was Mrs. Popják who brought whatever uniformity in style may exist, and who painstakingly checked many references; for her services we are greatly appreciative.

It is a pleasure also to acknowledge the assistance of Larry Tabata, Judy Wong, and Susan Y. Anderson, who have prepared most of the original illustrations. We also thank the several colleagues who permitted our use of their illustrations.

Finally, it is hoped that this book will stimulate interest and suggest research in lipid chemistry, biochemistry, metabolism, and nutrition, as well as indicate possible relationships between lipids and certain diseases; and that it may lead to improved understanding of the properties of lipids and to an appreciation of the way in which they contribute to the properties of all living organisms.

James F. Mead
Roslyn B. Alfin-Slater
David R. Howton
George Popják

CONTENTS

CHAPTER 1
Introduction 1
CHAPTER 2
Solubility and the Definition of Lipids 5
CHAPTER 3
Nature of the Fatty Acids: The Carboxyl and Hydrocarbon Moieties . . . 23
CHAPTER 4
Fatty Acids: Crystals, Monomolecular Films, and Soaps 49
CHAPTER 5
Distribution of Fatty Acids in the Tissue Lipids 69
Appendix: A note on the structure and naming of the fatty acids 77
CHAPTER 6
Peroxidation of Fatty Acids 83
CHAPTER 7
Catabolism of the Fatty Acids 101
CHAPTER 8
Biosynthesis of Fatty Acids 115
CHAPTER 9
Desaturation of Fatty Acids—The Essential Fatty Acids 133
CHAPTER 10
Prostaglandins, Thromboxanes, and Prostacyclin 149
CHAPTER 11
Eicosanoids: Leukotrienes and Slow-Reacting Substances of
Anaphylaxis 217
CHAPTER 12
Digestion and Absorption of Lipids 255
CHAPTER 13
Transport and Reactions of Lipids in the Blood 273
CHAPTER 14
Triacylglycerol Metabolism and the Reactions of Adipose Tissue 285
CHAPTER 15
Biosynthesis of Cholesterol and Related Substances 295
CHAPTER 16
The Amphiphilic Lipids: Structure, Properties, and Conformation 369
CHAPTER 17
Phosphoglyceride Metabolism 405

CHAPTER 18
  Sphingolipid Metabolism 429
CHAPTER 19
  Nutritional Value of Lipids 459
INDEX 475

1
INTRODUCTION

1.1. Historical Background

It is very probable that lipids were the first of the major classes of edible substances to be prepared in a fairly pure state by early man. The primitive hunter, roasting his meat over an open fire, must surely have noted the melted fat dripping onto the coals, some of it catching fire, which may have suggested its future application as an illuminant. Some of the melted fat may have mixed with the hot ashes and formed a substance that, on standing, was not very palatable, but had other interesting and potentially useful properties, one of which was the ability to lather with water and thereby remove grease and other dirt from skin, clothes, and utensils. Gradually, the great potential of these substances became apparent and, in addition to their being a source of energy, light, heat, and cleanliness, fats were found to be slippery and therefore useful as lubricants. Certain of their properties suggested a use in cosmetics and hair grooming, and urns have been found in the Egyptian tombs which contained palmitic and oleic acids, evidently hydrolytic products (produced in some obscure reaction) of palm and olive oils. There are also biblical references to balming effects of anointing the body with scented oils. The laxative properties of the vegetable oils must have been immediately apparent to anyone who tried to use them as a major food source, but their usefulness as protective coatings was apparently less quickly recognized; the earliest recorded use in this way seems to have been in the sixth century, while their extensive employment in the arts occurred much later still.

The scientific investigation of the nature of lipids probably began with Scheele, who in about 1779 isolated glycerol as a hydrolytic product of fats. The full impact of this important discovery was perhaps tempered somewhat by his belief that fats and oils were composed of phlogiston, fire, and earth. Among the many investigators associated with advances in the knowledge of lipid chemistry during the early years, the outstanding contributor was undoubtedly Chevreul, whose major research was carried out during the years 1813–1825. This remarkable man, who remained active to the age of 103, among other achievements determined the structure of fats, the nature of the hydrolytic process, and the structure and properties of the major saturated fatty acids from C_4 through C_{18}. Aside from the understandable error of believing a mixture of palmitic and stearic acids to be a single, seventeen-carbon acid, named margaric, his conclusions are still valid.

PROVISIONAL CLASSIFICATION OF LIPIDS

I. "Simple" carboxylic esters
   A. Fats (esters of fatty acids with glycerol, e.g., acylglycerols)
   B. Waxes (including sterol esters; esters of fatty acids with alcohols other than glycerol)
II. Complex carboxylic esters
   A. Glycerophospholipids
     1. Cholineglycerophospholipids
       (a) Diacyl (phosphatidyl)
       (b) O-alkenyl, acyl (plasmenyl)
       (c) O-alkyl, acyl (plasmanyl)
       (d) Di-O-alkyl
     2. Ethanolamineglycerophospholipids
       (a) Diacyl (phosphatidyl)
       (b) O-alkenyl, acyl (plasmenyl)
       (c) O-alkyl, acyl (plasmanyl)
       (d) Di-O-alkyl
     3. Serineglycerophospholipids
       (a) Diacyl (phosphatidyl)
       (b) O-alkenyl, acyl (plasmenyl)
       (c) O-alkyl, acyl (plasmanyl)
       (d) Di-O-alkyl
     4. Inositolglycerophospholipids
       (a) Diacyl (phosphatidyl)
       (b) O-alkenyl, acyl (plasmenyl)
       (c) O-alkyl, acyl (plasmanyl)
       (d) Di-O-alkyl
     5. Phosphatidylglycerols
   B. Glycoglycerolipids
   C. Glycoglycerolipid sulfates
III. Complex lipids (containing amides)
   A. Sphingolipids
     1. Sphingomyelins (containing phosphocholine)
   B. Glycosphingolipids
     1. (a) Ceramide monohexosides (cerebrosides)
       (b) Ceramide monohexoside sulfates (sulfatides)
     2. Ceramide polyhexosides
     3. Sialoglycosphingolipids (gangliosides) containing a sialic acid and hexosamine; for definition of globosides and hematosides, see Chapter 19
IV. Precursor and "derived" lipids
   A. Acids (including phosphatidic acid and bile acids)
   B. Alcohols (including sterols)
   C. Bases (sphinganines, etc.)
V. Hydrocarbons
   A. Straight-chain
   B. Simple branched
   C. Polyisoprenoid
VI. Lipid vitamins and hormones with multiple functional groups not clearly falling into any of the above classifications

INTRODUCTION

The much more difficult characterization of the unsaturated fatty acids and the more complex lipids has occupied the research careers of many scientists during the last hundred years and is still continuing. As space and pertinence permit, these discoveries will be presented in later chapters.

1.2. Classification of the Lipids

The need to classify is a human weakness (or strength) and in the case of the lipids has been indulged in frequently with varying degrees of success. For many years, the system originally devised by Bloor has continued to be widely used, and it is the basis of the classification presented here.

At this point, the question is, what use is intended of this classifcation of most of the common species of lipids? The chapters in this book will not follow the order of this classification, nor will all the species included in it be given equal treatment. Perhaps the most useful function of such a classification is to challenge the reader to improve upon it, or to devise a better way to classify the structurally wide variety of substances properly designated "lipids."

SELECT BIBLIOGRAPHY

Annual Review of Biochemistry, 1952–, Annual Reviews, Inc., Stanford University Press, Palo Alto, California.

Chapman, D., 1964, *The Structure of Lipids by Spectroscopic and X-ray Techniques,* Methuen and Co., Ltd., London.

Chapman, D., 1965, *The Structure of Lipids,* John Wiley and Sons, Inc., New York.

Dietschy, J. M., Gotto, A. M., and Ontko, J. A. (eds.), 1978, *Disturbances in Lipid and Lipoprotein Metabolism,* Am. Physiol. Soc., Bethesda, Maryland.

Gatt, S., Freyst, L., and Mandel, P. (eds.), 1978, *Enzymes of Lipid Metabolism, Adv. in Exp. Med. and Biol.,* Vol. 101, Plenum Press, New York.

Goodwin, T. W. (ed.), 1974, *MTP International Review of Science: Biochemistry of Lipids,* Butterworths, London.

Gurr, M. I., and James, A. T., 1971, *Lipid Biochemistry: An Introduction,* Cornell University Press, Ithaca, New York.

Hilditch, R. P., and Williams, P. N., 1964, *The Chemical Constitution of the Natural Fats,* 4th ed., Wiley, New York.

Holman, R. T. (ed.), 1953–1979, *Progress in the Chemistry of Fats and Other Lipids,* Pergamon Press, Oxford.

Holman, R. T. (ed.), 1979–, *Progress in Lipid Research,* Pergamon Press, Oxford.

Johnson, A. R., and Davenport, A. R. (eds.), 1971, *Biochemistry and Methodology of Lipids,* Wiley Interscience, New York.

Kates, M. (ed.), 1972, *Lipids-Analysis,* North-Holland Pub. Co., Amsterdam.

Kuksis, A. (ed.), 1978, *Handbook of Lipid Research: Fatty Acids and Glycerides,* Plenum Press, New York.

Lowenstein, J. M. (ed.), 1969, *Methods in Enzymology,* Vol. XV, *Lipids,* Academic Press, New York.

Marinetti, G. V. (ed.), 1976, *Lipid Chromatographic Analysis,* 2nd ed., Vols. 1–3, Marcel Dekker, New York.

Masoro, E. J., 1968, *Physiological Chemistry of Lipids in Mammals,* Saunders, Philadelphia.
The Nomenclature of Lipids (Recommendations 1976), IUPAC–IUB Commission on Biochemical Nomenclature, 1978, *Biochem. J.* **171**:21.
Paoletti, R., and Kritchevsky, D. (eds.), 1963–, *Advances in Lipid Research,* Academic Press, New York.
Schettler, F. G. (ed.), 1967, *Lipids and Lipidoses,* Springer-Verlag, Berlin.
Wakil, S. J. (ed.), 1970, *Lipid Metabolism,* Academic Press, New York.

2
SOLUBILITY AND THE DEFINITION OF LIPIDS

2.1. Introduction

Lipids differ strikingly from the other two major classes of tissue components, the proteins and carbohydrates, in their solubility characteristics—a distinction that is attributable to structural features responsible for the nature and strength of physical interactions of these substances with solvents.

The structural relationship between a solute and a solvent that readily dissolves it is not a simple one, however, and it is therefore not surprising that two widely used definitions of lipids—one based on structural similarities, the other on solubility characteristics—are sometimes at odds with one another.

In 1943, Bloor defined lipids as being "actually or potentially compounds of the fatty acids." Although many lipids are esters or amides of fatty acids, it was necessary to add the qualification " ... or potentially ..." in order to include others having more extensively altered fatty acyl groups. Some substances having the requisite solubility criteria, such as steroids and other polyisoprenoids, are difficult to embrace by this definition. Lipids may indeed be regarded as distinctive in terms of their widely disparate structural features, and Bloor's approach, however extensively qualified, is not completely applicable.

Lipids are also defined as "tissue components that are soluble in lipid solvents." This empirical definition has the advantage of permitting recognition of a lipid simply on the basis of conditions employed in its isolation from tissue, independent of any knowledge of its structure. This tautological definition, however, resembles that of Gertrude Stein (1922): "a rose is a rose is a rose;" and just as her "definition" is meaningless to anyone who does not know the flower, so the significance of the solubility definition of lipids depends (i) on knowing what the lipid solvents are and (ii) on understanding the implications of the fact that such solvents dissolve lipids well, and carbohydrates and proteins poorly. However, it should be pointed out that certain derivatives of fatty acids—gangliosides and soaps, for example—are Bloor lipids, but, being more readily soluble in water than in lipid solvents, have "inappropriate" solubility characteristics.

In order to understand how the two definitions of lipids are related, i.e., how the structures of lipids are responsible for their having distinctive solubility characteristics, it is desirable to define "ready" and "poor" solubility in a given solvent. The solubility of some proteins and carbohydrates in water, for example, is so low that they are correctly described as being insoluble in this solvent, and more importantly, so low that it would be difficult to demonstrate that they are, in fact, considerably less soluble still in lipid solvents. Such complications are readily accommodated in terms of procedures actually used to separate lipids from other tissue components: to the extent that all tissue components may be brought into solution in two-phase mixtures of water and a lipid solvent that has limited solubility in water, distribution of the lipids favors the lipid solvent phase, while that of non-lipids favors the aqueous phase. The functionality of some lipids is such that their distribution in such systems may be reversed by changes in the pH of the aqueous phase; strongly acidic lipids (such as the phosphoric acid esters of phosphatidylinositols, for example) tend to favor the aqueous phase unless their dissociation is suppressed by imposition of low pH. The low water solubility of certain proteins and carbohydrates is easily attributable to their being of high molecular weight (a structural feature rare in lipids); in such cases the criterion of low water solubility would be equivocal, but the complementary test of ready solubility in lipid solvents would not. (The tertiary structure of many water-soluble cytoplasmic proteins and the firm anchorage of others in biomembranes is attributable to the presence and location of certain amino acid moieties, the "R groups" of which impart lipid solubility characteristics to sections of such otherwise hydrophilic molecules.)

Why lipids dissolve much more readily in lipid solvents than in water would appear to be understandable in terms of the old saying "like dissolves like" if the question "Like in what aspect?" could be answered satisfactorily. "Polarity" is a simple answer that is popular because it leads to a variety of empirically useful corollaries. For example, if water is taken to be highly polar, then lipid solvents are less so; proteins and carbohydrates are relatively highly polar substances that tend to be water-soluble; and lipids are distinctive in possessing comparatively low polarity and are therefore more readily soluble in lipid solvents than in water. But having discerned and named a vague property of substances (including solvents) that determines whether or not they mix readily amounts to little more than a reiteration of empirical observation, and urges inquiry into the nature of "polarity," how the possession of this property to varying degrees is related to the structures of the substances involved, and why solution of a solute in a solvent is favored by the two substances having similar polarity.

2.2. Chemical Dynamics of Solubility

If partial dissolution of one substance (the solute, gas, liquid, or solid) in a liquid solvent can be regarded as a reversible process resulting in formation of a solution:

$$\text{solute} + \text{solvent} \rightleftarrows \text{solution}$$

then the condition of equilibrium, which defines the solubility of the solute in the solvent, is related to changes in enthalpy (or heat content, H) and in entropy (or randomness, S) that occur on mixing the two substances:

$$\Delta H° - T\Delta S° = -RT \ln K$$

where $\Delta H°$ and $\Delta S°$ represent changes in these properties of the system realized in the process taking place between arbitrarily standardized states (neat → unimolar solution), and K, the equilibrium constant, is a quotient of the product of the activities (concentrations, relative to one molar) of the substances in the equilibrated solution divided by the product of the activities of the unmixed substances. All terms in K are unitless, but the function is easily translated into any convenient units of solubility (e.g., grams or moles solute per unit weight, volume, or moles of solvent). Solubility is favored by an increase in entropy ($\Delta S° > 0$) and/or by a decrease in enthalpy ($\Delta H° < 0$, i.e., when dissolution is exothermic). Since enthalpy and entropy are essentially independent properties of state, a favorable change in one may offset an unfavorable change in the other.

With application of these principles to solubility of lipids in various solvents, three distinctly different situations may be encountered: (i) If the lipid is miscible in all proportions with a solvent, or if the solution is not saturated, no equilibrium is involved (i.e., neither solute nor solvent exist as discrete phases), but the free energy of the mixture ($H - TS$) must be considerably below that of the separated components. (ii) If a liquid lipid is incompletely dissolved in a solvent (e.g., castor oil in n-pentane; most other natural oils in methanol), some solvent must be presumed to be dissolved in the lipid-rich phase, and the system therefore involves two solutions in equilibirum; the partial free energies of each component must be identical in both phases, and their different concentrations attributed, therefore, to differences in the enthalpies and/or entropies of their two situations. Except for the increased randomness associated simply with mixing the two components, little difference in entropy would, in general, be expected, leaving ΔH primarily responsible for the immiscibility. (iii) If undissolved crystalline lipid is in equilibrium with a solution of it, solvent is presumably excluded from the solid, which is therefore in standard state; in this case the orderliness of the crystal lattice results in a substantial increase in entropy as the substance dissolves but the aggregate intermolecular forces responsible for stability of the crystalline phase (i.e., the low enthalpy of the state) act in opposition and thus limit solubility.

From the above discussion it might be concluded that dissolution is always driven by entropy increases, and solubility thus determined primarily by the magnitude and sign of enthalpy changes. A notable exception to this generalization is provided by the "hydrophobic effect," which deserves mention not only because of its importance in a variety of biochemical phenomena, but also because its correct rationalization requires recognition of the fact that the changes in enthalpy and entropy determining the position of the solubility equilibrium are those of the entire system, and may therefore be inadequately (or even incorrectly) explained in terms of characteristics of individual molecules of solute and solvent and of intermolecular interactions amongst them. The hydrophobic effect is manifest in a strong tendency for direct contact between apolar molecules (e.g., lipids) (or apolar portions of molecules) and water to be spontaneously minimized. The solubility of such substances in water would thus, for example, be expected to be low. The possibility that they are repelled by water is implied by the adjective, "hydrophobic." Such low solubility might be ascribed to increased enthalpy due to stronger interaction between water molecules than between those of water and solute. But water does not repel such molecules (or portions of molecules). The interaction is indeed considerably weaker than that between water molecules, but the dissolution of at least some apolar substances (e.g., methane) in water is actually exothermic ($\Delta H < 0$). The extremely low solubility of methane in water must therefore be due to a decrease in entropy that substantially overcompensates the favorable loss of enthalpy. Although substantial loss of entropy of the "iced in" apolar group or molecule may also be involved, the dominant changes in both enthalpy and entropy of such systems are those occurring in liquid water in close proximity to apolar moieties, which are uniquely able to reinforce (instead of weakening) the native ice-like structure of liquid water. The increased orderliness of proximate water is a direct consequence of this strengthening of hydrogen bonding of water molecules in contact with the apolar molecules or groups, but the cost of reducing the entropy of the system in this way is high in terms of its being paid for by the concomitant net decrease in enthalpy. The extensiveness of the phenomenon is therefore severely limited and the solubility of apolar substances in water, exemplifying a manifestation of the hydrophobic effect, is extremely low.

The unique hydrogen-bonding characteristics of water are essential to operation of the hydrophobic effect and the chemical dynamics of solubility in systems other than those of apolar solutes in water may be confidently expected to be more direct and therefore simpler. The dominant role of a large decrease in entropy in limiting the solubility of apolar substances in water must be recognized as extraordinary. In most if not all other cases (in which water is not involved) entropy may be assumed to increase, and thus to favor dissolution, and solubility must therefore be determined primarily by changes in enthalpy (ΔH) associated with mixing the solute with the solvent, related, in turn, to the magnitudes of intermolecular forces of attraction in the neat (solid or liquid) vs. mixed states.

In general, then, solubility should be favored if solute molecules are attracted more strongly to those of the solvent than the molecules of the individual components are to themselves ($\Delta H < 0$), and vice versa. It is clear, for example, that if more energy is required to separate molecules of a highly associated solvent than is regained by insertion of a solute molecule attracted weakly to the solvent molecules now surrounding it, the heat content of the resulting solution would be higher than that of the components (i.e., $\Delta H > 0$) and the solubility therefore low. Similar qualitative consideration of consequences of solute and solvent having other relative magnitudes of attraction leads to the conclusion that any given substance will be most soluble in a solvent having most nearly identical magnitudes of intermolecular attraction, i.e., for cases in which dissolution involves no change in enthalpy. To the extent that strengths of attraction of the components are disparate—regardless of whether solute or solvent is more strongly associated—mixing them is thus expected to result in an increase in enthalpy and suppression of solubility.

These considerations may be summarized as follows: strongly associated substances dissolve best in strongly associated solvents, and weakly best in weakly associated solvents. "Like dissolves like" now assumes more specific meaning and focuses interest on kinds and relative magnitudes of intermolecular forces associated with structural features of proteins, carbohydrates, lipids, and common solvents.

2.3. Intermolecular Forces

With rare exceptions, all molecules and atoms attract each other through the action of a variety of forces; lipids are not repelled by water!

2.3.1. Electrostatic Interactions

Most of the different kinds of intermolecular forces are essentially electrostatic (coulombic) in nature and involve interaction of electric fields associated with formally charged atoms in molecules (e.g., oxygen atoms in carboxylate and phosphate groups, nitrogen in ammonium groups, sulfur in S-adenosylmethionine) or with monoatomic ions (Na^+, Cl^-, Mg^{++}, etc.) or fields associated with partially developed charges on pairs of covalently bonded atoms of different electronegativity. Ion–ion, ion–dipole, and dipole–dipole classes of interactions are thus possible. Ion–ion interactions may be either attractive or repulsive, depending on the sign of charge of the ions, but if two interacting bodies are free to assume any orientation with respect to each other, ion–dipole and dipole–dipole interactions are always attractive.

Whether or not any given covalent bond is polarized (and thus has a permanent electrical dipole associated with it) and the extent of its polarization depend upon intrinsic properties of the atoms involved: their electronegativities, directly related to their relative tendencies to attract electrons, including those shared with another atom, toward their nuclei. The electronegativities of atoms of various elements increase quite consistently from the lower left to the upper right corner of the Periodic Table. Thus fluorine is the most and cesium the least electronegative (or most "electropositive") element. Relative electronegativities of elements of most direct interest, in Periodic Table arrangement, are shown in Table 2.1. Electrons shared by atoms of different electronegativity are therefore not shared equally; the molecular orbitals holding electrons of the bond(s) are distorted toward and thus create a partial negative charge on the more electronegative atom, and a partial positive charge of equal magnitude on the other. The

TABLE 2.1. Relative Electronegativities[a] of Selected Elements

H 2.20			
C 2.55	N 3.04	O 3.44	F 3.98
	P 2.19	S 2.58	Cl 3.16
			Br 2.96
			I 2.66

[a]Calculated by Allred (1961).

presence of a polarized covalent bond in a molecule and the sense or direction of polarization is denoted in either of two ways (where atom A is more electronegative than B):

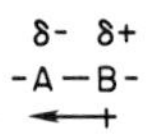

$$\overset{\delta-}{-A}\overset{\delta+}{-B-}$$

The notation, δ^-, indicates the presence (e.g., on atom A) of a charge anywhere between zero and the full unit charge of an isolated electron; and of a positive charge (on atom B) of magnitude equal to δ^+. The alternative "crossed arrow" notation indicates direction of displacement of bonding electrons toward the more electronegative atom.

Since the strength of the electric dipole associated with a polarized covalent bond will affect the magnitude of its attraction to a center of formal charge or to another such dipole (and, to an even greater extent, to a second dipole induced by it), this property and its relationship to the difference in electronegativities of the atoms involved are of immediate interest. The strength of the dipole (dipole moment) is the product of the absolute magnitude of partial charge (δ) on either atom times the internuclear distance (the covalent bond length). The charge can be shown to be directly related, in the case of diatomic molecules, to the exothermicity of the reaction $A-A + B-B \rightarrow 2 A-B$, and more generally (as shown by Pauling, 1960), to be calculable by use of the equation, $\delta = 1 - e^{-\Delta X^2/4}$, where ΔX is the difference in electronegativity of the two atoms involved. Values calculated in this way for covalent bonds likely to occur in substances of interest are shown in Table 2.2, grouped arbitrarily as being polarized strongly, polarized weakly, or unpolarized on the basis of the extensiveness of charge (δ) separations or polarizations being greater than 0.1, between 0.01 and 0.1, or less than 0.01 electron charge, respectively. (Polarization of covalent bonds is often expressed as "percent ionic character" $\equiv 100\delta$.)

The most highly polarized bonds occurring in proteins, carbohydrates, and lipids are those between oxygen or nitrogen and carbon or hydrogen, which therefore contribute importantly to the polarity of these substances.

The weak though appreciable polarity of the C−H bond may seem at odds with alkanes (e.g., "petroleum ether") being among the least polar of common solvents. The tetrahedral geometry of the four covalent bonds of sp^3 hybridized

TABLE 2.2. Polarization of Selected Pairs of Covalently Bonded Atoms

Pair[a]	Electronegativity difference (ΔX)	Polarization ($\delta \pm$)	
H—O	1.24	0.32	⎫
C—O	0.89	0.18	⎬ strongly polarized
H—N	0.84	0.16	⎭
C—Cl	0.61	0.09	⎫
C—N	0.49	0.06	⎪
C—Br	0.41	0.04	⎬ weakly polarized
H—S	0.38	0.035	⎪
H—C	0.35	0.03	⎭
C—I	0.11	0.003	⎫
C—S	0.03	0.0002	⎬ unpolarized
C—C[b]	0	0	⎭

[a]More electronegative atom to the right; sense of polarity: L$\overset{\leftrightarrow}{-}$R.
[b]Or any other pair of identical atoms.

carbon, however, results in the dipole moments of individual C—H bonds cancelling (by vector addition), such substances exhibiting no net dipole moment. More important perhaps, from the standpoint of understanding this apparent anomaly, the structure of such molecules precludes antiparallel orientation of pairs of C—H bonds, thus limiting the magnitude of their mutual attraction. Qualitative assessment of the importance of the effects of these "3% ionic character" bonds is afforded by recognizing that the exterior "surface" of alkane molecules is studded with hydrogen atoms, each of which bears a small positive charge; that such molecules exhibit weak but appreciable intermolecular attraction must therefore be attributable to the superiority of net aggregate London forces (see below) over weak electrostatic repulsions.

A characteristic of all electrostatic interactions that is of major significance in living (aqueous) systems is the inverse dependence of their magnitude on the dielectric constant (ϵ) of the medium separating the interacting bodies. (Interactions involving induced dipoles are inversely proportional to the square of the medium dielectric constant.) The dielectric constant of water is very high (about 80 times that of a vacuum), and its presence between the interacting species therefore reduces the strength of their interaction to about 1% of that in its absence.

2.3.2. London Forces

London forces (also called dispersion forces because the dispersion of light by a given molecule is a related phenomenon) have their origin in the fact that the motion of electrons in two atoms that are close influence each other in such a way as to result in creation of an attractive force (requiring quantum mechanics for rigorous explanation). Such London forces are solely responsible for the attraction of noble gas atoms, but, since they stem from characteristics shared by all atoms, give rise to attractive forces between atoms of molecules as well, regardless of what other kinds of interactions, attributable to more complex structural features, may be involved. These attractive forces are weak but increase in strength between atoms of higher atomic number and are maximal when the atoms are in van der Waals contact (the van der Waals radii of atoms are indeed defined on this basis); they fall off rapidly with greater separation, tending to be inappreciable at distances larger than one or two atomic diameters. As the internuclear distance between two atoms approaching each other begins to be less than that of the sum of their van der Waals radii, the interatomic force reverses sign and quickly becomes strongly repulsive, as the filled valence electron orbitals begin to overlap. This "repulsion" is grossly manifest in the "incompressibility" of liquids, molecules of which are in virtual van der Waals contact. London forces can be very substantial between large molecules, particularly when their structures involve regions of high topographical complementarity, providing the possibility of a number of pairs of atoms of the two molecules being simultaneously in intimate contact. Multiple London-force interactions of this sort are involved importantly in the highly specific and strong but purely physical complexing of enzymes with substrates, of antigens with antibodies, etc.

That London and electrostatic forces are qualitatively different in nature has an important consequence that is manifest, for example, in "hydrophobic bonding" (see below): uncharged atoms of a molecule may, if steric conditions permit close proximity, reinforce rather than compromise a preexistent electrostatic interaction.

London forces, moreover, are not sensitive to dielectric characteristics of the medium. This might appear to be of little consequence, since London forces would already have become negligible at any separation sufficient to permit interposition of even a single dielectric molecule. However, significant involvement of certain potentially strong interactions (e.g., the ion–ion "salt linkages") that would be greatly weakened in the presence of water appears to depend, in many cases, on cooperative action of a shield of London-interacting atom pairs unaffected by the highly dielectric medium.

2.3.3. Hydrogen Bonds

Hydrogen bonds are associated with the sharing of a hydrogen atom by two more electronegative atoms, to one of which the hydrogen atom is nominally covalently bonded; the hydrogen bond acceptor must be a monoatomic anion or have an atom more electronegative than that of other(s) to which it is covalently bonded (thus with some partial negative charge) and must bear at least one unshared electron pair. H-bond donors are therefore Lowry–Brønsted acids, and acceptors are bases (Lewis or Lowry–Brønsted), but the interaction is not of a simple acid–base type. The proton is not actually transferred; moreover, the relative strengths of H-bonds between O and/or N atoms clearly depend on the acidity of the donor, but appear to be insensitive to the basicity of the acceptor. It will be recognized that many (but not all: cf. NH_4^+) good H-bond donors are also good acceptors, while the reverse is not the case, since acceptors need not have hydrogen attached to their electronegative atoms. Although the functional groups involved in this kind of molecular interaction must both contain polarized covalent bonds, H-bonding is clearly atypical with respect to the usual dipole–dipole interaction in two ways: (i) it may be much stronger and (ii) the preferred orientation of the two dipolar bonds is not antiparallel. In their maximized-interaction orientation, the shared H atom lies on a straight line connecting the nuclei of the two electronegative atoms, the angle between vectors of head-to-tail dipoles thus being about 110° or 120° instead of the antiparallel 180°.

The strength of a hydrogen bond is dependent on the electronegativities of the atoms thus bound (e.g., $F \cdots H-F > O \cdots H-O > N \cdots H-N > S \cdots H-S$); on its linearity; and, of course, inversely on the "length" of the bond, which is weakened if the structure of the molecule precludes the hydrogen atom attaining optimal proximity to the acceptor atom.

With respect to judging the likelihood of strong, specific association in biological (i.e., aqueous) systems between two molecules (or parts of a molecule) having structural features amenable to strong H-bonding, the excellence of water as either H-bond donor or acceptor must not be overlooked. As pointed out elsewhere (see Chapter 3), the double H-bond association of pairs of carboxylic acid molecules is very strong indeed in non-H-bonding solvents, but short-chained carboxylic acids are monodisperse in aqueous solution.

2.3.4. Hydrophobic Bonding

Hydrophobic bonding gets its name from recognition of the strong apparent co-attraction of apolar molecules or parts of molecules in water that is responsible for the inherent stability of biomembranes and other micellar aggregates of amphiphilic substances in aqueous media; for maintenance of the native tertiary structure of many water-soluble, globular proteins; and, more generally, for the low solubility of apolar substances in water. As discussed earlier, in a chemical dynamics analysis of the hydrophobic effect (see page 7), such molecules or parts of molecules tend strongly to be aggregated rather than monodisperse in water because this minimizes the area of direct contact, in close proximity to which the icelikeness of the water is considerably enhanced. Such aggregation is thus entropy driven, quite literally by the melting of water that attends reduction in the area of contact with apolar moieties.

2.4. Relative Strengths of Intermolecular Forces

Electrostatic interactions between molecular centers bearing formal charge are very strong when the charged atoms are in van der Waals contact. The force of attraction between Na^+ and Cl^- ions at such minimum separation is, for example, 118 kcal/mol, equivalent to a strong covalent bond. However, if a solvent of high dielectric constant is interposed between the charged atoms, the attraction is greatly weakened. Water (dielectric constant about 80), for example, reduces the attraction to about 1% of its magnitude in the absence of water (dielectric constant of void $\equiv 1$).

The fact that water is the best of common solvents for salts (even though some have limited solubility in it) is attributable in part to this dielectric constant effect, but also to substitution of strong ion–dipole forces for the ion–ion and H-bond interactions that are disrupted. The magnitudes of various ion–dipole interactions are expected to vary widely, being dependent on the field strength of the formally charged atom (magnitude of charge vs. size of atom or ion) and on the moment of the dipole. In the case of ion hydration, however, it is clear that water molecules in contact with the ion interact more strongly in this manner than with other water molecules. In such ion–dipole interactions, the attraction is presumably greater than 5 kcal/mol, the average strength of liquid water H-bonds.

It serves present purposes simply to point out that intermolecular interactions, except for those involving formally charged centers and H-bonds, involve attractions that are of strength no greater than about 1 kcal/mol, and thus are quite vulnerable to disruption by thermal collision. These individually weak interactions are additive, of course, and, as pointed out above, may contribute to quite stable association of molecules of suitably complementary structure and/or conformation. Comparison of boiling points and dipole moments (see Table 2.3) of a selection of common organic solvents of similar molecular weight leads to some useful conclusions regarding the relative strengths of intermolecular forces of attraction attributable to the polarity of several functional groups of interest. The close similarity in boiling points of *n*-pentane and diethyl ether, the latter bearing strongly polarized C—O bonds (see Table 2.2), clearly suggests that

TABLE 2.3. Boiling Points and Dipole Moments of Some Common Organic Solvents

Solvent	Molecular weight	Dipole moment[a]	Boiling point (°C at atmospheric pressure)
n-Pentane	72.15	0	36.1
Diethyl ether	74.12	1.15	34.6
Butanone	72.10	2.8	79.6
Methyl acetate	74.08	1.86	56.9
1-Butanol	74.12	1.66	117.3

[a]Debyes, of vapors (except butanone, in apolar solvents).

dipole–dipole interactions in this case are not markedly different in strength from those of the London type. The greater polarity of the C=O group, resulting in stronger dipole–dipole interactions, is reflected in the considerably higher boiling point of butanone. The longer but much less highly polarized dipole of the carbalkoxy (ester) functional group of methyl acetate and many lipids results in a more modest elevation of boiling point. The relative strength of dipole interactions of C−O bond containing functional groups in these substances is, therefore, ether < ester < ketone.

The boiling point of 1-butanol (Table 2.3), almost 83° higher than that of the isomeric diethyl ether and 60° above that of methyl acetate (despite the somewhat higher dipole moment of the ester), illustrates dramatically that H-bond interactions can be much stronger than dipole–dipole attractions. Among the hydrogen bonds of primary interest are those of greatest strength: those in which the electronegative atoms of the donor and acceptor are oxygen and/or nitrogen. The strength of such H-bonds is usually in the neighborhood of 5 kcal/mol, i.e., an order of magnitude stronger than other intermolecular interactions. Intermolecular attractions of this magnitude are of prime pertinence in judging, on the basis of structural features, whether solute and solvent are "like" each other in polarity and hence tend to form solutions readily.

Since H-bonds between oxygen and/or nitrogen atoms are thus seen to play a major role in determining solubility characteristics of tissue components, it is of interest to compare certain features of the four types of such interactions that are possible, each of which is involved in biological systems: O · · · H−O, O · · · H−N, N · · · H−O, and N · · · H−N. Direct quantitative comparison of the strengths of these species of H-bonds is difficult, insofar as values obtained by calculation from various physical measurements or under different conditions vary widely, although all fall in the approximate range of 4–7 kcal/mol (see Pimentel and McClellan, 1960). The lengths of such bonds are, however, more consistent, and the extensiveness of overlap can be taken to represent their relative strengths. Results of such calculations and some comparisons with measured strengths are shown in Table 2.4. The consistent dependence of the strengths of

TABLE 2.4. Relative Strengths of H-Bonds Between O and/or N Atoms, in Terms of Overlap of Hydrogen and Acceptor Atoms

Type of bond	Length, Å[a]	Overlap, Å[b]	ΔH kcal/mol H-bond[a]
O−H · · · O	2.76 (liquid H_2O)	0.74	5.0
	2.63 ± 0.1 (RCOOH)	0.87	7.0
O−H · · · N	2.80 ± 0.09	0.80	—
N−H · · · O	2.88 ± 0.13 ($R_3\overset{+}{N}H$)	0.62	—
	2.93 ± 0.10 (amides)	0.57	(3.9)[c]
	3.04 ± 0.13 (R_2NH)	0.46	—
N−H · · · N	3.10 ± 0.13 (NH_3)	0.50	4.4

[a]Data from Pimentel and McClellan (1960).
[b]Minimal, using 1.1 Å as the van der Waals radius of H.
[c]In CCl_4 solution; others neat.

these bonds on acidity of the donor is shown by comparison of those in water with those in carboxylic acids; of those formed between oxygen acceptors and ammonium ions (full formal charge), amides (in which, by resonance, the nitrogen atoms bear a partial positive charge), or amines (uncharged N); and of O—H···A vs. N—H···A bonds. The basicity of the acceptor atom (N considerably greater than O), however, appears to have little if any effect on the strength of such interactions. In summary, the four species of strong hydrogen bonds that may, because of functionality of various tissue components, be involved in living systems have the following relative strengths (range: from 7 down to about 4 kcal/mol):

$$O-H \cdots O \cong O-H \cdots N > N-H \cdots O \cong N-H \cdots N.$$

This survey of the relative strengths of various types of interactions between atoms and/or functional groups of atoms within molecules leads to the broader generalization: covalent bonds (and ion–ion interactions in low dielectric media) > ion–dipole interactions ≥ H-bonds between O and/or N atoms > all other types of interactions (dipole–dipole, ion– and dipole–induced dipole, and London). The potential for participation of tissue constituents in strong ion–dipole interactions (as well as in strong H-bonds) is associated, again, with the presence of O and/or N atoms, which are the usual bearers of formal charge (e.g., in $-COO^-$, in phosphate mono - and diesters, and in $-\overset{+}{N}R_3$ moieties).

2.5. Relationship of Polarity to Solubility

Having concluded that substances are most soluble in solvents of like polarity, and that this property is imparted to both solvents and solutes by the presence of functional groups capable of involvement in strong physical intermolecular interactions, it is important to point out that the presence of such a functional group in a molecule may be insufficient to make the whole molecule polar, as might be judged, for example, by ready solubility in a highly polar solvent (e.g., water). Many lipids bear excellently hydrophilic (highly polar) substituents, yet exhibit characteristically low water solubility. The polarity of a molecule is the net result of counteracting effects of number and/or polarity of its functional groups vs. extensiveness of the apolar remainder of the molecule. This is readily understandable in terms of the chemical dynamics of dissolution: passage of the polar groups into a polar solvent involves a favorable loss in free energy, but this is countered or may easily be greatly exceeded by increases in free energy associated with the obligatory, concomitant dissolution of the apolar part.

Using solubility in water as a semiquantitative gauge of polarity, the limits of miscibility with water of homologues bearing the same hydrophilic substituent reflect the relative abilities of different substituents of this kind to prevail over countereffects of apolar n-alkyl groups of various size. Good H-bonders are about equally effective: acetone is miscible with water, but methyl ethyl ketone is not (nor is butyraldehyde); n-propanol is the largest normal alcohol that is miscible with water, and n-butyric acid the largest normal carboxylic acid; the primary amino group is more hydrophilic, n-amylamine (but not n-hexyl) being soluble in water in all proportions. The ability of these good H-bonding functional groups to solubilize n-alkyl groups is quite limited. The clear superiority of formally charged hydrophilic substituents is seen in the greatly enhanced solubility of amines (i.e., of the corresponding ammonium ions) in water at low pH, and of carboxylic acids (converted to carboxylate anions) in water at high pH. Interpretation of the effectiveness of conversion of $-COOH$ to $-COO^-$ is clouded by micelle formation by soap anions (see Chapter 4), a hydrophobic-effect phenomenon that might appear to indicate that the carboxylate group has a virtually unlimited ability to solubilize alkyl groups; octanoate ions, however, are essentially monodisperse in aqueous solution up to about two-molar concentration (when micelle formation begins) and are thus considerably more soluble than the corresponding free carboxylic acid (see below).

The progressive loss in overall polarity (again, as reflected in water solubility) of normal saturated carboxylic acids with increasing size of the apolar alkyl group is shown in Table 2.5. It may be noted that for acids no longer miscible with water ($>C_4$), presence of an additional methylene group results in an approximately 75% loss in water solubility, decreasing to about 40% and remaining quite constant for all succeeding (crystalline) acids ($>C_9$, m.p. $>20°C$), showing that each added methylene group contributes actively, through its interaction with the solvent, to the growing positive ΔG of solution rather than simply passively diluting out the polarity contribution of the carboxyl group.

TABLE 2.5. Solubilities of Normal Saturated Carboxylic Acids, $CH_3(CH_2)_m COOH$, in Water

Total carbon atoms ($m + 2$)	Solubility[a] (g/100 g at 20°C)
2–4	miscible
5	3.7
6	0.97
7	0.24
8	0.068
9	0.026
10	0.015
11	0.0093
12	0.0055
13	0.0033
14	0.0020
15	0.0012
16	0.00072
17	0.00042
18	0.00029

[a] From Ralston and Hoerr (1942).

The presence of a limited number of highly polar or hydrophilic functional groups in a large, otherwise apolar molecule results in such a substance having typical lipid solubility characteristics, the resistance of the polar moieties to immersion in a solvent of low polarity being overwhelmed by the contrary propensity of the rest of the molecule. The inherent solvent affinities of both parts of such an amphiphilic solute can, however, be realized at an interface between polar and apolar media, and molecules of this kind therefore have a strong tendency to be concentrated and oriented in regions of discontinuous polarity, e.g., sterols at surfaces of biomembranes.

"Like dissolves like" having been recognized as referring to similarities in polarity, and polarity of molecules as representing an averaging of the polarity of their parts, high polarity being associated with strong H-bonding or formally charged oxygen and/or nitrogen substituents, it is now possible to reconcile the distinctive solubility of lipids vs. that of proteins and carbohydrates in broad terms of their structural differences: lipid molecules have a low ratio of O $\pm$ N to C, while those of proteins and carbohydrates have high values of this ratio. In carbohydrates, the ratio is equal to or not much below 1(O):1(C). In proteins (average wt. % N, 16; number of N and O approximately equal), the ratio is about 1 (O + N):2.5(C). In lipids, the ratio is from zero (for hydrocarbons, e.g., squalene) to about 1:4 (cytolipin H, 1:3.9, is soluble in chloroform, but not in water, while ganglioside G_{M1}, 1:2.2, is insoluble in lipid solvents, although it forms micellar aqueous solutions). It will be noted that the upper "limit" of the ratio corresponds roughly to that of water miscibility of homologous solvents.

On the basis of the conclusion that any given substance will be most soluble in a solvent of matching polarity, it might be easily predicted that the best common solvent for, say, squalene would be petroleum ether (preferably containing some alkenes). More importantly, however, this quite apolar solvent might not be best for lipids that are less apolar by virtue of bearing polar substituent(s). Comparison of solubilities of a substance in a variety of solvents having different polarities would be of interest in providing a test of the indicated relationship between polarities of solutes and solvents and the ability of one to dissolve the other, and indeed might be used to rate solutes with respect to polarity. However, as indicated by the above discussion of the diversity of the kinds of intermolecular interactions that may be involved in determining how extensively one substance will spontaneously mix with another, assignment of specific polarities to solvents is no simple matter, and cannot, in any case, be meaningfully equated to any single specific property of solvents, since properties of the solute are also crucial. (For various approaches to developing generally applicable ratings of solvents with respect to polarity, see Cosaert (1971) and Krygowski and Fawcett (1975).)

That no single property of solvents is consistently correlatable with their effectiveness in dissolving a given solute is shown in Table 2.6, listing solubilities of stearic acid (taken as an example of a lipid) in an assortment of solvents having a wide range of dielectric constants and thus, although manifest in a quite different way, of widely different polarity. The correlation between these two polarity-

TABLE 2.6. Solubilities of Stearic Acid in Selected Common Solvents Compared with Their Dielectric Constants

Solvent	Dielectric constant[a]	Solubility[b]
Water (HOH)	78.5	0.00029
Acetonitrile (CH_3CN)	36.7	<0.008
Methanol (CH_3OH)	32.6	0.08
Ethanol (C_2H_5OH)	24.3	1.8
Acetone (CH_3COCH_3)	20.7	1.2
Butanone ($C_2H_5COCH_3$)	18.5	2.4
2-Propanol (($CH_3)_2CHOH$)	18.3	1.6
1-Butanol (C_4H_9OH)	17.8	1.3
Methylene chloride (CH_2Cl_2)	9.1	1.7
Ethyl acetate ($CH_3COOC_2H_5$)	6.0	0.45
Chloroform ($CHCl_3$)	4.8	8.9
Trichloroethylene ($Cl_2C=CHCl$)	3.4	5.6
Toluene ($C_6H_5CH_3$)	2.4	1.7
Benzene (C_6H_6)	2.3	2.2
Carbon tetrachloride (CCl_4)	2.2	3.8
Cyclohexane (C_6H_{12})	2.0	1.9
n-Heptane (C_7H_{16})	1.9	0.3

[a]Defined as the ratio of coulombic forces between two charged bodies at fixed distance *in vacuo* to that with solvent between them, i.e., $\epsilon_{vac} \equiv 1$; all values at 25°C, somewhat lower than at 20°C (e.g., $\epsilon_{H_2O}^{20}$ 80.37).

[b]g/100 ml at 20°C, based on data reported by Ralston and Hoerr (1942); Hoerr and Ralston (1944); Hoerr *et al.* (1946); Kolb and Brown (1955); Preckshot and Nouri (1957); and Brandreth and Johnson (1971).

related characteristics of the solvents is obviously very poor for solvents of dielectric constant less than that of methanol (all solvents in the list of dielectric constant equal to or less than that of ethanol would usually be considered to qualify as "lipid solvents;" most natural oils, i.e., liquid triacylglycerol mixtures, are miscible with absolute ethanol, but not with methanol). The low solubility of stearic acid in heptane is in accord with the view that alkanes are too apolar to serve as good solvents for lipids bearing even a single respectably polar substituent. The general excellence of chlorinated hydrocarbons as lipid solvents (viz. chloroform, trichloroethylene, and carbon tetrachloride, Table 2.6) is widely recognized but difficult to rationalize.

Stearic acid was the lipid of choice here because its solubilities in a large number of different solvents have been measured. However, this and other carboxylic acids are atypical with respect to their strong tendency to "dimerize" by double H-bonding (see Chapter 3) in the absence of alternative H-bond donors and/or acceptors. Although this characteristic has no clear impact on the relationship between solubility and solvent dielectric constant, the solute is largely monomeric in butanol and solvents of greater polarity (and of high H-bonding potential), but dimeric in the others, which, with the possible exception of ethyl acetate, are constitutionally incapable of disrupting this very strong intermolecular association.

2.6. Anomalous Solubilities

Generalizations correlating phenomena as complex as those involved in determining the solubility characteristics of the structurally diverse lipids are subject to exception. The anomalous low solubility of phospholipids in acetone, an otherwise excellent lipid solvent, is perhaps the best and longest known of these. Before the advent of highly superior chromatographic techniques, empirically discovered differences in solubilities of certain classes of lipids were used extensively to effect separations that would now be regarded as quite crude. Treatment of the acetone-insoluble phospholipid mixture with ethanol, for example, served to separate the relatively soluble lecithins (phosphatidylcholines) from the "cephalins" (phosphatidylethanolamines and -serines), which are less soluble in this solvent. Similarly, sphingolipids (sphingomyelins and cerebrosides) were found to have low solubility in both acetone and diethyl ether. Separations based on such differential solubilities are notoriously capricious, and failure to appreciate this undoubtedly led in many cases to faulty conclusions. The solubility of phospholipids in acetone, for example, is now known to be greatly enhanced by the presence of triacylglycerols, which, in most tissues, occur in substantial quantity. The "hydrolecithins" (those having no unsaturated fatty acyl moiety) usually occur in quite small amounts, but their true abundance was undoubtedly frequently underestimated before recognition of the fact that such phosphatidylcholines have a distinctly low solubility in diethyl ether and thus tend to follow the sphingolipids in classical solubility fractionation of lipid mixtures.

2.7. Solvent Extraction of Lipids from Tissues

Complete extraction of lipids from any biological sample must take into account the fact that tissue structure may present formidable barriers to solvent access. It is therefore customary, first, to disrupt the native structure of the material mechanically (i.e., to homogenize it), and then to employ a mixture of solvents, one of high and one of low polarity (but not so disparate in polarity as to preclude their being miscible), in order to favor ready penetration of persistent biomembrane-enclosed realms. In his pioneering studies of lipids, Bloor (1943) used diethyl ether–ethanol mixtures for this purpose, but mixtures of chloroform and methanol (as originally proposed by Folch et al. (1957), then modified in detail by Bligh and Dyer (1959)) proved superior and are still used extensively for this purpose. Hara and Radin (1978) have pointed out that a mix of hexane and isopropyl alcohol is similarly effective, and has the advantage of being less toxic. However, hexane may be too apolar to match the excellence of chloroform as a general lipid solvent; indeed the hexane–2-propanol mixture does not completely extract gangliosides. In the Folch–Bligh–Dyer procedure, a mixture of chloroform and methanol (2:1; v/v) is employed in quantity sufficient to give a homogeneous liquid phase when the water of the tissue has been taken into it.

After removal of insoluble solid debris, addition of more water and chloroform yields a two-phase system, the more dense (chloroform-rich) of which may, with considerable confidence, be expected to contain virtually all of the lipids, and the less dense aqueous phase all of the soluble non-lipid components of the sample under investigation. (As mentioned earlier, extraction of "acidic" lipids, e.g., phosphorylated phosphatidylinositols, tends to be incomplete unless the aqueous phase is acidified.) Without fear of losing lipids from it, the chloroform-rich phase may then be freed of methanol by shaking with water, and finally of traces of water by application of an inert, insoluble drying agent, e.g., finely divided anhydrous magnesium sulfate.* The resulting dry chloroform extract is suitable for application of such techniques as silicic acid-column chromatography, which serves to distinguish classes of lipids on the basis of differences in hydrogen-bonding potentialities of their functional groups (see Sweeley, 1969).

2.8. Relationship of Distinctive Lipid Solubility to Life Processes

The fact that polymeric structure is typical of proteins and common among carbohydrates, but rare in lipids, may be related indirectly to differences in solubility characteristics that distinguish these classes of substances. In the complex, living, essentially aqueous organism, it is clearly necessary to maintain close control over the facility with which the myriad substances involved in the life of the organism move about in it. Compartmentalization obviously plays an important role in such regulation, but control of the solubility of various kinds of substances in the aqueous medium provides another, quite different way of limiting their mobility. Conversion of water-soluble amino acids and monosaccharides into high-order polymeric proteins and polysaccharides greatly reduces what might otherwise be too facile solubility in the solvent of living systems. In contrast, very few of the inherently water-insoluble lipids are polymeric; they are innately immobile, and elaborate mechanisms have therefore had to be evolved to facilitate their transport *in vivo*.

The essential effectiveness of the compartmentalization of living matter, preventing the casual mixing of aqueous solutions of different composition, also involves the fundamental principles discussed here. The continuous apolar central region of biomembranes responsible for maintenance of the integrity of these compartments represents a formidable barrier to adventitious passage of water-soluble, highly polar substances. Replacement of the strong association of such substances with water—even briefly—with the much weaker interactions with the hydrocarbon moieties of the interior of the biomembrane is highly unfavorable

---

*This frequently used desiccant absorbs up to seven mole equivalents of H_2O quite strongly and has a negligible tendency either to absorb or to react chemically with most lipids. In our experience, however, it binds variable amounts of cholesterol (and hence possibly, by implication, other hydroxylic lipids); chloroform extracts dried with $MgSO_4$ cannot, therefore, be relied upon with respect to containing all of the cholesterol originally present in the tissue sample.

thermodynamically and thus unlikely to occur in the absence of mechanisms specifically designed to effect such transpositions.

REFERENCES

Allred, A. L., 1961, Electronegativity values from thermochemical data, *J. Inorg. Nucl. Chem.* **17**:215.
Bligh, E. G., and Dyer, W. J., 1959, A rapid method of total lipid extraction and purification, *Can. J. Biochem. Physiol.* **37**:911.
Bloor, W. R., 1943, *Biochemistry of the Fatty Acids and Their Compounds, the Lipids,* Reinhold, New York.
Brandreth, D. A., and Johnson, R. E., 1971, Solubility of stearic acid in some halofluorocarbons, chlorocarbons, ethanol, and their azeotropes, *J. Chem. Eng. Data* **16**:325.
Cosaert, E., 1971, Oplosbaarheidsparameters om de oplosbaarheid van een vaste stof in een systeem van solventen te voorspellen, *Chim. Peintures,* **34**:169 (*Chem. Abstr.,* 1971, Solubility parameters allowing the prediction of the solubility of a solid in a solvent system, **75**:204, Abstract Number 113248d).
Folch, J., Lees, M., and Sloane Stanley, G. H., 1957, A simple method for the isolation and purification of total lipids from animal tissues, *J. Biol. Chem.* **226**:497.
Hara, A., and Radin, N. S., 1978, Lipid extraction of tissues with a low-toxicity solvent, *Anal. Biochem.* **90**:420.
Hoerr, C. W., and Ralston, A. W., 1944, The solubilities of the normal saturated fatty acids. II. *J. Org. Chem.* **9**:329.
Hoerr, C. W., Sedgwick, R. S., and Ralston, A. W., 1946, The solubilities of the normal saturated fatty acids. III. *J. Org. Chem.* **11**:603.
Kolb, D. K., and Brown, J. B., 1955, Low temperature solubilities of fatty acids in selected organic solvents, *J. Am. Oil Chem. Soc.* **32**:357.
Krygowski, T. M., and Fawcett, W. R., 1975, Complementary Lewis acid–base description of solvent effects. I. Ion–ion and ion–dipole interactions, *J. Am. Chem. Soc.* **97**:2143.
Pauling, L., 1960, *The Nature of the Chemical Bond,* 3rd ed., Cornell University Press, Ithaca, New York, p. 98.
Pimentel, G. C., and McClellan, A. L., 1960, *The Hydrogen Bond,* W. H. Freeman, San Francisco.
Preckshot, G. W., and Nouri, F. J., 1957, Phase behavior of fatty acids–chlorinated solvent systems, *J. Am. Oil Chem. Soc.* **34**:151.
Ralston, A. W., and Hoerr, C. W., 1942, The solubilities of the normal saturated fatty acids, *J. Org. Chem.* **7**:546.
Stein, G., 1922, "Sacred Emily", in: *Geography and Plays,* The Four Seas Company, Boston.
Sweeley, C. C., 1969, Chromatography [of lipids] on columns of silicic acid, in: *Methods in Enzymology,* Vol. XIV (J. M. Lowenstein, ed.), Academic Press, New York, pp. 254–267.

SELECT BIBLIOGRAPHY

Hildebrand, J. H., Prausnitz, J. M., and Scott, R. L., 1970, *Regular and Related Solutions,* Van Nostrand Reinhold, New York.
Joesten, M. D., and Schaad, L. J., 1974, *Hydrogen Bonding,* Marcel Dekker, New York.
Ralston, A. W., 1948, *Fatty Acids and Their Derivatives,* John Wiley, New York.
Tanford, C., 1980, *The Hydrophobic Effect: Formation of Micelles and Biological Membranes,* 2nd ed., John Wiley, New York.

3
NATURE OF THE FATTY ACIDS: THE CARBOXYL AND HYDROCARBON MOIETIES

Because of their low polarity, lipids can be separated almost quantitatively from the non-lipid components of tissue by extraction with certain organic solvents (Chapter 2). Development of a variety of chromatographic methods has made it possible now to separate classes of lipids from one another on the basis of what are, in many cases, quite subtle differences in their physical properties. These advances have led to a marked acceleration of progress in knowledge and understanding of the chemistry and metabolism of lipids.

Many of the chemical as well as physical and solubility properties of the lipids are determined by the nature of the fatty acids they contain either in ester or amide linkages. It is not an uncommon practice to deduce as much as possible about the composition of a class of lipids by analysis of the mixture of fatty acids obtained from it by hydrolysis. The properties of the fatty acids are therefore of much importance, even though they occur in free form only in trace amounts in natural lipid mixtures.

The fatty acids are carboxylic acids, and are represented by the general formula RCOOH, in which R is usually an unbranched straight-chain or normal hydrocarbon group, frequently containing one or more *cis*-olefinic linkages. Most fatty acids contain an even total number of carbon atoms, commonly 16 or 18, reflecting their elaboration from two-carbon units (see Chapter 8). Structural features such as cycloalkyl, conjugated polyene, and hydroxyl substituents are more rarely encountered, almost always in plant rather than in animal lipids.

The discussion that follows focuses first on characteristics of the carboxyl group, then on certain features of the hydrocarbon moieties (R of RCOOH), which are of particular interest since these are retained unmodified in the lipids and thus contribute importantly to many of their properties.

3.1. Dissociation of Carboxylic Acids

The strength of the fatty acids is of practical as well as theoretical interest. Establishing just how strong (or weak) these extremely slightly water-soluble acids

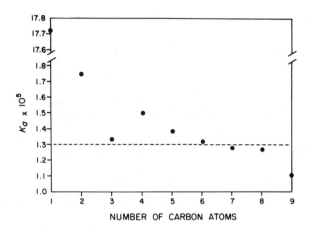

FIGURE 3.1. Variation in acid dissociation constant (K_a) with number of carbon atoms in short (appreciably water-soluble) normal carboxylic acids (after Dippy (1938)).

are presents some technical difficulties, however, since this quality of an acid is expressed in terms of the dissociation constant nominally representing the equilibrium constant for the hypothetical* reaction $HA \rightleftarrows H^+ + A^-$. K_a expresses the extensiveness with which proton release by a Brønsted acid takes place in water which, by virtue of its interactions with all three species involved, plays an important role in determining the position of the equilibrium.

Two fairly obvious approaches to determination of the strength of the nearly water-insoluble fatty acids have been used. One method measures the strengths of the lowest homologues in the series, which are soluble in water, with the expectation that studies of this rather limited number of substances would reveal a convergence of K_a values supporting confident extrapolation. This is a reasonable expectation on theoretical grounds: hyperconjugation effects should not differ, since C_3 and higher homologues have the same number of α hydrogen atoms, and the weak inductive effect of the terminal methyl group should die out quickly with increasing chain-length. Dissociation constants of the lower, appreciably water-soluble homologues, measured by Dippy (1938), are shown in Fig. 3.1.

The outcome of this approach is not completely satisfactory. Although the dissociation constants appear to be settling down nicely at about 1.3×10^{-5} with

---

*The equilibrium reaction actually involved is $HA + H_2O \rightleftarrows H_3O^+ + A^-$, the position of the equilibrium depending on the readiness with which the acid HA yields its proton to the standard base, H_2O, i.e., on the relative basicities of H_2O and A^-. The equilibrium constant for this reaction includes the concentration of water as a parameter: $K_{eq} = [H_3O^+][A^-]/[H_2O][HA]$. However, since these relationships apply most accurately to dilute solutions, $[H_2O]$ may be taken to be unchanging—equal to that of pure water, i.e., 55.5 mol/liter at room temperature—and may therefore be incorporated into the constant. Thus the acid dissociation constant $K_a = K_{eq}[H_2O]$ or 55.5 K_{eq} at room temperature. If it is recognized, further, that any proton not associated with A^- is associated with H_2O, i.e., that free protons, *per se*, do not exist under such conditions, then $[H^+]\equiv[H_3O^+]$, and the meaning of the expression $K_a = [H^+][A^-]/[HA]$ is understood and convenient, even though it is certainly not the equilibrium constant for the reaction $HA \rightleftarrows H^+ + A^-$ in the absence of water.

increasing chain-length, the significantly lower K_a of *n*-nonanoic acid is disconcerting. The relationship between chain-length and K_a is clearly at least triphasic. The tenfold weakening of acetic over formic acid is readily ascribable to hyperconjugation (Eq. 3.1):

$$\underset{\overset{|}{\text{H}}}{\overset{\text{H}}{\text{C}}}\!\!-\!\!\underset{\text{OH}}{\overset{\ddot{\text{O}}:}{\text{C}}} \longleftrightarrow \underset{\overset{|}{\text{H}}}{\overset{\text{H}^{\oplus}}{\text{C}}}\!\!=\!\!\underset{\text{OH}}{\overset{:\ddot{\text{O}}:^{\ominus}}{\text{C}}} \quad \text{hybrid:} \quad \underset{\overset{|}{\text{H}}}{\overset{\text{H}^{\delta+}}{\text{C}}}\!\!=\!\!=\!\!\underset{\text{OH}}{\overset{\text{O}^{\delta-}}{\text{C}}} \tag{3.1}$$

but this effect should be less influential in the higher homologues, having two instead of three α hydrogen atoms. The relatively low acidity of propionic acid may be rationalized by the inductive effect of the terminal methyl group which compensates for the loss of an α hydrogen atom. As expected, butyric is a somewhat stronger acid than propionic; but the trend, as the methyl becomes more remote from the carboxyl group, does not continue. Yet another consideration now sustains the apparent further drop in K_a with increasing chain-length: the hydrophobic effect. Increasingly appreciable aggregation results in overestimation of equilibrium concentrations of undissociated, monomeric RCOOH (HA) and hence in spuriously low K_a values. (The aggregation of carboxylic acids in aqueous solution due to the hydrophobic effect is further discussed below.)

The other approach to determination of dissociation constants of the nearly water-insoluble fatty acids involves use of a solvent (or mixture of solvents) in which these substances are adequately soluble. Any such solvent will, of course, be less polar than water and thus have a lower dielectric constant. The resulting increase in electrostatic force suppresses dissociation by making it more difficult for the proton to leave the carboxylate anion. Any difference between the basicity of the solvent and that of water will also affect the observed equilibrium constant. However, corrections for these effects can be made by comparing the K_a values of a water-soluble acid determined in water and in the less polar solvent employed. Noerland (cited by Schmidt-Nielsen, 1946) found, on the basis of studies using 95% ethanol–5% H_2O, that the variation in pK_a values of normal carboxylic acids containing 6–16 carbon atoms was less than 0.01 pK_a unit. Similarly, Garvin and Karnovsky (1956), using β-ethoxyethanol ($C_2H_5OCH_2CH_2OH$) containing 1 vol % H_2O and 0.001 M KCl, found that the pK_as of the C_{12}–C_{18} fatty acids were all within 0.1 pK_a unit of that of acetic acid.

For present purposes it can therefore be taken as having been demonstrated experimentally that the fatty acids are virtually of identical acid strength, regardless of chain-length: K_a about 1.3×10^{-5} (pK_a 4.8) at 25°C, i.e., somewhat less than that of acetic acid. Commonly occurring structural variations are not expected to affect the acidities of fatty acids appreciably. Chain-length effects have been discussed. Branching of the alkyl group (unless this occurs at the α carbon, where it would reduce hyperconjugation) should also have a negligible effect. The usual location of carbon–carbon unsaturation is far too remote from the carboxyl group to influence its dissociation; even when the olefinic center is immediately adjacent to (i.e., conjugated with) the carboxyl group, as it is in acids corresponding to the α,β-unsaturated intermediates of fatty acid synthesis and β-oxidation, the effect is small: cf. K_a 2.0 × 10^{-5} for *trans*-crotonic (2-butenoic) acid with K_a 1.50 × 10^{-5} for butyric (butanoic). The powerful inductive (electron-withdrawing) effect of the hydroxyl group elevates the acid strength of fatty acids bearing such a substituent when it is close to the carboxyl group: β-hydroxy acids (corresponding, for example, to the intermediates of fatty acid metabolism) are about twice as strong, and the α-hydroxy acids that occur commonly in cerebrosides are about ten times as strong as analogues having no hydroxyl group.

The acidity typical of fatty acids is of interest from several practical and physiological standpoints. Although customarily classified as "weak" (in comparison with the common mineral acids), carboxylic acids are (with the exception of some containing sulfur) the most strongly acidic of organic substances. That they release CO_2 from bicarbonate, reflecting acidity greater than that of carbonic acid, is a virtually unique property serving to identify them qualitatively. The chemical change RCOOH → RCOO⁻ is accompanied by a dramatic increase in water solubility, the much more hydrophilic, formally charged carboxylate group being able to overcome the hydrophobicity of alkyl groups considerably larger than those solubilizable by the undissociated carboxyl group (see Chapter 2). Such conversion of fatty acids to the corresponding salts (soaps), together with the equally facile reversal of the process (by acidification of the aqueous solution), provides a simple way of isolating fatty acids from other lipid hydrolysis products or from unhydrolyzable lipids, which are therefore termed "unsaponifiables" (meaning literally not convertible to soaps, or, in practice, not rendered soluble by action of aqueous alkali).

Although soaps are more soluble in water than in lipid solvents, their water solubility—particularly that of those involving sodium cations and/or saturated fatty carboxylate moieties—may be quite small.

The other very important practical consequence of the acidity of the fatty acids is that they are easily and precisely titratable in 95% ethanol by standard aqueous alkali, with phenolphthalein as indicator to an end-point at pH > 7; such titration yields a neutralization equivalent (g of acid ÷ moles of alkali required). Since the quantity of base consumed per unit weight of the acidic substance is inversely related to its molecular weight (and thus to its chain-length), the size of the hydrocarbon group of an isolated fatty acid (or the average size of such groups in mixtures of fatty acids) is readily determined in this simple way.

Because of the great differences in the natures of free (undissociated) fatty acids and of the corresponding carboxylate anions, and hence in the manner of their interaction with aqueous media (as, for example, in digestion of lipids and in fatty acid transport in blood), it is important to bear in mind that these substances are about 99% dissociated (i.e., in the carboxylate form) at pH 7.

An important change occurs when the carboxyl group is deprotonated, insofar as its two oxygen atoms become equivalent—a fact that might easily be overlooked in the usual simplified way of representing the change (3.2):

$$-C\begin{matrix}\\^{\displaystyle O}\\_{\displaystyle OH}\end{matrix} \xrightarrow{-H^{\oplus}} -C\begin{matrix}\\^{\displaystyle O}\\_{\displaystyle O^{\ominus}}\end{matrix} \qquad (3.2)$$

Since no more than a redistribution of electrons is required to convert one form of the anionic carboxylate group to another that is the exact equivalent of the first, the structure of the group is actually that of a resonance hybrid of the two (δ denotes a partial charge, anywhere between 0 and 1; in this case each of the two δs has a value of ½—exactly equal in magnitude and totalling 1) (Eq. 3.3):

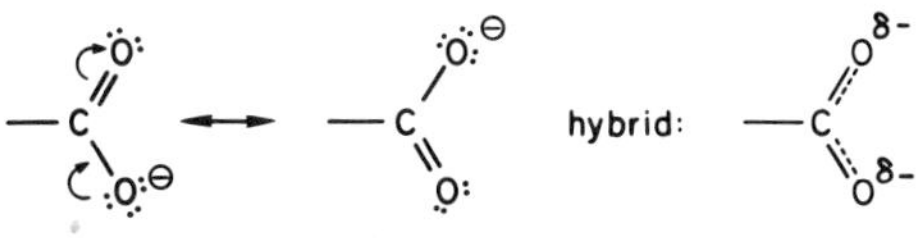

$$\begin{array}{c}\text{(resonance structures shown)}\end{array} \qquad (3.3)$$

The resonance energy associated with this hybridization provides driving force for the dissociation of the carboxyl group, contributing very substantially to the strength of such acids (compare, for example, the acid strengths of alcohols, the deprotonated anionic forms of which are not stabilized in this way: $K_a \sim 10^{-17}$, i.e., some 10^{12}-fold smaller than for the analogous carboxylic acids). The symmetry of the resonance-stabilized carboxylate group is manifest in experimental observations showing, for example, that the carbon–oxygen bond length and vibrational frequency are those of a single type of group (i.e., $C \mathop{=\!=\!=} O^{\delta-}$), intermediate between those expected for the isolated or noninteracting $C=O$ and $C-O^-$ groups.

The fact that the oxygen atoms of the carboxylate ion carry half negative charges (as opposed to either bearing a full charge) also has consequences of possible physiological importance with respect to the effects of the group on the arrangement of water molecules near it in aqueous solutions.

3.2. Association of Carboxylic Acids

Methyl acetate (molecular weight 74) boils at 60°C (at atmospheric pressure), and since boiling point is usually directly related to molecular weight, the parent acetic acid (molecular weight 60) would be expected to have a boiling point below that of its methyl ester. The acid boils at 118°C! The fact that the volatility of fatty acids is increased by conversion to the methyl esters (despite the nominal increase in molecular weight) is used to great practical advantage in the fractional distillation of fatty acid mixtures and in their analysis by gas–liquid chromatography. But why do the carboxylic acids behave as though their molecular weights were larger than they really are? A simple explanation is provided by the data of Brocklesby (1930) on the relative molar freezing point depressions of methyl oleate and oleic acid in several solvents. The data in Table 3.1 show that in the apolar solvent, cyclohexane, the effective molecular weight of the free acid is exactly twice the formal value, and that the completeness of the dimerization falls off as the polarity (or, more explicitly, the hydrogen-bonding potentiality) of the solvent increases.

TABLE 3.1. Ratio (α) of Molar Freezing
Point Depression of Methyl Oleate to That
of Oleic Acid in Several Solvents[a]

Solvent	α
Cyclohexane	2.00
Benzene	1.77
Dioxane	1.10
Acetic acid	1.01

[a]Brocklesby (1930).

The reason carboxylic acids display a very strong tendency to associate, generally in dimeric form, is readily apparent on detailed inspection of the structure of the functional group. Since the carbonyl group is a good H-bond acceptor and the hydroxyl group a good H-bond donor, it is expected that carboxyl groups would tend to bind strongly to each other by H-bonding (Eq. 3.4):

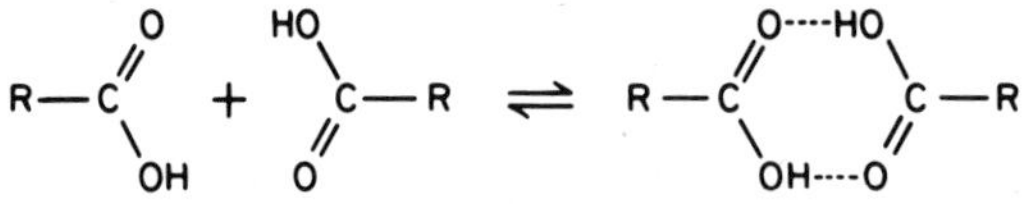

$$R-C(=O)(OH) + (HO)(O=)C-R \rightleftharpoons R-C(O\cdots HO)(OH\cdots O)C-R \qquad (3.4)$$

Actually, the presence in the carboxyl group of both carbonyl oxygen and hydroxyl groups on the same carbon atom gives rise to an electron delocalization that enhances both the H-bond-accepting potential of the carbonyl oxygen and the H-bond-donating ability of the hydroxyl group (Eq. 3.5):

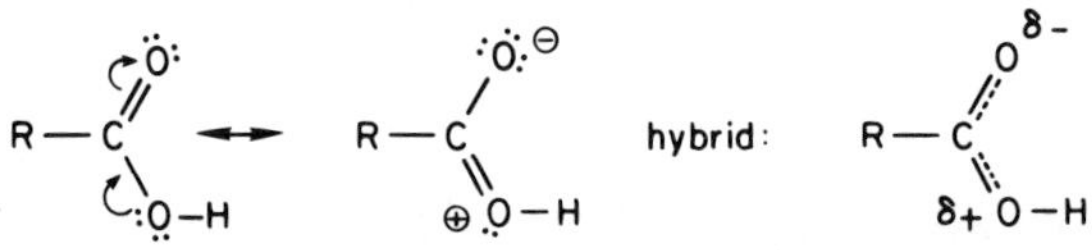

$$R-C(\ddot{\ddot{O}}:)(\ddot{\ddot{O}}-H) \leftrightarrow R-C(:\ddot{O}:^{\ominus})(\overset{\oplus}{\ddot{O}}-H) \quad \text{hybrid:} \quad R-C(O^{\delta-})(\overset{}{O}-H_{\delta+}) \qquad (3.5)$$

The striking stability of carboxylic acid dimers (in the absence of competitive hydrogen-bonding substances) invites consideration of resonance hybridization (see Eq. 3.6).

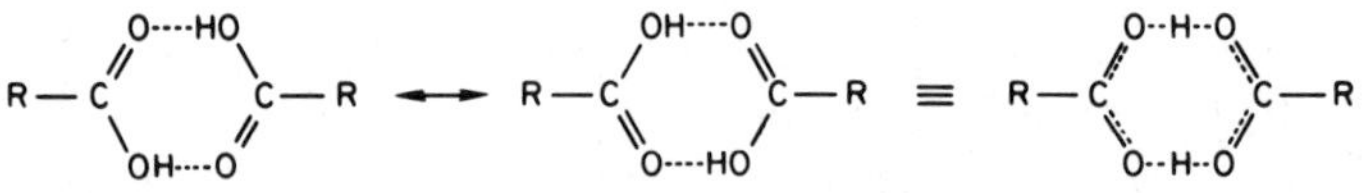

$$R-C(O\cdots HO)(OH\cdots O)C-R \leftrightarrow R-C(OH\cdots O)(O\cdots HO)C-R \equiv R-C(O\cdots H\cdots O)(O\cdots H\cdots O)C-R \qquad (3.6)$$

which would place the shared hydrogen atoms equidistant between oxygen atoms, and would make all four C—O bonds equal in length. Determination of the actual placement of atoms in the acetic acid dimer (see below) shows, however, that this is not the case. Hydrogen exchange between carboxylic acid molecules is, of course, rapid, but a true equilibrium (rather than resonance hybridization) is involved, presumably via an acid–base reaction and an ion-pair transition state (Eq. 3.7):

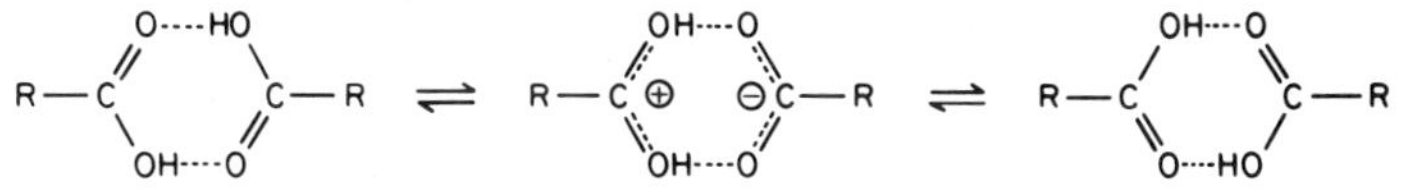

$$R-C(\text{O}\cdots\text{HO})(\text{OH}\cdots\text{O})C-R \rightleftharpoons R-C^{\oplus}(\text{OH}\cdots\text{O})(\text{OH}\cdots\text{O})\ominus C-R \rightleftharpoons R-C(\text{OH}\cdots\text{O})(\text{O}\cdots\text{HO})C-R \quad (3.7)$$

An analogous transfer of hydrogen from H-bond donor (hydronium ion) to acceptor (water) is responsible for the remarkably great apparent mobility of protons in ice.

That the H atoms in such dimers are not midway between the pairs of oxygen atoms, and that there are two (instead of one) carbon–oxygen bond lengths is shown, for example, by electron diffraction study of the vapor-phase dimer of acetic acid, disclosing the interatomic distances of interest shown in the diagram below:

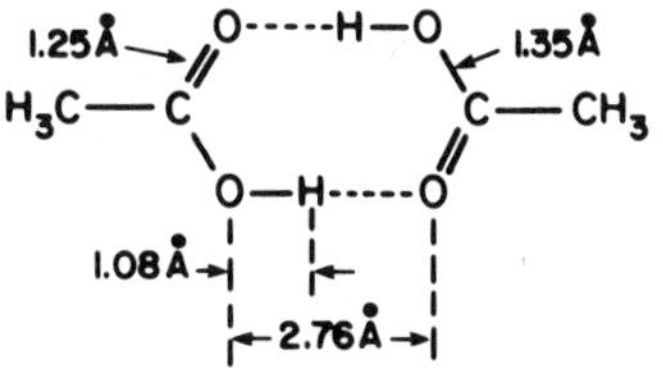

$$H_3C-C\underset{1.25\text{Å}}{\overset{\text{O}\cdots\text{H}-\text{O}}{\diagup}}\underset{1.35\text{Å}}{\diagdown}C-CH_3 \atop \text{O}-\text{H}\cdots\text{O} \atop 1.08\text{Å} \quad 2.76\text{Å} \quad (3.8)$$

Although such data clearly exclude the resonance hybrid from serious consideration, they do serve to substantiate other features of the carboxyl group: resonance polarization, reflected in shortening of the C—O distance—normally 1.44 Å—and lengthening of the normally 1.24-Å C=O distance; and strong H-bonding (stretching of the normally 0.96-Å H—O bond length, and characteristic O—H ··· O distance). It will be recalled (see Table 2.4) that the H-bonds in liquid water are of the same length, but the O—H contribution to this distance is 0.99 instead of 1.08 Å.

The hydrogen bonds involved in the carboxylic acid dimers are among the strongest known for those between oxygen atoms: 7 kcal/mol each (cf. 5 kcal/mol for those in ice), a total of 14 kcal/mol favoring formation of the dimer, measurable in terms of the effect of temperature on the equilibrium constant for the process 2 monomers $\rightleftharpoons$ dimer. Consideration of specific values of such equilibrium constants serves to emphasize the quantitative importance of this association process and thus to underscore its influence on various aspects of the behavior of the fatty acids. The vapor density of propionic acid (C_2H_5COOH), for example, yields a value of 1258 atm^{-1} at 27°C for $K_{eq} = P_D/P_M^2$ (P_D and P_M are partial pressures of dimer and monomer, respectively), from which it may be calculated that over 82% of the molecules of this substance in the gas phase at this temperature are associated. Similarly, by taking advantage of the fact that the hydroxyl group involved in hydrogen bonding absorbs light (in the infrared range) of appreciably longer wavelength (3.2 µm) than does the unbonded group (2.8 µm), Wenograd and Spurr (1957) showed that C_2 through n-C_{12} carboxylic acids are about 90%

associated in carbon tetrachloride solution at 24°C (formal concentration, 0.02 mol/liter). Except for formic and acetic acids, such substances are completely dimeric in the crystalline state (see Chapter 4). (In crystals of formic and acetic acids, each carboxyl group accepts an H-bond from a second and donates an H-bond to a third such group, and these substances are therefore polymeric rather than dimeric in the solid state.)

In addition to the cryoscopic measurements discussed earlier, Brocklesby (1930) also investigated the extensiveness of association of oleic acid as reflected in its effects (as solute, compared with those of methyl oleate) on the boiling points of a number of common solvents. His findings are summarized in Table 3.2 in which α has a connotation analogous to that in Table 3.1, a value of 1 being expected for complete monomolecular dispersion, and 2 for complete association as dimers. In view of the fact that association is suppressed independently by elevated temperature and/or by H-bond-accepting or -donating potential of solvent, association is appreciable ($\alpha > 1$) only in solvents having little or no ability to be involved in H-bonding, and is most extensive in the lowest boiling of these, carbon disulfide. On the other hand, association is weak or inappreciable, regardless of boiling point, in the other solvents, which, by virtue of containing strongly polarized carbon–oxygen bonds, are good hydrogen bond acceptors. (The unusual behavior of chloroform may be taken to reflect its appreciable ability to function as an H-bond donor; Wenograd and Spurr (1957) also observed carboxylic acid dimer dissociation constants to be about ten times greater in chloroform than in carbon tetrachloride.) These considerations illustrate the competitive characteristics of the hydrogen-bonding phenomenon and emphasize the fact that the association of carboxylic acids is strongly influenced by the presence or absence of alternative H-bond donors and/or acceptors in their environment. In the absence of other substances, in the gaseous, liquid, or solid states, or dissolved in apolar solvents, the fatty acids are extensively associated as doubly H-bonded dimers.

TABLE 3.2. Ratio (α) of Molar Boiling Point Elevation of Methyl Oleate to That of Oleic Acid in Selected Solvents of Different Boiling Point and Polarity[a]

Solvent	Boiling point[b]	α
Dioxane	101	1.00
Cyclohexane[c]	81	1.56
Benzene[c]	80	1.51
Carbon tetrachloride[c]	77	1.53
Chloroform	62	1.18
Acetone	56	0.97
Carbon disulfide[c]	46.5	1.89
Diethyl ether	34.5	1.03

[a]After Brocklesby (1930).
[b]Of pure solvent, °C at atmospheric pressure.
[c]Solvents of very low polarity.

NATURE OF THE FATTY ACIDS

But in dilute solution in a solvent of even modest H-bonding potentialities (such as diethyl ether), such association is highly suppressed. The state of stearic acid dissolved in pentane, for example, is thus quite different from that of the substance dissolved in ether.

It is of considerable practical importance to stress an obvious extension of these conclusions: that in the presence of water, the carboxyl–carboxyl-group association of fatty acids will be virtually completely eliminated by access to this particularly effective H-bond-accepting and -donating substance. (Note the disposition of long-chain carboxylic acids in laterally compressed monomolecular films on water surfaces, in which the apolar hydrocarbon moieties have escaped the inimically highly polar solvent, but each molecule is firmly attached to the liquid phase by carboxyl-group–water H-bonds. See Chapter 4.)

It should be pointed out here that studies of certain short-chain carboxylic acids by Martin and Rossotti (1959, 1961) revealed that such substances also dimerize appreciably in aqueous solution. Scheraga and coworkers (see Schrier *et al.*, 1964) have pointed out that such association is of an entirely different sort, involving hydrophobic bonding of the alkyl (rather than H-bonding of carboxyl) moieties of these solutes, reflecting the strong tendency of such highly apolar entities to minimize their contact with water (see also Smith and Tanford, 1973). In agreement with explanations of these two quite different dimerization processes is the fact that, while the stabilities of dimers formed in apolar media are virtually independent of alkyl-group chain-length, those of dimers in aqueous solution are directly and linearly related to this structural feature.

Although Scheraga has proposed that the orientation of the two molecules in such hydrophobically bound dimers is one that might be termed parallel, involving a single H-bond between the carboxyl groups of the paired molecules, the alternative antiparallel arrangement:

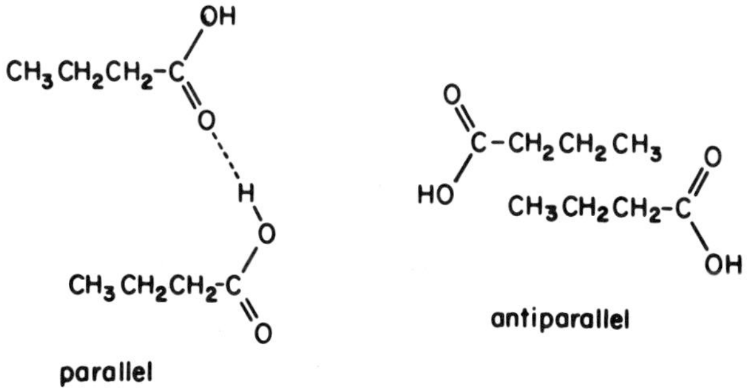

$$(3.9)$$

would result in superior hydrophobic bonding. (Geometry of the carboxyl group and of H-bonding are such that either strength of the H-bond or ability of methylene groups near the carboxyl groups of the two molecules to be in van der Waals contact would have to be seriously compromised in the parallel arrangement.) It also seems improbable that hydrogen bonding between the carboxyl groups would be significantly competitive with that between the separate carboxyl groups and water.

3.3. Conformation of the Carboxyl-C–C_α Bond in Carboxylic Acids

Before discussing the conformation of the R groups of fatty acids, it is appropriate to consider the conformation at the juncture of the R and carboxyl groups, i.e., about the carboxyl–C–C_α bond. Except in α,β-unsaturated species, the syncoplanar conformation of the C_β–C_α–C=O moiety is preferred in fatty acids as well as in the corresponding esters and amides (see Dunitz and Strickler, 1968). The apparent "eclipsing" of the double bond by a substituent of the sp^3-hybridized carbon (C_α) in such groups is very common and can be rationalized as actually involving a virtual staggered conformation, the double bond being considered to consist of two splayed, curved single bonds (τ type) instead of a $\sigma + \pi$ combination. Eclipsing of the carbonyl oxygen by the considerably larger β-CH_2 instead of by an α-H may be attributable to superior coulombic interaction between the negatively biased oxygen atom (via resonance; see p. 28) and slightly positive β-hydrogens (via C–H bond polarization), which are closer than α-hydrogens.

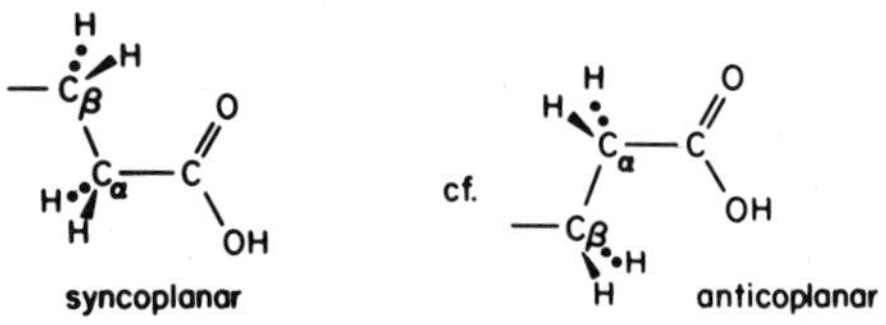

syncoplanar   cf.   anticoplanar

Similarly, the preferred anticoplanar conformation (**1**) of the first C–C bond of *trans*-α,β-unsaturated carboxylic acids is easily rationalized in terms of its correspondence to a staggered rather than to an eclipsed (syncoplanar, **2**) conformation of the C_α–COOH bond:

Another array involving a nominally single bond between sp^3- and sp^2-hybridized carbon atoms—one that is of considerable interest—is that in which a methylene group is joined to an olefinic center:

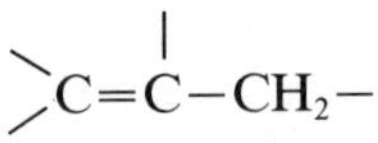

$$\rangle C=\overset{|}{C}-CH_2-$$

In accord with arguments presented above, the preferred (staggered) conformation about the bond in question places one substituent of the α-carbon atom in the plane of the olefinic center; as will be shown later, this, for steric reasons, will be an α-hydrogen (if available), even when the *cis* substituent is H (i.e., minimally small) (compare structure **3**, better than **4**):

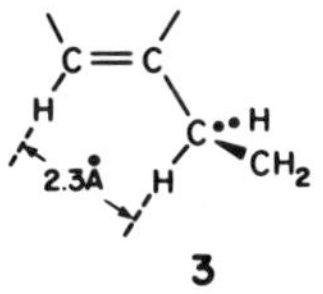

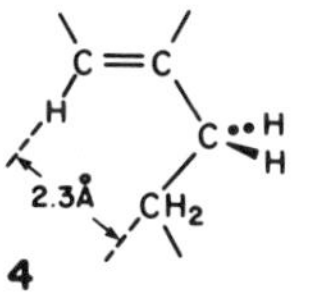

(The calculated internuclear distances should be compared with van der Waals radii sums: 2.4 Å for H · · · H and 3.2 Å for H · · · CH$_2$.) It will be noted that such conformations add an additional one or two atoms to the six already present, but exclude any β-carbon atoms from the plane of the olefinic array.

3.4. Conformation of the Hydrocarbon Moiety of Fatty Acids

Conformations of the fully saturated (and thus simplest) of the R groups of the fatty acids, RCOOH, are considered first, then those containing one or more olefinic centers, in order to decide to what extent it is meaningful and useful to define the conformation of these inherently flexible molecules and to discuss effects that various types of environment might have on the shapes they are most likely to possess.

Many years ago, after he had pulled down the widely held shibboleth that large cycloalkanes are difficult to synthesize because they are highly strained, Ruzicka is said to have dramatized his lectures on the subject by likening the conformational restraints (or lack of them) in such substances to that of a string of beads. This greatly oversimplified analogy was faithful insofar as both beads and atoms occupy closely defined minimal amounts of space that are not easily violated, and in that the individual units are fairly closely fixed distances apart; but the analogy was faulty in failing to show restrictions imposed by two other parameters: the rigidity of the C—C—C bond angle and limitations to free rotation of groups at each end of carbon–carbon bonds with respect to each other.

In considering what shapes a saturated hydrocarbon chain may assume with ease (i.e., without excessive strain), the fundamental dimensions exemplified in propane will be taken to be virtually fixed. Several features of these data warrant emphasis. (i) Since the monosubstituted carbon atom is rarely encountered and therefore seldom of interest, the 2.0-Å van der Waals radius of the entire methyl (or methylene) group is more useful; this represents an average, effective radius of groups that are not, of course, spherically symmetrical. (ii) The usual 1.2-Å van der Waals radius of the H atom is used, although the apparent size of this atom is unusually variable—up to about 1.5 Å, depending on circumstances. Finally, (iii) the 112° C—C—C bond angle characteristic of normal acyclic hydrocarbon chains represents an appreciable departure from the tetrahedral angle (109.5°),

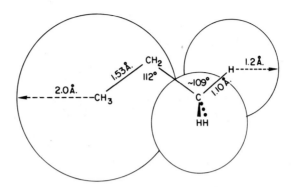

FIGURE 3.2. Dimensional parameters of a molecule of propane having one hydrogen atom of one methyl group anticoplanar with the carbon atoms.

reflecting that the four substituents of a saturated carbon atom are quite crowded and that the two bulky methylene (or methyl) groups will be accommodated by a more facile (ca. 0.1 kcal/mol) bond bending (rather than a stretching) adjustment; the spreading of the C−C−C bond angle by the two carbon substituents would occasion a concomitant compression of the H−C−H angle separating the two less bulky hydrogen substituents to less than 109.5°. The drawing of the propane molecule (Fig. 3.2) includes an indication, to scale, of the sizes of atoms and groups of atoms. These parameters of the shapes of molecules are heavily deemphasized in most drawings and models in order to make bond lengths and bond angles more easily discernible; but the fact that space in such molecules is at a high premium cannot be ignored, since this conclusion plays a decisive role in determining the shape they assume preferentially.

There is yet another variable that is of considerably greater latitude and therefore of much greater significance in determining the shape of such molecules: the dihedral angle describing the orientation with respect to each other of the two arrays of substituents situated on carbon atoms joined by a single covalent bond. This has to do with answering the question: how "free" is rotation about such a single bond? and constitutes the basis of the powerful approach of conformational analysis to the rationalization of many organic chemical phenomena. As any nonspherically symmetrical part of a molecule is rotated about the axis of a single bond joining it to the rest, the molecule assumes successively an infinitude of shapes, each characterized (conveniently in terms of the dihedral angle between the two parts, measured from some arbitrarily chosen starting orientation), as one of a family of "conformers" (or "rotamers"). Each may also be differentiated in terms of changes in strain or potential energy arising from the bulkiness of the various substituents that must be accommodated in the limited amount of space around the central single bond; in some cases, other types of so-called "nonbonding interactions" may be involved as well. These potential energy differences determine quantitatively which of the possible conformations predominate at any given temperature, and to what extent.

Even in ethane, in which all six of the hydrogen atoms on its two singly bonded carbon atoms are minimally bulky, rotation of one methyl group with respect to the other is not free. Comparison of the calculated with experimentally determined entropy and enthalpy values for this simplest of conformationally active alkanes reveals a 2.9-kcal/mol barrier to rotation; other considerations show that the minimum-energy (more stable) conformation of these molecules is that in which the hydrogen atoms on different carbon atoms are staggered (most remote; dihedral angle 60°) rather than eclipsed (in closest apposition; reference dihedral angle 0°).

In the system of immediate interest, that involving any four consecutive carbon atoms in the alkyl moiety of a saturated fatty acid, e.g., stearic acid, one hydrogen atom on each of the two central carbon atoms is replaced with a considerably more bulky methylene (or methyl) group, and the obstacles to free rotation would be expected to be substantially greater than in ethane. Before detailed consideration of the conformations of a saturated fatty acid, it is informative to examine those of n-butane. This alkane has been studied extensively, the relative potential energies of its cardinal conformations have been established with some precision, and it serves as a good model for any unit of four consecutive carbon atoms or "butane array" within the alkyl group of a saturated fatty acid (see Fig. 3.3).

As the dihedral angle between the two central-carbon substituent arrays is varied from 0° to 360°, starting arbitrarily from the conformation in which the methyl groups are as close as possible to each other, several extreme (or cardinal) conformations are encountered that represent states of either minimal or maximal potential energy; these, depicted conveniently as Newman projections, representing views along the central C—C axis, are shown in Fig. 3.4.

Conclusions reached in considering potential energies of the cardinal conformations of ethane immediately imply that the staggered (anti and gauche) conformations of n-butane represent states of minimal potential energy (i.e., of greater stability or less strain), the anti (or *"trans"*) being more stable (since the bulky methyl groups are further apart) than the gauche (or skew), encountered at dihedral angles of ±60°; the gauche and gauche' conformers are nonsuperimposable mirror images of each other, and therefore of identical energy. (The size of the methylene or methyl substituents and their close proximity in the gauche conformation are such that they are spread to an appreciable extent, at the expense

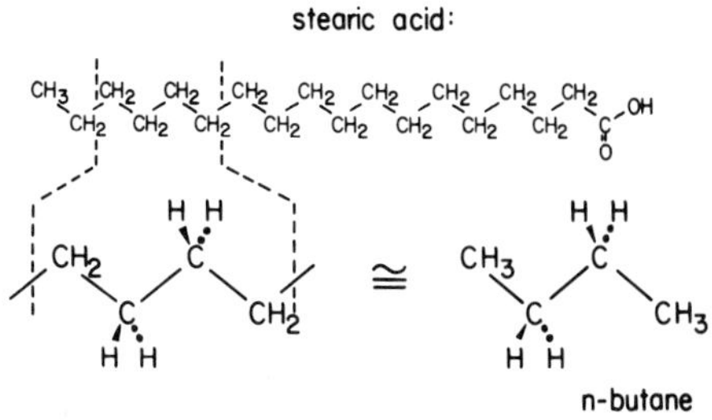

FIGURE 3.3. Relationship of a butane array (e.g., that with C14–C15 central bond) in stearic acid to n-butane.

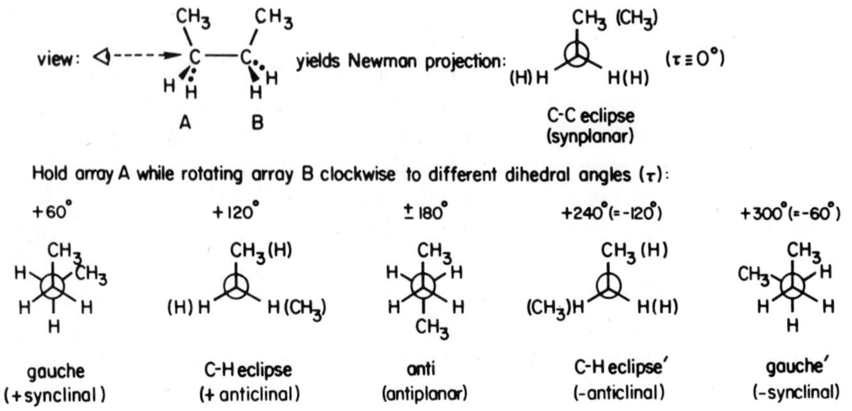

FIGURE 3.4. Newman projections of cardinal conformations of *n*-butane. (Substituents of the more remote carbon atoms, hidden in the eclipsed conformations, are shown in parentheses.) By convention (Klyne and Prelog (1960); see Bentley (1969)), the dihedral (or torsion) angle (τ) is taken to be zero when the distinctive substituents (methyl groups in this case) on the two singly bonded carbon atoms of interest are in eclipse (i.e., in the synplanar conformation). Clockwise rotation of the more remote array (B) yields positive dihedral angles, and counterclockwise, negative. When the two substituents are most remote ($\tau = \pm 180°$), the conformation is antiplanar (or anti). Other conformations are termed ±synperiplanar (from, but not including 0° = synplanar, to $\tau = \pm 30°$); ±synclinal ($\tau = \pm 30°$ to 90°, including gauche and gauche'); ±anticlinal ($\tau = \pm 90°$ to 150°, including the C—H eclipse conformations); and ±antiperiplanar ($\tau = \pm 150°$ up to but not including 180°-antiplanar [antiplanar]).

of partial eclipsing with smaller H substituents, the gauche potential energy minima thus occurring actually at dihedral angles of about ±68°.)

The eclipsed conformations represent potential energy maxima, the one with the methyl groups eclipsed (the C—C eclipse or "*cis*" conformation, at $\tau \equiv 0°$) being more highly strained than those (C—H eclipse, at ± 120°) in which both methyl groups are as close as possible to hydrogen substituents of the other carbon atom, crowding the smaller hydrogen substituents.

Changes in the potential energy of a molecule of *n*-butane, as the full complement of its central-bond conformations is explored through 360° of change in dihedral angle, are shown in Fig. 3.5. Quantitation of the differences in energy of these cardinal conformations of butane is of great interest because the C—H-eclipse potential energy (3.6 kcal/mol above that of the anti conformation) represents the activation energy required to permit interconversion of the anti and gauche forms and thus determines the rate of this interconversion. Since this is only about 4.5 times the average kinetic or thermal energy of the molecules (equal to RT or about 0.60 kcal/mol at room temperature), the interconversion is quite rapid and one would therefore not expect to be able to isolate samples of pure gauche or pure anti butane under ordinary conditions—the change from one to the other is simply too facile. Moreover, the 0.8 kcal/mol difference in energy of the two staggered conformers (anti and gauche) permits, by resort to the Boltzmann equation, calculation of the relative populations of these two most prevalent species at any given temperature. This equation states that the number (n_g)

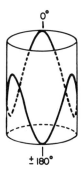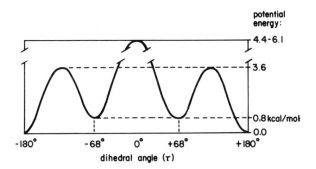

FIGURE 3.5. Potential energies (kcal/mol) of conformations of n-butane (plotted against dihedral angle, τ) in excess of that of the most stable, antiplanar conformation ($\tau = \pm 180°$). If the right cylinder (at left), upon which the actually continuous curve is drawn, were cut along the 180° coordinate and flattened out, the more usually seen two-dimensional curve (at right) would result. The gauche minima are shown at their true dihedral angles of $\pm 68°$ (that of the synplanar conformer being defined as 0°), that between bonds to the methyl groups being somewhat larger than those between a methyl group and smaller opposite hydrogen substituents.

of molecules in either gauche conformation is related to that (n_a) of anti molecules in thermal equilibrium at temperature T (°K) as follows:

$$\frac{n_g}{n_a} = 2e^{-(E_g - E_a)/RT}$$

where the E's represent the potential energies of the two conformations (the factor of 2 accounts for the involvement of the two energy-equivalent but distinguishable mirror-image gauche conformers); as noted above, $E_g - E_a = 0.8$ kcal/mol.

It is thus possible to calculate that at room temperature (25°C) about 66% of the molecules in a sample of n-butane are, at any instant, in the anti and about 34% in the gauche conformation. (Similarly, prevalence of the C—H-eclipsed conformers is calculated to be about 0.5% at this temperature; in general—even in the case of ethane—eclipsed conformations can be safely assumed to be inappreciable at "ordinary" temperatures.) It should be noted that the temperature dependence of the ratio is such that the relative quantitative importance of the more highly strained gauche conformers increases with temperature; at 37°C, for example, prevalence of the gauche conformation is about 1% higher (35%).

On the basis of the reasonableness of the assumption that in higher alkanes or alkyl groups having more than one four-carbon "butane array" (see Fig. 3.3), each is essentially conformationally independent of the others,* conclusions

---

*Certain facts are ignored in the simplifying assumption involved here. The terminal butane array, having a methyl in place of one methylene substituent, is more likely to be gauche than the others, in apparent conflict with expectation, insofar as $E_g - E_a$ is appreciably higher in n-butane (0.8 kcal/mol) than in higher normal homologues (0.5 kcal/mol). This smaller difference in potential energy is, however, offset by some limitation of conformational freedom peculiar to higher homologues of n-butane (notably the improbability of adjacent butane arrays being in enantiomeric gauche confor-

reached in the specific study of *n*-butane can be applied to the conformationally more complex situation extant in saturated fatty acids. In stearic acid, for example, there are 15 distinguishable butane arrays, there being 17 C−C bonds, the terminal two of which are not centers of butane arrays. (For simplicity it is assumed, certainly with considerable liberty, that the nonbonding interactions of the carboxyl group—see Section 3.3, p. 32—are equivalent in net effect to those of the methyl and/or methylene groups; including this with the 14 true butane arrays should not, however, seriously compromise the validity of the conclusions reached.) The conformation of the alkyl group of the isolated stearic acid molecule at room temperature is therefore expected to have the following characteristics:

(i) Individual butane arrays in molecules of this (or any other such substance) are gauche about 34% of the time and anti the rest of the time; or

(ii) On time average, about $\frac{1}{3}$ (i.e., 5) of the 15 butane arrays present in the stearic acid molecule are gauche and the remaining 10 anti.

It should be noted, since it might be easy to come to an erroneous, hasty conclusion in this regard, that even though the fully extended, longest, all-anti conformation of stearic acid has the lowest potential energy and must therefore exist in largest amount, the probability of such a molecule having none of its 15 butane arrays in gauche conformation is very small. At room temperature, for example, only 0.2% (i.e., $0.66^{15} \times 100$) of a large number of such molecules would, at any instant, have this shape. Even though no single alternative conformation, having at least one gauche butane array, would be present to the extent of the all-anti, the number of such alternative conformations is, of course, very large.

To this extent, then, the "shape" or conformation of the typical, rather lengthy, very flexible saturated fatty acid molecule (or fatty acyl moiety) may be defined. It is important, however, to remember that change from one conformation to another takes place easily and rapidly and that, for example, any elevation in temperature will be accompanied by an essentially instantaneous increase in gaucheness. It should be expected, moreover, that the conformation of such substances will be very sensitive to environmental conditions: the conclusions reached above are applicable to molecules in the gas phase, in the neat liquid state, or in dilute solution in a nonpolar solvent, but may be far off the mark under other circumstances. Two of these are of particular interest:

(i) In the solid (crystalline) state, such substances tend strongly to assume the all-anti, "zigzag," fully extended conformation (see the depiction of the stearic acid molecule in Fig. 3.3); molecules having this particular shape pack well in

---

mations—the so-called "pentane effect"). Predictions based on the assumed equivalence of *n*-butane and butane arrays are, when these are experimentally verifiable, substantially correct. Conformational characteristics of *n*-alkanes (and other long-chain substances) have been rigorously and elegantly analyzed by Flory (1969).

crystal lattices and thus provide the greatest possible integrated intermolecular forces of attraction. The uniquely maximal length of such molecules in this conformation was recognized early in X-ray diffraction crystal structure analysis as well as in laterally compressed monomolecular film studies, and has been confirmed in detail by more recent, much more highly refined crystal structure determinations. The literal "freezing out" of gauche conformations accompanying the passage from the liquid to the crystalline state is dramatically evident in infrared absorption spectra, those of liquid-state substances being much more complex, every conformer having its own set of characteristic vibrational frequencies; but only one set of these (that of the all-anti conformers) survives reduction of temperature below the melting point. The infrared spectrum of liquid n-butane, for example, exhibits two sets of absorption peaks that are readily distinguishable insofar as one set (that of the gauche conformers) diminishes in intensity while the other (that of the anti conformer) grows as the temperature is lowered; the fading of the gauche peaks with decrease in temperature is, however, discontinuous, suddenly becoming complete at the freezing point of the substance.

(ii) In an aqueous medium, on the other hand, it may be argued (albeit difficult to demonstrate) that the amount of gaucheness should be enhanced. Since the amount of surface presented to the water envelope by a long-chain saturated hydrocarbon moiety is inversely related to the extensiveness of involvement of gauche conformations, such groups (in dilute aqueous soap solutions, for example—see Chapter 4) would be expected to be significantly distorted in the sense that they would be more "balled-up" or "crumpled" than in nonaqueous media.

In summary, then, the shape of a long, straight-chain, saturated hydrocarbon moiety of the sort frequently encountered in lipids can be conceptualized as involving a composite of four-carbon units (distinguishable in having different central C—C bonds), each in either anti or gauche conformation, in facile thermal equilibrium, and therefore highly sensitive to temperature changes and to environmental constraints and other effects imposed by the environment; these conclusions are brought together in Fig. 3.6.

Of very great practical interest is the fact that the gaucheness of an n-alkyl group inversely affects its most probable length, L, conveniently taken to be the time-average distance between nuclei of the terminal carbon atoms of the chain. Such considerations have a direct bearing, for example, on the thickness of biomembranes. As shown in Fig. 3.7, a change of any butane array in such a chain from anti to gauche conformation not only foreshortens the distance between the

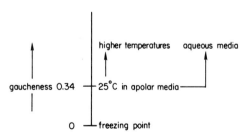

FIGURE 3.6. Effects of various conditions on the gaucheness (time-average number of butane arrays in gauche or gauche' conformation ÷ total number of butane arrays) of n-alkanes and n-alkyl groups.

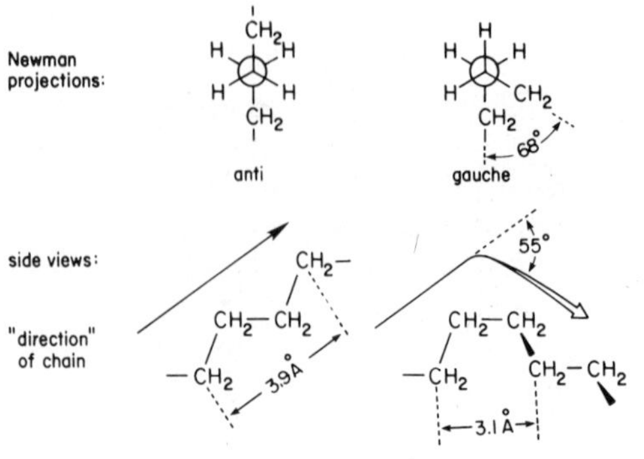

FIGURE 3.7. Shortening effects of the anti → gauche change in conformation of a butane array.

terminal carbon atoms of the array, but also, by bending the chain, brings its terminal carbon atoms closer together (the more so, of course, the nearer the gauche array is to the center of the chain). Taking the change in dihedral angle to be 112° (i.e., 180 − 68°—see Fig. 3.5), the terminal carbon atoms of the butane array are thus brought about 0.64 Å closer; on the same basis the change in direction of the chain is calculated to be 55°.

Since the C_{18} chain of the stearoyl group (for example) is expected to have, on average, about five gauche butane arrays at room temperature (see above), such a fatty acyl group has an effective length considerably shorter than maximal (all-anti). Figure 3.8 illustrates schematically this conformational effect on the probable length ($L_{a,g}$) of a stearoyl moiety of a phosphatidylcholine molecule. Taking the C−C bond length to be 1.53 Å and the C−C−C bond angle to be 112° yields a contribution of 1.28 Å per C−C bond to the axial length (L_a) of the all-anti chain—21.6 Å for the C_{18} stearoyl group. On the basis of Flory's statistical analyses and the "quite approximate" temperature corrections of Tanford (1974), the probable length ($L_{a,g}$) of the partially gauche chain at room temperature is

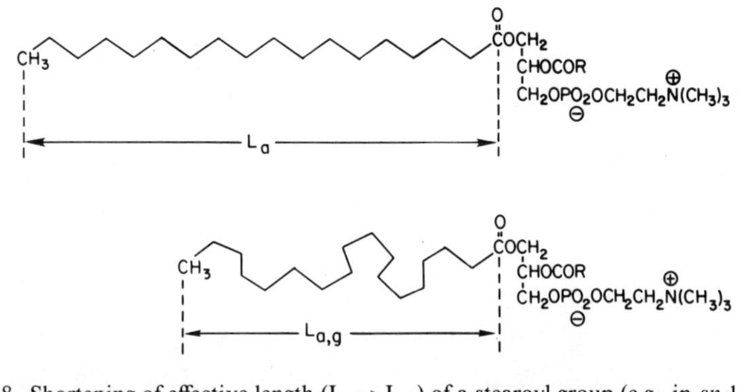

FIGURE 3.8. Shortening of effective length ($L_a \rightarrow L_{a,g}$) of a stearoyl group (e.g., in sn-1-position of a phosphatidylcholine) by change of several butane arrays from anti to gauche conformation. (L_a: distance between nuclei of first and last carbon atoms of the stearoyl group in all-anti conformation; $L_{a,g}$: shortened distance resulting when several of the butane arrays of the stearoyl group change from anti to gauche conformation.)

estimated to be about 63% of L_a, i.e., 13.6 Å. (The "gauche shortening"—to 63% of L_a in the case of the stearoyl moiety—is, of course, less extensive for shorter chains: e.g., to 65% for the C_{16} palmitoyl; to 71% for the C_{14} myristoyl; etc.) The stearoyl (or other saturated fatty acyl group) would have its maximum length (L_a) in crystalline fatty acids (or esters or amides), but would suffer a sudden and substantial reduction in length (to $L_{a,g}$ characteristic of the temperature) at the melting points of such substances. Attention is called to the curious—and possibly important—fact that since gaucheness increases with temperature, fatty acyl groups actually become shorter (and thus, for example, biomembranes thinner) as temperature is raised.

Seelig (1977) discusses experimental evidence that in amphiphilic lipid bilayers (e.g., biomembranes), certain saturated fatty acyl group conformations may be favored in which pairs of gauche butane arrays of opposite configuration occur, separated by one ("kinks") or by three or more anti arrays ("jogs"), presumably because such arrangements minimize overall "change in direction" (departure from essential straightness) of the alkyl chains.

3.4.1. Effects of Unsaturation

Since fatty acyl moieties frequently contain one or more *cis*-olefinic moieties (usually spaced 1,4 if polyunsaturated), it is of interest to consider the effects such structural features have on the shapes of hydrocarbon chains containing them.

The olefinic center occurring anywhere in (except at the end of) the hydrocarbon chain of the fatty acyl group may have either of two configurations: *cis (Z, zusammen)* or *trans (E, entgegen)* (see also Appendix, Chapter 5).

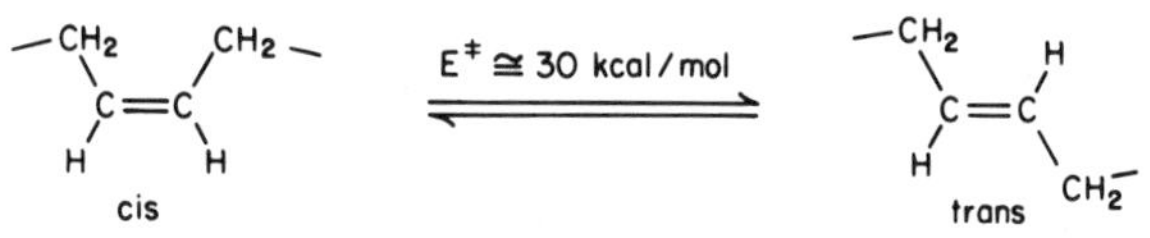

$$\text{cis} \quad \overset{E^{\ddagger} \cong 30 \text{ kcal/mol}}{\rightleftharpoons} \quad \text{trans}$$

The activation energy ($E^{\ddagger}$) necessary to permit the change from one configuration to the other is very high—about ten times as high as that separating the anti and gauche conformations of the butane arrays of normal saturated hydrocarbons. The probability that any individual olefinic molecule would acquire this amount of additional potential energy under ordinary conditions is negligible—which is to say that olefins having either configuration are thermally stable: they can be separated from one another, have distinctly different physical (and chemical) properties, of course, and only under special conditions (all of which presumably involve a transitory conversion of the double to a single bond) do they have any appreciable tendency to interconvert. This stability is clearly shown by the fact that oleic acid isolated from hydrolysis products of most natural lipids is found to contain no elaidic acid (the *trans* isomer of oleic), even though the substance (or derivatives of it) may have been submitted in the process to quite high temperatures (e.g., during gas chromatography of its methyl ester). As readily determined, for example, by measurement of differences in heats of combustion or of hydrogenation, *trans* alkenes are about 0.7 kcal/mol more stable than the *cis* isomers (coincidentally very nearly the same as for the anti vs. gauche conformers of *n*-butane). In the presence of certain catalysts, oleic acid can be "elaidinized," i.e., converted to equilibrium mixtures of oleic and elaidic acids in which the *trans* isomer predominates in ratios of about 3:1 (if the isomerization is induced by exposure to NO_2 at room temperature) or 2:1 (selenium at 200°C), in good agreement with Boltzmann equation calculations.

Thus, in sharp contrast to singly bonded carbon atoms, those joined by a double bond do not rotate with respect to each other; moreover, they are rigidly constrained to an arrangement in which six atoms (the doubly bonded carbon atoms and the initial atoms of their four substituents) are coplanar.

3.4.1.1. cis-Unsaturation

Presence of a double bond in an otherwise saturated fatty acyl chain therefore has important consequences with respect to its most likely shape: whereas the polymethylene sections of the group retain their conformational freedom (subject to limitations discussed above), the *cis* enoyl group contains a four-carbon sequence fixed in a configuration analogous to the $C-C$ eclipse (*"cis"* or synplanar) conformation, shortening the internuclear distance between the terminal carbon atoms of the array from 3.84 Å to 3.16 Å, and imposing a minimal bend of 44° in the carbon chain (based on dimensions of oleic acid in the crystalline state*, determined by Abrahamsson and Ryderstedt-Nahringbauer (1962)) (see Fig. 3.9 and Chapter 4). Since, in oleic acid, the *cis* double bond occurs in the center of the molecule, the effective length of the oleoyl group must, if unperturbed by intermolecular interactions, be substantially less than that of the corresponding saturated (stearoyl) group.

The work of Abrahamsson and Ryderstedt-Nahringbauer reveals details of the geometry of the four-carbon *cis* olefinic group, some of which are not likely to have been modified by intermolecular forces of the crystal lattice and may therefore be taken to be generally characteristic of this structural group. The lengthening of the $C=C$ bond, together with the shortening of the $=C-C_\alpha$ single bonds, are changes that can be ascribed jointly to hyperconjugation; the fact that the single bonds join sp^2- and sp^3- hybridized carbon atoms, however, also contributes to their shortness. The $C=C-C$ bond angle (nominally 120°) is clearly enlarged by steric interaction of the two α-methylene groups (calculated internuclear distance, 3.2 Å; cf. the 2.0-Å van der Waals radii of the methylene groups). (The fact that, in the crystal, the C_βs are not coplanar with the other carbon atoms

---

*The spreading of the α-methylene groups to give a $C=C-C$ angle of 126° (in crystalline oleic acid) was used in calculating the internuclear $C_\alpha-C_{\alpha'}$ distance; however, calculation of the minimal change in direction of the carbon chain is based on assumption that, when free of intermolecular interactions, the preferred conformation of oleic acid has anticoplanar $C_\alpha-C_\beta$ bonds, by analogy with conformational characteristics of *cis*-2-butene (see Karabatsos and Fenoglio, 1970). As pointed out earlier, the preferred conformation of a single α-methylene group involves a syncoplanar α-H; but α-Hs on both methylene groups of a *cis* alkene cannot be accommodated simultaneously in this way.

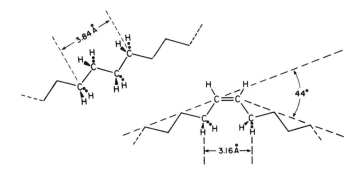

FIGURE 3.9. Change in shape of a normal carbon chain (all-anti) caused by insertion of a *cis* double bond.

of the group might, by bringing α-hydrogens into closer apposition, tend to spread this angle still further; the substantial reduction of the $=C-C_\alpha-C_\beta$ angle may be a related lattice effect.)

Since the oleoyl group occurs in lipids of a preponderance of all living organisms, and is, indeed, the major fatty acyl group in lipids of many plants and animals, it seems worthwhile reiterating conclusions reached with respect to its most likely shape in the liquid state and out of contact with water, as it would be, for example, in the interiors of biomembranes or of fat vacuoles: although the butane arrays of its two polymethylene segments retain their facility for alternation between anti and gauche conformations, the central "*cis*-butene array" imposes a fixed bend in the carbon chain that cannot be straightened out and very substantially reduces the time-average length of the group. Calculations carried out by Karabatsos and Lande (see Karabatsos and Fenoglio, 1970) indicate that the enthalpy of *cis*-2-butene (with one α-H anticoplanar and the other syncoplanar with the double bond) is only 250 cal/mol above that of the doubly anticoplanar conformer (that of the doubly syncoplanar is 2.5 kcal/mol higher). On the assumption that this simplest *cis* alkene is a valid model in this respect for the oleoyl group, one concludes that at 37°C this important fatty acyl group is, at the olefinic center, syncoplanar–anticoplanar about 40% of the time, and doubly anticoplanar about 60% of the time. It was pointed out above that the more probable doubly anticoplanar conformation imposes a 44° redirection or bend in the chain; note that the syn–anticoplanar form would create an even greater redirection (it is 67° in crystalline oleic acid, where the conformation is close to doubly syncoplanar, with the two polymethylene segments "*cis*," i.e., extending from the same side of the plane of the olefinic center).

3.4.1.2. trans-Unsaturation

The effect of the presence of a *trans* olefinic center (although relatively rare in natural lipids) on the shape of an otherwise normal fatty acyl chain is quite different, but in a possibly unexpected way. "All-anti" elaidic acid, with both $C_\alpha-C_\beta$ bonds anticoplanar with the double bond, is almost exactly isosteric with all-anti stearic acid, the shorter C=C bond compensating for the larger C=C−C

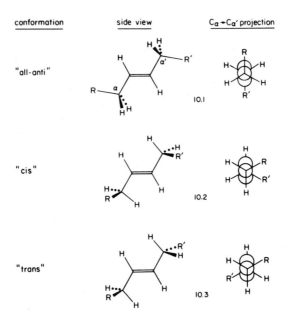

FIGURE 3.10. Cardinal conformations of a normal carbon chain near a *trans* double bond. (For clarity, hydrogen atoms of the olefinic center are not shown in the projections.)

angles; neither the overall length nor the directions of the chain are therefore affected appreciably. However, this is not the preferred conformation of the *trans* olefinic center. In contrast to the *cis* configuration, both α-methylene groups can now assume their best (i.e., lowest energy) arrangements, with α-Hs syncoplanar with the double bond. The plane of the olefinic center now embraces eight atoms (by addition of the two α-Hs). But the other substituents of the α-carbon atoms, including the β-carbons (and thus both extensions of the chain), are out of the plane, and since each α-carbon atom bears two hydrogen atoms, either of which may be in the syncoplanar position, two minimal-energy conformations are possible: one with both β-carbons on the same side of the plane (*"cis"* or syn), and another in which they are on opposite sides of the plane (*"trans"* or anti). From the fact that the energy barrier for rotation of the methyl groups of *trans*-2-butene is 1.95 kcal/mol (see Karabatsos and Fenoglio, 1970), it may be estimated that the energy of the "all-anti" conformation of the *trans* olefinic fatty acyl group is almost 4 kcal/mol above that of the *"cis"* and *"trans"* conformers. Differences between these and the "all-anti" conformation are shown in Fig. 3.10, which includes (at right) Newman projections along the Cα—Cα' axis, illustrating the close analogy of the *"cis"* and *"trans"* to gauche and anti conformations, respectively, of butane arrays. These considerations lead to the following conclusions with respect to the comparative shapes of "free" (i.e., unaffected by intermolecular interactions) elaidoyl (*trans*-enoyl) and analogous stearoyl (saturated) moieties: except for about 2% of the time (at 25°C), when one or the other C_α—C_β bond is anticoplanar with the double bond, the elaidoyl group has equally high (49%) probabilities of being either (i) in the *"trans"* conformation, which imposes

NATURE OF THE FATTY ACIDS 45

no change in the general direction of the chain except for a "jog" or offset of about 1.2 Å that reduces the effective overall length negligibly (by about 0.2 Å), compared to that of the all-anti stearoyl group; or (ii) in the *"cis"* conformation, which fixes the $C_\beta - C_\beta$ internuclear distance some 1.2 Å less than the maximum length of the corresponding segment in the stearoyl group, and imparts a 67° change in chain direction (compare the 55° "turn" imposed by a single gauche array on an otherwise all-anti polymethylene chain), reducing the effective length of the chain very considerably.

The conclusion that the "all-anti" conformation of elaidic acid, essentially isosteric with all-anti stearic acid, is, in fact, substantially strained provides a possible explanation for the melting point of elaidic acid (44°C) being considerably below that of stearic acid (70°C), and suggests that the crystal structure of elaidic acid (to our knowledge not yet determined) will prove to contain molecules in the *"trans"* conformation described above (packing of molecules in the *"cis"* conformation would presumably give a solid of considerably lower melting point; cf. that of oleic acid).

3.4.1.3. *Conformation of all-cis-1,4-Polyenes*

From the conformational analysis of oleic acid, it would appear reasonable to consider—at least initially—possible arrangements of the *cis,cis*-1,4-diene system (that occurs, for example, in the linoleoyl group) in which all seven carbon

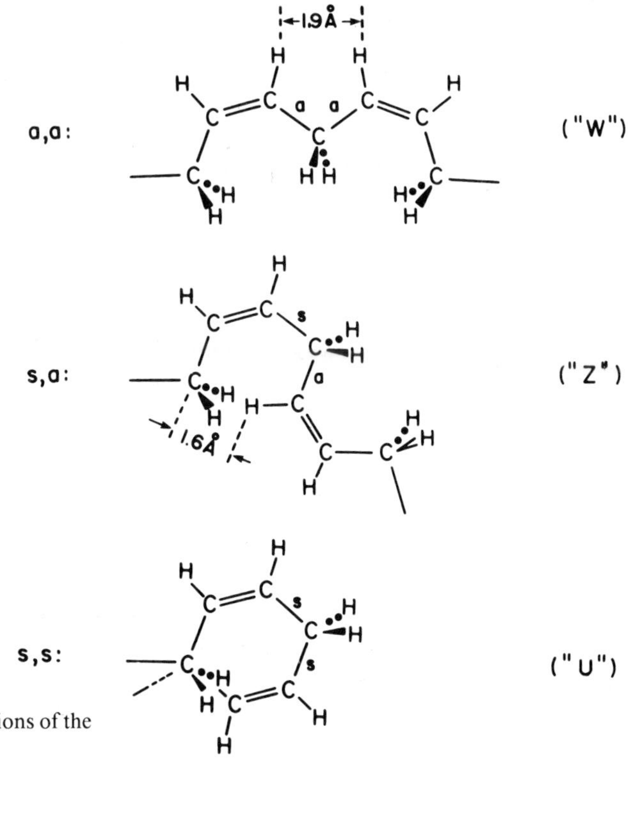

FIGURE 3.11. Planar conformations of the *cis,cis*-1,4-diene group.

atoms (including those of the external α-CH$_2$ groups) are coplanar (see Fig. 3.11). The three possible arrangements of this kind can be distinguished by designating them as a,a; s,a; and s,s, in terms of whether the internal C$_\alpha$–C$_\beta$ single bonds are anti-(a)- or syn-(s)-coplanar with the more remote double bond. (In discussing the analogous configurations of free radicals theoretically or actually obtainable from 1,4-pentadiene by hydrogen atom abstraction, Sustmann and Schmidt (1979) have designated these as W, Z, and U forms, respectively.) Of these postulated planar conformations, only the a,a (W) appears to be reasonably free of steric strain, its two close hydrogen atoms easily accommodable by slight spreading of the central =C–C$_\alpha$–C= bond angle (this could also be effected by rotations out of anticoplanarity). Overlap in the s,a (Z) form is already quite serious (cf. sum of van der Waals radii of CH$_2$ and H $\simeq$ 3.2 Å) while the s,s (U) form requires superposition of the outer α-methylene groups. (Note, however, that the cyclo-oxygenase initiating conversion of certain all-*cis*-1,4-polyenoic fatty acids to prostaglandins presumably requires holding a *cis,cis*-1,4-diene group in a conformation quite close to the s,s type—see Chapter 10.)

Very strong presumptive evidence that, in the absence of perturbation by intermolecular interactions, anticoplanar conformations of the central single bonds of *cis,cis*-1,4-dienes (conformation a,a) are strongly preferred is provided by structures of hydroperoxides formed by autoxidation of linoleic acid (see Chapter 6) either neat or dissolved in solvents of low polarity. Only four are formed: the *cis*- and *trans*-9,*trans*-11-dienyl-13-hydroperoxides and the *trans*-10, *cis*- and *trans*-12-dienyl-9-hydroperoxides. The important observation is that, although double bonds of these products in original positions (9 or 12) have either *cis* or *trans* configurations, those in "new" positions (10 or 11—corresponding to the single bonds of interest in the starting material) are *trans*. This implies, first, that the configurations of the partially double bonds (9,10,11, and 12) of the resonance hybrid radical formed by abstraction of a hydrogen atom from the central methylene group are frozen (i.e., once established, do not change); and second, that the conformation of the 10 and 11 single bonds prior to formation of the radical must be anticoplanar (or close to it) in order to become fixed in the *trans* configuration as a partially double bond in the intermediate radical, then as a fully developed double bond in the eventual hydroperoxides. The fact that double bonds in original positions may have, in the final products, either *cis* or *trans* configuration has been shown very elegantly by Porter *et al.* (1980) to be a consequence of the reversibility of the reaction of molecular oxygen with the dienyl radical, giving a conjugated dienyl hydroperoxy radical in which a bond that was double in the starting substance is now single, and able to change conformation before reverting to a fixed partial-double-bond configuration on departure of the oxygen molecule.

Based on the conclusions (see above) (i) that (e.g., at 37°C and in the absence of intermolecular effects) the oleoyl group, with respect to its *cis*-olefinic center, would be in its favored doubly anticoplanar conformation about 60% of the time; and (ii) that this conformation involves a chain redirection of about 44°, it may be of interest to explore their implications with respect to certain aspects of the probable shape of some important all-*cis*-1,4-polyenes. Thus, under the same conditions, if the two olefinic centers are assumed to behave independently, the linoleoyl group is expected to possess a fully anticoplanar segment of nine consecutive carbon atoms (C7–C15) about $(60\%)^2$ or 36% of the time, imposing an 88° bend in the chain. Similarly, the linolenoyl (cis-6,9,12- or -9,12,15-octadecatrienoyl) group would have a 12-carbon planar array $(60\%)^3$ or about 22% of the time, and a 132° bend; and arachidonoyl a 15-carbon planar array 13% of the time, with a 176° redirection of its C_{20} chain—very nearly assuming a true "U" shape. The arachidonoyl moiety (for example) could, of course, easily affect modifications of its conformation in this region (as it presumably does, independently of intermolecular interactions, about 87% of the time) by $=C-C_\alpha$ rotations, sacrificing the coplanarity of the array at a small cost in enthalpy. van Soest (see discussion by van Dorp (1974)) has shown, for example, by study of X-ray diffraction patterns of crystalline (powder) 2-arachidonoyldipalmitoylglycerol, that in response to demands imposed by the crystal structure of this substance, the arachidonoyl group is much more gently curved, in a conformation about 19 Å long. This effective length of the arachidonoyl group should be compared with the 24-Å length of the all-anti conformation of a saturated C_{20} chain; a shortening of about 0.75 Å per cis-olefinic group reduces this to 21 Å, and the slight overall curvature accounts for the remainder.

REFERENCES

Abrahamsson, S., and Ryderstedt-Nahringbauer, I., 1962, The crystal structure of the low-melting form of oleic acid, *Acta Cryst.* **15:**1261.

Bentley, R., 1969, *Molecular Asymmetry in Biology,* Vol. 1, Academic Press, New York, pp. 60–61.

Brocklesby, H. N., 1930, Fatty acids and their esters. I. Some observations on the cryoscopy and ebullioscopy of fatty acids, *Can. J. Res.* **14B:**222.

Dippy, J. F. J., 1938, Chemical constitution and the dissociation constants of monocarboxylic acids. Part X. Saturated aliphatic acids, *J. Chem. Soc. London,* **1938:**1222.

Dunitz, J. D., and Strickler, P., 1968, Preferred conformation of the carboxyl group, in: *Structural Chemistry and Molecular Biology* (A. Rich and N. Davidson, eds.), W. H. Freeman, San Francisco, pp. 595–602.

Flory, P. J., 1969, *Statistical Mechanics of Chain Molecules,* Interscience, New York.

Garvin, J. E., and Karnovsky, M. L., 1956, The titration of some phosphatides and related compounds in a non-aqueous medium, *J. Biol. Chem.* **221:**211.

Karabatsos, G. J., and Fenoglio, D. J., 1970, Rotational isomerism about sp^2–sp^3 carbon–carbon single bonds, in: *Topics in Stereochemistry,* Vol. 5 (E. L. Eliel and N. L. Allinger, eds.), Wiley-Interscience, New York, pp. 167–203.

Klyne, W., and Prelog, V., 1960, Description of steric relationships across single bonds, *Experientia* **16:**521.

Martin, D. L., and Rossotti, F. J. C., 1959, The hydrogen-bonding of monocarboxylates in aqueous solution, *Proc. Chem. Soc. London,* **1959:**60.

Martin, D. L., and Rossotti, F. J. C., 1961, The structure of dimeric acetic acid in aqueous solution, *Proc. Chem. Soc. London,* **1961:**73.

Porter, N. A., Weber, B. A., Weenen, H., and Khan, J. A., 1980, Autoxidation of polyunsaturated lipids. Factors controlling the stereochemistry of product hydroperoxides, *J. Am. Chem. Soc.* **102:**5597.

Schmidt-Nielsen, K., 1946, Investigations on the fat absorption in the intestine, *Acta Physiol. Scand.* **12**(Suppl. 37):38.

Schmidt-Nielsen, K., 1946, Investigations on the fat absorption in the intestine, *Acta Physiol. Scand.* **12**(Suppl. 37):38.

Schrier, E. E., Pottle, M., and Scheraga, H. A., 1964, The influence of hydrogen and hydrophobic bonds on the stability of the carboxylic acid dimers in aqueous solution, *J. Am. Chem. Soc.* **86**:3444.

Seelig, J., 1977, Deuterium magnetic resonance: theory and application to lipid membranes, *Quant. Rev. Biophys.* **10**:353.

Smith, R., and Tanford, C., 1973, Hydrophobicity of long chain n-alkyl carboxylic acids, as measured by their distribution between heptane and aqueous solutions, *Proc. Nat. Acad. Sci. U.S.A.* **70**:289.

Sustmann, R., and Schmidt, H., 1979, Pentadienyl-Radikale—Struktur und Spindichteverteilung, *Chem. Ber.* **112**:1440.

Tanford, C., 1974, Theory of micelle formation in aqueous solutions, *J. Phys. Chem.* **78**:2469.

Van Dorp, D. A., 1974, Essential fatty acids and prostaglandins, in: *XXIVth International Congress of Pure and Applied Chemistry,* Vol. 2, Butterworths, London, pp. 117–136.

Wenograd, J., and Spurr, R. A., 1957, Characteristic integrated intensities of bands in the infrared spectra of carboxylic acids, *J. Am. Chem. Soc.* **79**:5844.

4
FATTY ACIDS: CRYSTALS, MONOMOLECULAR FILMS, AND SOAPS

4.1. Crystallization

Fatty acids typically possess "high crystallizing power"—which is to say that they crystallize very readily at temperatures below their melting points and show little tendency to persist as supercooled melts or as supersaturated solutions. This characteristic implies that when the temperature of a melt is lowered or the concentration of a solution is raised to levels no longer favoring the liquid or dissolved states of such a substance, there is a high probability that some minimal number of molecules will aggregate in an arrangement representative of—or closely similar to—that of its characteristic crystal lattice. This prerequisite for ready crystal nucleus formation is provided in the case of the fatty acids by the strongly dominant inter-carboxyl-group forces and, particularly for full saturated members of the class, by the inherent flexibility of the hydrocarbon moiety, which can, following initial union at the carboxyl-group termini, easily and quickly assume conformations that pack well and thus contribute maximally to the sum of intermolecular forces stabilizing the growing crystal. Early events in the initiation of crystallization from the liquid state (melt) or from solution in an apolar solvent, in which the units eventually assembled into the crystal lattice will actually consist of H-bonded "dimers" (see Chapter 3, Section 2), may be pictured as shown in Fig. 4.1. The process will, of course, be somewhat more complicated when H-bonding solvents are involved, since in this case the initial units are solvated monomers that are converted to dimers during crystallization.

As additional units are added to the assemblage, the resulting microcrystal more perfectly assumes its eventual architecture, in which the H-bonded pairs of carboxyl groups are arranged side by side, forming a central plane sandwiched between two brush-like layers of parallel, all-anticoplanar-conformation hydrocarbon chains (see Chapter 3, p. 38). Since the dipole–dipole interactions called

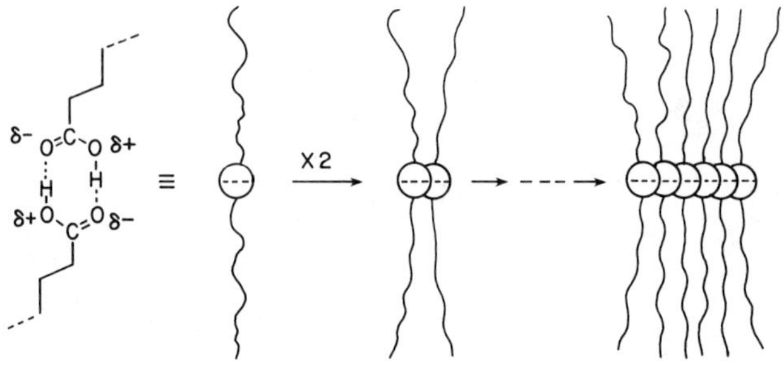

FIGURE 4.1. Schematic representation of incipient crystallization of a dimeric fatty acid.

into play by addition of further units to the edges of these early crystals are much stronger than the London forces of methyl group interaction involved in growth of additional parallel planar arrays on the faces of the first, the accretion process is predominantly two-dimensional. It seems quite probable that although the initial bimolecular platelet may well grow by addition of single units to its edges, additional thickness is realized by stacking of such platelets rather than by accumulation of units on their faces. In any case, this tendency to grow well in only two directions is manifest grossly in the characteristically flaky crystals of such substances (Fig. 4.2).

It is readily appreciated that, in the fully developed fatty acid crystal, envisioned as a stack of such bimolecular platelets, midway between successive parallel paired-carboxyl-group planes will be others in which the methyl-group termini of the fatty acid molecules are in contact. Because of the weakness of intermolecular forces here, these planes resist shearing stresses poorly and the facile basal cleavage, "greasy feel," and lubricant qualities of such solids are readily understandable.

Although fatty acids crystallize very readily, it has not, for reasons discussed above, been easy to obtain single crystals of the three-dimensional high quality necessary for detailed X-ray diffraction crystal structure analysis. On the other hand, however, the nature of the easily obtained flaky material is such that powdered samples serve well for determination by X-ray diffraction of the so-called major or long spacing, representing the distance between successive carboxyl-

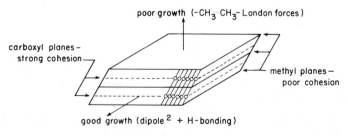

FIGURE 4.2. Schematic representation of a fully developed fatty acid crystal. The open circles represent pairs of H-bonded carboxyl groups.

FIGURE 4.3. Typical inclination of the hydrocarbon chains with respect to the plane of the carboxyl groups in fully developed fatty acid crystals.

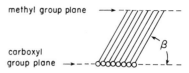

group planes in the crystalline array. This distance generally turns out to be considerably shorter than the length of two all-anti-conformation fatty acid molecules laid end to end (compare $d = 40$ Å for the C form of stearic acid with the 25-Å length of a fully extended molecule of this substance), revealing the further typical structural feature of inclination of the "direction" of the hydrocarbon chain with respect to the reference carboxyl-group planes. This angle of inclination (β) is usually in the neighborhood of 60° (see Fig. 4.3). This feature of typical saturated fatty acid crystal structure presumably reflects attainment of optimal inter-carboxyl-group-pair register (with respect to dipole–dipole interactions). Improved aggregate London-force interaction between hydrocarbon moieties may also be involved, although long-chain normal saturated hydrocarbons assume lattice structures in which the fully extended chains are perpendicular (i.e., not inclined) to the terminus planes.

The striking depression of melting point effected by substitution of a *cis*-olefinic center for a pair of methylene groups in a normal fatty acid, particularly when the substitution occurs near the center of the long chain (e.g., 18:0, m.p. 69.5° vs. *cis*-9-18:1, m.p. 13 or 16°C, depending on polymorph) has been known for a long time, and ascribed truistically to weakening of aggregate intermolecular forces by the presence of the rigid *cis*-configuration center of unsaturation. (*trans*-Unsaturation has a similar but not nearly so impressive effect.) It was not possible, however, to develop an understanding of this effect until the structure of the lower-melting (13°C) crystalline oleic acid was determined by Abrahamsson and Ryderstedt-Nahringbauer (1962). Certain salient features of the arrangement of molecules in this crystal lattice are shown in Fig. 4.4. As in crystals of the simpler saturated fatty acids, the molecules assemble in planes of doubly hydrogen-bonded pairs; planes of apposed terminal methyl groups alternate with those in which the carboxyl groups are situated; and the initial section (C-2 through C-8) of the hydrocarbon chain is tilted (56.5°) with respect to the carboxyl-group plane. The intervening *cis*-olefinic center, however, reverses the tilt of the remaining C-11 through C-18 saturated segment. Both saturated segments are (again, as usual) all-anti in conformation, and the planes in which their carbon atoms lie are both very nearly normal to that defined by their intersecting axes (note their apparently straight linear arrangement in Fig. 4.4a). Because the C-2 to C-8 segment is anticoplanar, alternate C—C bonds are parallel, and a Newman projection may be used to advantage to show that neither the carboxyl- nor olefinic-group planes are coincident with that of the intervening methylene groups (see Fig. 4.4b). The carboxyl groups are 26° out of the otherwise preferred syncoplanar conformation (see Chapter 3, Section 3). Similarly, the olefinic array is 47° out of the plane of the methylene groups, resulting in a 67° bend (see Fig. 4.4a) that is considerably larger than that (44°) expected for consistent coplanarity through the center of unsatu-

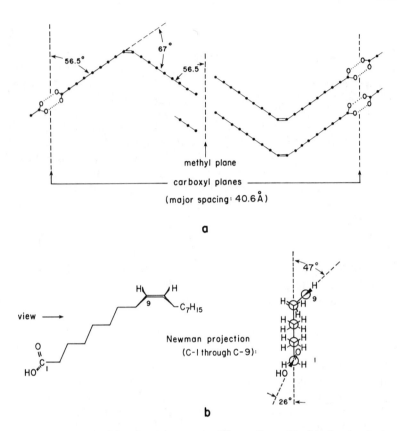

FIGURE 4.4. Arrangement (a) and certain aspects of the conformation (b) of molecules in crystalline oleic acid (m.p. 13°C) [after Abrahamsson and Ryderstedt-Nahringbauer (1962)].

ration (Chapter 3, Section 3); this rotation of the olefinic group also brings the C_α and C_β substituents (α and β with respect to the double bond, i.e., C7–C8 and C11–C12) to but 13° from C–H eclipsing. The departures from minimal intramolecular potential energies associated with these structural features contribute to the low melting temperature of the substance.

The crystal structure of elaidic acid (the *trans* isomer of oleic), which could, with minimal departure from its most stable conformation, be virtually isosteric with stearic acid, is not yet known in detail, although it has been shown that the long spacing of elaidic acid (i.e., the separation of carboxyl-group planes) is considerably larger (50.6 Å) than that of stearic acid (see below), in accord with the conclusion that the hydrocarbon moieties are normal rather than tilted with respect to the carboxyl-group planes (Bailey *et al.*, 1975).

The readiness with which fatty acids crystallize from solution invites attempts to isolate individual components from mixtures of such substances (as would be obtained, for example, by saponification of natural fats) by fractional crystallization. The general usefulness of this approach is, however, severely limited by the usual occurrence in such mixtures of closely homologous saturated fatty acids, the solubilities of which (like their melting points) do not differ greatly. Moreover, such homologues readily form mixed crystals; close homologues fit each other's lattices very comfortably, particularly so since the units laid down are H-bonded dimers as likely to be mixed as not. Before the existence of techniques that would easily have revealed such errors, mixtures of palmitic (16:0) and stearic (18:0) acids were repeatedly misidentified as the actually quite rare margaric acid (17:0). The very substantial melting-point-depressing (hence solubility-increasing) effect of the presence of *cis* double bond(s), however, suggests that saturated, mono-, and multiply *cis*-unsaturated fatty acids (all of approximately the same chain-length) could be satisfactorily separated from each other in this way. The low-temperature crystallization technique developed by J. B. Brown (see Brown and Kolb, 1955) indeed serves as the method of choice for isolation (on a large scale if desired) of certain unsaturated fatty acids that are strongly dominant components of fatty acid mixtures obtained by hydrolysis of suitably chosen natural (triglyceride) oils; e.g., oleic from olive oil and linoleic from safflower, sunflower, or poppy seed oils. In the course of cooling (by means of a dry-ice–acetone bath) of a 5-wt. % acetone solution of mixtures of such fatty acids, palmitic and stearic acids crystallize out fairly completely at $-20°C$; oleic at $-65°C$; and linoleic at the lowest temperature (about $-78°C$) obtainable with such a cooling bath. Small amounts of less unsaturated fatty acids precipitated by further cooling are efficiently removed by recrystallization, since acids differing in content of *cis*-unsaturation (unlike homologous saturated fatty acids) do not co-crystallize readily.

4.2. Polymorphism

Fatty acids (and other types of lipids as well) have a marked tendency to crystallize in more than one form, depending on differences in conditions under which the substances are induced to solidify. For example, when stearic acid crystallizes slowly from solution in a nonpolar solvent, the flaky solid obtained (polymorph B) may be difficult to distinguish by visual inspection from that (polymorph C) prepared by crystallization from a polar solvent (e.g., ethanol or acetic acid) or by cooling the neat molten substance. Moreover, both will be found to have the same melting point when that property is determined in the usual way. But X-ray diffraction analysis readily reveals that the two are in fact distinctly different, the first having a major spacing of 44.1 Å, and the other, 40.0 Å, indicating that the tilt of the molecules in the lattices of these two polymorphs of stearic acid are quite different (66° and 63.5°, respectively). Careful thermal analysis shows that when the solid crystallized from an apolar solvent is heated slowly and constantly, a brief hesitation in temperature elevation occurs at about 46°C (the B → C transition temperature). Although the B polymorph may, on this basis, be regarded as having a considerably lower melting point than the other, the recrystallization of the substance as the higher-melting polymorph C takes place so readily and rapidly that the observed melting point is that of form C

(69.6°C), and these forms are thus not distinguishable by this means. In a few cases polymorphs exhibiting different melting points—as determined in the usual way—are encountered; oleic acid, for example, may be obtained in either of two forms of m.p. 13°C or 16°C.

The general tendency of crystallizable lipids to exhibit polymorphism is ascribable to features of the more or less extensive polymethylene moieties most of them contain; to the effective topography of such groups, determining the sum of London forces involved when they lie side by side in van der Waals contact in the crystal lattice; and to the inherent flexibility of such hydrocarbon chains. When crystallization starts, the first few molecules that come and remain together may do so in several slightly different ways involving small shifts along their lengths, affecting the "tilt" of the essentially linear molecules with respect to the carboxyl planes but representing very small differences in potential energy or stability of the assemblage. Additional variation in the aggregate interactions between normal saturated hydrocarbon moieties of crystalline fatty acids (and other lipids) is provided by differences in the orientation with respect to each other of the planes of adjacent, axis-parallel anticoplanar polymethylene chains (see Abrahamsson et al., 1978). Once the "seed" of the eventual crystal has formed, it serves as a template to which additional molecules deposited on it as the crystal grows readily conform. The register of the first few molecules that unite to form the seed crystal could easily be influenced by their solvation and/or by their being either monomeric or dimeric.

4.3. Melting Point Alternation

Insofar as the melting point represents the temperature at which the kinetic energy of the molecules concerned becomes sufficient to overcome the forces responsible for retaining them in the crystal lattice, this property is related in a very complex way to molecular structure. Melting points of series of homologous substances tend to be related in a way that may be taken to reflect significant contribution to total intermolecular forces of additional methylene groups. The unbranched saturated fatty acids having an even number of carbon atoms that are commonly encountered in natural lipids, for example, exhibit a regular progression of increasing melting point with longer chain-length: cf. palmitic acid (16:0), m.p. 62.9°C, and stearic acid (18:0), m.p. 69.6°C. But the rarer acids having an odd number of carbon atoms do not conform to this generalization: margaric acid (17:0), for example, melts at 61.1°C, lower than palmitic acid. Plotting melting points of all unbranched (normal) saturated fatty acids against chain-length reveals a pattern in which (except for very short-chain homologues in which effects of the variable-size alkyl group play a less important role) melting points of even-carbon-number homologues fall nicely on the higher of two essentially parallel curves, and those of the odd homologues on the lower. Malkin (1952) has attributed this curious relationship to differences in crystal lattice forces in the planes of apposition of terminal methyl groups. On the basis of the

FATTY ACIDS

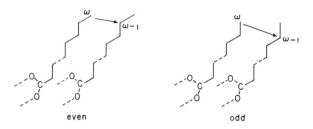

FIGURE 4.5. Possible different orientations of terminal methyl groups in crystals of fatty acids with even and odd numbers of carbon atoms.

discussion of early events involved in the initiation of crystallization of fatty acids, it seems reasonable to assume that the orientation of carboxyl groups, which ultimately determines the architecture of the crystal, should be virtually independent of whether the long hydrocarbon chain contains an even or odd number of carbon atoms (a structural difference far too remote to be sensed at the carboxyl-group end of the molecule), but that the final disposition of the terminal methyl groups would be quite different in the two cases, as indicated in Fig. 4.5. These greatly oversimplified pictures are meant only to suggest, as well as this can be done in two dimensions, that acids otherwise identically oriented with respect to each other in the crystal lattice may be expected to have their terminal methyl groups in quite different situations, depending on their having either an even or an odd total number of carbon atoms. It should be noted that although the distance between methyl groups is essentially the same in both cases, these may be much closer to the penultimate carbon atom of the adjacent chain in the even than in the odd, and that the resulting additional London-force contribution (note that the magnitude of such forces falls off very rapidly with distance) to the overall lattice stabilization might well be reflected in an appreciable elevation in melting point. This explanation of the alternating melting point phenomenon is dependent on the essentially linear hydrocarbon chains being tilted in the crystal lattice. It will be readily recognized that if the hydrocarbon chains are situated perpendicular to the plane of their methyl termini, the environments of the methyl groups will be identical, irrespective of whether the chain contains an even or odd number of carbon atoms. Certain classes of homologues (e.g., hydrocarbons, fatty alcohols, and the ethyl esters of fatty acids) that are orthorhombic in crystal habit indeed do not exhibit such melting point alternation.

4.4. Monomolecular Films

The arrangement of fatty acid molecules in monomolecular films under lateral compression on aqueous surfaces exposed to gases or in contact with liquids of low polarity (hence low miscibility) resembles in many respects that in individual layers of such molecules in their crystal lattices. Differences in detail and the origins of those differences, however, warrant discussion.

The classical, Nobel prize studies of Langmuir (1917) on such systems are remarkable both for their simplicity and for the insight they provided into certain aspects of the behavior of matter at the molecular level. Using an apparatus consisting essentially of a trough of water provided with barriers defining a rectangular surface area variable by movement of one side, and a sensitive means of measuring the amount of force required to reduce the confined area, Langmuir added known amounts of fatty acids dissolved in volatile, low-polarity solvents; permitted the solvent to evaporate, leaving the nonvolatile, insoluble solute thinly spread and floating on the water surface; and then noted the force required to compress the film (thus reducing the area to which the molecules on the surface were confined) by gradually diminishing the distance between the movable and opposite fixed barriers. It was observed that very little pressure was required initially, but that when a certain minimum area (directly proportional to the weight of fatty acid that had been applied to the surface) was reached, the pressure required for further reduction of the surface area abruptly increased. By comparing results obtained with fatty acids of different chain-length, it was revealed that these discontinuities in the force:area relationship occur at an area per molecule that is independent of the molecular weight (or length) of the acid involved. The thickness of the film under these conditions therefore varies directly with (and is, in fact, equal to) the lengths of the floating molecules, which could easily be calculated on the basis of the density of the neat substances. It has been pointed out that Benjamin Franklin, on noting the surprisingly large area ultimately occupied by a single drop of olive oil placed on the surface of a pond, was close to a discovery very similar to Langmuir's but a century earlier.

These observations are readily understandable in terms of the carboxyl groups of the floating fatty acids being very strongly bonded to the water surface and thus resisting tenaciously any tendency of the molecule as a whole to depart from the surface; conversely, the hydrophobicity of the hydrocarbon chain of the fatty acid strongly resists submersion of this part of the molecule (see Chapter 2, Section 3.4). Initially, when the floating molecules are widely dispersed over an ample surface, it may be presumed that the hydrocarbon chains lie on the surface, retained by weak London forces between methylene (or methyl) groups and water molecules. Contrary to commonly seen illustrations, there is no reason to believe that the hydrocarbon chains of the fatty acid are normal to the water surface before lateral compression. But as the area to which the fatty acid molecules are constrained is reduced, the hydrocarbon chains are forced to leave the surface, inter-methylene- (or -methyl-) group interactions of approximately the same magnitude being substituted for those previously involved, and assume ever more completely all-anti (fully extended) conformations as the necessity to conserve space and pack together more efficiently is heightened. The sudden increase in resistance to further compression is identifiable with attainment of a situation in which the hydrocarbon moieties of all the fatty acid molecules are standing erect, perpendicular to the water surface, essentially fully extended (giving the film its molecular-length thickness), and in close lateral contact. The fact that the final area per molecule (20–21 $Å^2$) is somewhat greater than that in the crystal lattice (18.2 $Å^2$) reveals that the packing is not quite as orderly, although very nearly so.

It will be noted that the arrangement of molecules in these laterally compressed, monomolecular films is closely similar to that in the crystal lattice insofar as in both cases the hydrocarbon chains are in the all-anti conformation and lie side by side, essentially parallel to each other. However, the chains are not tilted (with respect to the carboxyl-group "plane") in the film as they usually are in the crystal. The discrepancy in cross-sectional area suggests that the untilted chains do not nest as efficiently and thus that the molecular inclination typifying fatty acid crystal structure may be dictated by features of the hydrocarbon chain rather than—or in addition to—those of optimal carboxyl-group interactions. These monomolecular films collapse under application of pressure sufficient to tear the carboxyl groups away from the water surface, giving rise to films that are more than one molecule thick.

4.5. Soaps in Aqueous Solution

A vast amount of study has been devoted over a period of many years to mixtures of water and salts of fatty acids (soaps) in attempts to understand better a variety of intriguing properties exhibited by such solutions. Much of the support and motivation for this work is attributable to soaps being important articles of commerce, although much of what was discovered later proved to be of considerable biochemical interest. Soaps are rarely encountered in living systems (except during the digestion of fats—see Chapter 12) simply because the free fatty acids themselves almost never occur in greater than trace quantities. But the soaps are good and relatively simple models for an extensive, complex group of substances (to which a variety of names apply: amphipaths, amphiphiles, surfactants, detergents, emulsifiers) embracing many, of which the phospholipids are representative, that are of very great biochemical interest. Although such compounds vary widely in detailed structure, they have in common the presence of both highly polar and apolar moieties covalently linked together. In the presence of water, therefore, the avidity of the polar portions of such molecules for the solvent draws the apolar (hydrophobic) portions into an environment essentially inimical to them; the solute is thus literally forced to assume an assortment of nonrandom dispositions, all of which reflect a strong tendency to minimize, in one way or another, exposure of the apolar parts of the solute to the aqueous phase. Much of what has been learned about the nature of soaps in aqueous solution is applicable, by generalization, to understanding the behavior of all amphiphilic lipids in the aqueous systems of living organisms.

Before considering each in detail, it is worthwhile summarizing the various ways in which soaps dissolved in water typically depart from "ideal solution" behavior (defined as uniform distribution throughout the system, with the hydrocarbon chain in its normal room-temperature conformation), as the amounts of the two components of the system are varied over the complete range from no soap to no water. The "critical micellar concentration" (c.m.c.) and that at which "dimerization" becomes appreciable, tabulated below, are those pertaining specifically to potassium laurate. Although these critical concentrations differ for and are characteristic of each particular soap, concentrations at which the higher aggregates—micelles, rods, and lamellae—predominate are, on the other hand, essentially identical for any soap at temperatures above its melting or "Krafft" point (also called, for this reason, its critical micellar temperature see p. 392).

Soap concentration:

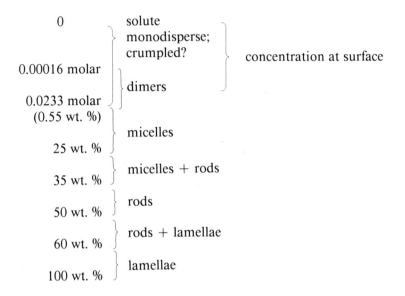

4.5.1. Surface Concentration

Addition of even very small amounts of a soap to a sample of water leads to a marked reduction in surface tension that is readily manifest in the tendency of such solutions to form "suds" on mechanical agitation—a reflection of their ability to stabilize emulsions of air. Quantitative measurements show that the surface tension of such solutions diminishes continuously with increasing soap concentration until—but not after—micelles begin to form (at 0.55 wt. % in the case of potassium laurate). It can be shown in other ways that at this point, when the surface can no longer accommodate additional soap molecules, these occupy an area (per molecule) of about 45 $Å^2$ (cf. the 20–21 $Å^2$ area per molecule of fatty acid in laterally compressed monomolecular films). Since each soap anion carries a formal negative charge, this limitation is presumably imposed by heightened electrostatic repulsion forces as the ions are packed more closely together.

The extent to which soap molecules prefer being at the air–water interface, rather than in the bulk of the solution, may be gathered by the fact that a cube of 0.0233-molar potassium laurate in water, one centimeter on a side and having a 1 cm^2 surface exposed to air will, at saturation (45 $Å^2$ per molecule or anion), contain about 2×10^{14} molecules in that surface, while of the total 1.4×10^{19} molecules in the whole sample, the number statistically expected to be present in that surface at any instant is $(1.4 \times 10^{19})^{2/3}$ or 5.8×10^{12}; the surface concentration is thus enhanced about 34.5-fold.

This marked tendency to crowd into a surface exposed to a less polar phase (air or oil) represents one way (albeit limited in capacity) in which the intrusion of the hydrocarbon moieties of the amphiphilic substance into the aqueous medium may be lessened, i.e., by literally escaping from it—to the extent that continued immersion of the polar moiety in the aqueous phase will permit. Supposedly a few methylene groups may remain submerged by virtue of the strong tendency of the carboxylate anionic group to remain fully hydrated (the relatively poor surface tension-reducing powers of short-chain carboxylic acid salts reflect this), but the bulk of the typical long-chain hydrocarbon moiety of a soap presumably lies on the water–air surface, held there by the weak London forces until the surface population density no longer permits this. At saturation (45 Å^2 per molecule) much of the hydrocarbon chain must of necessity be directed away from the water surface, in van der Waals contact with other protruding hydrocarbon moieties, closely akin to the situation prevailing in laterally compressed monomolecular films of free fatty acids on water, but considerably less constrained to the space-conserving all-anticoplanar conformation.

4.5.2. Monodisperse Solute

Osmotic pressure and conductivity measurements show that the solute in very dilute soap solutions is essentially monodisperse. It has been pointed out earlier, however, that the flexibility of the saturated hydrocarbon chain, totally immersed in the water envelope, permits some reduction of the effective water–hydrocarbon contact area by change to conformations of greater gaucheness (see Chapter 3, p. 39). It can be shown, for example, that conversion of a room-temperature 36%-gauche–64%-anti hydrocarbon chain into a fully gauche helix involves an increase in strain (or potential energy) that would be more than offset by the energy provided by reunion of water molecules in the volume increment resulting from this more compact arrangement of the hydrocarbon chain. Although difficult to demonstrate experimentally—at least in part because of the diluteness of the solutions involved—the hydrocarbon chains of isolated soap molecules in aqueous solution probably tend to be "crumpled" (i.e., to assume formations of enhanced gaucheness) in response to such advantages of minimizing hydrocarbon–water contact.

4.5.3. Low-Order Aggregates of Solute

As the concentration of soap in water is increased, approaching that at which micelles begin to form, the anions of the solute would be expected to have a pronounced tendency to form, via hydrophobic bonding, antiparallel dimers of configuration closely resembling those known to exist in aqueous solutions of undissociated carboxylic acids (Chapter 3, p. 31). This process represents yet another way in which entropically unfavorable apolar group–water contact can be lessened. The two formally negatively charged carboxylate groups would, because of electrostatic repulsion, be expected to be as far from one another as possible without sacrifice of lateral contact of the hydrocarbon chains.

In his classical studies of free fatty acid transport in blood as albumin complexes, Goodman (1958; see also Smith and Tanford (1973)) noted that as the total amount of acid distributed between heptane and water (buffered at pH 7.45) was increased, the partitioning of the solute between the two immiscible solvents became more and more inadequately quantitatable in terms solely of equilibria involving dimerization (via H-bonding of carboxyl groups) in the heptane phase, distribution of the monomeric acid between heptane and water, and dissociation of the acid in the aqueous phase. The experimentally observed discrepancies could be accounted for by postulating that soap anions tend significantly to unite in pairs:

$$
\begin{array}{c}
\text{dimer} \\
\updownarrow \\
\text{heptane} \quad\quad \text{RCOOH} \\
\hline
\quad\quad\quad\quad\quad\quad \updownarrow \\
\text{water} \quad\quad \text{RCOOH} \quad\quad \text{—— additional equilibrium} \\
\quad\quad\quad\quad \updownarrow \quad \times 2 \\
\text{H}^+ + \text{RCOO}^- \rightleftharpoons (\text{RCOO}^-)_2
\end{array}
$$

In addition to the dimeric ion $[(\text{RCOO}^-)_2 \equiv A_2^=]$, other low-order aggregates (the "acid soap," $\text{HA} \cdot \text{A}^-$ or HA_2^-, and dimers of the complex, $\text{H}_2\text{A}_4^=$) also appear to be present in appreciable quantity. Eagland and Franks (1965) showed, for example, that at a formal concentration of 2.6×10^{-4} molar and at 25°C, aqueous sodium palmitate contains solute in forms A^-, $H_2A_4^=$, $A_2^=$, and HA_2^- in relative amounts of about 6:6:1:1.

It is a remarkable fact that aggregates of order intermediate between that of those discussed above and that of others (the micelles) containing 25–50 times as many soap anions do not exist in appreciable quantity.

4.5.4. Micelles*

It was noted many years ago that as the concentration of a given soap in water is increased, a point is reached at which the specific osmotic activity of the

---

*There is a growing tendency to use the term "micelle" generically, applying it to all types of aggregates formed by amphiphilic substances in aqueous solution. The term is used here, however, in the original sense, signifying the first (presumably spherical) aggregates of high order that begin to form when the gradually increased concentration of a soap, for example, in aqueous solution reaches a certain critical value.

solute drops dramatically (and its electrical conductivity somewhat less so)—an obvious indication that some quite high-order aggregation of the solute molecules is occurring. These aggregates, the micelles, were believed—largely on philosophical grounds, little in the way of tangible pertinent evidence being available at the time—to be essentially spherical in structure, with the anionic carboxylate groups of the soap molecules constituting an exterior shell or skin enclosing the hydrocarbon portions, and of radius equal to that of the length of the constituent soap anions. Subsequent study showed these views to be essentially correct.* The critical micellar concentration (c.m.c.) at which micelles begin to be formed is highly characteristic of the particular soap involved; in the case of potassium laurate, for example, the c.m.c. at room temperature is about 0.0233 molar or 0.55 wt. %. Variations of the c.m.c. of soaps with changes in the structure of R in $RCOO^-M^+$ are discussed below.

The fact that micellar soap solutions are optically clear implies that the micelles have no dimension exceeding the shortest wavelength of visible light (about 4000 Å). This property in itself made it difficult to establish experimentally any further detail of the structure of micelles, although logic suggests, on the basis of the presumed radial spherical structure, that the radius of the body could not be greater than the length of the fully extended constituent molecule, since this would require either that the sphere be incompletely filled or that highly polar carboxyl-group heads be pulled into the apolar interior. Early X-ray diffraction analyses of aqueous soap solutions gave results compatible with the anticipated spherical shapes and radii of micellar aggregates. A further clue to an important feature of the structure of micelles came from early conductometric studies of soap solutions, which indicated that only a small fraction of the soap molecules incorporated in the micelle retain their ability to contribute normally to the overall conductivity of such solutions.

The fact that micellar soap solutions readily dissolve apolar solutes (of extremely limited solubility in water alone) is consistent with the concept of the interior of the micelle representing, in essence, a packet of liquid hydrocarbon. The ability of any given soap to form aggregates of the types discussed here is, in fact, critically dependent on the liquidity of the internal region; as mentioned above, such aggregates are not formed at temperatures below the "Krafft point," which corresponds closely to the melting point of the soap. This characteristic of micelles provides a means of finding out more about them, in that they can be rendered visible by dissolving intensely colored nonpolar dyes in them. Thus Hoyer and Mysels (1950) were able, by adding small amounts of Sudan IV (a water-insoluble dye that dissolves readily in the soap micelle) to micellar aqueous solutions of potassium laurate, to determine self-diffusion coefficients, electrophoretic mobilities, and ultracentrifugal sedimentation characteristics. Sharpness of boundaries observed during ultracentrifugation reveals that the micelles are

---

*Although it is easy to support the conclusion presented here that the micelle should be spherically symmetrical (or difficult to understand why it should not have such symmetry), Tanford (1980, pp. 54–55) has presented arguments favoring a somewhat oblate shape for these initially formed, high-order aggregates.

quite uniform in size and shape. These measurements also served to show that under the conditions employed, the potassium laurate micelle is indeed essentially spherical, with a radius of about 25 Å, and has an effective molecular weight of about 40,000 (corresponding to about 170 soap molecules per micelle). Each micelle, however, bears only about 12 (far less than 170) negative charges.

With this information, a number of important conclusions regarding the construction of the micelle could now be drawn. First, the curiously low net charge (confirming conductivity evidence) indicates that the micelle has a marked ability to bind a large fraction of the total number of counterions (K^+ in the present case) which would otherwise (i.e., in the monodisperse situation) be expected to be virtually completely dissociated. As the number of negatively charged soap anions increases in the growing micelle, the field strength at the surface of the body (which may be expressed as $F = q/4\pi r$, where q is the charge and r the radius of the essentially spherical assemblage) builds up, favoring reassociation of counterions with soap anions already incorporated into the micelles. This process has the important consequence of reducing electrostatic repulsion of carboxyl groups as these become more crowded in the micelle surface. The extensiveness of binding of counterions by the anionic carboxylate groups in the surface of the micelle represents, of course, an equilibrium state, which can be imagined to have been reached by changes in a micelle initially composed of intact (i.e., undissociated) soap molecules which would lose cations to the aqueous medium. As the number that succeeded in dissociating increased, however, so would the electric field tending to inhibit escape of additional counterions, resulting in establishment of an equilibrium which, as Hoyer and Mysels's experiments showed, involves strong binding of a large fraction of the total counterion complement of the micelle.

The observed effective radius (25 Å) of the potassium laurate micelle seems large in comparison with the conclusion that this parameter should not exceed the length of the fully extended normal alkyl (potassium) carboxylate units comprising the body. The length of the lauric acid molecule in its crystal lattice is 18.5 Å, of which about 1.4 Å represents half the length of the H-bonds joining carboxyl groups; the potassium ion contribution presumably lies somewhere between its "bare" and hydrated radii, i.e., 1.33–3.31 Å. The contribution of the shell of hydration of the micelle to its effective radius is difficult to assess precisely, but it should also be remembered that every micelle involved in these measurements carries at least one molecule of the dye, Sudan IV (an azo dye, $C_{22}H_{20}N_4O$, including one naphthalene and two benzene moieties), which is quite large and, if present at the center of the micelle, would permit the body to attain a significantly greater radius. If this is so, then the number of soap molecules per micelle is also greater, and the extensiveness of counterion dissociation less, than in the dye-free body—a clear case of a system being modified by the procedures used to study it.

All 170 of the carboxyl groups of this micelle being in the surface of a sphere of radius 25 Å, the area per carboxyl group is easily calculated to be about 47 Å^2, quite close to that (45 Å^2) found for air–water interfacial surfaces saturated with soap molecules (see above). This value is, however, over twice as great as the close-packing cross-sectional areas per molecule found in the saturated fatty acid crystal lattice (18.2 Å^2) or in laterally compressed monomolecular films (20–21 Å^2), and it is thus clear that about half the surface of the micelle must consist of hydrophobic methylene and/or methyl groups of the hydrocarbon chain of the soap in contact with the aqueous medium. This represents, of course, a compromise with the raison d'être of the micelle—the minimization of hydrocarbon-water contact; but the price is a small one to pay for complete insulation of the large remainder of hydrophobic groups. The presence of portions of these groups at the surface of the micelle implies their having substantial conformational freedom; the apparent density of the micelle interior corresponds closely to that of liquid hydrocarbons, and the hydrocarbon moieties of the soap anions are presumed to possess similar conformational characteristics, except for limitations imposed by one end of each of them being anchored, by attachment to the polar head group, in the surface of the micelle. Although the maximum radius of the micelle is determined by the all-anti-conformation lengths of these groups, only three or four of the many can possess this conformation at any given instant, or need to do so.

The formation of a micelle is a truly remarkable process. As the critical micellar concentration is approached, suddenly the monomeric and dimeric soap anions, together with the completely dissociated counterions, form full assemblages of about 100 molecules. These then continue to represent essentially the sole state of additional solute over a great range of concentrations from near 0 to about 25 wt. %. A close analogy to the process of crystallization from a supersaturated solution suggests itself, but with the crucial distinction that, unlike the crystal, which is not limited with respect to the size it may attain and thus the number of molecules it may accommodate, the maximum size of the micelle is strictly limited—ultimately by the maximum extendability of the hydrocarbon moiety of the soap. Thus, as solute continues to be added to the system, additional micelles are formed.

Although the process, once begun, is so rapid that intermediates between the soap anion dimer and the completely assembled micelle may probably never be verifiable by examination, it can readily be imagined to involve steps of the sort pictured in Fig. 4.6. Hydrophobic bonding of a third anion to the antiparallel dimer is electrostatically unfavorable at concentrations below the c.m.c.; but accommodation of higher concentrations of solute leads to formation of trimers, in which coulombic repulsion of the carboxylate anion heads keeps them as remote from each other as possible, hence about 120° apart and at ends of leading sections of tails as lengthy as their hydrophobically resisted extension will permit. Similarly, the carboxylate groups of tetramers would tend to be equidistant (about 110° apart), at ends of hydrocarbon moieties somewhat shortened by growth of the central core of closely nested methyl termini of the tails. Further growth involves increasingly strong binding of counterions (i.e., reassociation of soap molecules), reducing repulsion of the increasingly crowded head groups, until the assemblage (the completed micelle) can accommodate no additional soap molecules without attaining a volume of radius too great for the tails, even at maximal extension, to reach the assemblage's center. The speed of the overall, multi-step process and the absence of discernible intermediates suggest that each successive step is increasingly exergonic and that none involves substantial activation energy.

Formation of micelles from molecularly dispersed, dissociated soap mole-

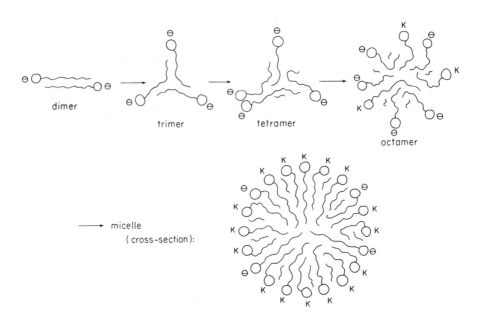

FIGURE 4.6. Possible early events in formation of an alkali metal soap micelle by stepwise aggregation of anions, then binding of counterions (e.g., K^+), starting with the anion dimer.

cules might appear at first blush to involve a substantial increase in order and therefore unlikely to occur spontaneously; but the process is actually strongly driven by the substantial entropy increases of the hydrophobic effect, involving increased conformational freedom of the hydrocarbon chains and melting of the ice-like sheaths enclosing them, concomitant with their being incorporated into the micelle. These aggregate-favoring changes, together with those of substitution of strong intermolecular forces between water molecules for the weak ones between water and hydrocarbon moieties, easily exceed the contrary endergonic effects of crowding the anionic carboxylate groups (substantially offset by counterion binding) and of ion dehydration that must accompany such reassociation of the soap ions. The thermodynamics of this complex phenomenon have been discussed in detail by Kresheck *et al.* (1966) and by Tanford (1980, pp. 60–78).

4.5.4.1. Effects of Differences in the Hydrophobic Group of Soaps on Micelle Formation

Critical micellar concentrations are, as pointed our earlier, characteristic of the particular soap, varying with both the cation and the structure of the fatty-carboxylate anion. The c.m.c.'s of soaps with the same cation vary inversely with the length (ℓ) of the normal saturated fatty acid soap and thus with its number of carbon atoms according to the empirical relationship:

$$\log_{10} \text{c.m.c. (moles/liter)} = 2.26 - 0.231\ \ell(\text{Å}).$$

This indicates an approximately 50% decrease in c.m.c. per added methylene group. The c.m.c. of a stearate soap, for example, is about one-fourth that of the corresponding palmitate soap. Increasing the ionic strength of the aqueous solvent (e.g., by adding KCl to a potassium soap solution) markedly depresses the c.m.c. and amplifies the homology effect. Presence of cis-unsaturation causes a two- to fourfold increase in c.m.c. (e.g., stearate as compared with oleate soaps; see Klevens (1953)) that would appear to be attributable, at least in part, to the smaller increase in entropy accompanying the passage of the more rigid olefinic hydrocarbon group from solution to micelle. Presence of a hydroxy substituent significantly reduces the magnitude of the hydrophobic effect, resulting in a further three- to fourfold elevation of c.m.c. (e.g., in the case of ricinoleate = 12-hydroxyoleate as compared with oleate soaps).

4.5.4.2. The Model Micelle

Deductive reasoning alone is sufficient to suggest that micelles could be spherical, space-conserving bodies of radius equal to the length of the fully extended constituent anions and of charge determined by a constant limiting field strength produced by dissociations of a fraction of the soap cations. If this picture is essentially correct, a number of interesting relationships between various features of the micelle and the number (n) of carbon atoms contained in the saturated fatty acid soap molecules of which it is composed may be formulated on the basis of simple geometric considerations.

The micelle is imagined to be composed of N flexible cylindrical elements of length proportional to the number, n, of carbon atoms in the soap anions (and of fixed cross-sectional area, a); the radius of the micelle is of course also proportional to n.

The volume of the micelle may then be expressed as $(4/3)\pi(kn)^3$ (where k is a proportionality factor representing the linear contribution of each carbon to the length of the molecule), but is also equal to $Nkna$ (the summed volumes of the elements), from which it is apparent that the number (N) of soap molecules per micelle is proportional to n^2. The stearate soap micelle, for example, is thus expected to contain $18^2/16^2$ or 1.265 times as many molecules as that of a palmitate micelle.

Since the surface area of the micelle, $4\pi(kn)^2$, is also proportional to n^2, it follows that the fraction of this total area occupied by the soap molecule carboxyl groups, all of which are in the surface of the body, is independent of n. Moreover, from the volume equation, $V = (4/3)\pi(kn)^3 = Nkna$, it is apparent that the total head-(carboxyl-) group area, Na, equals $(4/3)\pi(kn)^2$, and thus that the termini of the cylindrical units occupy one-third of the surface area. This is in good agreement with observations on potassium laurate (see above), if one takes into account the fact that the head groups are actually of somewhat greater cross-sectional area than the tails.

If the field strength of a sphere of charge q and of radius r ($= kn$) may be expressed as $F = q/4\pi kn$, and the limiting field strength preventing further net dissociation of counterions bound to the micelle is independent of its size, it follows that q/n is constant (i.e., that q is directly proportional to n), and, since N is proportional to n^2, that the fraction of dissociated soap molecules in the micelle, q/N, is inversely proportional to n. This leads to the expectation, for example, that dissociation of soap molecules in a stearate micelle would be about 100 × 16/18 or 89% as extensive as in one of palmitate.

4.5.5. Rods

When the concentration of soap in water reaches about 25 wt. % (irrespective of the structure of the soap, providing only that it is sufficiently soluble), the population of spherical micelles is so great that they in turn begin to coalesce, growing in one direction to form hemispherically capped cylindrical or rod-shaped aggregates of variable length, but retaining essentially the same radius as that of the spherical precursor (see Fig. 4.7). The dimensional characteristics of these bodies are revealed by X-ray scattering measurements and are reflected grossly in a marked increase in solution viscosity, as might be expected. As solute is added to the system, these rods grow in average length and population density, and eventually tend strongly to line up in extensive parallel, hexagonally packed arrays responsible for the remarkably high viscosity of the so-called "middle soaps." Curiously, as the soap concentration is increased still further, the viscosity of the solution begins to diminish again as a new type of aggregate, discussed more fully below, begins to be formed.

From earlier discussions of the spherical micelle, it is clear that the radius of these rod-shaped aggregates cannot exceed the length of the fully extended soap molecule. However, since the liquid-hydrocarbon interior of the rod is virtually incompressible, and assuming that all carboxyl groups must remain on the surface, in contact with the aqueous environment, some adjustment is required to accommodate the fact that the cylindrical (formerly spherical) unit inserted into the growing rod has a considerably diminished exposed-area-to-volume ratio. Thus in the spherical body the ratio $A/V = 3/r$, while in the cylindrical unit of the interior of the rod, exposed $A/V = 2/r$. If the radius of the rod did not change, the population density of carboxyl groups along the sides of the rod would therefore have to increase by 50%. Since the rods form spontaneously under sufficient concentration pressure, they cannot be highly strained structures and therefore there must be some compensation for the heightened electrostatic repulsion of the more highly crowded carboxyl groups. Relief could be provided by additional counterion binding and/or by reducing the radius of the cylindrical section of the

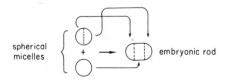

FIGURE 4.7. Growth of a spherical micelle to a rod-shaped assembly.

body, the latter being accompanied by some increase in gaucheness of the hydrocarbon moieties.

4.5.6. Lamellae

Soaps having sufficient water solubility to permit preparation of solutions containing as much as 50 wt. % of the solute yield solutions that, at about this concentration, begin to lose their high viscosity, signaling the onset of formation of a new type of aggregate. These bodies, which can be pictured as arising by lateral union of rods, are (as indicated by X-ray scattering studies) of indefinite extent in two dimensions but of a thickness approximately equal to the diameter shared by the spherical and rod-shaped aggregations. These plate-like aggregates (lamellae) are "bimolecular leaflets" (closely analogous in structure to that of the arrangement of amphiphilic lipid molecules in biomembranes), in which all the soap molecules are in one or the other of two parallel planar arrays with the hydrophobic hydrocarbon groups inside, in terminal methyl-group contact, and all carboxyl groups in the surfaces, exposed to the aqueous medium and extensively associated with counterions. Arguments presented in detail about rod-shaped aggregates apply here as well, which suggests that the thickness of lamellae should be somewhat less than the diameter of the spherical micelles of the same soap and/or that counterion binding should be more extensive.

In crystalline soaps, the molecules are arranged in parallel planes, in each successive one of which the orientation of the individual molecules (more or less normal to the planes) alternate, giving rise to planar interfaces of carboxylate groups and counterions parallel to and alternating with others in which the methyl termini of the molecules are in apposition. Such crystals therefore consist essentially of stacked lamellae, with the important distinction, however, that the apolar layers consist of hydrocarbon groups in typical crystalline order (in all-anti conformation and parallel to each other) rather than in liquid-like disorder and conformational freedom. At temperatures below their "Krafft points" (see above), crystalline soaps are appreciably soluble in water, but the dissolved molecules are (except for formation of anion dimers) molecularly dispersed, dissociated, and in direct equilibrium with crystals of the soaps (if the solution is saturated), no intermediate micelles and/or rod-type aggregates being formed.

As discussed in Chapter 16, amphiphiles of the type that appear to be obligate components of biomembranes (differing from soaps in having two or more tails attached to each hydrophilic head group) assume, like soaps, a stacked-lamellar molecular arrangement in the neat phase. Because of their more dominant hydrophobicity, these substances are virtually insoluble in water. In contact with water, however, the membrane amphiphiles exhibit the remarkable characteristic of swelling at temperatures below their "Krafft points," in some cases manifest (under microscopic observation) as the "growth of dendritic figures," shown by closer examination to consist of coaxial cylinders of lamellae. Such substances, the so-called "swelling amphiphiles," thus tend spontaneously to disperse in water directly as lamellae (although the molecular arrangement of these aggregations pre-exists in the neat phase), with no involvement of intermediate monodispersity or formation of spherical micellar or rod-like aggregates. Substances exhibiting such behavior are essential components of all biomembranes.

REFERENCES

Abrahamsson, S., and Ryderstedt-Nahringbauer, I., 1962, The crystal structure of the low-melting form of oleic acid, *Acta Cryst.* **15**:1261.
Abrahamsson, S., Dahlén, B., Löfgren, H., and Pascher, I., 1978, Lateral packing of hydrocarbon chains, in: *New Concepts in Lipid Research* (R. T. Holman, ed.), Pergamon Press, Oxford, pp. 125–143.
Bailey, A. V., Mitcham, D., French, A. D., and Sumrell, G., 1975, Unit cell dimensions of some long chain fatty acid polymorphs, *J. Am. Oil Chem. Soc.* **52**:196.
Brown, J. B., and Kolb, D. K., 1955, Applications of low temperature crystallization in the separation of the fatty acids and their compounds, in: *Progress in the Chemistry of Fats and other Lipids,* Vol. 3 (R. T. Holman, W. O. Lundberg, and T. Malkin, eds.), Pergamon Press, London, pp. 57–94.
Eagland, D., and Franks, F., 1965, Association equilibria in dilute aqueous solutions of carboxylic acid soaps, *Trans. Faraday Soc.* **61**:2468.
Goodman, D. S., 1958, The distribution of fatty acids between n-heptane and aqueous phosphate buffer, *J. Am. Chem. Soc.* **80**:3887.
Hoyer, H. W., and Mysels, K. J., 1950, A method for determining the properties of micelles, *J. Phys. Colloid Chem.* **54**:966.
Klevens, H. B., 1953, Structure and aggregation in dilute solutions of surface active agents, *J. Am. Oil Chem. Soc.* **30**:74.
Kresheck, G. C., Hamori, E., Davenport, G., and Scheraga, H. A., 1966, Determination of the dissociation rate of dodecylpyridinium iodide micelles by a temperature-jump technique, *J. Am. Chem. Soc.* **88**:246.
Langmuir, I., 1917, The constitutional and fundamental properties of solids and liquids. II. Liquids, *J. Am. Chem. Soc.* **39**:1848.
Malkin, T., 1952, The molecular structure and polymorphism of fatty acids and their derivatives, in: *Progress in the Chemistry of Fats and Other Lipids,* Vol. 1 (R. T. Holman, W. O. Lundberg, and T. Malkin, eds.), Pergamon Press, London, pp. 1–17.
Smith, R., and Tanford, C., 1973, Hydrophobicity of long chain n-alkyl carboxylic acids, as measured by their distribution between heptane and aqueous solutions, *Proc. Nat. Acad. Sci. U.S.A.* **70**:289.
Tanford, C., 1980, *The Hydrophobic Effect: Formation of Micelles and Biological Membranes,* 2nd ed., John Wiley, New York.

5
DISTRIBUTION OF FATTY ACIDS IN THE TISSUE LIPIDS

Before discussion of the metabolism of the fatty acids, it will be useful to outline briefly their distribution in potential dietary sources and in the tissue lipids, since these represent precursors and products of the degradative and synthetic processes.

For this purpose, the most useful reference is still the compilation by T. P. Hilditch and P. N. Williams, *The Chemical Constitution of the Natural Fats* (1964). Although much of the analytical work reported therein was done before the advent of gas–liquid chromatography and other now indispensable techniques, identifications and relative amounts of the fatty acyl groups in these fats are undoubtedly substantially correct, and the aims of this chapter are thus well served by reference to Hilditch's book.

The analyses by Hilditch and many of his students were performed over a lifetime by tedious methods requiring large amounts of material for accuracy. They included fractional crystallization, which separates the fatty acids with similar ratios of unsaturation to chain-length; low-pressure distillation of methyl or ethyl esters, which distinguishes those of different chain-length but does not distinguish esters of different unsaturation; lead (and other heavy metal) salt precipitation, which separates in bulk the saturated from unsaturated fatty acids; and, in some cases, alkaline isomerization, which quantitates the number of double bonds per molecule. Location of double bonds was deduced by various oxidative methods and examination of the fragments thus produced. Modern techniques of column and simple thin-layer and argentation chromatography and reversed phase and gas–liquid chromatography, including the coupling of the last technique with mass spectrometry, are now in common use. These newer, much more powerful techniques permit even the neophyte to perform accurate analyses on microgram quantities of fatty acids. Moreover, fatty acids present in amounts too small for detection and much too small for quantitation by the older methods are now readily identified and quantitated. Thus, modern fatty acid analyses have become so complete that salient points may be hidden in the mass of data. For this reason, the analyses reported in this chapter will, by and large, be simplified to illustrate points. For this purpose, the analyses found in Hilditch's work are usually quite adequate.

The mixtures of fatty acids found in the triacylglycerols of most plant fats are comparatively simple (Table 5.1). Except for trace components, the fruit fats (e.g., olive and the very similar avocado) consist largely of oleates and palmitates. The seed fats, on the other hand, contain major amounts of linoleic and, in some cases, linolenic acids. The fatty acids of green leaves (largely from chloroplast and other membrane lipids) consist of the usual mixture, with sizable proportions of linolenate and significant amounts of t3–16:1, a finding almost unique to chloroplasts (cf. Klenk and Knipprath, 1962). In contrast to the composition of typical plant fats, coconut oil contains largely a mixture of saturated medium-chain (C_8–C_{12}) acids with very little oleic and linoleic acids.

Certain families of plants produce large quantities of rare or bizarre fatty acids. Examples of these are the enothera with all $c6,9,12$–18:3 (γ-linolenic acid), and hydnocarpus with the cyclopentene-containing acids, chaulmoogric, $(2\text{-}C_5H_7)(CH_2)_{12}COOH$ and hydnocarpic, $(2\text{-}C_5H_7)(CH_2)_{10}COOH$. Other more common but interesting plant fatty acids are the cyclopropene-containing acids, sterculic, $CH_3(CH_2)_7-C_3H_2-(CH_2)_7COOH$ and malvalic, $CH_3(CH_2)_7-C_3H_2-(CH_2)_6COOH$,* and the 22-carbon monounsaturated fatty acid, erucic ($c13$–22:1), of the *Brassica* family, a major constituent of most species of rapeseed, an important food fat of many northern countries, and ricinoleic acid, 12h-$c9$–18:1, present in a significant proportion in castor oil.

Since the fatty acids of microorganisms will be discussed to some extent in later chapters, they will not be considered here. In passing, however, it is of interest to note that the bacterial fatty acids are often branched rather than unsaturated, possibly to fill a function similar to that of unsaturated fatty acids in the membrane lipids, since a branching also lowers the melting point of the fatty acid.

In most animals, the tissue from which the lipids are derived seems to be of more importance in determining their structure than the animal species. In other words, fatty acids from the same tissue in different animals are more likely to be similar than are those from different tissues in the same animal. For example, adipose tissue is composed mainly of triglycerides containing a rather simple mixture of fatty acids, largely synthesized in the tissue. Oleate and palmitate are usually the major components. The lipids of muscle and of most of the organs are, for the most part, typical membrane lipids with high proportions of phospholipids containing relatively large amounts of polyunsaturated long-chain acids and the saturated acid, stearate. Nervous tissue contains little neutral lipid and consists largely of phospholipid, sphingolipid, and sterol. The phospholipids have the typically high amounts of long-chain polyunsaturates, while the sphingolipids usually contain saturated or monounsaturated acids of exceptionally great chain-length, with both odd and even chains. The otherwise unusual α-hydroxy acids are also common components of the sphingolipids.

The fatty acyl groups of these lipids arise in three main processes:

(i) From the diet: In some lower animals, this may be the sole source of fatty acids, and these animals are consequently at the mercy of the environment in maintaining functionally optimal composition of their membrane lipids (Dayton *et al.*, 1967).

(ii) By synthesis: Most organisms possess this capacity, which is, however,

*$(2\text{-}C_5H_7)$ = cyclopent-2-enyl; —C_3H_2— = 1, 2-cyclopropenylene.

TABLE 5.1. Typical Major Fatty Acids of Some Fats of Plant Origin[a]

Vegetable oils	Saturated								Unsaturated				
	C_4–C_8	10:0	12:0	14:0	16:0	18:0	20:0	22:0	16:1	18:1	18:2	18:3	20:1	22:1
Corn					13	3	1			28	53	1		
Peanut				1	12	3	2	3		53	26			
Cottonseed					26	3			1	18	52			
Soybean				2	12	4				24	51	7		
Olive				1	13	3			1	73	9	1		
Coconut	7	6	49	20	9	2				6	2			
Pasture grass				3	15	3				8	23	46		
Rapeseed[b]														
Colza					3					11	14	8	7	51
Canbra					5					60	21	7	2	1

[a]From Hilditch and Williams (1964).
[b]See Chapter 19 for nutritional discussion.

TABLE 5.2. Typical Fatty Acid Composition of Some Fats of Animal Origin[a]

Animal fats	Saturated								Unsaturated					Other eicosapolyenoic acids	Docosapolyenoic acids
	C_4–C_8	10:0	12:0	14:0	16:0	18:0	20:0	22:0	16:1	18:1	18:2	18:3	20:4		
Lard				2	27	14			3	44	11	1			
Chicken			2	7	25	6			8	36	14		3		
Egg				3	25	10				50	10	2	1		
Beef				3	29	21	1		3	41	2	1			
Butter	6	3	4	12	28	13			3	29	1				
Human milk		2	7	9	21	7	1		3	36	7	1	1		
Menhaden				9	19	6			16		25[b]			15	9

[a]From Hilditch and Williams (1964).
[b]Plus other C_{18} polyenes.

inhibited by large amounts of dietary fatty acids. The fatty acids from this source are largely the usual oleate, palmitate, stearate mixture.

(iii) Alteration of fatty acids from these sources: This consists largely of chain elongation and desaturation, processes that will be discussed in Chapters 8 and 9.

In Table 5.2 the fatty acids of some typical fats of animal origin are listed. Such comparisons are actually not completely valid since, in some cases, the fats analyzed are not derived from the same tissues (an impossibility in many cases such as fish and mammals) and the influence of diet may not be considered. However, certain generalizations can be made. Almost without exception, palmitate (16:0) is the major saturated fatty acid. In the milk fats of many mammals shorter-chain fatty acids occur in significant amounts, but even in these cases, 16:0 represents more than 20% of the total. Oleic acid (9–18:1) is the major unsaturated fatty acid and, indeed, the major fatty acid of most fats. Polyunsaturated fatty acids having more than two double bonds are generally major constituents of fats of marine origin only. The probable origin of the high polyunsaturated content of fish fatty acids is shown in Table 5.3, which reports the results of an

TABLE 5.3. Fatty Acid Composition of *Chaetoceros, Artemia,* and Guppy Oils (Weight Percent of Total Ester)[a,b]

Fatty acid	*Chaetoceros*	*Artemia salina*	Guppy 17 ± 1°C	Guppy 24 ± 1.5°C
Shorter chain	Trace	0.4	Trace	
12:0	0.4	Trace	0.2	Trace
13:0	0.7	Trace	Trace	Trace
14:0	13.0	5.0	2.0	0.9
14:2	0.6	Trace	0.6	0.5
15:0	2.0	2.0	Trace	0.2
16:0	18.0	12.0	23.0	36.0
16:1	48.0	45.0	16.0	9.0
16:2	3.0	Trace	0.2	0.2
16:3	4.0	2.0		0.6
16:4	Trace			0.5
18:0	0.5	2.0	8.0	10.0
18:1	9.0	18.4	18.3	15.0
18:2	2.0	0.7	Trace	Trace
18:3	Trace	0.5	1.0	0.8
20:1		0.9		Trace
18:4 & 20:2		0.8	0.3	
20:3			0.2	Trace
20:4		Trace	2.0	2.0
20:5		12.0	5.0	5.0
22:4			1.0	1.0
22:5			6.0	7.0
22:6			17.0	12.0

[a]From Kayama et al. (1963).
[b]See text for discussion of feeding experiment.

experiment by Kayama *et al.* (1963). In this study, brine shrimp *(Artemia salina)* were maintained on a diet of the diatom *Chaetoceros,* and the fatty acids of the total lipids of each were analyzed. It can readily be seen that the *Artemia* fatty acids reflect those of *Chaetoceros* but that, in addition, further elongation and desaturation have taken place. In the case of the guppies maintained solely on the *Artemia,* it is possible to find a reflection of the *Artemia* lipids, but in addition, extensive breakdown, resynthesis, and alteration have occurred. The effect of temperature can also be seen in that more highly unsaturated fatty acids are deposited at 17°C than at 24°C. This widespread phenomenon will be dealt with in Chapter 9.

In Table 5.4 can be seen the fatty acids of some typical fish body fats. The muscle fats of fish vary widely and a comparison between different families may not be justified, since the storage of fat in fish is accomplished in several different ways and a comparison of adipose tissue may not be possible. The differences in the fatty acids of marine and freshwater fish may again be a reflection of the diet, since the marine plankton synthesize more highly unsaturated fatty acids than do the freshwater varieties. The effect of temperature may be important but cannot be assessed in this case. In any event, it can be seen that the young salmon deposit fat with a mixture of fatty acids fairly typical of freshwater fish while they reside in streams and that this changes to the marine type of mixture after their migration to the ocean.

In modern gas–liquid chromatographic analysis, so much information is obtained that trace constituents assume an importance out of proportion to their concentrations. One such analysis, a *tour de force* by a group extensively involved in the development of this now indispensable tool, is the analysis of menhaden oil fatty acids by Stoffel and Ahrens (1960). For assessing general trends some simplification is necessary.

In Table 5.5 are presented the fatty acid spectra of some animals phylogenetically between fish and mammals, as well as those of two marine mammals. The fatty acids of the latter undoubtedly reflect their marine diet in that long-chain polyunsaturated acids form a large proportion of the lipid component. The short-chain acids (largely isopentanoic) of the porpoise represent a material synthesized for a special purpose, possibly echo-location. Reptiles (as typified by the

TABLE 5.4. Percentage of Component Acids of Body Fats of Fresh- and Saltwater Fish[a]

	Fatty acids							
	Saturated			Unsaturated				
	C_{14}	C_{16}	C_{18}	C_{14}	C_{16}	C_{18}	C_{20}	C_{22}
Typical marine fish (tuna)	4	19	4		6	26	24	18
Typical freshwater fish (carp)	4	15	2	1	18	46	15	Trace
Young salmon (fresh water)	3	18	3	3	22	30	13	10
Mature salmon (salt water)	5	11	1	1	9	26	27	21

[a]From Hilditch and Williams (1964).

TABLE 5.5. Percentage of Component Acids of Body Fats of Marine Mammals, a Reptile, and a Bird[a]

	Fatty acids									
	Saturated					Unsaturated				
	C_5	C_{12}	C_{14}	C_{16}	C_{18}	C_{14}	C_{16}	C_{18}	C_{20}	C_{22}
Whale oil			6	18	2	4	13	38	11	6
Porpoise	14	4	12	5		5	27	17	11	7
Lizard			4	29	10		12	40	5	
Chicken			1	24	4		7	64[b]		

[a]From Hilditch and Williams (1964).
[b]43% 9–18:1 plus 21% 9,12–18:2.

lizard) do not show any unusual features, except, possibly, the slightly higher proportion of palmitate, ostensibly indicating a greater contribution from synthesis rather than diet. The chicken, on the other hand, has a relatively high content of 9,12–18:2, reflecting a diet high in this fatty acid, probably derived from grains (see Table 5.1).

As an additional complication in fish, lipids are stored in various organs and serve a variety of functions. Thus, the high squalene content of certain shark livers and the glyceryl ethers of other species may be for buoyancy rather than energy storage. The same may be true of the very high wax ester content of certain vertically migratory fish (Nevenzel, 1970).

When we turn to typical mammals, adipose tissue fatty acids can be compared although here, too, the effect of the diet is difficult to assess. In Table 5.6 can be seen the fatty acid composition of the adipose tissue of the rat on a fat-free diet and it is remarkable how similar it is to the composition of the adipose

TABLE 5.6. Typical Fatty Acid Composition of Some Mammalian Depot Fats[a]

	Fatty acid (average percentage of composition)										Higher unsaturated acids
	12:0	14:0	16:0	18:0	20:0	14:1	16:1	18:1	18:2	18:3	
Horse (pasture-fed)		2	27	4		1.8	4	38	5	17	0.9
Cow (pasture-fed)		4	32	22		0.5	5	35	0.6	0.3	3
Lion (fed in captivity)		5	29	18		0.6	2	40			
Rat (fed fat-free diet in laboratory)		7	24	5	1	1	6	49	5		0.5
Human subject	1	5	26	5	3	0.5	8	48	3		Trace

[a]From Hilditch and Williams (1964).

tissue fat of man. Typically 16:0 and 18:1 predominate (73% of the total for the two). The polyunsaturated fatty acid content of rat fat is presumably acquired from the mother during the suckling period. Other nonruminants tend to show the same distribution of fatty acids and do not vary far from it unless some specialized diet has been imposed over a long period of time. Ruminants, on the other hand, have an unusually high proportion of 18:0, which is probably the result of hydrogenation of dietary 18-carbon unsaturated fatty acids by rumen bacteria. In these animals some of the unsaturated fatty acids have the *trans* configuration, a result of partial hydrogenation and isomerization of dietary polyunsaturated fatty acids. The lion, although a carnivore, is listed with the herbivores to illustrate the effect of the lipid composition of the prey on that of the predator. In this case, the "prey" was probably horse meat, beef, and mutton fed in the zoo rather than the prey more usual in the African plains.

As mentioned earlier, the effect of temperature on fatty acid unsaturation is almost universal. Even in those mammals in which anatomical and environmental features permit temperature differences, local fatty acid composition may vary accordingly. Thus, in the classic experiment with pigs carried out by Henriques and Hansen in 1901 (Table 5.7), it can be seen that the pig-back fats differ in degree of unsaturation, depending on the temperature at which they are deposited. Of course the difference is much greater in poikilotherms, as will be discussed in Chapter 9.

Finally, it is of interest to consider the lipid composition of a specialized secretion product, mammalian milk (Table 5.8). The large proportion of short-chain fatty acids in the milk of many mammals, particularly ruminants, reflects not only the dietary fats and the usual synthetic fatty acids but also the imposition on these of two reactions that are unusual for most animals. The first is the result of metabolism of long-chain fatty acids by rumen bacteria, resulting in the formation and absorption of short-chain fatty acids. The second is a special deacylase in mammary tissue that hydrolyzes the acyl thioester before completion of palmitate synthesis on the fatty acid synthetase. The result of the action of this deacylase is the release of short- and medium-chain fatty acids at the expense of the production of the usual palmitate. This subject will be covered in Chapter 8.

TABLE 5.7. Relationship of Temperature to Unsaturation in Pig-Back Fats[a]

Depth of fat	Temp. (°C)	I.V.[b]	% Oleic	% Polyene
1 cm	33.7	60	48	10
4 cm	39.0	54	41	9
rectal	39.9			
	Ambient temp.			
	0	72.3		
	0	67.0		
	(in a sheepskin coat)			

[a]From Henriques and Hansen (1901).
[b]Iodine value—a measure of unsaturation given as the number of grams of iodine absorbed per 100 g of fat.

TABLE 5.8. Percentage of Component Acids of Milk Fat of Cow and Man[a]

	C_4	C_6	C_8	C_{10}	C_{12}	C_{14}	C_{16}	C_{18}	C_{20}	C_{14}	C_{16}	C_{18} (−2H)	C_{18} (−4H)	$>C_{18}$
Cow	4	2	1	3	2	9	25	11	4	1	5	32	4	
Human subject				2	6	8	22	9	1	1	4	37	8	3

[a]From Hilditch and Williams (1964).

Appendix: A Note on the Structure and Naming of the Fatty Acids

Many of the physical properties of the fatty acids are reflected in those of the complex lipids. A consideration of the structures and naming of some of the common members of the fatty acid family and the lipids which contain them is therefore important. In this Appendix, only the fatty acids will be treated; the complex lipids will be considered in the chapters devoted to their properties.

Although it may seem to the gas chromatographer that every conceivable fatty acid structure can be found in some natural source, certain simplifying rules can be stated, which, though not invariably true, will hold for the vast majority of lipids in nature.

(i) Fatty acids commonly have straight (unbranched) chains.
(ii) Most fatty acids contain an even number of carbon atoms.
(iii) The chain-lengths of the most abundant fatty acids in nature are in the C_{16}–C_{20} range (actually mostly C_{16}–C_{18}).
(iv) In unsaturated fatty acids, the double bonds are usually of the *cis* configuration (see Chapter 9 and below).
(v) If two or more double bonds are present, they are usually in the divinyl methane or 1,4-relationship to each other.

Interesting exceptions will be encountered for each rule.

The naming of the fatty acids may follow several different conventions (see Table 5.9). First, their trivial names, by which the most common members have been known for a very long time, are usually sufficient to indicate their exact structure to the initiated. These are used in this book where they can be used conveniently and without ambiguity. Second, the Geneva Convention, somewhat modified, is completely descriptive and may serve to identify the more common fatty acids as well as those with unusual structures. Third, for the sake of simplicity, several shorthand designations are current and will be used in many chapters to save space. In the commonly accepted shorthand, the chain-length is indicated by a number separated by a colon from a second number designating the number of double bonds. Thus, the saturated fatty acids will have 0 for the second number (e.g., palmitic acid: 16:0). In the case of the unsaturated fatty acids, a complication is introduced by the need to know the position and configuration of

TABLE 5.9. Structures and Names of Some Common Saturated Fatty Acids

Structure (R in RCOOH)	Trivial name (if commonly used)	Geneva system	Shorthand
H	Formic	—	1:0
CH_3	Acetic	—	2:0
C_2H_5	Propionic	Propanoic	3:0
C_3H_7	Butyric	Butanoic	4:0
C_4H_9	Valeric	Pentanoic	5:0
C_5H_{11}	Caproic	Hexanoic	6:0
C_7H_{15}	Caprylic	Octanoic	8:0
C_9H_{19}	Capric	Decanoic	10:0
$C_{11}H_{23}$	Lauric	Dodecanoic	12:0
$C_{13}H_{27}$	Myristic	Tetradecanoic	14:0
$C_{15}H_{31}$	Palmitic	Hexadecanoic	16:0
$C_{16}H_{33}$	Margaric	Heptadecanoic	17:0
$C_{17}H_{35}$	Stearic	Octadecanoic	18:0
$C_{19}H_{39}$	Arachidic	Eicosanoic	20:0
$C_{23}H_{47}$	Lignoceric	Tetracosanoic	24:0

the double bond. Thus, 18:1 designates an octadecenoic acid, but not necessarily oleic. The problem is solved in several ways, two of which will be used here.

In the first, the position of the double bond (or of that nearest the carboxyl group if more than one is present) is indicated, as a prefix, by the lower of the numbers of the two doubly bonded carbon atoms, counting from the carboxyl carbon. Thus oleic (*cis*-9-octadecenoic) acid is *c*9–18:1. Alternatively, for reasons that will become apparent in Chapter 9, it may be desirable to indicate the position of the double bond nearest the methyl terminus of the molecule. In this case, the numbering of the carbon atoms starts from the methyl end, and the location of the first (or only) double bond is indicated by a single number as before (usually a different number) either as a suffix preceded by "ω" or in parentheses and preceded by *n*-. Oleic acid might, therefore, be designated as *c*18:1ω9, which informs the reader that the most distal double bond is nine carbons from the methyl terminus. It is assumed that, unless otherwise specified, all multiple double bonds in polyunsaturated fatty acids have the 1,4-relationship to each other. Another convention subtracts the number of carbons separating the distal double bond and terminal methyl from the number of carbons in the chain—i.e., *n*. By this convention, oleic acid is 18:1(*n*−9), revealing that the double bond is also nine carbons from the carboxyl group. The somewhat shorter ω terminal-numbering shorthand is favored in this test (see Table 5.10). (Because the double bond of oleic acid is in the exact center of the molecule, it is in position 9 counting from either end. *Palmitoleic* acid, on the other hand, is *c*9–16:1 or *c*16:1ω7.)

Although the *cis* and *trans* convention has been presented in this chapter, it is possible that in some cases it may be ambiguous and new designations of *Z* (from the German *zusammen*) and *E* (from the German *entgegen*) have, in recent

TABLE 5.10. Structures and Names of Some Common Unsaturated Fatty Acids

Structure	Trivial name (if generally used)	Shorthand (carboxyl numbering)	Shorthand (terminal numbering)
Monoenoic			
$CH_3-(CH_2)_5-\overset{H}{C}=\overset{H}{C}-(CH_2)_7-COOH$	Palmitoleic	c 9-16:1	c 16:1ω7
$CH_3-(CH_2)_7-\overset{H}{C}=\overset{H}{C}-(CH_2)_7-COOH$	Oleic	c 9-18:1	c 18:1ω9
$CH_3-(CH_2)_7-\overset{H}{C}=\underset{H}{C}-(CH_2)_7-COOH$	Elaidic	t 9-18:1	t 18:1ω9
$CH_3-(CH_2)_7-\overset{H}{C}=\overset{H}{C}-(CH_2)_{11}-COOH$	Erucic	c 13-22:1	c 22:1ω9
$CH_3-(CH_2)_7-\overset{H}{C}=\overset{H}{C}-(CH_2)_{13}-COOH$	Nervonic	c 15-24:1	c 24:1ω9
Dienoic			
$CH_3-(CH_2)_4-\overset{H}{C}=\overset{H}{C}-CH_2-\overset{H}{C}=\overset{H}{C}-(CH_2)_7-COOH$	Linoleic	c,c 9,12-18:2	c,c 18:2ω6
Trienoic			
$CH_3-CH_2-\overset{H}{C}=\overset{H}{C}-CH_2)_3-(CH_2)_6-COOH$	α-Linolenic	All c 9,12,15-18:3	All c 18:3ω3

(*continued*)

TABLE 5.10. Structures and Names of Some Common Unsaturated Fatty Acids (*Continued*)

Structure	Trivial name (if generally used)	Shorthand (carboxyl numbering)	Shorthand (terminal numbering)
$CH_3-(CH_2)_4-\overset{H\ \ \ H}{(C=C-CH_2)_3}-(CH_2)_3-COOH$	γ-Linoleic	All c 6,9,12–18:3	All c 18:3ω6
$CH_3-(CH_2)_4-\overset{H\ \ \ H}{(C=C-CH_2)_3}-(CH_2)_5-COOH$	Bishomo-γ-linolenic	All c 8,11,14–20:3	All c 20:3ω6
$CH_3-(CH_2)_7-\overset{H\ \ \ H}{(C=C-CH_2)_3}-(CH_2)_2-COOH$		All c 5,8,11–20:3	All c 20:3ω9
Tetraenoic			
$CH_3-(CH_2)_4-\overset{H\ \ \ H}{(C=C-CH_2)_4}-(CH_2)_2-COOH$	Arachidonic	All c 5,8,11,14–20:4	All c 20:4ω6
Pentaenoic			
$CH_3-CH_2-\overset{H\ \ \ H}{(C=C-CH_2)_5}-(CH_2)_2-COOH$		All c 5,8,11,14,17–20:5	All c 20:5ω3
$CH_3-(CH_2)_4-\overset{H\ \ \ H}{(C=C-CH_2)_5}-CH_2-COOH$		All c 4,7,10,13,16–22:5	All c 22:5ω6
Hexaenoic			
$CH_3-CH_2-\overset{H\ \ \ H}{(C=C-CH_2)_6}-CH_2-COOH$	Clupanodonic	All c 4,7,10,13,16,19–22:6	All c 22:6ω3

years, replaced the *cis* and *trans* notations. The origin of these new designations can be readily understood with reference to structures (**a**) and (**b**).

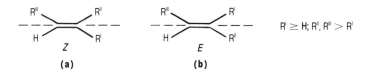

If, on bisecting the double bond by a straight line, the substituents of highest priority are on the same side of the line, the structure has the Z configuration (corresponding often to *cis*); when they are on opposite sides of the line bisecting the double bond, the *E* designation is used. For further discussion and definition of priority see Alworth (1972).

REFERENCES

Alworth, W. L., 1972, *Stereochemistry and its Application to Biochemistry,* Wiley-Interscience, New York, p. 148, footnote.
Dayton, S., Hashimoto, S., and Pearce, M. L., 1967, Adipose tissue linoleic acid as a criterion of adherence to a modified diet, *J. Lipid Res.* **8**:508.
Henriques, V., and Hansen, C., 1901, *Skand. Arch. Physiol.* **11**:151.
Hilditch, T. P., and Williams, P. N. (eds.), 1964, *The Chemical Consitution of the Natural Fats,* 4th ed., Wiley, New York.
Kayama, M., Tsuchiya, Y., and Mead, J. F., 1963, A model experiment of aquatic food chain with special significance in fatty acid conversion, *Bull. Japan Soc. Scientific Fisheries* **29**:452.
Klenk, E., and Knipprath, W., 1962, Über das Vorkommen der *trans*-Δ^3-Hexadecensäure in den Lipoiden einer Süsswasseralge *(Scenedesmus obliquus)* und deren Isolierung, *Hoppe-Seyler's Z. Physiol. Chem.* **327**:283.
Nevenzel, J. D., 1970, Occurrence, function and biosynthesis of wax esters in marine organisms, *Lipids* **5**:308.
Stoffel, W., and Ahrens, E. H., Jr., 1960, The unsaturated fatty acids in menhaden body oil: the C_{18}, C_{20} and C_{22} series, *J. Lipid Res.* **1**:139.

SELECT BIBLIOGRAPHY

Holman, R. T., 1971, Chapter I, in: *Progress in the Chemistry of Fats and Other Lipids,* Vol. 9 (R. T. Holman, ed.), Pergamon Press, Oxford, pp. 3–6.
Mead, J. F., and Fulco, A. J., 1976, *The Unsaturated and Polyunsaturated Fatty Acids in Health and Disease* (I. Newton Kugelmass, ed.), Charles C. Thomas, Springfield, Illinois.
The Nomenclature of Lipids (Recommendations 1976), IUPAC–IUB Commission on Biochemical Nomenclature, 1978, *Biochem. J.* **171**:21.

6
PEROXIDATION OF FATTY ACIDS

One of the most typical properties of the unsaturated fatty acids, particularly the polyenoic acids, is their susceptibility to peroxidation, chiefly the type termed autoxidation (or autocatalytic oxidation). Autoxidation is often (but probably incorrectly) defined as oxidation by molecular oxygen under usual conditions of temperature and pressure. Perhaps a better definition would be a radical chain reaction involving molecular oxygen as a reactant in one of the steps. The term "peroxidation," which includes autoxidation, refers to production of peroxides and their degradation products.

In any event, most organic compounds, including most tissue constituents, are thermodynamically unstable in the presence of oxygen. Fortunately, because of both energy barriers and electron-spin barriers to the direct reaction of ordinary triplet-state oxygen with methylene groups, the reaction is generally very slow and living organisms, therefore, do not burn on exposure to air—at least not immediately! However, certain substances present in the tissues react relatively readily with oxygen. Among these are the unsaturated fatty acids.

6.1. Chemistry of Lipid Peroxidation

Although saturated and monounsaturated fatty acids can undergo autoxidation, especially at high temperatures in the presence of certain catalysts, the rates of the reactions are much lower than are those for the polyunsaturates of the commonly occurring 1,4-polyene type. Thus, at 100°C, the ratios of the rates of oxygen absorption by methyl oleate, linoleate, and linolenate have been found to be 1:12:24 (Gunstone and Hilditch, 1945). We will consider linoleate as a model for fatty acid autoxidation since the most definitive studies on the mechanism of autoxidation, performed largely by the scientists of the British Rubber Producers Research Association, were carried out with this substance.

Empirically, it was found that the rate of autoxidation is directly proportional to substrate (or rather, double bond) concentration and to the partial pressure of oxygen above 100 mm. The rate also increases with the extent of oxidation, indicating a chain reaction and the autocatalytic nature of the process. The principal initial products are hydroperoxides. Equation 6.1 expresses these relationships, where K_O and λ are empirical constants, RH is unoxidized substrate, and pO_2 the partial pressure of oxygen:

$$-\frac{dO_2}{dt} = \frac{K_O[RH][ROOH]}{1 + \frac{\lambda[RH]}{pO_2}} \quad (6.1)$$

The generally accepted mechanism of autoxidation of linoleate, leading to the same relationships, is shown in Fig. 6.1.

The initiation reactions are of particular importance since it has been shown that highly purified polyunsaturated fatty acids are stable for long periods of time in the presence of oxygen (Privett and Blank, 1962). The reaction $RH + O_2 \rightarrow R\cdot + HO_2\cdot$ is endothermic to the extent of about 64 kcal/mol and is unlikely with RH in the ground state and O_2 in the triplet state. However, traces of peroxides or transition metals, and ultraviolet or ionizing radiation, in addition to several other factors, are known to bring about initiation. Singlet oxygen, formed photochemically, reacts directly with pure methyl linoleate (Rawls and Van Santen, 1970). Once initiated, the reaction continues by a chain mechanism involving resonance-stabilized free radicals that react readily with oxygen to form peroxy radicals, which can then initiate new chains by the slower abstraction of a hydrogen atom from another molecule of substrate as the peroxy radicals are converted to hydroperoxides. These latter compounds decompose under certain conditions by homolysis, giving radicals that can initiate new chains (Eq. 6.2):

$$\text{At high ROOH:} \quad ROOH\cdots OOR \rightarrow H_2O + RO\cdot + RO_2\cdot \atop | \atop H \quad (6.2a)$$

$$\text{At low ROOH:} \quad ROOH \rightarrow RO\cdot + \cdot OH \quad (6.2b)$$

but a much more likely decomposition mechanism is the metal-catalyzed reaction in which certain metal ions readily convert hydroperoxides to alkoxy radicals and hydroxyl ion (Eq. 6.3) (Pryor, 1973):

$$ROOH + M^{n+} \rightarrow RO\cdot + OH^- + M^{(n+1)+} \quad (6.3)$$

The structures of the hydroperoxides formed during autoxidation have been investigated extensively. Reduction of the hydroperoxides of oleic acid usually gives the 8-, 9-, 10-, and 11-hydroxystearic acids in about equal amounts. From linoleic acid, mainly the 9- and 13-hydroxystearic acids are obtained despite the initial attack on the susceptible diallylic position 11 (Eq. 6.4):

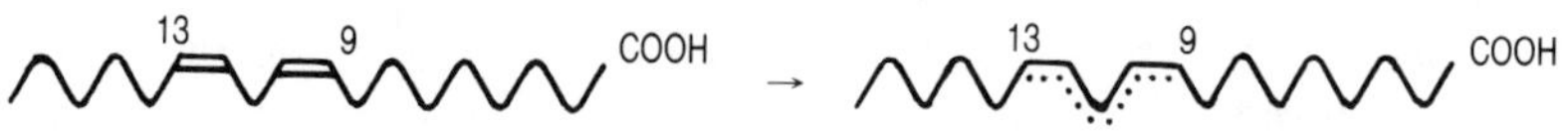

$$\begin{array}{c}\text{13}\quad\text{9}\\\diagup\!\diagdown\!\diagup\!\diagdown\!=\!\diagdown\!=\!\diagup\!\diagdown\!\diagup\!\diagdown\!\diagup\!\diagdown\text{COOH}\end{array} \rightarrow \begin{array}{c}\text{13}\quad\text{9}\\\diagup\!\diagdown\!\diagup\!\diagdown\!\cdots\!\diagdown\!\cdots\!\diagup\!\diagdown\!\diagup\!\diagdown\!\diagup\!\diagdown\text{COOH}\end{array} \quad (6.4)$$

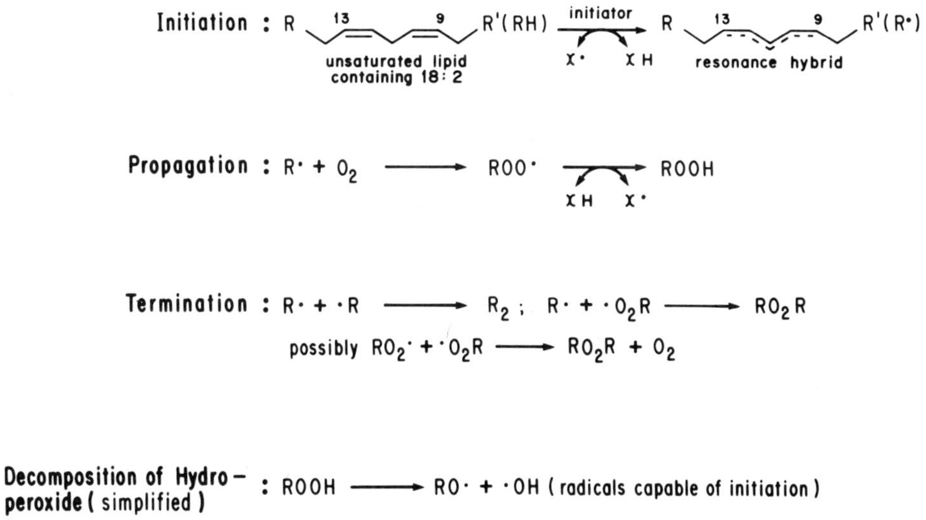
Initiation : R 13 9 R'(RH) initiator R 13 9 R'(R•)
unsaturated lipid containing 18:2
X• XH
resonance hybrid
Propagation : R• + O2 ROO• ROOH
XH X•
Termination : R• + •R R2 ; R• + •O2R RO2R
possibly RO2• + •O2R RO2R + O2
Decomposition of Hydroperoxide (simplified) : ROOH RO• + •OH (radicals capable of initiation)

FIGURE 6.1. The initial reactions in the autoxidation of a diene (e.g., linoleic acid). R• = alkyl radical; RO• = alkoxy radical; χH is any hydrogen donor, possibly the starting substance, RH.

This is due to the fact that the species that reacts with oxygen is the resonance-stabilized pentadienyl radical, the odd electron of which is delocalized almost completely to carbons 9 and 13, yielding 9-hydroperoxy-10-*trans*,12-*cis*-octadecadienoic acid (**a**) and 13-hydroperoxy-9-*cis*,11-*trans*-octadecadienoic acid (**b**) (Fig. 6.2). Substantial quantities of the corresponding *trans, trans* isomers (**c** and **d**, Fig. 6.2) are also formed as a consequence of the reversibility of the reaction of pentadienyl radicals with oxygen (Porter *et al.*, 1980). Since formation of the stable final products (the hydroperoxides) by abstraction of hydrogen from a second linoleic acid molecule represents a fate of the intermediate peroxy radicals

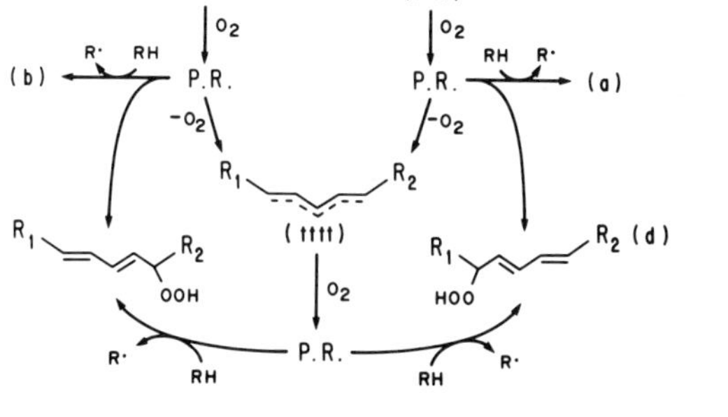
13 9
R1 R2
(cttc)
O2
RH R•
P.R.
-O2
R1 R2 (a)
OOH
OOH
R1 R2 (b)
R1 R2 R1 R2
(cttt) (tttc)
O2 O2
R• RH RH R•
(b) P.R. P.R. (a)
-O2 -O2
R1 R2
(tttt)
R1 R2 R1 R2 (d)
OOH HOO
O2
R• P.R. R•
RH RH

FIGURE 6.2. Formation of geometric isomers of linoleate hydroperoxides. From Porter *et al.* (1980). c = *cis*; t = *trans*; thus cttc: *cis,trans,trans,cis*. P.R. = peroxy radical. RH is a hydrogen-containing substance, possibly the starting substance.

that is a competitive alternative to loss of O_2 accompanied by configurational change, the *cis, trans* isomers (**a** and **b**, Fig. 6.2) predominate in autoxidation of neat linoleic acid, but the product composition shifts in favor of the *trans, trans* isomers (**c** and **d**) if the diene is diluted with an inert solvent before exposure to oxygen.

The hydroperoxides are reasonably strong oxidizing agents and are reduced by thiol-containing proteins, glutathione, cysteine, and, particularly, by a glutathione-dependent factor in the cytosol which apparently has this function as part of the antioxidant defense of the cell. If not reduced to the alcohol, the hydroperoxide can undergo homolysis, usually catalyzed by transition metal ions, to form a hydroperoxy or alkoxy radical, depending upon the oxidation state of the metal (Eqs. 6.3, 6.5):

$$ROOH + M^{(n+1)+} \rightarrow ROO\bullet + M^{n+} + H^+ \tag{6.5}$$

The alkoxy radical formed, as in Eq. 6.3, reacts with any susceptible molecule in its vicinity, usually by abstraction of a hydrogen atom and creation of another radical likely to initiate further reactions. The alkoxy radical may also undergo a fragmentation, termed β scission, in which either C—C bond β to the oxygen is broken, giving an aldehyde and an alkyl radical, as shown in Eq. 6.6:

$$R-\overset{\overset{\bullet}{\overset{|}{O}}}{C}H-R' \begin{matrix} \nearrow RCHO + R'\bullet \\ \searrow R\bullet + R'CHO \end{matrix} \tag{6.6}$$

Indeed, it is this type of reaction that is responsible for the unpleasant odor and flavor of oxidatively rancid fat, the odor being that of unsaturated aldehydes produced in this manner from unsaturated fats. For example, hydroperoxides formed by autoxidation of oleic acid would be expected to decompose in this way to give aldehydes shown in Eq. 6.7:

8-Hydroperoxy-9-octadecenoic acid → 2-undecenal		(6.7a)
9-Hydroperoxy-10-octadecenoic acid → 2-decenal		(6.7b)
10-Hydroperoxy-8-octadecenoic acid → nonanal		(6.7c)
11-Hydroperoxy-9-octadecenoic acid → octanal		(6.7d)

Additional reactions that alkoxy radicals may undergo are shown in Eqs. 6.8a–6.8d:

Abstraction of a hydrogen atom, initiating new chains:

$$\underset{R_2}{\overset{R_1}{\diagdown}}CHO\bullet + RH \rightarrow \underset{R_2}{\overset{R_1}{\diagdown}}CHOH + R\bullet \tag{6.8a}$$

Disproportionation with another radical:

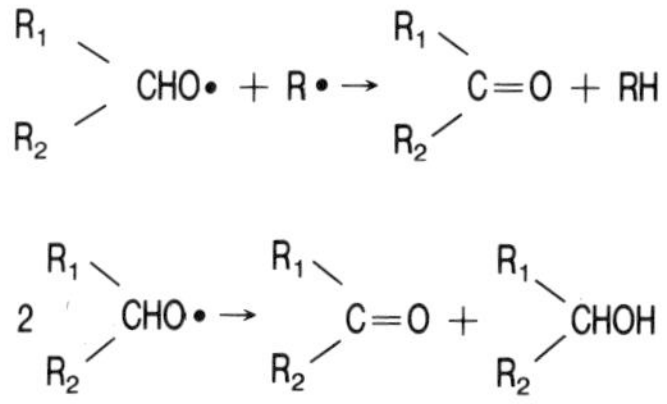

$$\begin{array}{c}R_1\\ \\R_2\end{array}\!\!\!\!\!>\!CHO\bullet + R\bullet \rightarrow \begin{array}{c}R_1\\ \\R_2\end{array}\!\!\!\!\!>\!C\!=\!O + RH \qquad (6.8b)$$

$$2\,\begin{array}{c}R_1\\ \\R_2\end{array}\!\!\!\!\!>\!CHO\bullet \rightarrow \begin{array}{c}R_1\\ \\R_2\end{array}\!\!\!\!\!>\!C\!=\!O + \begin{array}{c}R_1\\ \\R_2\end{array}\!\!\!\!\!>\!CHOH$$

Coupling:

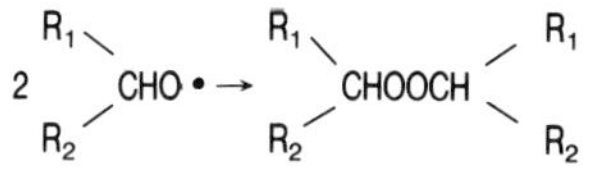

$$2\,\begin{array}{c}R_1\\ \\R_2\end{array}\!\!\!\!\!>\!CHO\bullet \rightarrow \begin{array}{c}R_1\\ \\R_2\end{array}\!\!\!\!\!>\!CHOOCH\!\!<\!\!\begin{array}{c}R_1\\ \\R_2\end{array} \qquad (6.8c)$$

Addition to double bonds:

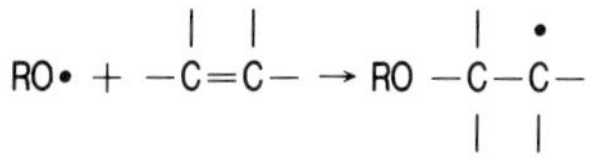

$$RO\bullet + -\underset{|}{\overset{|}{C}}\!=\!\underset{|}{\overset{|}{C}}- \rightarrow RO-\underset{|}{\overset{|}{C}}-\underset{|}{\overset{\bullet}{C}}- \qquad (6.8d)$$

The last reaction can lead to polymeric products of high order of the types commonly found among the products of lipid peroxidation.

6.2. Measurement of Lipid Peroxidation

Analysis of the products of peroxidation is accomplished by a variety of methods, none of them completely satisfactory. Conventional iodimetry is useful for moderate concentrations of hydroperoxides but fails with very small amounts because of a high background stemming from oxidation of I^- by oxygen. Other colorimetric reducing agents have been used with varying success. However, as is apparent from the following discussion, hydroperoxide content represents the extent of peroxidation only in the initial stages. Spectrophotometric detection of the conjugated double bond system formed in peroxidation of polyunsaturated acids is particularly useful since chromophoric properties of the group are little affected by some changes in the hydroperoxy group, at least before fragmentation of the initially formed products. Unfortunately, the maximum absorption of a conjugated diene is in the neighborhood of 235 nm, a region in which other components of the reaction mixture show high end absorption.

The thiobarbituric acid (TBA) test has been used by many investigators, particularly when seeking signs of lipid peroxidation in tissues. This test presumably depends on the formation of malondialdehyde and its reaction with thiobarbituric acid to give an intensely colored product ($\lambda_{max.}$ 530 nm) (Eq. 6.9):

$$\text{(structure)} + 2 \text{ thiobarbituric acid} \longrightarrow \text{(colored adduct)} \qquad (6.9)$$

There are two major difficulties with this reaction. First, malondialdehyde is a very minor peroxidation product. Second, it is uncertain that malondialdehyde is the only substance quantitated in this way; other aldehydes as well as products that are converted to aldehydes under the conditions of the test also give colored products. In addition, malondialdehyde is not formed from linoleate, the major polyunsaturated fatty acid of most tissues, but only from those with three or more double bonds. This distinction was initially rationalized by Dahle et al. (1962), in terms of a reaction mechanism requiring presence of a β,γ-double bond in the peroxy intermediate (possible only with tri- and higher 1,4-polyenes). α-Linolenic (9,12,15-octadecatrienoic) acid, for example, in a process initiated by abstraction of H· from C-11, would ultimately yield malondialdehyde containing carbons 13, 14, and 15 via a cyclic peroxide formed by intramolecular addition of a (C-13)−OO· group to the 15–16 (β–γ) double bond (Eq. 6.10):

$$\text{(reaction scheme)} \qquad (6.10)$$

(Malondialdehyde derived from carbons 10, 11, and 12 of linolenic acid would presumably be formed in about equal probability by an analogous sequence of events initiated by loss of H· from C-14 instead of C-11).

The validity of the Dahle hypothesis is uncertain insofar as it is not clear that the stability of malondialdehyde would compensate for concomitant formation of two high-energy free radicals. Moreover, an alternative explanation is provided by recent findings: (i) that tri- (and higher, but not di-) 1,4-polyenes react spon-

PEROXIDATION OF FATTY ACIDS

taneously (i.e., nonenzymatically) with oxygen to give very small yields of prostaglandins of the PGG type (cf. Chapter 10); and (ii) that such bicyclic endoperoxides give malondialdehyde on being heated (Eq. 6.11), in a reaction closely analogous to reversal of Diels–Alder condensation (Pryor et al., 1976):

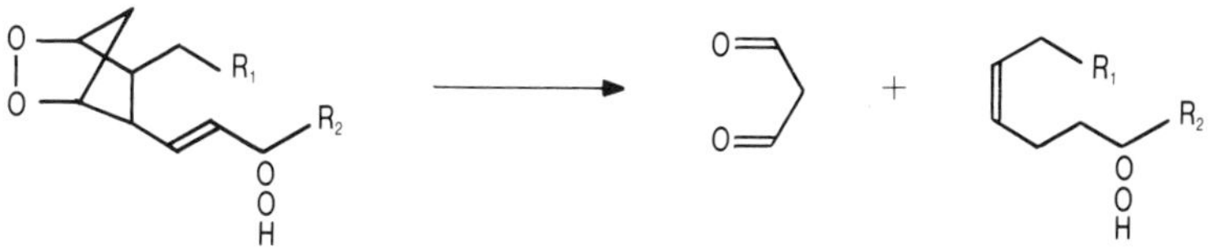

$$\text{(6.11)}$$

Malondialdehyde production is also measurable by the fluorescence analysis proposed by Tappel and his associates (Malshet and Tappel, 1973) as a sensitive means of detection of lipid peroxidation. "Schiff base" reaction of malondialdehyde with free amino groups of tissue components (e.g., proteins, PE, PS, and other amino-containing substances) yields products bearing a chromophore (thought to be $-N=CH-CH=CH-NH-$) with an absorption maximum at 350–360 nm and emission at 430–450 nm. Although the reaction, as it occurs in tissues, is not completely understood, it is a useful and sensitive assay for the peroxidation of tissue constituents.

Probably the least invasive means of detecting the occurrence of a peroxidative event in the living animal is by measurement of expired hydrocarbons in the breath. The formation of ethane from the terminal two carbons of $\omega 3$ polyunsaturated fatty acids (e.g., α-linolenic) and of pentane from the terminal five carbons of $\omega 6$ species (e.g., linoleic) is thought to involve alkoxy radicals formed from hydroperoxides by reactions of types illustrated in Fig. 6.1 and Eq. 6.3, from the 16-hydroperoxide in the case of linolenic acid (Eq. 6.12a) or from the 13-hydroperoxide in the case of linoleic acid (Eq. 6.12b):

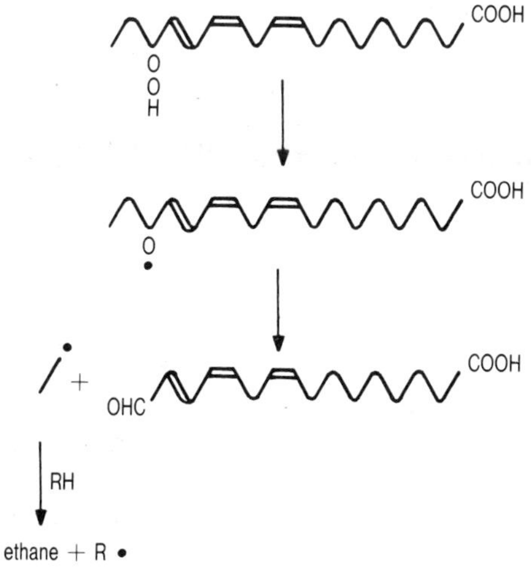

$$\text{(6.12a)}$$

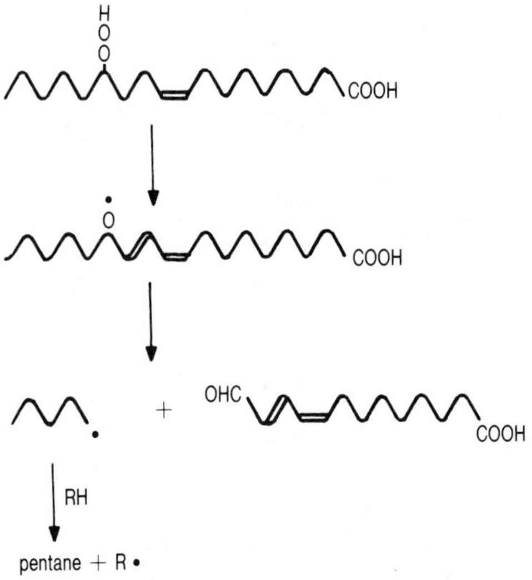

$$\text{(diagram: hydroperoxide} \rightarrow \text{alkoxy radical} \rightarrow \text{alkyl radical} + \text{aldehyde} \rightarrow \text{pentane} + R\cdot\text{)} \qquad (6.12b)$$

β-Scission of the intermediate alkoxy radicals yields alkyl radicals that may abstract hydrogen atoms from neighboring molecules.

Tappel and his associates (Tappel and Dillard, 1981) have shown that conditions presumably leading to a peroxidation, such as ingestion of CCl_4, CCl_3Br, and excess iron, exposure to ozone, or vitamin E deficiency, lead to large increases in expired pentane and ethane in rats. Administration of vitamin E and certain other antioxidants usually decreases hydrocarbon expiration to normal values. Applicability of the method to human subjects is of considerable interest but remains to be explored, though it has been reported (Gelmont *et al.*, 1981) that—at least in rats—much of the pentane is produced by the action of intestinal bacteria on dietary hydroperoxides.

Considerable use has been made in the past of yet another approach to assessing the extensiveness of peroxidation, by measurement of decrease of tissue lipid polyunsaturated fatty acids. In tissue preparations, the most highly unsaturated fatty acids, particularly 22:6 and 20:4, are usually found to decrease under conditions expected to enhance peroxidation. Quantitative interpretation of results obtained in this way is limited not only by technical disadvantages, but also because such interpretation does not give careful distinctions about the mechanism of disappearance of these metabolically active substances.

6.3. Lipid Peroxidation in Model Membranes

It would seem from the nature of the autoxidation reaction that it would occur most readily in systems such as cell membranes, in which the lipids are in an ordered arrangement that would facilitate the propagation reactions. However, the complexity of such membranes has made the investigation of their peroxidation difficult and has led to the use of several model systems in order to gain insight into the reactions involved.

One such system consists of monomolecular films of unsaturated fatty acids adsorbed on silica gel. Autoxidation of linoleic acid in this way, induced by small amounts of peroxides, leads to disappearance of linoleic acid by apparent first-order kinetics. Equation 6.1 indicates that autoxidation of polyunsaturated fatty acids (neat) would exhibit much more complex kinetics, as has been verified for both neat and emulsified unsaturated lipids. The monomolecular film thus appears to exhibit a different mechanism of peroxidation, and this is confirmed by the finding that major products are the fatty acid epoxides, presumably formed by transfer of an oxygen from the initially formed peroxy radical to a neighboring double bond, possibly as illustrated in Fig. 6.3. This reaction is of interest since some of the most potent chemical mutagens and carcinogens are epoxides. The rate of reaction for this system is maximal and constant over a wide range of linoleate to silica gel ratios, probably involving variable but fractional surface coverage of the particles by patches of linoleic acid monolayer. At very low ratios of linoleate to silica gel, or if saturated fatty acids are included as diluents, the reaction rate decreases markedly and products of intramolecular oxygen transfer, such as 11-hydroxy-9,10- (or 12,13)- epoxy-octadecadienoic acids, become prominent. Formation of such products can be rationalized in terms of suppression of the second propagation step (cf. Fig. 6.1: abstraction of allylic H• from a "new" 1,4-polyene by the intermediate peroxy radical) by reduced access to a readily abstractable hydrogen. Alternative intramolecular reactions of the peroxy radical thus become competitive, e.g., as shown for the case of linoleate in Fig. 6.4.

In another model system, the unilamellar liposome, formed by sonication of an unsaturated phosphatidylcholine in aqueous buffered solution, the products of autoxidation are the expected 9- and 13-hydroperoxyoctadecadienoic acids. However, with inclusion of saturated phosphatidylcholine in the liposome, the product mixture becomes much more complex, containing also epoxides, hydroxy-epoxides, and di- and trihydroxy acyl moieties (see also Fridovich and Porter, 1981). A simple membrane system from *Acholeplasma laidlawii,* with a fatty acid composition of the membrane phospholipids consisting of a 50:50 mixture of 18:2 and 16:0, gave very similar fatty acid products, indicating that results with the model systems are indeed representative of those occurring in biomembranes.

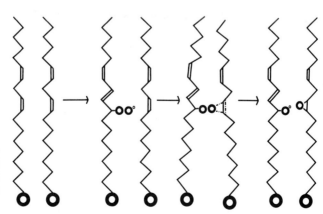

FIGURE 6.3. Proposed formation of fatty acid epoxide in a monolayer of linoleic acid adsorbed on silica gel. From Sevanian *et al.* (1979).

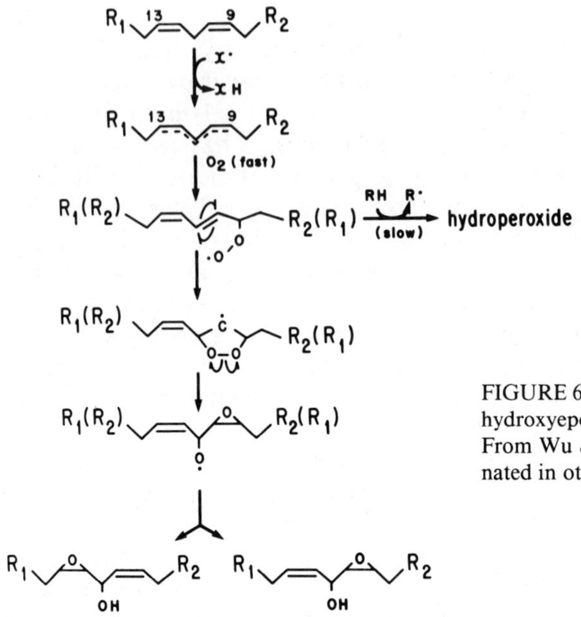

FIGURE 6.4. Proposed formation of contiguous hydroxyepoxides in model membrane systems. From Wu *et al.* (1977). RH and R• are as designated in other figures.

6.4. The Action of Antioxidants

As can be seen in Fig. 6.1, the autoxidation can be terminated by reaction of radicals with each other (Eq. 6.13):

$$2R\bullet \rightarrow RR \tag{6.13a}$$
$$R\bullet + ROO\bullet \rightarrow ROOR \tag{6.13b}$$
$$2ROO\bullet \rightarrow RO_4R \rightarrow ROOR + O_2 + \text{degradation products} \tag{6.13c}$$

Reaction 6.13a is unlikely since it could occur only under oxygen pressures too low to support the autoxidation reaction. Reaction 6.13c would require partial pressures too high for *in vivo* conditions, with production of singlet oxygen, a potential initiating agent. Of the three, therefore, reaction 6.13b would appear to be most likely in living organisms. In any case, all three reactions produce dimeric products that are difficult to isolate and identify.

Lipid peroxidation can also be limited through termination reactions of several types of antioxidants or radical trapping agents operating at different stages of the process.

Photosensitized oxidation, initiated by absorption of light by chromophoric impurities, probably involves intermediacy of singlet oxygen and can be distinguished from autoxidation *per se* by the products. For example, the photosensitized oxidation of methyl oleate yields only the 9- and 10-hydroperoxides whereas autoxidation yields the 8-, 9-, 10-, and 11-hydroperoxides. Moreover, photosensitized oxidation is inhibited by singlet oxygen quenchers such as carotenes, triethylamine, and nickel chelates, which are not effective in suppression of autoxidation.

However, autoxidation is prevented by several types of antioxidants. Metal chelators, such as DTPA (diethylenetriaminepentaacetate) or EDTA (ethylenediaminetetraacetate), limit the formation of chain initiators by preventing the metal-assisted homolysis of hydroperoxides (Eqs. 6.3 and 6.5). Reducing agents can convert the hydroperoxides (potential initiators, see Eq. 6.2b) into alcohols. In particular, glutathione-mediated enzymatic reductions, such as that of glutathione peroxidase, or, more likely, the GSH-dependent cytosolic factor proposed by McCay *et al.* (1981) can function in such reduction of hydroperoxides.

The most prevalent type of antioxidant is the hindered phenol. This class of antioxidants or radical quenchers includes the tocopherols, BHA [3-(or 2-)t-butyl-4-hydroxyanisole], BHT (3,5-di-t-butyl-4-hydroxytoluene), PG (n-propylgallate), and NDGA (nordihydroguaiaretic acid). In general, these substances act by furnishing a hydrogen atom to a chain-carrying radical while they themselves are initially converted to considerably more stable radicals that are less effective or ineffective as chain carriers, or that can dismutate to form non-radical products (Eq. 6.14):

$$\begin{aligned} &R\bullet + AOH \rightarrow RH + AO\bullet \\ &2AO\bullet \rightarrow AO^- + A=O \quad \text{(if AOH is a hydroquinone)} \\ &AO\bullet + R\bullet \rightarrow \text{non-radical products} \end{aligned} \qquad (6.14)$$

A characteristic of the action of antioxidants of this type is that, at least in *in vitro* reactions, they introduce a lag period that is proportional in length to their concentration relative to that of the polyunsaturated lipid and that continues until about 90% of the antioxidant has been destroyed. During this lag period, lipid peroxidation proceeds at a very low rate, but at the end, oxidation resumes at a rate equal to that of the unprotected lipid, or even greater. It is significant that at the relative concentration of α-tocopherol usually found in tissues (about 0.04 mol % of polyunsaturated fatty acyl groups present), one molecule of antioxidant can prevent the peroxidation of a great many fatty acid molecules, probably by destruction of some mobile chain-carrying radicals (see Logani and Davies, 1980, for complete discussion).

6.5. *Effects of Lipid Peroxidation in Living Organisms*

A question of considerable interest to biochemists and physiologists is whether the reactions discussed above are of significance in the living organism. This question has been approached in several ways: by studying the effects of ingestion of peroxidized fat in the diet; by examining the results of peroxidation of tissue homogenates and subcellular fractions; and by attempting to find evidence that such reactions occur in living organisms (see Wills, 1969).

The first effect attributable to peroxidized lipid in the diet is the destruction of oxidizable components, including essential nutrients, both in the diet before ingestion and in the gastrointestinal tract (of rats: Burr and Barnes, 1943). Even when this difficulty is avoided by administering a vitamin supplement separately or by a different route, peroxidized fat gives rise to toxic symptoms. Several investigators have reported that peroxidized, highly unsaturated marine oils are quickly fatal to rats and that toxicity is proportional to their peroxide content. On the other hand, others have shown that toxic substances are produced by heating polyunsaturated fatty acids in the absence of air (Crampton *et al.*, 1956). In these cases, peroxides are, of course, not involved. Indeed, when fats are heated to fairly high temperatures, even in the presence of air, polymeric rather than peroxidic substances eventually result; the initially formed hydroperoxides rapidly decompose at such temperatures, giving rise to a variety of polymeric products that may or may not include added oxygen.

Andrews *et al.* (1960) fed rats, after they were weaned, on diets containing 20% soybean oil catalytically oxidized to various peroxide numbers (meq peroxide per kg). In addition, samples of fresh oil were brought to the same peroxide number by addition of highly concentrated peroxide. Either oil, providing it was equally peroxidic, had the same effect, showing that peroxides, rather than the polymeric products, were involved. As can be seen in Fig. 6.5, a peroxide number (PN) of 100 had no effect on the growth of the rats, whereas with PNs of 400 or 800, growth was depressed and a PN of 1200 was quickly fatal. Since PNs even as high as 100 are very unlikely to occur in the human diet, this possibility may not represent a virtual hazard. Moreover, in this same study it was shown that the peroxides themselves, at least in small amounts, are absorbed only after reduction in the intestine and thus that the toxicity was manifest in the lumen of the intestine.

Studies in many laboratories have demonstrated that tissue homogenates, when incubated in the presence of air or oxygen, form lipid peroxides (as measured in a variety of ways). This is particularly true if the tissue is from vitamin E-deficient animals or is high in polyunsaturated acids or if such acids are added *in vitro*. The rate of formation depends on the tissue used, being generally high in homogenates of brain, liver, and kidney and low in those of testis and intestine. It is also different for different subcellular particles: highest in microsomes and

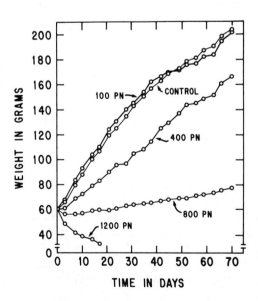

FIGURE 6.5. Growth of female rats receiving various levels of oxidized fat. From Andrews *et al.* (1960). PN = peroxide number (number of milliequivalents of thiosulfate used to titrate the iodine liberated from iodide by the peroxide in 1 kg fat).

lowest in nuclei. In the case of *in vitro* peroxidation, the reaction seems to depend on interactions of iron, oxygen, possibly ascorbic acid, and the polyunsaturated fatty acids found in the membrane lipids, in the absence of adequate amounts of antioxidants such as thiols or tocopherols (Barber and Bernheim, 1967). Such a simple mixing would be unlikely to occur in the intact cell and may depend on the disruption associated with the homogenates or cell fractions.

In any case, use of such model systems has shown that peroxidation may be very damaging to enzymes, membranes, other proteins—in short, to all susceptible cell constituents. There is also some indication, through *in vitro* studies with subcellular particles, that peroxidation may be both enzymatic and nonenzymatic.

Nonenzymatic peroxidation is probably a consequence of disruption of tissues or of exposure to amounts of pro-oxidants that overwhelm normal protective mechanisms. Normally, the radical-generating substances, such as Fe^{2+} and ascorbate, are kept separate from the membrane polyunsaturated fatty acids, but tissue disruption can co-mingle these substances with drastic results.

In the case of the enzymatically induced peroxidation, most oxidation–reduction systems ultimately result in the stepwise reduction of oxygen through the O_2^- and $O_2^=$ ($\cdot OH$) oxidation states. In all tissues of aerobic organisms, specific protective substances and enzymes exist to prevent damage by these substances. The superoxide (radical) anion ($O_2^- \cdot$) is converted by the selenium-containing or non-selenium-containing superoxide dismutases (SOD) to hydrogen peroxide and molecular oxygen (Eq. 6.15):

$$2O_2^- \cdot + 2H^+ \xrightarrow{SOD} H_2O_2 + O_2 \qquad (6.15)$$

Hydrogen peroxide is dismutated by catalase or reduced by glutathione peroxidase (Eq. 6.16):

$$2H_2O_2 \xrightarrow{catalase} 2H_2O + O_2 \qquad (6.16)$$

The hydroxyl radical, however, is so reactive that it reacts with virtually any tissue constituent it encounters and is thus unlikely to survive long enough to react specifically with a polyunsaturated fatty acylated constituent. Since $\cdot OH$ can be produced by reaction of the superoxide ion with hydrogen peroxide (Eq. 6.17),

$$O_2^- \cdot + H_2O_2 \to O_2 + OH^- + \cdot OH \qquad (6.17)$$

it is conceivable that superoxide dismutase could be involved in the formation of $\cdot OH$ and thus in promoting peroxidation. Indeed, this has been reported by McCay and his coworkers (Fong *et al.*, 1973). Aust and his coworkers (Tien *et al.*, 1981), on the other hand, consider that neither microsomal NADPH-dependent lipid peroxidation nor xanthine oxidase-promoted lipid peroxidation involves hydroxyl radicals; since iron (chelated by ADP) appears to be required by these systems, they may initiate peroxidation through agency of a ferryl ion which can be considered equivalent to $\cdot OH$ (Eq. 6.18):

$$ADP = Fe^{3+} + 3e^- + 2H^+ + O_2 \to (ADP-[FeO]^{2+}) + H_2O \qquad (6.18)$$

Although these reactions have been shown to occur in tissue homogenates (and thus potentially in intact tissues), to date it has been very difficult to show unequivocally that destructive lipid peroxidation takes place in living tissue. The enzyme-controlled formation of prostanoids, however, involves mechanisms very similar in other respects to those of peroxidation (see Chapter 10) and, as a matter of fact, it has been reported that certain of the prostaglandins that are normally produced in specific enzymatic reactions can also be found as products of autoxidation of arachidonic acid (Pryor and Stanley, 1975). Nevertheless, in general, experiments designed to demonstrate occurrence of destructive peroxidative reactions in living tissue have proved to be contradictory and confusing. Part of the difficulty is attributable to analytical techniques, which must be so sensitive to measure the small amounts of peroxides involved that spurious effects, such as those due to atmospheric oxygen, may interfere seriously. Moreover, since the organism or tissue is necessarily disrupted before the measurement is made, there is always the question of whether the peroxide might have formed during the manipulations of isolation. Living tissues, of course, are plentifully supplied with protective mechanisms specifically designed to prevent such eventualities. As discussed above, these involve water-soluble antioxidants, such as thiols; the lipid-soluble compounds, such as tocopherols; a variety of metals, especially heme compounds; and specific enzyme systems, such as the soluble glutathione-dependent factor of McCay *et al.* (1981), that reduce hydroperoxides to alcohols. Nevertheless, there is considerable evidence that, even if appreciable concentrations of peroxides do not actually accumulate in the living organism, they do exist transiently. It has been noted that in rats raised on vitamin E-deficient diets, depot fat became yellow-brown in color (Dam and Granados, 1945) and incisors lost pigmentation (Granados and Dam, 1945). Destruction of certain oxidizable factors and susceptible enzymes in livers of vitamin E-deficient animals have also been noted. Membrane breakdown in lysosomes has been proposed as a mechanism by which small amounts of peroxides can bring about extensive damage by consequent release of hydrolytic enzymes. In any case, the fact that more peroxide can be measured *in vitro* in tissues of animals subjected to a variety of conditions such as low antioxidant, high polyunsaturated fat, or high oxygen pressure, leads to the belief that an oxidative reaction has taken place *in vivo* and that protective mechanisms have been disturbed or overwhelmed.

Several other procedures appear to bring about lipid peroxidation. Administration of ethanol (DiLuzio, 1973) or carbon tetrachloride (Recknagel and Glende, 1973) induces peroxidation of mitochondrial or microsomal membrane lipids and antioxidants can prevent such damage. Nitrogen dioxide and ozone induce peroxidation of the lung lipids of rats, and ionizing radiation has been implicated in peroxide formation in membrane lipids in many tissues, including those of lysosomes.

Admittedly these are all abnormal conditions and, in most cases, have been contrived to emphasize the expected result. Nevertheless, there is also considerable evidence—indirect, to be sure—that a similar reaction takes place continuously in the living animal and is, indeed, a consequence of living in our oxygen-containing environment. As discussed above, it is obvious and has been shown experimentally that many of the energy-producing reactions of cells involve free radicals. These are prevented from damaging susceptible cell substances by compartmentalization, rapid reaction with the normal substrate, generally low oxygen tension, and presence of various antioxidants and radical quenchers. However, no protective system is perfect and it is reasonable to postulate that occasional errors may result in slight but cumulative damage. Factors such as antioxidant deficiency or exposure to radiation or hyperbaric oxygen would increase the frequency of such events, but continuous exposure to cosmic radiation and airborne oxidants is normal. The many similarities between aging, radiation damage, and vitamin E-deficiency symptoms have led several investigators, particularly Harman (1981), to postulate that cumulative free radical and peroxide damage is an important component of the aging process. Indirect evidence for this comes from the steady accumulation in the less metabolically active tissues of most animals of brown, fluorescent particles of a pigment called lipofuscin, ceroid or age pigment, which increase in these tissues with age in human beings, to the extent of 0.6% of the cell volume per decade according to Strehler and his coworkers (Hendley *et al.*, 1963). Although the exact nature of these particles is still obscure, they bear some resemblance to lysosomes and may arise from these organelles as a consequence of ingestion of peroxidized and disrupted membranes. In any event, they seem to be composed of peroxidized polymerized lipid and denatured, cross-linked protein and evidently arise from some type of peroxidative process. The ceroid pigment in atherosclerotic lesions may have similar properties and has also been reported to be high in peroxide (Glavind *et al.*, 1952). Similar polymeric products have been produced *in vitro* from peroxidized lipids and various proteins, and experiments by Tappel (1973) have shown that the principal chromophore leading to fluorescence in all cases is $-N=CH-CH=CH-NH-$, a product of the reaction of malondialdehyde (a 1,4-polyene autoxidation product, see Section 2) or other dialdehyde with free amino groups in lipids or proteins.

The conclusion at the present time appears to be that with a high dietary (and hence tissue) proportion of polyunsaturated fatty acids, there is a possibility that peroxidic reactions may occur, particularly if concentrations of various antioxidants and other protective devices are deficient. The cumulative result of such reactions may be a variety of cellular degradations, including those associated with aging.

If it is true that free radical reactions contribute to the symptoms of aging, including limitation of longevity, then it follows that prevention of these reactions should diminish the symptoms and increase longevity. Experiments designed to test this idea have, however, not been entirely satisfactory. The average (but not the maximal) life span of mice is increased about 30% by 1% dietary 2-mercaptoethylamine, but not by the natural antioxidant, tocopherol (Harman, 1981). Examples of increased longevity in other organisms have been recorded but so far not consistently. Some diseases associated with aging, such as certain types of cancer, also seem to respond to antioxidant administration; and the accumulation of aging pigments, lipofuscin and ceroid, has, in some cases, been slowed by antioxidants, including tocopherols. The inescapable conclusion is that free radical reactions, including lipid peroxidation, are intimately involved in life processes, and that understanding and controlling them could reap rewards in improved treatment of degenerative disease and in lengthening life.

REFERENCES

Andrews, J. S., Griffith, W. H., Mead, J. F., and Stein, R. A., 1960, Toxicity of air-oxidized soybean oil, *J. Nutrition* **70**:199.

Barber, A. A., and Bernheim, F., 1967, Lipid peroxidation: its measurement, occurrence and significance in animal tissues, *Advanc. Gerontol. Res.* **2**:355.

Burr, G. O., and Barnes, R. H., 1943, Non-caloric functions of dietary fats, *Physiological Reviews* **23**:256.

Crampton, E. W., Common, R. H., Pritchard, E. T., and Farmer, F. A., 1956, Studies to determine the nature of the damage to the nutritive value of some vegetable oils from heat treatment. IV. Ethyl esters of heat-polymerized linseed, soybean, and sunflower seed oils, *J. Nutrition* **60**:13.

Dahle, L. K., Hill, E. G., and Holman, R. T., 1962, The thiobarbituric acid reaction and the autoxidations of polyunsaturated fatty acid methylesters, *Arch. Biochem. Biophys.* **98**:253.

Dam, H., and Granados, H., 1945, Peroxidation of body fat in vitamin E deficiency, *Acta Physiol. Scand.* **10**:162.

DiLuzio, W. R., 1973, Antioxidants, lipid peroxidation and chemical-induced liver injury, *Fed. Proc.* **32**:1875.

Fong, K., McCay, P., Poyer, J. L., Keele, B. B., and Misra, H., 1973, Evidence that peroxidation of lysosomal membranes is initiated by hydroxyl free radicals produced during flavin enzyme activity, *J. Biol. Chem.* **248**:7792.

Fridovich, S. E., and Porter, N. A., 1981, Oxidation of arachidonic acid in micelles by superoxide and hydrogen peroxide, *J. Biol. Chem.* **256**:260.

Gelmont, D., Stein, R. A., and Mead, J. F., 1981, The bacterial origin of rat breath pentane, *Biochem. Biophys. Res. Commun.* **102**:932.

Glavind, J., Hartmann, S., Clemmeson, J., Jesson, K., and Dam, H., 1952, Studies on the role of lipoperoxides in human pathology. II. The presence of peroxidized lipids in the atherosclerotic aorta, *Acta Pathol. Microbiol. Scand.* **30**:1.

Granados, H., and Dam, H., 1945, Role of fat in incisor depigmentation of vitamin E-deficient rats, *Science* **101**:250.

Gunstone, F. D., and Hilditch, T. P., 1945, Union of gaseous oxygen with methyl oleate, linoleate and linolenate, *J. Chem. Soc*, p. 836.

Harman, D., 1981, The aging process, *Proc. Nat. Acad. Sci. U.S.A.* **78**:7124.

Hendley, D. D., Mildvan, A. S., Reporter, M. C., and Strehler, B. L., 1963, The properties of isolated human cardiac age pigment. I. Preparation and physical properties, *J. Geront.* **18**:144.

Logani, M. K., and Davies, R. E., 1980, Lipid oxidation: biologic effects and antioxidants—a review, *Lipids* **15**:485.

Malshet, V. G., and Tappel, A. L., 1973, Fluorescent products of lipid peroxidation. 1. Structural requirements for fluorescence in conjugated Schiff bases, *Lipids* **8**:194.

McCay, P. B., Gibson, D. D., and Hornbrook, K. R., 1981, Glutathione-dependent inhibition of lipid peroxidation by a soluble, heat-labile factor not glutathione peroxidase, *Fed. Proc.* **40**:199.

Porter, N. A., Weber, B. A., Weenen, H., and Khan, J. A., 1980, Autoxidation of polyunsaturated lipids. Factors controlling the stereochemistry of product hydrocarbons, *J. Am. Chem. Soc.* **102**:5597.

Privett, O. S., and Blank, M. L., 1962, The initial stages of autoxidation, *J. Am. Oil Chem. Soc.* **39**:465.

Pryor, W. A., 1973, Free radical reactions and their importance in biochemical systems, *Fed. Proc.* **32**:1852.

Pryor, W. A., and Stanley, J. P., 1975, A suggested mechanism for the production of malonaldehyde during the autoxidation of polyunsaturated fatty acids. Nonenzymatic production of prostaglandin endoperoxides during autoxidation, *J. Org. Chem.* **40**:3615.

Pryor, W. A., Stanley, J. P., and Blair, F., 1976, Autoxidation of polyunsaturated fatty acids: II. A suggested mechanism for the formation of TBA-reactive materials from prostaglandin-like endoperoxides, *Lipids* **11**:370.

Rawls, H. R., and Van Santen, P. J., 1970, A possible role for singlet oxygen in the initiation of fatty acid autoxidation, *J. Am. Oil Chem. Soc.* **47**:121.

Recknagel, R. O., and Glende, E. A., Jr., 1973, Lipid peroxidation in acute carbon tetrachloride liver injury, *CRC Critical Rev. Toxicol.* **2**:268.

Sevanian, A., Stein, R. A., and Mead, J. F., 1979, Lipid epoxide hydrolase in rat lung preparations, *Biochim. Biophys. Acta* **614**:489.

Tappel, A. L., 1973, Lipid peroxidation damage to cell components, *Fed. Proc.* **32**:1870.

Tappel, A. L., and Dillard, C. J., 1981, *In vivo* lipid peroxidation: measurement via exhaled pentane and protection by vitamin E, *Fed. Proc.* **40**:174.

Tien, M., Swingen, B. A., and Aust, S. D., 1981. Superoxide-dependent lipid peroxidation, *Fed. Proc.* **40**:179.

Wills, E. D., 1969, Lipid peroxide formation in microsomes, *Biochem. J.* **113**: p. 315, General considerations; p. 325, The role of non-haem iron; p. 333, Relationship of hydroxylation to lipid peroxide formation.

Wu, G.-S., Stein, R. A., and Mead, J. F., 1977, Autoxidation of fatty acid monolayers adsorbed on silica gel. II. Rates and products, *Lipids* **12**:971.

SELECT BIBLIOGRAPHY

Lundberg, W. O. (ed.), 1961 and 1962, *Autoxidation and Antioxidants,* Vols. I and II, Interscience, New York.

7
CATABOLISM OF THE FATTY ACIDS

7.1. Historical Introduction

It is a historical curiosity that the outline of the mechanism of fatty acid oxidation proposed by some of the earliest workers was essentially correct although the details were not worked out for another fifty years.

The designation of the principal oxidation system for fatty acids as β-oxidation is due to the experiments of F. Knoop (1905), who reported what was probably the first metabolic tracer experiment. Realizing that the phenyl group is metabolically inert, he synthesized a number of ω-phenyl-substituted fatty acids and fed them to dogs. From their urine he isolated glycine conjugates of benzoic or phenylacetic acids—hippuric and phenylaceturic acids, respectively. From fatty acids of the type $C_6H_5(CH_2)_n COOH$ in which n was even, hippuric acid was derived, whereas those with n odd gave phenylaceturic acid. It appeared from these results that carbon atoms were lost in pairs, presumably via oxidative attack on the β carbon atom (C-3). Shortly after these findings, starting in 1908, Dakin carried out experiments, including the oxidation of fatty acids with H_2O_2 and liver perfusion, and from the results was able to propose the following as the most likely mechanism of oxidation of caproic acid by the β-oxidation process (see Dakin, 1923):

$$CH_3-(CH_2)_2-CH_2-CH_2-COOH \rightarrow CH_3(CH_2)_2-CH=CH-COOH \rightarrow \quad (7.1)$$
$$CH_3-(CH_2)_2-CHOH-CH_2-COOH \rightarrow CH_3(CH_2)_2-\overset{O}{\underset{\|}{C}}-CH_2-COOH \rightarrow \text{short-chain products}$$

As will be seen, a very closely analogous sequence of chemical events is involved in metabolic β-oxidation of fatty acids. During the half century after Dakin's experiments, investigators continued to study fatty acid oxidation, first in perfused organs and tissue slices and then in homogenates, and to interpret their results in favor of diverse mechanisms. Highlights of this period included the following:

(i) The observation that acetoacetate and β-hydroxybutyrate are formed during fatty acid oxidation;

(ii) The finding that acetoacetate can be formed from acetate as well as from long-chain fatty acids and that more than one mole of acetoacetate can be derived from one mole of fatty acid of chain-length greater than four;

(iii) The finding by Weinhouse et al. (1944) that acetoacetate formed from carboxy-labeled octanoic acid was labeled in both the carboxyl and keto carbons, which showed that fatty acids are oxidized to acetate (or another 2-carbon compound) which is condensed to form acetoacetate, for example:

$$CH_3-CH_2-CH_2-CH_2-CH_2-CH_2-CH_2-COOH \rightarrow 4\ CH_3-\overset{}{C}- \rightarrow 2\ CH_3-\overset{O}{\overset{\|}{C}}-CH_2-COOH \qquad (7.2)$$

(iv) The studies, in the 1940s and early 1950s, by Lehninger and his coworkers (e.g., Kennedy and Lehninger, 1949) of the *in vitro* oxidation of long-chain fatty acids, with the finding that oxidation takes place in the mitochondria and requires, as cofactors, Mg^{2+}, ATP, H_3PO_4, a tricarboxylic acid cycle intermediate and, in aged preparations, NAD. An active "two-carbon fragment" was produced which condensed to acetoacetate, the major product of β-oxidation;

(v) The finding by Lynen (1954) and others that the active "two-carbon fragment" is acetyl–coenzyme A and that the fatty acids are oxidized as their CoA-thioester derivatives;

(vi) The isolation, purification, and characterization of the enzymes of β-oxidation from mitochondria by Green and his coworkers at the Wisconsin Enzyme Institute (Beinert et al., 1953).

Thus, detailed knowledge of this important energy-producing process was finally elucidated and the stage was set for studies of its control and other refinements.

7.2. Translocation of Fatty Acids

The fatty acids, derived either from the diet or from adipose tissue, are probably delivered to the cell physically complexed with albumin or other proteins. They then transfer from binding sites on the albumin to those on the cell membrane. Goodman (1958) has estimated the dissociation constants (for oleate) of the two strongest albumin sites at 2×10^{-8} (see Chapter 13), while those for the erythrocyte membrane are of the order of 2×10^{-6}. However, with about 3×10^7 sites per erythrocyte, for example, transfer is efficient provided that rates of transfer of fatty acid from albumin to the cell and of transport through the cell

CATABOLISM OF THE FATTY ACIDS

membrane are rapid. This latter process requires energy and may involve activation of the fatty acid and its transfer to some transport vehicle within the membrane or possibly reversible incorporation into membrane lipids.

In their transport through the cytoplasm, the fatty acids are bound to proteins of small molecular weight termed, aptly, fatty acid-binding proteins (FABP) with molecular weight about 12,000.

The fatty acids, in this form, are presented to the outer membrane of the mitochondria, where they are converted to the coenzyme A derivatives by the long-chain acyl-CoA synthetase.

Since the inner mitochondrial membrane is poorly permeable to long-chain acyl-CoA derivatives, this translocation is accomplished by the action of two enzymes (carnitine acyltransferases I and II) located near the two surfaces of this membrane, with the function of transferring the acyl chain from coenzyme A to carnitine and back to coenzyme A. The overall reaction can be represented in the following simplified way:

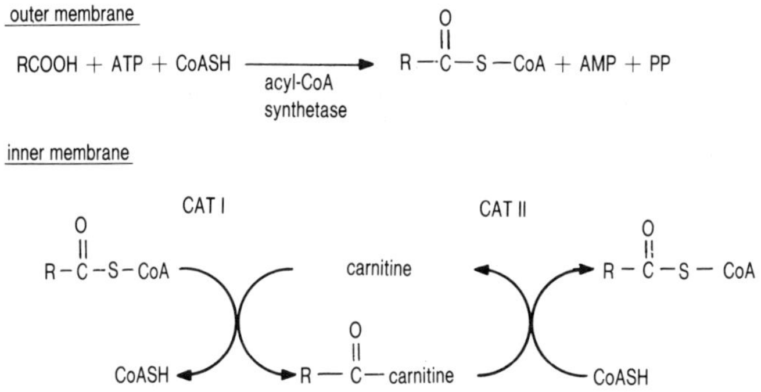

outer membrane

$$RCOOH + ATP + CoASH \xrightarrow{\text{acyl-CoA synthetase}} R-\overset{\overset{O}{\|}}{C}-S-CoA + AMP + PP$$

inner membrane

$$\begin{array}{ccc} & \text{CAT I} & \text{CAT II} \\ R-\overset{\overset{O}{\|}}{C}-S-CoA & \text{carnitine} & R-\overset{\overset{O}{\|}}{C}-S-CoA \\ CoASH & R-\overset{\overset{O}{\|}}{C}-\text{carnitine} & CoASH \end{array}$$

(7.3)

Carnitine, first discovered as an insect vitamin and only later recognized as a promoter of mitochondrial fatty acid oxidation, has an interesting structure involving a hydroxyl group between two oppositely charged ends of the molecule:

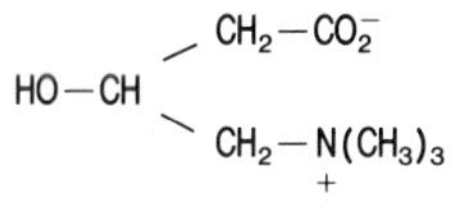

$$HO-CH \begin{array}{c} CH_2-CO_2^- \\ CH_2-\overset{+}{N}(CH_3)_3 \end{array}$$

The ester bond formed by acyl groups with this hydroxyl group has an equilibrium constant (hydrolysis) similar to that of acyl-CoA, since the exchange reaction has a K of about one. The two enzymes, acyl-CoA–carnitine acyltransferases I and II, appear to catalyze the exchanges in opposite directions. The ability of carnitine to aid in translocation of the long-chain acyl groups may reside in its net positive charge, compared with coenzyme A's negative charge, or in its comparatively small size, which furnishes little hindrance to passage of the ester through the lipophilic interior of the membrane (compare structures of CoA, below, and of carnitine, above):

7.3. Mechanism of β-Oxidation

The outline of fatty acid oxidation, as elucidated largely by Green and his coworkers (Beinert *et al.*, 1953) is shown in Fig. 7.1. The thiokinase (or acyl-CoA synthetase) reaction, which has already been mentioned, is carried out in the presence of five or more chain-length-specific enzymes, with those responsible for

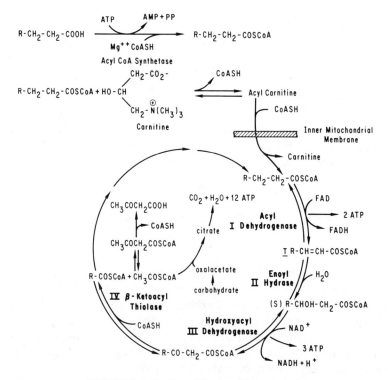

FIGURE 7.1. Outline of the β-oxidation spiral.

activation of long-chain fatty acids located in the outer mitochondrial membrane (presumably the outer face). Enzyme-bound acyl adenylate is an intermediate, which explains the requirement for ATP and Mg^{2+}.

Following the transfer of the acyl group from carnitine to CoA at the inner face of the inner mitochondrial membrane, the first oxidative step (I) takes place, mediated by FAD and catalyzed by acyl dehydrogenase. This dehydrogenation involves two one-electron transfer steps, two hydrogen atoms reducing the oxidized form of the flavin via a semiquinone intermediate. The chain of electron transfer is next to a second flavoprotein, electron transfer protein (ETF) and then to cytochrome b. The subsequent electron transfer to oxygen may be accompanied by phosphorylation of two molecules of ADP.

Three chain-length-specific enzymes have been reported: a green enzyme, containing copper, when isolated, and with a specificity for about C_4, and two yellow enzymes, Y_1 and Y_2, with specificities for about C_{10} and C_{16}, respectively.

The product of this reaction, *trans*-α,-β-unsaturated acyl-coenzyme A, is hydrated in the presence of the enzyme enoyl hydrase to form the (S) β-hydroxyacyl-CoA (step II). This hydroxy acyl derivative then undergoes the second dehydrogenation step to the β-ketoacyl-CoA in the presence of NAD and the enzyme β-hydroxyacyl dehydrogenase. The transfer of a hydride ion, a two-electron transfer, is followed by one-electron transfers to a flavoprotein and thence to the electron transfer chain, with formation of three molecules of ATP. It can be seen from Eq. 7.4 that the reaction is sensitive to pH changes being shifted to the right by elevation of pH greater than seven:

$$\text{Hydroxyacyl-CoA} + \text{NAD}^+ \rightleftharpoons \text{Ketoacyl-CoA} + \text{NADH} + \text{H}^+ \qquad (7.4)$$

The last reaction in the spiral consists of the thiolytic cleavage of the β-ketoacyl-CoA in the presence of β-ketoacyl thiolase with the formation of acetyl-CoA and the CoA derivative of a fatty acid two carbon atoms shorter than the starting acid. Thus, with each turn of the spiral, two carbon atoms are lost, as acetyl-CoA, until the entire fatty acid chain has been degraded. Each turn of the spiral also results in an initial formation of about five molecules of ATP. If the acetyl-CoA is oxidized through the TCA cycle, a considerably higher yield of ATP (about 12 per turn) is realized.

This "high-yield" oxidation of fatty acids depends, of course, on the availability of oxalacetate and thus presupposes an adequate carbohydrate metabolism. If the supply of carbohydrate is inadequate, as in starvation, a high-fat diet, or diabetes, there is a shift toward a "low-yield" fatty acid oxidation which, however, occurs to some extent even in the presence of oxalacetate. This pathway involves the condensation of two molecules of acetyl-CoA to form acetoacetyl-CoA and a molecule of CoA. Thus it appears that, at least in the liver, the response to increasing or exclusive use of fat as an energy source results in a decreased energy yield per molecule.

The product of the reaction, acetoacetyl-CoA, may be deacylated by one of three mechanisms. In tissues other than liver, direct deacylation, with release of CoA or transfer to other acids (e.g., succinate) by a thiophorase, can occur. These reactions are readily reversible and acetoacetic acid can be reactivated in these tissues for use as an energy source. In the liver mitochondria, however, a major pathway for the generation of free acetoacetate involves intermediate formation of hydroxymethylglutaryl-CoA in the mitochondria (Eq. 7.5):

$$CH_3COCH_2COSCoA + CH_3COSCoA \rightarrow HOOCCH_2-\underset{\underset{OH}{|}}{\overset{\overset{CH_3}{|}}{C}}-CH_2COSCoA + CoASH$$

$$HOOCCH_2-\underset{\underset{OH}{|}}{\overset{\overset{CH_3}{|}}{C}}-CH_2COSCoA + H_2O \rightarrow CH_3COCH_2COOH + CH_3COSCoA \tag{7.5}$$

(net, overall: $CH_3COCH_2COSCoA + H_2O \rightarrow CH_3COCH_2COOH + CoASH$)

This reaction is irreversible and the acetoacetate generated by it could be used for energy and synthetic reactions only after reconversion to the CoA ester by an acetoacetyl-CoA synthetase or thiophorase. For this reason acetoacetate is not used efficiently as an energy source in liver mitochondria.

Other tissues, particularly those of ectodermal origin (brain, spinal cord, and skin) readily use acetoacetate and both (R)- and (S)-3-hydroxybutyrate as sources of energy and of carbon for synthetic reactions (Webber and Edmond, 1977). These processes are particularly important during conditions of low supply of carbohydrate such as a high-fat diet, starvation, or diabetes. However, they are also of importance during development (e.g., 9–11 days after birth in the rat) when they represent the major source of carbon for synthesis of both brain sterols and fatty acids. Tissues capable of such efficient utilization of these "ketone bodies" also differ from those of the liver and kidney in that they can readily activate acetoacetate and 3-hydroxybutyrate, as mentioned above, either through the exchange reaction via a thiophorase or by direct activation with acetoacetyl-CoA synthetase in the presence of ATP, CoASH, and Mg^{2+}. The importance of these reactions lies, of course, in the availability to certain vital organs of a substrate produced by the liver during periods of short supply of the more usual substrate, glucose, or during times of urgent need for rapid biosyntheses.

Processes discussed above account in detail for the complete metabolism of the even-chain saturated fatty acids. However, there are certain differences in the pathway if the acids are odd-chain or unsaturated.

In the case of the odd-chain fatty acids, the β-oxidation pathway ultimately leads to formation of propionyl-CoA, which is converted to succinyl-CoA in the following series of reactions (Eq. 7.6). The carboxylation reaction involves a biotin-containing peptide (see Chapter 8, "Biosynthesis of Fatty Acids"). Inversion of the absolute configuration of the methylmalonyl-CoA by methylmalonyl-CoA racemase is then followed by an intramolecular rearrangement involving transfer of the $-COSCoA$ to the methyl group, with vitamin B_{12} serving as coenzyme.

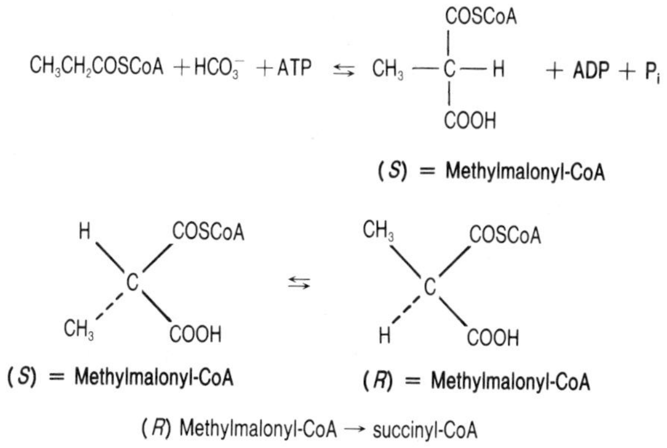

$$CH_3CH_2COSCoA + HCO_3^- + ATP \rightleftharpoons CH_3-\underset{COOH}{\overset{COSCoA}{\underset{|}{\overset{|}{C}}}}-H + ADP + P_i$$

(S) = Methylmalonyl-CoA

(7.6)

(S) = Methylmalonyl-CoA ⇌ (R) = Methylmalonyl-CoA

(R) Methylmalonyl-CoA → succinyl-CoA

The common unsaturated fatty acids (having a *cis* double bond nearest the carboxyl group in an odd position), on the other hand, are ultimately degraded to an unsaturated acyl-CoA with a *cis* β,γ double bond, rather than the normal intermediary *trans* α,β double bond. This potential difficulty may be handled in either of two ways: (i) via agency of a mitochondrial isomerase capable of converting the *cis* β,γ to the *trans* α,β bond, thus permitting the normal sequence of β-oxidation to proceed; (ii) it has been claimed that the *cis* β,γ bond can be hydrogenated to yield a saturated acyl-CoA, which can then be oxidized in the usual manner.

Certain features of the nature of the intact native enzyme complex that carries out the β-oxidation spiral in mammals are suggested by rationalization of the following initially curious observations. The isolated individual enzymes metabolize their respective substances normally, and reconstituted complexes bring about the full sequence of events as expected. However, when the same intermediate substances are added to the native mitochondrial system, they fail to contribute carbon to the final products. Stanley and Tubbs (1975) have explained these facts by comparing the operation of the β-oxidation system to that of a "leaky hosepipe." Under pressure some intermediates may leak out into the surrounding medium, but during normal operation the enzymes are too tightly coupled to permit appreciable entrance of intermediates added to the milieu.

7.4. Control of β-Oxidation

The factors controlling fatty acid oxidation appear to be more complex than implied by currently understood details of the process (Bremer and Wojtczak, 1972).

Although it would seem evident that availability of carnitine and CoA would be regulatory (as indeed they are *in vitro*), these cofactors are usually present in excess *in vivo*. It is true that under certain unusual circumstances, such as following ingestion of hypoglycin (an unusual amino acid from the akee fruit, giving rise to a nonmetabolizable fatty acid that ties up both carnitine and CoA), these factors become limiting, but under normal conditions they do not appear to be involved. A major control point appears to be the concentration of free fatty acid or, more accurately, the ratio of fatty acid to albumin or other fatty acid-binding proteins. This concentration (or ratio) is controlled by dietary or hormonal influences on adipose tissue (see Chapter 15). In brief, however, an increase of dietary fat, uncontrolled diabetes, or a period of fasting, brings about an increase in plasma-free fatty acids (FFA). This excess of metabolizable fatty acid has several effects. First, it provides a substrate for energy production in many tissues. Second, it inhibits several enzymes—possibly, in some cases, by the nonspecific surfactant action of acyl-CoA. In particular, the glycolytic enzymes as well as those of the citric acid cycle are inhibited, thus decreasing available oxalacetate for condensation with acetyl-CoA formed as an end product of fatty acid oxidation. This shifts fatty acid oxidation in the direction of the low-energy-yielding production of acetoacetate and permits oxidation in the liver of greater amounts of fatty acid without a concomitant disastrous increase in energy charge:

$$\frac{(ATP + \frac{1}{2} ADP)}{(ATP + ADP + AMP)}$$

The acetoacetate and 3-hydroxybutyrate resulting from this shift in metabolism can supply the energy needs of certain vital organs, such as heart and brain, despite the resulting acidosis associated with their overproduction.

Since free fatty acids, or, more accurately, their CoA derivatives, depress fatty acid biosynthesis, it would be expected that synthesis and oxidation would have a reciprocal relationship. This has been shown to be the case but the main mechanism behind the relationship may have a less direct cause. As will be discussed in Chapter 8, conditions that promote fatty acid biosynthesis, such as high dietary carbohydrate, result in increased levels of malonyl-CoA, which is a potent inhibitor of the enzyme carnitine acyltransferase (CAT I). Inhibition of this enzyme, which is essential for transport of fatty acids into the mitochondria, results in inhibition of fatty acid oxidation. On the other hand, conditions that result in decreased formation of malonyl-CoA, such as a high-fat diet, fasting, or uncontrolled diabetes, relieve this inhibition and promote fatty acid oxidation. Glucagon, which mimics fasting by decreasing availability of glucose, has the same effect (*Nutrition Reviews,* 1980).

Additional feedback inhibition is provided by the inhibitory effect of acetoacetyl-CoA on the enzyme crotonase (enoyl-CoA hydrase). The effect of such regulation, however, is not apparent.

Although most studies of fatty acid oxidation have been concerned with the β-oxidation system of mitochondria, a separate system, which appears to be initiated by a rate-limiting CN-insensitive fatty acid oxidase, exists in the peroxysomes. This system is effective in shortening fatty acids of unusually great chain-length and may assume importance with large intakes of these long-chain fatty acids or when the mitochondrial system is depressed.

7.5. ω- and α-Oxidation

Although the β-oxidation system is by far the most important means of deriving useful energy from fatty acids, certain other systems are capable of degrading fatty acids by other pathways under special conditions.

7.5.1. ω-Oxidation

The original observations of ω-oxidation were made in the 1930s, when Verkade and van der Lee (1934) found that fatty acids of intermediate chain-length (C_9–C_{11}), fed to human subjects, are excreted in the urine as the corresponding dicarboxylic acids. Obviously an oxidation of the terminal methyl (ω-carbon) had taken place, yielding a product resistant to further metabolism. It seemed significant that the oxidation took place at the site involved in desaturation of stearate to oleate but no connection could be established between the two mechanisms.

Considerably later, Wakabayashi and Shimazono (1961), using guinea-pig liver microsomes in the presence of NADPH and oxygen, showed that fatty acids of intermediate chain-length are first hydroxylated and, in the presence of the $100,000 \times g$ supernatant and NAD^+, are oxidized to the aldehyde and then to the acid (e.g., Eq. 7.7):

$$CH_3-(CH_2)_7-COOH \rightarrow HOCH_2-(CH_2)_7-COOH \rightarrow OHC(CH_2)_7-COOH \rightarrow HOOC-(CH_2)_7-COOH \quad (7.7)$$

These observations were confirmed and extended by M. J. Coon and his coworkers (Coon, 1978) using several systems from bacteria, yeast, or liver microsomes. In general it was found that, although maximum activity was manifest with C_{11}, ω-oxidation could take place with fatty acids of chain-length from C_6 to C_{16}. Moreover, the attack was not limited to the ω-carbon, but could have involved the ω-1-carbon, with formation of a secondary alcohol. In the case of microorganisms, not only fatty acids but also *n*-alkanes were oxidized analogously, particularly if the organism, *Pseudomonas oleovorans,* had been cultured in their presence.

Investigation of ω-oxidation in *P. oleovorans* revealed a requirement of two cofactors: rubredoxin (a non-heme, iron-containing, low-molecular weight (~20,000) protein with two binding sites for iron), and rubredoxin-NADPH reductase—in addition, of course, to the hydroxylase and oxygen. Investigation of the reaction mechanism by L. J. Morris and coworkers using a yeast *(Torulopsis groppengiesseri),* in which rubredoxin is replaced by cytochrome P-450, revealed that the reaction involves oxygen insertion rather than desaturation and double bond hydration. In the case of the ω-1 oxidation, it was shown that the oxygenation occurred with the stereospecificity typical of monooxygenases (Morris, 1970).

In the liver microsomal system, the requirements are somewhat different. As in the yeast, cytochrome P-450 (or P-448), NADPH, cytochrome *c* reductase, and oxygen are required. In addition, a heat-stable factor, which proved to be phosphatidylcholine, is necessary. The function of this compound is not clear but may be involved in enzyme conformation (see Chapter 16) or in substrate binding.

The function of such a system in the organism is also obscure, particularly in the higher animals. However, it may be concerned with oxidation of hydrocarbon chains under conditions in which other systems are unavailable. Preiss and Bloch (1964), for example, studied it in aged liver microsomes in which a CoAse and an ATPase prevented the normal activation of fatty acids. The enzyme may also serve as a more or less nonspecific hydroxylase involved in degradation or detoxification of a number of endogenous or exogenous substrates or xenobiotics.

7.5.2. α-Oxidation

Alpha oxidation was first studied in higher plants by Stumpf and his coworkers (Shine and Stumpf, 1974), who found that, in peanut cotyledons, fatty acids could be degraded by one-carbon steps in a system in which, in the presence of a peroxide-generating system, the fatty acid is converted to an aldehyde with one less carbon. The aldehyde is then oxidized to the acid by NAD–aldehyde dehydrogenase. An early proposal for the mechanism involved the action of the perferryl ion (Eq. 7.8):

$$\begin{aligned} Fe^{2+} + H_2O_2 &\rightarrow FeOOH^+ + H^+ \\ FeOOH^+ + H_2O_2 &\rightarrow FeO_2^{2+} + 2H_2O \\ FeO_2^{2+} + R-CH_2-CO_2^- &\rightarrow Fe^+OOCH_2-R + CO_2 \\ Fe^+OOCH_2-R &\rightarrow FeO + R-CHO + H^+ \end{aligned} \quad (7.8)$$

In the laboratory of A. T. James (see Hitchcock *et al.*, 1967), a similar system in mature leaves was reported to substitute oxygen for the H_2O_2 and to proceed through formation of an α-hydroxy acid (Eq. 7.9):

$$RCH_2COOH \rightarrow RCHOHCOOH \xrightarrow{-CO_2} RCHO \rightarrow RCOOH \quad (7.9)$$

In this case also, it was shown that the oxygen was added by direct hydroxylation of [2-^{3}H]substrates with retention of configuration. The D form was thought to accumulate in the leaves because the L form underwent further reaction (i.e., decarboxylation).

Stumpf has cast doubt on this interpretation and proposed that the D-2-hydroperoxypalmitate is the first product from palmitic acid and that this is either reduced to the D-2-hydroxypalmitate by glutathione peroxidase or decarboxylated to pentadecanal and CO_2 (see Shine and Stumpf, 1974). (The L-2-hydroxypalmitate is not formed and in fact inhibits the oxidative attack on palmitic acid.) This reaction sequence could be depicted as follows (Eq. 7.10):

$$CH_3(CH_2)_{13}CH_2COOH \xrightarrow{[O]} CH_3(CH_2)_{13}\underset{OOH}{CH-COOH}$$

$$\downarrow GSH \qquad \text{or} \searrow$$

$$CH_3(CH_2)_{13}CHOHCOOH \qquad CH_3(CH_2)_{13}CHO + CO_2 + H_2O \tag{7.10}$$

In the animal, it has been shown that the long-chain odd and α-hydroxy acids of the brain sphingolipids are also products of an α-oxidation, or one-carbon degradation process. When carboxy-labeled acetate was injected into 14-day-old rats and, after six months, the C_{23} and C_{24} acids of their brain cerebrosides were degraded stepwise, it was found that the label was largely in the odd carbons of the C_{24} acids and in the even carbons of the C_{23} acids. This could have resulted only from a one-carbon degradation process, since a synthetic process involving carboxy-labeled acetyl- or malonyl-CoA could result in label only in the odd carbons (Levis and Mead, 1964) (Eq. 7.11):

$$12\ CH_3-\overset{\bullet}{C}OOH \rightarrow CH_3-(\overset{\bullet}{C}H_2-CH_2)_{10}-\overset{\bullet}{C}H_2-CH_2-\overset{\bullet}{C}OOH \rightarrow CH_3-(\overset{\bullet}{C}H_2-CH_2)_{10}-\overset{\bullet}{C}H_2-CHOH-\overset{\bullet}{C}OOH \rightarrow$$
$$\overset{\bullet}{C}O_2 + CH_3-(\overset{\bullet}{C}H_2-CH_2)_{10}-\overset{\bullet}{C}H_2-COOH \tag{7.11}$$

The first step of α-oxidation, α-hydroxylation, was shown by Kishimoto *et al.* (1979) to occur in the brain mitochondrial fraction when free fatty acid substrates were used, but to a greater extent in the microsomal fraction when the CoA esters were the substrates. The system is specific for the very long-chain fatty acids, and requires NADP(H), oxygen, and two cytosolic factors—a heat-stable and a heat-labile factor, at least for the free acids. The reaction involves direct replacement of the pro-R-hydrogen with cleavage of this C—H bond as the rate-limiting step. All of the resulting α-hydroxy acids are incorporated into ceramides or cerebrosides and the hydroxylation system is intimately involved in the synthesis of these lipids (see Chapter 18).

The decarboxylation of the α-hydroxy acids has been shown to occur in the brain microsomal fraction and to involve the free hydroxy acids. It also appears to be selective for the longer-chain fatty acids. Requirements were found to be oxygen, Fe^{2+}, ascorbic acid, and a soluble protein that was probably an ascorbate reductase (Mead and Hare, 1971). The α-keto acid could be shown to be an intermediate in some cases but under circumstances that indicated a reaction sequence with no release of free intermediates (Levis and Mead, 1964).

Although liver microsomal systems are not active in decarboxylation of long-chain α-hydroxy acids, a separate α-oxidation system appears to exist in the liver mitochondria. This system came to light when it was found that, in Refsum's disease, the accompanying nervous disorders could be attributed to an accumulation of phytanic acid in the lipids of the nervous system (Steinberg, 1972).

$$CH_3-\underset{\underset{CH_3}{|}}{CH}-CH_2-CH_2-(CH_2-\underset{\underset{CH_3}{|}}{CH}-CH_2CH_2)_2-CH_2-\underset{\underset{CH_3}{|}}{CH}-CH_2-COOH$$

Since β-oxidation cannot proceed with a β-methyl acid, an α-oxidation step is necessary before the molecule can be degraded further (Eq. 7.12):

$$R-CH_2-CH(CH_3)-CH_2-COOH \xrightarrow{\alpha\text{-oxidation}} R-CH_2-CH(CH_3)-COOH \xrightarrow{\beta\text{-oxidation}} R-COOH, \text{etc.} \quad (7.12)$$

In Refsum's disease, the α-hydroxylation enzyme appears to be missing, and the undegradable phytanic acid therefore accumulates. That this system is indeed different from that concerned with the brain long-chain fatty acids has also been shown by the finding that, in Refsum's disease, the brain odd-chain and α-hydroxy acids are present in normal quantities.

The function of the α-oxidation system thus appears to be the degradation of fatty acids, the β-oxidation of which, for structural reasons, is difficult or impossible. With phytanic acid, the β-methyl blocks the formation of a β-keto acid. With the long-chain fatty acids of the brain, the acyl-CoA synthetase appears to have low activity. However, this consideration does not explain the accumulation of the products of the process (odd-chain and hydroxy acids) in the brain sphingolipids and, although it is difficult to assign any special attributes to the odd-chain fatty acids, those with α-hydroxyl groups may play an important role in favoring optimal conformations of biomembrane lipids containing them (see Chapter 17). Thus this essentially degradative process may ultimately serve to impart essential specific properties to structures such as myelin.

REFERENCES

Beinert, H., Brock, R. M., Goldman, D. S., Green, D. E., Mahler, H. R., Mii, S., Stansly, P. G., and Wakil, S. J., 1953, The reconstruction of the fatty acid oxidizing system of animal tissues, *J. Am. Chem. Soc.* **75**:4111.

Bremer, J., and Wojtczak, A. B., 1972, Factors controlling the rate of fatty acid β-oxidation in rat liver mitochondria, *Biochim. Biophys. Acta* **280**:515.

Coon, M. J., 1978, Oxygen activation in the metabolism of lipids, drugs and carcinogens, *Nutr. Rev.* **36**:319.

Dakin, H. D., 1923, Experiments on the catabolism of caproic acid and its derivatives, *J. Biol. Chem.* **56**:43.

Goodman, DeW. S., 1958, The interaction of human erythrocytes with sodium palmitate, *J. Clin. Invest.* **37**:1729.

Hitchcock, C., James, A. T., and Morris, L. J., 1967, The stereochemistry of α-oxidation of fatty acids in plants. *Biochem. J.* **103**:8.

Kennedy, E. P., and Lehninger, A. L., 1949, Oxidation of fatty acids and tricarboxylic acid cycle intermediates by isolated rat liver mitochondria, *J. Biol. Chem.* **179**:957.

Kishimoto, Y., Akanuma, H., and Singh, I., 1979, Fatty acid α-hydroxylation and its relation to myelination, *Molecular and Cellular Biochemistry*, **28**:93.

Knoop, F., 1905, Der Abbau aromatischer Fettsäuren im Tierkörper, *Beitr. Chem. Physiol. Pathol.* **6**:150.

Levis, G. M., and Mead, J. F., 1964, An alpha-hydroxy acid decarboxylase in brain microsomes, *J. Biol. Chem.* **239**:77.

Lynen, F., 1954, Participation of coenzyme A in the oxidation of fat, *Nature* **174**:962.

Mead, J. F., and Hare, R. S., 1971, Alpha oxidation of cerebronic acid in brains from scorbutic and ascorbic acid-supplemented guinea pigs, *Biochem. Biophys. Res. Commun.* **45**:1451.

Morris, L. J., 1970, Mechanisms and stereochemistry in fatty acid metabolism, *Biochem. J.* **118**:681.

Nutrition Reviews, 1980, The regulation of fatty acid synthesis and oxidation by malonyl-CoA and carnitine, **38**:25.

Preiss, B., and Bloch, K., 1964, ω-Oxidation of long-chain fatty acids in rat liver, *J. Biol. Chem.* **239**:85.

Shine, W. E., and Stumpf, P. K., 1974, Fat metabolism in higher plants. LVIII. Recent studies on plant α-oxidation systems, *Arch. Biochem. Biophys.* **162**:147.

Stanley, K. K., and Tubbs, P. K., 1975, The role of intermediates in mitochondrial fatty acid oxidation, *Biochem. J.* **150**:77.

Steinberg, D., 1972, Phytanic acid storage disease: Refsum's syndrome, in: *The Metabolic Basis of Inherited Disease* (J. B. Stanbury, J. B. Wyngaarden, and D. S. Fredrickson, eds.), McGraw-Hill, New York, pp. 688–706.

Verkade, P. E., and van der Lee, 1934, Researches on fat metabolism, *Biochem. J.* **28**:31.

Wakabayashi, K., and Shimazono, N., 1961, Studies *in vitro* on the mechanism of ω-oxidation of fatty acids, *Biochim. Biophys. Acta* **48**:615.

Webber, R. J., and Edmond, J., 1977, Utilization of L(+)-3-hydroxybutyrate, D(−)-3-hydroxybutyrate, acetoacetate and glucose for respiration and lipid synthesis in the 18-day-old rat, *J. Biol. Chem.* **252**:5222.

Weinhouse, S., Medes, G., and Floyd, N. F., 1944, Fatty acid metabolism. The mechanism of ketone body synthesis from fatty acids with isotopic carbon as tracer, *J. Biol. Chem.* **155**:143.

8
BIOSYNTHESIS OF FATTY ACIDS

8.1. Historical Introduction

It has been known for a very long time that fat can be synthesized in the body from non-lipid precursors, but the mechanism of such a synthesis remained obscure. The likelihood that two-carbon units were important intermediates was recognized by Raper as early as 1907 (Raper, 1907), but whether these were in the form of acetaldehyde, ethanol, or acetic acid could not be determined. Tracer experiments ultimately showed that all the carbons of the fatty acids could be furnished by acetate, but the nature of the "active acetate" required for such a process was not immediately apparent.

The first supportive evidence for the speculative ideas of Raper was provided by Rittenberg and Schoenheimer (see Schoenheimer, 1949) who were the first to introduce deuterium for the study of the flux of body constituents, including the fatty acids. They found that when the concentration of D_2O in the body water of the animals was maintained constant (usually at the 2–3% level) for an extended period, the saturated fatty acids in the lipids of the animals contained deuterium at a concentration equal to about one-half of that present in the body water. Rittenberg and Schoenheimer interpreted their results to mean that the fatty acids were formed by coupling of small units and that the deuterium was introduced into the fatty acids through reductive reactions of intermediates at alternate carbon atoms (Schoenheimer, 1949). When ^{13}C became available, Bloch and Rittenberg (1945) showed, with the aid of $CDH_2^{13}COOH$, that both carbon atoms of acetate were used for fatty acid biosynthesis. Because aromatic amines, such as p-aminobenzoic acid and sulfanylamide, when administered to animals, are excreted as the N-acetamido derivatives, Bloch and Rittenberg calculated from the isotope content of the acetylsulfanilamide and p-acetamidobenzoic acid, after administration of $CDH_2^{13}COOH$, that an adult rat produced 15–20 mmole (0.9–1.2 g) of acetate per 100 g body weight per day. These observations revealed for the first time the important role of acetate in metabolism. It was suspected very early that acetate as such was not the true metabolic intermediate but that some "activated" form of it existed. Kaplan and Lipmann (1948) discovered that the donor of acetyl groups in the acetylation of aromatic amines was acetyl-CoA.

Although it was thought that acetyl-CoA was also the source of C_2-units in fatty acid biosynthesis, more than ten years elapsed between the discovery of acetyl-CoA and the full appreciation of its participation in fatty acid biosynthesis.

After the elucidation of the reaction sequence of β-oxidation of fatty acids, it was not an unnatural assumption that the biosynthesis would be a mere reversal of β-oxidation. Stansly and Beinert (1953), in Green's laboratory, succeeded in converting acetyl-CoA to butyrate, while Seubert et al. (1957), in Lynen's laboratory, were able to convert hexanoate to octanoate. Evidently the mitochondrial system could function in elongation of fatty acids but very poorly in their total synthesis. As will be seen below, elongation can take place in both mitochondria and endoplasmic reticulum and it is apparent that both Green's and Lynen's groups had, in fact, investigated a major elongation system.

The first glimmer of evidence that fatty acids are synthesized by an enzyme system independent of the mitochondrial β-oxidation enzymes became available between 1952 and 1954 when Gurin and his colleagues (Brady and Gurin, 1952) found that homogenates of avian and rat liver free of mitochondria and microsomes could synthesize fatty acids from acetate. Similarly Popják and Tietz (1955) reported that the soluble supernatants of homogenates made from the mammary gland of lactating rats and sheep synthesized a whole range of fatty acids (C_4–C_{18}) from acetate when the preparations were supplemented with ATP, CoA, crude preparations of NAD(H) and NADP(H), and with α-keto glutarate or malonate. Two further observations set the scene for the final elucidation of fatty acid biosynthesis and established unequivocally that the process was not a simple reversal of β-oxidation. The first of these was the demonstration by Langdon (1957) that the electron donor in fatty acid biosynthesis was NADPH and not NADH. The second crucial observation, made in 1958 in Green's laboratory at the Enzyme Institute in Wisconsin, was that fatty acid biosynthesis in avian liver preparations from acetate, CoASH, ATP, and NADPH was enormously stimulated in HCO_3^- buffers (Gibson et al., 1958). It is of interest to recall, however, that the stimulation of fatty acid biosynthesis by bicarbonate buffers in mammary gland slices had been observed in the early 1950s, but the significance of the observation was not immediately apparent.

8.2. The Soluble or Total Synthesis System

In Fig. 8.1 the overall stoichiometry of the liver system is the one depicted, since the final product of this system is the free acid rather than the CoA derivative as is the case with yeast. The first step of the reaction shown is the activation of acetate by acetyl thiokinase. However, as will be seen below, this means of formation of acetyl-CoA is involved primarily in the in vitro reaction and is probably not important in the intact animal. The acetyl-CoA is then converted to malonyl-CoA by the enzyme acetyl-CoA carboxylase, which contains biotin as the prosthetic group. The location of the active CO_2 on one of the "ureido" nitrogens, as proved by Lynen (1967), is shown. The malonyl group is thought to be trans-

BIOSYNTHESIS OF FATTY ACIDS

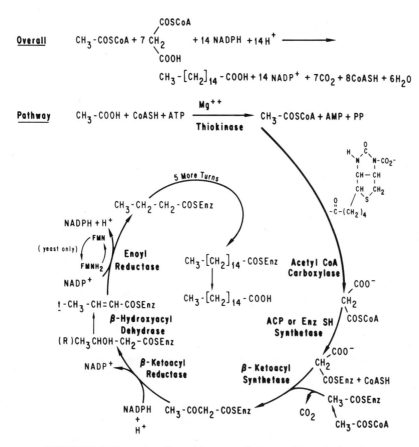

FIGURE 8.1. General outline of the cytosolic fatty acid synthesis spiral.

ferred to another thiol group, either on the low-molecular weight acyl carrier protein (ACP) in the system of most microorganisms or on a similar peptide that is part of the complex fatty acid synthetase in higher organisms and yeast. Acetyl-CoA is also transferred to the thiol group of the ACP or enzyme and condensation takes place with elimination of CO_2.

It might be appropriate, at this point, to call attention to the importance of this series of reactions in directing the pathway toward biosynthesis. In the case of the mitochondrial β-oxidation system, reversal was difficult, partly because the equilibrium constant of the β-ketothiolase reaction favors thiolysis over condensation (Eq. 8.1):

$$CH_3COCH_2COCoA + CoASH \rightleftharpoons 2\ CH_3COCoA$$

$$K = \frac{[\text{Acetoacetyl-CoA}][\text{CoASH}]}{[\text{Acetyl-CoA}]^2} = 1.6 \times 10^{-5}(25°) \tag{8.1}$$

In the synthetase, this unfavorable equilibrium is overcome in two ways. First, the α-carboxylate group of the malonyl-CoA renders its methylene group more nucleophilic than the methyl group of acetyl-CoA and thus facilitates attack at the carbonyl carbon of the acetyl-CoA. Second, release of CO_2 during the condensation is substantially endergonic. In any event, the reaction is changed from one strongly favoring thiolysis to another much more favorable to condensation (Eq. 8.2):

$$K = \frac{[\text{Acetoacetyl-Enz}][\text{CoA}]^2[CO_2]}{[\text{Acetyl-CoA}][\text{Malonyl-CoA}][\text{Enz}]} = 2 \times 10^{-2} \text{M}; \quad \text{pH 7.0} \qquad (8.2)$$

Reduction of the β-keto group to give the (R)-3-hydroxyacyl enzyme is accomplished with NADPH. This is followed by dehydration to give the *trans* α,β-unsaturated acyl enzyme and, finally, the double bond is reduced by NADPH + H^+. In this second reduction, in microorganisms and yeast, $FMNH_2$ is the initial hydrogen donor and the resulting FMN is then reduced by NADPH + H^+. In the liver system, however, NADPH appears to function directly.

Thus, the saturated 4-carbon acyl enzyme is produced and is recycled through the same series of reactions until a chain-length of about 16 carbons is attained, at which point the product is released from the enzyme. A detailed examination of this process and of the enzymes involved reveals that it is not nearly so straightforward as may appear at first sight, as will be evident from a consideration of the individual reactions.

The first reaction, formation of acetyl-CoA, takes place as shown in the scheme mainly under *in vitro* conditions with acetate as substrate. In the cell, acetyl-CoA is normally produced directly by oxidation of various substrates within the mitochondria rather than by activation of free acetate. Since the fatty acid synthetase is extramitochondrial, and since mitochondrial membranes appear to be impermeable to acetyl-CoA generated within these organelles, a transport problem exists. Figure 8.2 presents a summary of the findings from several laboratories on how such transport is managed.

Since acyl-CoAs pass the mitochondrial membrane only with difficulty (pathway **a**), another means of transport must be involved. Hydrolysis to acetate, which more easily diffuses through the membrane and may then be activated by extramitochondrial acetyl thiokinase (pathway **b**), is considered to be too slow for the known rate of fatty acid synthesis. Furthermore, the carnitine transport system (see Chapter 7) does not appear to be active in translocation of acetyl groups. The most likely pathway (**c**) appears to be the conversion of acetyl-CoA to citrate inside the mitochondrion, transport of citrate through the membrane, and cleavage of citrate by ATP citrate lyase (citrate cleavage enzyme) to yield acetyl-CoA and oxaloacetate.

Evidence for the function of citrate as a source of substrate for fatty acid synthesis is derived from several sources. The formation and transport of citrate is rapid enough to account for fatty acid synthesis; citrate cleavage enzyme is a member of the group of enzymes (citrate cleavage, acetyl-CoA carboxylase, "malic enzyme"—malate dehydrogenase, decarboxylating, NADPH—and fatty acid synthetase) that are located in the cytosol and respond similarly to dietary changes; and citrate has been shown to stimulate fatty acid synthesis in moderate

BIOSYNTHESIS OF FATTY ACIDS

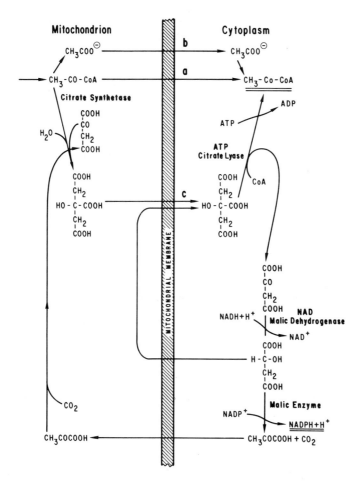

FIGURE 8.2. Schematic representation of sources and substrates for fatty acid synthesis in a mitochondrion–cytosol system.

concentrations but to depress incorporation of label from acetate at higher concentrations. The experiments of Lowenstein appear to provide convincing support for the contention that citrate is an obligatory intermediate in the normal pathway of fatty acid synthesis. Using a preparation of mitochondria and 105,000 × g supernatant, Watson and Lowenstein (1970) provided labeled alanine, which is converted to pyruvate within the mitochondria as substrate for fatty acid synthesis in the cytosol. Unlabeled citrate inhibited the incorporation of label into fatty acid by dilution but promoted the incorporation of tritium from tritiated water, showing that fatty acid synthesis in general was not inhibited. Furthermore, hydroxycitrate, an inhibitor of citrate cleavage enzyme, inhibited fatty acid synthesis to the same extent.

Additional evidence came from studies using the same system, with the methyl group of the alanine labeled with tritium. A little reflection reveals that, of the various proposed methods of transport, only the one involving the intermediacy of citrate would result in loss of tritium (i.e., Eq. 8.3):

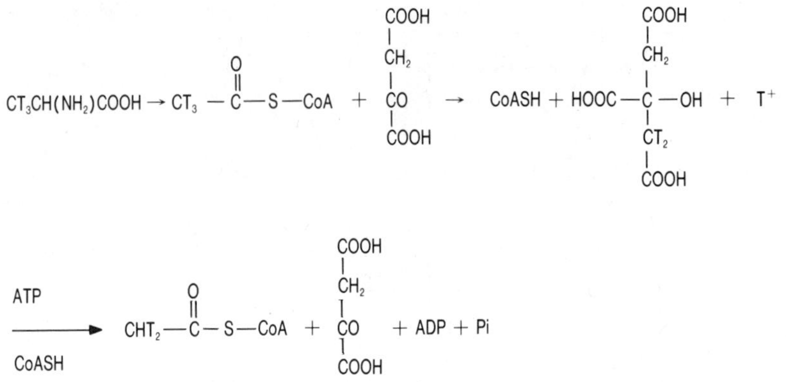

$$CT_3CH(NH_2)COOH \rightarrow CT_3-\overset{O}{\underset{\|}{C}}-S-CoA + \begin{matrix} COOH \\ | \\ CH_2 \\ | \\ CO \\ | \\ COOH \end{matrix} \rightarrow CoASH + HOOC-\underset{|}{\overset{|}{C}}-OH + T^+ \atop {CT_2 \atop COOH}$$

$$\begin{matrix} ATP \\ \\ \downarrow \\ CoASH \end{matrix} CHT_2-\overset{O}{\underset{\|}{C}}-S-CoA + \begin{matrix} COOH \\ | \\ CH_2 \\ | \\ CO \\ | \\ COOH \end{matrix} + ADP + Pi$$

(8.3)

When fatty acid synthesis was carried out starting with labeled alanine, loss of tritium from the terminal methyl of the product palmitate was observed, thus further substantiating the proposed mechanism.

The next reaction is carried out by the key enzyme, acetyl-CoA carboxylase, which catalyzes the following reactions (Eq. 8.4):

$$ATP + HCO_3^- + Enz.\text{-biotin} \xrightarrow{Mn^{2+}} ADP + Pi + Enz.\text{-biotin}-CO_2^-$$
$$Enz.\text{-biotin}-CO_2^- + Acetyl\text{-}CoA \longrightarrow Malonyl\text{-}CoA + Enz.\text{-biotin}$$

(8.4)

The first step is probably a concerted reaction. No phosphorylated intermediate has been isolated. As shown by Lynen (1967), the carbon dioxide is "activated" by attachment to one of the nitrogen atoms of the biotin moiety (see below):

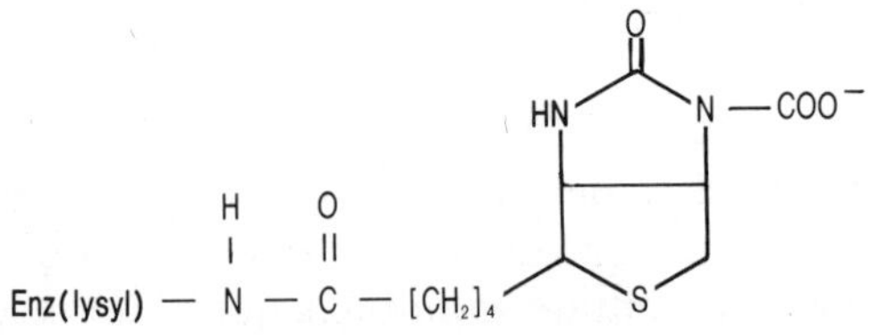

$$Enz(lysyl) - \underset{H}{\overset{|}{N}} - \underset{\|}{\overset{O}{C}} - [CH_2]_4 \begin{matrix} \text{biotin ring with } HN, N-COO^- \end{matrix}$$

The participation of biotin in this and other carboxylation reactions has been demonstrated in many ways, including its unchanging proportion during enzyme purification and the inhibition of such carboxylation reactions by avidin, the biotin-binding protein of egg white.

Insight into the nature of the reaction has come from studies by Vagelos (1971) on the *Escherichia coli* enzyme. It was found that the acetyl-CoA carboxylase of this organism could be separated into two subunits—one containing biotin and the other catalyzing the transfer of CO_2 from biotin to acetyl-CoA. The biotin-containing protein could be further subdivided into biotin carboxylase and a low-molecular weight (22,000) peptide, biotin carboxy-carrier protein (BCCP), the Enz.-biotin in Eq. 8.4 above. The transfer of CO_2 from biotin to acetyl-CoA is catalyzed by a transcarboxylase.

BIOSYNTHESIS OF FATTY ACIDS

the Enz.-biotin in Eq. 8.4 above. The transfer of CO_2 from biotin to acetyl-CoA is catalyzed by a transcarboxylase.

The enzyme from the liver of higher animals appears to be considerably more complex. Electron micrographs show it as a long ribbon-like aggregate (overall dimensions about 1000×10 Å) with four nonidentical protomeric units, about 130–140 Å in length, each with a molecular weight (M_r) of about 100,000. Three have binding sites for biotin, citrate, and acetyl-CoA, respectively. The fourth may be the CO_2 transferase. It is interesting that synthesis of this enzyme complex is controlled at least in part by dietary biotin, increasing markedly in the absence of this vitamin, but decreasing as dietary biotin permits completion of the assembly.

The study of the fatty acid synthetase itself has been difficult in such higher organisms as yeast and mammalian or avian liver. The multi-enzyme complexes involved have not been amenable to separation into their individual active components. For this reason, study of fatty acid synthesis in microorganisms such as *E. coli* has been valuable since, in these organisms, synthesis is accomplished by a group of enzymes that can be separated and that function separately. For example, it has been possible to isolate a low-molecular weight ($\sim$9,000), heat-stable peptide that binds the condensing acyl groups and the growing chain during subsequent operations. This peptide, acyl carrier protein (ACP), contains one thiol group per molecule, provided by 4'-phosphopantetheine, to which the acyl groups remain attached during elongation. The proposed sequence is given in Eq. 8.5:

$$\text{Malonyl}-S-\text{CoA} + \text{ACP}-\text{SH} \underset{\text{transacylase}}{\overset{\text{malonyl}}{\rightleftarrows}} \text{Malonyl}-S-\text{ACP} + \text{CoASH}$$

$$\text{Acetyl}-S-\text{CoA} + \text{ACP}-\text{SH} \overset{\text{acetyl}}{\rightleftarrows} \text{Acetyl}-S-\text{ACP} + \text{CoASH} \quad (8.5)$$

$$\text{Acetyl}-S-\text{ACP} + \text{Malonyl}-S-\text{ACP} \xrightarrow{\text{transacylase}} CO_2 + \text{Acetoacetyl}-S-\text{ACP} + \text{ACP}-\text{SH}$$

The isolation of acetoacetyl$-$S$-$ACP marked it as the initial condensation product. With NADPH and the remaining enzymes, the further reactions of fatty acid synthesis, as shown in Fig. 8.3, take place with recycling until C_{16} or C_{18} thioesters are produced. It is interesting that the ACP appears to be nonspecific for acyl group (malonyl, acetyl, and higher acyl) and for species (it is interchangeable between plants and microorganisms). It also can serve as an activating group for formation of glycerides and for other reactions requiring an active acyl moiety.

The multi-enzyme complexes of liver and yeast have particle weights in the neighborhood of 5×10^5 and 2.5×10^6, respectively, and contain all the component enzymes of fatty acid synthesis. For example, a heat-stable ACP-like peptide chain is a component of each complex, furnishing 4'-phosphopantetheine groups attached to peptides with the same amino acid sequence as occurs in bacterial ACP. There appear to be two such moieties per complex of M_r 5×10^5 in the mammalian or avian liver systems and six per complex of M_r 2.5×10^6 in the yeast system.

Originally, amino-terminal analysis of the liver complex indicated seven or eight polypeptide units, perhaps fortuitously corresponding to the seven enzyme

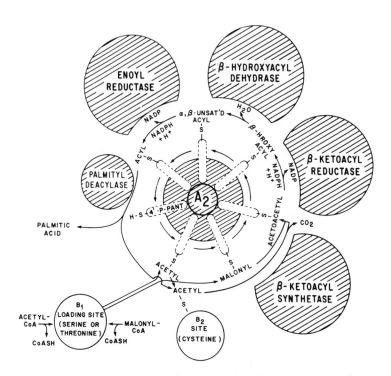

FIGURE 8.3. Schematic representation showing proposed spatial arrangements of enzymatic functions in the fatty acid synthetase system. From Phillips *et al.* (1970).

activities. This information was interpreted to mean that the component enzymes were bound noncovalently and could be separated and recombined. However, it was later shown that these separate peptides resulted from the action of proteases during isolation. Careful isolation and separation techniques now appear to indicate a homodimer of two identical protomers of molecular weight 250,000 for the liver enzyme complex and two nonidentical units, α and β, with molecular weights 213,000 and 203,000, respectively, for the yeast enzyme complex; thus the latter complex must have the formula $\alpha_6\beta_6$. Both monomers of the liver system have been reported to possess all the component enzyme activities in covalently bound polypeptide chains, leading to the possibility of one (or possibly two) genomes coding for all the different enzymes. However, the first step in the pathway, condensation of malonyl and acetyl moieties to form the acetoacetyl thioester, requires the presence of both monomers, indicating that interaction between two identical chains is necessary for the condensation step.

In the yeast system, the α-chain appears to contain the β-ketoacyl synthetase and reductase and the 4'-phosphopantetheine center, while the β-chain has acetyl and malonyl transacylases, β-hydroxyacyl dehydrase, enoyl reductase, and the palmitoyl transferase (or, as discussed below, thioesterase I). Stoops and Wakil (1980) have envisaged the yeast complex as consisting of plates (α-chains) and arches (β-chains), as shown in the schematic representation, Fig. 8.4. In connec-

BIOSYNTHESIS OF FATTY ACIDS

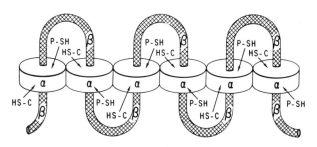

FIGURE 8.4. Schematic representation showing possible structure–function relationships in the yeast fatty acid synthetase system. From Stoops and Wakil (1980). P–SH = pantotheinyl sulfhydryl; C–SH = cysteinyl sulfhydryl.

tion with a proposed mechanism of condensation (see below), it may be noted that this arrangement permits juxtaposition of the pantetheine and cysteine thiol groups.

Before the discovery of the polymeric nature of the synthetase complexes, similar mechanisms of synthesis had been proposed by Phillips *et al.* (1970) and by Lynen (1970), as shown in Fig. 8.3. The primer (acetyl or butyryl) was thought to be transferred from CoA to a non-thiol binding site in the complex (B_1 in the diagram). Similarly, malonyl groups were also transferred to the B_1 site and both primer and malonyl were transferred to a B_2 site (cysteinyl thiol). Transfer of both from cysteine groups to the A_2 (4′-phosphopantetheine) sites was pictured as being followed by condensation to form the β-ketoacyl thioester, which was then transferred in sequence to sites of the various enzyme activities involved in elongation. With each trip around the spiral the growing acyl moiety was transferred to a B_2 site where an additional malonyl moiety was fed into the system. This spiral was thought to continue until, at a chain-length of 16–18 carbons, the acyl group was transferred to CoA and released (yeast) or hydrolyzed to the free acid (liver) by acyl thioesterase I, also a component of the synthetase complex.

However, in a continuing study of the liver synthetase, Wakil and his coworkers, using the proteolytic functions of several enzymes, specific labeling of the different enzymatic sites by labeled substrates or inhibitors and monoclonal antibodies binding the synthetase, have mapped the locations of the seven catalytic centers and the acyl carrier site. Three domains are recognized for each synthetase subunit. Domain I contains the NH_2-terminal end and the β-ketoacyl synthetase and acetyl and malonyl transacylase sites. Domain II contains the β-ketoacyl and enoyl reductases, probably the dehydratase, and the acyl carrier site to which the growing acyl chain remains attached. Domain III contains the thioesterase activity and the carboxyl terminus.

A functional model (Fig. 8.5), shows the subunits arranged in head-to-tail fashion such that a functional unit consists of Domain I of one subunit and Domains II and III of the other subunit. This would explain the necessity of two identical subunits containing all the catalytic centers. These relationships are shown in Fig. 8.5 (Tsukamoto *et al.*, 1983).

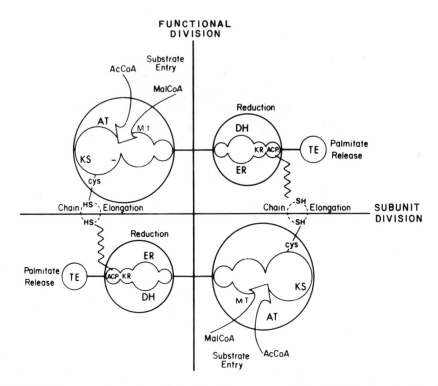

FIGURE 8.5. Two subunits are drawn in head-to-tail arrangement (subunit division) so that two sites of palmitate synthesis are constructed (functional division). From Tsukamoto *et al.* (1983). The abbreviations for partial activities used are *AT*, acetyl transacylase; *MT*, malonyl transacylase; *KS*, β-ketoacyl synthetase; *KR*, β-ketoacyl reductase; *DH*, dehydratase; *ER*, enoyl reductase; *TE*, thioesterase; and *ACP*, acyl carrier protein. The wavy lines represent the 4′-phosphopantetheine prosthetic groups.

At this point a word about the nature of the product of the synthetase is in order. Under conditions of normal nutrition, the principal product is palmitate (16:0), which, if formed from carboxy-labeled acetate, is evenly labeled—that is, every odd carbon has equal activity. The implication is that during fatty acid synthesis the same pool of acetyl-CoA is used, with no release and reincorporation of shorter-chain fatty acids, an eventuality that would be accompanied by dilution at each stage, evidencing stepwise synthesis. Also pertinent is the very rapid incorporation of malonyl-CoA once initiation of synthesis by acetyl-CoA occurs.

It is possible to replace the initiating and elongating molecules leading to production of unusual fatty acids. For example, initiation by propionyl-CoA results in production of odd-chain fatty acids, albeit at a reduced rate. Thus, the ratio of odd-chain to even-chain fatty acids produced is generally very low. Butyryl-CoA can also initiate synthesis and, indeed, appears to be the preferred initiator. Other intermediate-chain-length CoA esters can also serve as initiators. Thus, it is of interest that certain tissues, such as skin, accumulate sizable amounts of branched-chain fatty acids that must obviously have been formed by initiation by acyl-CoAs arising from oxidation products of amino acids. Horning and coworkers (1961), for example, found the following processes to be effected by adipose tissue synthetase (Eq. 8.6):

$$\begin{aligned}
&\text{Acetyl-CoA} \rightarrow C_{16}\ (C_{14},\ C_{18}) \\
&\text{Propionyl-CoA} \rightarrow C_{15}\ (C_{13},\ C_{17}) \\
&\text{Valine} \rightarrow \text{isobutyryl-CoA} \rightarrow \text{iso-}C_{16}\ (C_{14},\ C_{18}) \\
&\text{Leucine} \rightarrow \text{isovaleryl-CoA} \rightarrow \text{iso-}C_{15}\ (C_{17}) \\
&\text{Isoleucine} \rightarrow \alpha\text{-methylbutyryl-CoA} \rightarrow \text{anteiso-}C_{15},\ C_{17}
\end{aligned} \qquad (8.6)$$

Thus the iso- and anteiso fatty acids are derived from the correspondingly branched amino acids.

It will be noted that the chain-lengths of major products, regardless of the initiator, tend to be about C_{16}. Two characteristics of the synthetase enzyme complex contribute to this result. First, thioesterase I is increasingly active with acyl thioesters that have chain-lengths greater than C_{14}. Second, the rate of elongation of the acyl chain becomes slower after C_{16} so that the formation and release of the 16-carbon product is more likely. However, there are obviously other factors involved. For example, it has been known for a long time that milk fats (particularly in the milk from ruminants) contain fatty acids with much shorter chains than do the fats of other tissues. In experiments with the lactating mammary gland (see Chapter 5) it was also found that incorporation of carboxy-labeled acetate does not give evenly labeled fatty acids, in contrast to results obtained with systems of other tissues (see above). However, experiments by several groups have demonstrated that the fatty acid synthetase isolated from rat mammary gland is the same as that from liver of the same animal, usually producing the 16-carbon product. Other factors must therefore be involved in the formation of short- and medium-chain fatty acids by mammary gland tissue *in vivo*. One such factor is an independent cytosolic thioesterase II in mammary gland that enhances release of medium- and short-chain acyl moieties from the synthetase (Libertini and Smith, 1979).

Another factor in the formation of shorter-chain-length fatty acids appears to be the ratio of malonyl-CoA to acetyl-CoA. With high ratios, the product from either synthetase has the usual 14–18-carbon chain-length. As the ratio decreases, however, the chain-length shortens until, as it approaches zero, butyrate may represent as much as 50% of the product. One factor in this case may be the competition between acetyl or butyryl and the growing acyl chain for the A_2 site. Lynen (1970) attempted to put these relationships on a quantitative basis by considering (i) the probability that an enzyme-bound acyl group will transfer to CoA (in yeast) as a function of the relative velocities of the condensing and transferring reactions and (ii) that the growing acyl chain interacts with the transferase with increasing velocity after a length of 13 carbon atoms. An equation based on these considerations showed close agreement with experimental results.

As mentioned above, not only the initiating, but also the elongating moieties are replaceable. Evidence for this came first from the structure of certain fatty acids isolated from the preen gland of the goose or from the tubercle bacillus in which a regular methyl branching is present, as in that of the following (preen gland) fatty acid:

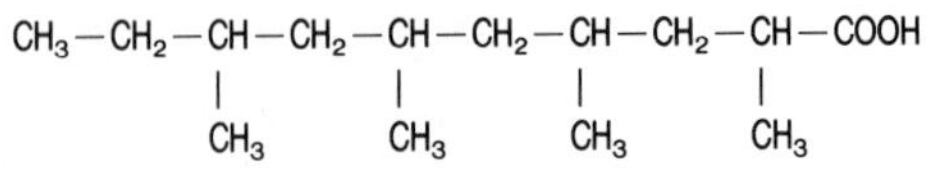

$$CH_3-CH_2-CH-CH_2-CH-CH_2-CH-CH_2-CH-COOH$$
$$\,\,|\,\,|\,\,|\,\,|$$
$$CH_3CH_3CH_3CH_3$$

It seems evident that in the case of fatty acids of this type, after initiation with acetyl-CoA, elongation was continued with methylmalonyl-CoA. This was confirmed by Lederer (1963), who showed the incorporation of labeled methylmalonyl-CoA into certain bacterial fatty acids. Considering the function of vitamin B_{12} coenzyme in the metabolism of methylmalonyl-CoA (see Chapter 7), it seems probable that a deficiency of this vitamin might result in an increase in the formation of branched-chain fatty acids of this type. This supposition has been borne out by the work of Abeles and his coworkers (Barley *et al.,* 1972). The incorporation of such branched-chain fatty acids into neural membranes may account for some of the symptoms of B_{12}-deficiency disease.

8.3. *Control of Fatty Acid Synthesis*

The essential steps in fatty acid biosynthesis and the nature of the enzymes involved having been reviewed, it is now possible to consider certain aspects of control of the overall process. Since fatty acid biosynthesis is of course intimately interrelated with carbohydrate metabolism, it is expected that control will be complex and imposed at several different levels and sites.

Three longstanding observations must be adequately explained by any tenable regulatory system: fatty acid synthesis is (i) depressed toward zero by starvation, (ii) greatly increased by feeding carbohydrate to the starved animal, and (iii) usually depressed by feeding fat.

Effects of concomitant carbohydrate metabolism can be considered to be results of several interactions. Carbohydrate represents a principal source of carbons for fatty acid biosynthesis via pyruvate, acetyl-CoA, and citrate, as discussed above. It is also a source of reducing power as NADPH from, among other sources, NADP isocitrate dehydrogenase, glucose-6-phosphate dehydrogenase and, particularly, the malic enzyme (malate dehydrogenase, decarboxylating (NADP)EC1 1.1.40). As a matter of fact, one group of enzymes—fatty acid synthetase, acetyl-CoA carboxylase, citrate cleavage enzyme, and malic enzyme—are known as the fatty acid synthesis group and, in most cases, respond to dietary changes in parallel fashion. Figure 8.2 indicates some relationships among some of these enzymes. The formation and cleavage of citrate as a source of extramitochondrial acetyl-CoA for fatty acid synthesis have already been discussed. It can be seen that the citrate cleavage enzyme also produces extramitochondrial oxalacetate, which can be reduced by NAD malate dehydrogenase and then oxidatively decarboxylated by the malic enzyme with production of pyruvate and extramitochondrial NADPH, thus resulting in a transhydrogenation. The pyruvate can be carboxylated within the mitochondria to oxalacetate, which can then condense with acetyl-CoA to form citrate. Meanwhile, the malate produced extramitochondrially also activates the citrate permease (or other mechanism of citrate transport across the mitochondrial membrane) as well as the transport of other di- and tricarboxylic acids. Thus the system can act in a cyclic manner, producing substrate and reducing power and facilitating transport.

However, control does not seem to reside in this system under normal conditions. In the diminished fatty acid synthesis characteristic of starvation, neither NADPH nor substrate appear to be limiting and their addition to the depressed *in vitro* system does not restore synthesis.

A logical control point for synthesis is acetyl-CoA carboxylase. Not only is it the first committed step in the pathway from acetyl-CoA to fatty acid, but the activity of this enzyme seems to be limited when compared to those of citrate cleavage or the synthetase. It has been found that acetyl-CoA carboxylase, as usually separated from the tissues, is in an inactive form that is activated allosterically by citrate. The inactive protomeric form of the enzyme is not only activated but also induced to polymerize by citrate, isocitrate, and many other acids. The reaction is not simply an associated activating polymerization, however. Polymerization can be accomplished by tricarballylic acid without activation. On the other hand, the enzyme can be activated slowly by incubation with Mg^{2+} at 37°C or more rapidly with trypsin. Greenspan and Lowenstein (1973) propose a stepwise activation of the inactive low-molecular weight protomer as follows:

$$I_{light} \xrightarrow{Mg^{2+}} P_{light} \xrightarrow[\text{(short)}]{\text{citrate}} A'_{light} \xrightarrow[\text{(long)}]{\text{citrate}} A_{heavy} \qquad (8.7)$$

where I is inactive, P is preactive, A and A' are active forms of the enzyme. Actually, the active polymeric enzyme has been isolated from the tissues by homogenization at 38°C. Apparently the polymer is dissociated during isolation at the usually employed lower temperatures.

The logic of the activation of this key enzyme by citrate is, of course, that citrate activates the enzyme that uses it as a substrate. However, the concentrations necessary for rapid activation are considerably greater than those normally present in the tissues, so it is evident that this type of control is not ordinarily important or that concentrations may be different in different compartments.

Another type of control of acetyl-CoA carboxylase has been proposed by Lynen (1970). All the fatty acid synthesis enzymes, particularly acetyl-CoA carboxylase, are markedly depressed by fatty acids and especially by their immediate derivatives, the long-chain acyl-CoAs. Moreover, these substances increase during starvation and decrease with carbohydrate feeding. Thus, a feedback inhibition system appears to be a logical control mechanism for acetyl-CoA carboxylase and hence for fatty acid synthesis (Numa *et al.*, 1972). However, it is well known that fatty acids and acyl-CoAs are nonspecific detergent-type inhibitors of many enzymes and that this inhibition is reversed by protein, e.g., albumin or microsomal protein. Thus, this type of inhibition may not be important under usual *in vivo* conditions. Another result of an increase in concentration of free fatty acids would be their competition for CoA, for example with citrate cleavage enzyme, thus increasing fatty acid oxidation and decreasing synthesis. This short-term control of fatty acid synthesis is readily demonstrated in chicks because of their rapid digestive process. When chicks are starved for very short periods and then refed, fatty acid synthesis responds more rapidly to both conditions than do the activities of the fatty acid synthesis enzymes. Here, as well as in fat feeding, free CoA changes in a manner parallel to that of fatty acid synthesis. The result of these changes may be interpreted as stemming from the competition for utilization of free CoA to form acyl-CoA from the increased free fatty acids and for the requirement for CoA by the ATP citrate lyase, a necessary factor in fatty acid biosynthesis. CoA has also been shown to be a positive allosteric effector of acetyl-CoA carboxylase.

Additional regulation may be provided by the cyclic-AMP-dependent phosphorylation of acetyl-CoA carboxylase. Phosphorylation inactivates this enzyme and dephosphorylation by a phosphatase reactivates it.

It is also possible that the concentration rather than the activity of the enzyme is involved. Workers in several laboratories have found that prolonged fasting results in a decrease in the absolute amounts of the synthesis group of enzymes. Moreover, acetyl-CoA carboxylase is degraded (or is dissociated) considerably more rapidly in the starved than in the fed animal (Gibson *et al.,* 1970). Aging also decreases fatty acid synthesis, but this trend can be partially offset by action of insulin or thyroxine.

Thus it appears that control of fatty acid synthesis is exerted at several different sites and under different conditions. Actually, considering the complexity of the energy-producing reactions, it might have been expected that control of their interrelationships would not be simple.

Yet another mode of control of fatty acid synthesis in certain bacteria has been reported from the laboratory of K. Bloch (Flick and Bloch, 1974). Fatty acid synthesis in *Mycobacterium phlei* is carried out by two systems: by a multienzyme complex (I) similar to those of liver and yeast as well as by a "separated" enzyme system of the sort generally characteristic of microorganisms. Enzyme complex I is unstable at low ionic strength and dissociates with loss of activity. At low acetyl-CoA concentration, it depends for activity on a mixture of FMN and three polysaccharides, the function of which appears to be to lower than the K_m for acetyl-CoA and thus to control the rate of synthesis.

8.4. The Particulate Elongation Systems

One of the problems of studying a system such as the soluble fatty acid synthetase after isolation from other cell components is that these other components may have profound effects on it. Although the study of the isolated system serves to reveal effects of added factors on its activity, this approach obviates investigation of its function and control in the tissues from which it was isolated. Thus, for example, when microsomes from livers of starved animals are added to the isolated synthetase, inhibition results, probably because of strong ATPase and CoAase activities of the microsomes. Microsomes from livers of fed animals, on the other hand, stimulate synthesis, possibly by providing binding sites for inhibitory acyl-CoA (see Donaldson *et al.*, 1970). In any event, both microsomes and mitochondria have the capacity to alter the product of the soluble synthetase, and it is to such effects that attention is now addressed (Landriscina *et al.*, 1977).

Palmitate produced by the soluble synthetase, following injection of carboxy-labeled acetic acid into the whole animal or addition of labeled acetyl-CoA to the purified preparation, is found to be evenly labeled, i.e., all of the odd carbons of the chain, including the carboxyl group, are equally labeled, and the even carbons are label-free:

$$CH_3 - (\overset{*}{C}H_2 - CH_2)_7 - \overset{*}{C}OOH$$

In practice, this can be demonstrated simply by measuring the activity of the $^{14}CO_2$ obtained by decarboxylating the palmitic acid. It would, of course, have close to 12.5% of the total activity of the molecule. However, those fatty acids with chain-lengths longer than 16 carbons, which make up the majority of the fatty acids of the tissue lipids, are found to have a quite different distribution of activity: the carboxyl group has an activity considerably greater than that expected on the basis of "even labeling." Such observations imply existence of a system that can elongate unlabeled or lightly labeled fatty acids by addition of two or more carbons from the labeled acetate pool.

Such a system was actually known before the discovery of the soluble synthetase. When it was initially hypothesized that fatty acid synthesis might occur by reversal of the β-oxidation process, at least two groups of investigators attempted to effect synthesis of long-chain fatty acids from acetate or an intermediate-chain-length fatty acid in the presence of the purified mitochondrial enzymes and oxidative cofactors in the reduced states (see Seubert *et al.*, 1957; Stansly and Beinert, 1953). Although they were not successful in their original aim, both groups succeeded in elongating their starting acids by two carbons. Thus it became apparent that mitochondria contain systems capable of elongating fatty acids.

That this is indeed the case was shown by Boone and Wakil (1970), who carried out elongation of stearate to arachidate with whole mitochondria or with corresponding acetone powders or sonicates. They also reported some total synthesis but this was later denied by most workers in the field, who were able to achieve elongation systems containing mitochondria, an acyl-CoA, acetyl- (but not malonyl-) CoA, NADH, and NADPH. Little if any total synthesis or malonyl-CoA incorporation was demonstrable. Podack and Seubert (1972) reconstructed the mitochondrial elongation system, finding that it is not exactly the reverse of the oxidative system and that it requires, for example, an NADPH enoyl reductase and certain chain-length-specific enzymes such as β-keto thiolase.

Separation of the inner and outer mitochondrial membranes has somewhat confused the picture in that the elongation system seems to be located in the outer membrane, whereas the inner membrane may be able to carry out some *de novo* synthesis of medium-chain fatty acids. This total-synthesis capacity is masked by the substantially greater elongation activity of the intact mitochondrion. It is interesting that long-chain acyl-CoAs, which inhibit total synthesis by the soluble system, actually serve as substrates for the elongation system. Thus it would be expected that a high-fat diet would convert *de novo* synthesis to elongation (and, of course, oxidation). This has been found to be the case in several animals, such as the pig, and in certain tissues, such as that of the heart, which contains no citrate cleavage enzyme or acetyl-CoA carboxylase and incorporates acetate into fatty acids solely by elongation.

In addition to the mitochondrial elongation system, a quite different mechanism for elongation of somewhat different substrates occurs in the endoplasmic reticulum. This microsomal system will be treated in part in Chapter 9, since it is involved in the formation of the long-chain polyunsaturated fatty acids. However, it also appears to function importantly in elongation of the saturated fatty acids. Unlike the mitochondrial elongation system, which uses acetyl-CoA as the elongating substrate, the system in the endoplasmic reticulum uses malonyl-CoA in a reaction similar to that of the cytosolic palmitate synthetase. As a matter of fact, some total synthesis can occur in a microsomal system incubated anaerobically with malonyl- and acetyl-CoA (Bourre *et al.*, 1977).

Stemming from the finding in the laboratory of N. A. Baumann (Bourre *et al.*, 1973) of a deficiency of long-chain fatty acids in the brain sphingolipids of two mutant myelin-deficient strains of mice, "quaking" and "jimpy," Bloch and his coworkers (Goldberg *et al.*, 1973) found that three microsomal elongation systems exist in the brains of normal mice. These have optimal activity for coenzyme A derivatives of 16:0, 18:0, and 20:0, respectively. Only the 20:0 elongating enzyme appears to be decreased or absent in the mutants.

In other organs this division of the elongation systems may not be the same as in brain. Bernert and Sprecher (1979) have found that the rate of elongation of palmitoyl-CoA by rat liver microsomes is about 11 times that of stearoyl-CoA, largely because of a more rapid condensation step, which is rate-controlling in the elongation process.

REFERENCES

Barley, F. W., Sato, G. H., and Abeles, R. J., 1972, An effect of vitamin B_{12} deficiency in tissue culture, *J. Biol. Chem.* **247**:4270.

Bernert, J. T., Jr., and Sprecher, H., 1979, Factors regulating the elongation of palmitic and stearic acid by rat liver microsomes, *Biochim. Biophys. Acta* **574**:18.

Bloch, K., and Rittenberg, D., 1945, An estimation of acetic acid formation in the rat, *J. Biol. Chem.* **159**:45.

Boone, S. C., and Wakil, S. J., 1970, *In vitro* synthesis of lignoceric and nervonic acids in mammalian liver and brain, *Biochemistry* **9**:1470.

Bourre, J.-M., Pollet, S., Chaix, G., Daudu, O., and Baumann, N., 1973, Etude "in vitro" des acides gras synthétisés dans les microsomes de cerveaux de souris normales et "quaking", *Biochimie* **55**:1473.

Bourre, J.-M., Paturneau-Jouas, M. Y., Daudu, O. L., and Baumann, N. A., 1977, Lignoceric acid biosynthesis in the developing brain. Activities of mitochondrial acetyl-CoA-dependent synthesis and microsomal malonyl-CoA chain-elongating system in relation to myelination. Comparison between normal mouse and dysmyelinating mutants (Quaking and Jimpy), *Eur. J. Biochem.* **72**:41.

Brady, R. O., and Gurin, S., 1952, Biosynthesis of fatty acids by cell-free or water-soluble enzyme systems, *J. Biol. Chem.* **199**:421.

Donaldson, W. E., Wit-Peeters, E. M., and Scholte, H. R., 1970, Fatty acid biosynthesis in rat liver. Relative contributions of the mitochondrial, microsomal and non-particulate systems, *Biochim. Biophys. Acta* **202**:35.

Flick, P. K., and Bloch, K., 1974, In vitro alterations of the product distribution of the fatty acid synthetase from *Mycobacterium phlei*, *J. Biol. Chem.* **249**:1031.

Gibson, D. M., Titchener, E. G., and Wakil, S. J., 1958, Requirement for bicarbonate in fatty acid synthesis, *J. Am. Chem. Soc.* **80**:2908.

Gibson, D. M., Lyons, R. T., Scott, D. F., and Muto, Y., 1970, Synthesis and degradation of the lipogenic enzymes of rat liver, in: *Advances in Enzyme Regulation*, Vol. 10 (G. Weber, ed.), Pergamon, Oxford, pp. 187–204.

Goldberg, I., Schechter, I., and Gloch, K., 1973, Fatty acyl-coenzyme-A elongation in brain of normal and quaking mice, *Science* **182**:497.

Greenspan, M. D., and Lowenstein, M. M., 1973, Effects of magnesium ions, adenosine triphosphate, palmitoyl carnitine and palmitoyl coenzyme A on acetyl coenzyme A carboxylase, *J. Biol. Chem.* **243**:6273.

Horning, M. G., Martin, D. B., Karmen, A., and Vagelos, P. R., 1961, Fatty acid synthesis in adipose tissue. II. Enzymatic synthesis of branched chain and odd-numbered fatty acids, *J. Biol. Chem.* **236**:669.

Kaplan, N. O., and Lipmann, F., 1948, Assay and distribution of coenzyme A, *J. Biol. Chem.* **174**:37.

Landriscina, C., Gnoni, G. V., and Quagliariello, E., 1977, Fatty acid biosynthesis. The physiological role of the elongation system present in microsomes and mitochondria of rat liver, *Eur. J. Biochem.* **29**:188.

Langdon, R., 1957, Biosynthesis of fatty acids in rat liver, *J. Biol. Chem.* **226**:615.

Lederer, E., 1963, The biosynthesis of bacterial branched chain acids, in: *Proceedings of the Fifth International Congress of Biochemistry*, Vol. VII, *Biosynthesis of Lipids* (G. Popják, ed.), Pergamon Press, Oxford, pp. 90–103.

Libertini, L. J., and Smith, S. J., 1979, Synthesis of long-chain acyl-enzymes by modified fatty acid synthetases and their hydrolysis by a mammary gland thioesterase, *Arch. Biochem. Biophys.* **192**:47.

Lynen, F., 1967, The role of biotin-dependent carboxylations in biosynthetic reactions, *Biochem. J.* **102**:381.

Lynen, F., 1970, Comparative aspects of fatty acid synthesis, *Miami Winter Symposia*, Vol. 1, North-Holland Publishing Co., Amsterdam, p. 151.

Numa, S., Hashimoto, T., and Nakanishi, S., 1972, Regulatory mechanisms for liver acetyl-coenzyme A carboxylase, *Biochem. J.* **128**:2P.

Phillips, G. T., Nixon, J. E., Dorsey, J. A., Butterworth, P. H. W., Chesterton, C. J., and Porter, J. W., 1970, The mechanism of synthesis of fatty acids by the pigeon liver enzyme system, *Arch. Biochem. Biophys.* **138**:380.

Podack, E. R., and Seubert, W., 1972, On the mechanism of malonyl-CoA independent fatty acid synthesis. II. Isolation, properties and subcellular location of *trans*-2,3-hexenoyl-CoA and *trans*-2,3-decenoyl-CoA reductase, *Biochim. Biophys. Acta* **280**:235.

Popják, G., and Tietz, A., 1955, Biosynthesis of fatty acids in cell-free preparations. 2. Synthesis of fatty acids from acetate by a soluble enzyme system prepared from rat mammary gland, *Biochem. J.* **60**:147.

Raper, H. S., 1907, The condensation of acetaldehyde and its relation to the biochemical synthesis of fatty acids, *J. Chem. Soc.* **91**:1831.

Schoenheimer, R., 1949, *The Dynamic State of Body Constituents*, Harvard University Press, Cambridge, Mass.

Seubert, W., Greull, G., and Lynen, F., 1957, Die Synthese der Fettsäuren mit gereinigten Enzymen des Fettsäurecyclus, *Angew. Chem.* **69**:359.

Stansly, P. G., and Beinert, H., 1953, Synthesis of butyryl-coenzyme A by reversal of the oxidative pathway, *Biochim. Biophys. Acta* **11**:600.

Stoops, J. K., and Wakil, S. J., 1980, Yeast fatty acid synthetase: structure–function relationship and nature of the β-ketoacyl synthetase site, *Proc. Nat. Acad. Sci. U.S.A.* **77**:4544.

Tsukamoto, Y., Wong, H., Mattick, J. S., and Wakil, S. J., 1983, The architecture of the animal fatty acid synthetase complex. IV. Mapping of active centers and model for the mechanism of action, *J. Biol. Chem.* **258**:15312.

Vagelos, P. R., 1971, Regulation of fatty acid biosynthesis, *Current Topics in Cell Regulation* **4**:119.

Watson, J. A., and Lowenstein, J. M., 1970, Citrate and the conversion of carbohydrate into fat. Fatty acid synthesis by a combination of cytoplasm and mitochondria, *J. Biol. Chem.* **245**:5993.

9
DESATURATION OF FATTY ACIDS—THE ESSENTIAL FATTY ACIDS

The unsaturated fatty acids, which are the most abundant fatty acids in nature, were the last to reveal their detailed structure and the mechanism of their biosynthesis. In fact, the details of their formation are still not completely clear in some cases. The first real "breakthrough" came with the announcement by Bloomfield and Bloch (1960) that a particulate fraction from yeast converts stearic to oleic acid or palmitic to palmitoleic acid in a reaction requiring oxygen and NADPH (or NADH). Since that time, two major systems for formation of monounsaturated acids have been recognized—the aerobic system used by all higher organisms and most aerobic bacteria and the anaerobic system used by obligate anaerobes and certain other bacteria. In this chapter, the anaerobic system will be discussed first, since it is directly related to the fatty acid biosynthesis pathway discussed in Chapter 8.

9.1. The Anaerobic System

In view of the discovery that desaturation in yeast requires oxygen, it seemed curious that certain anaerobic bacteria could produce unsaturated fatty acids at all. This problem was settled by Bloch and his coworkers when they found that, in *Escherichia coli,* desaturation does not involve oxidative removal of two hydrogens but rather the dehydration of a hydroxyacyl derivative that is a normal intermediate in fatty acid biosynthesis in this organism. At the C_{10} stage, β-hydroxyacyl-ACP can be dehydrated (and thus desaturated) in two directions—to form *trans*-α,β-decenoyl-ACP, which is then reduced to the decanoyl-ACP and further elongated to the usual saturated products; or to form *cis*-β,γ-decenoyl-ACP in which the center of unsaturation is thus set aside and preserved while normal elongation processes result in formation of palmitoleic or vaccenic acid derivatives (see Fig. 9.1).

The hydroxydecanoyl-ACP desaturase has the unusual property of being fairly demanding with respect to chain-length specificity but relatively nonspecific

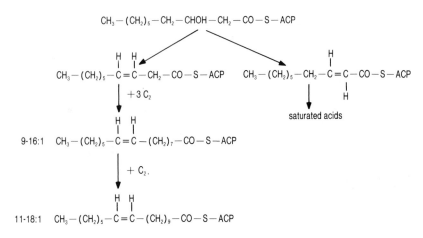

FIGURE 9.1. Mechanism of the anaerobic desaturation system.

with respect to the thiol involved: only derivatives with chain-lengths of C_{10} are desaturated, but they may be esters of CoA, ACP, pantetheine, or N-acetylcysteamine (NAC). When this 3-hydroxydecanoyl-thioester dehydrase is inhibited by 3-decenoyl-NAC, synthesis of unsaturated fatty acids ceases, although saturated fatty acids continue to be produced as usual. Thus two enzymes appear to be concerned with dehydration of β-hydroxyacyl thioesters: a *trans-α,β-dehydrase* not specific for chain-length, which leads to the synthesis of saturated fatty acids, and the β-hydroxydecanoyl-thioester dehydrase specific for chain-length.

The action of the latter enzyme is of particular interest since it appears to carry out reversibly several different types of reactions—dehydration in both the α,β and β,γ senses and isomerization of the unsaturated esters. Bloch (1969) has depicted the action of the enzyme in the manner shown in Eq. 9.1.

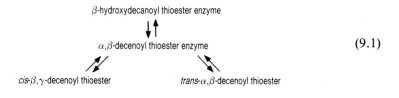 (9.1)

Thus, the intermediacy of the enzyme-bound α,β-unsaturated acid permits the isomerization of this acid to the β-γ-unsaturated acid, normally a thermodynamically unfavorable reaction.

9.2. The Aerobic System

The aerobic desaturase system, as revealed in the yeast experiments of Bloomfield and Bloch, is employed by all higher organisms that have been inves-

DESATURATION OF FATTY ACIDS

tigated and, with some modifications, by most microorganisms. It is a particulate system requiring oxygen and NADPH and resulting usually in the formation of a *cis* double bond at the 9-position of a saturated fatty acid, regardless of chain-length; for example see Eq. 9.2.

$$\text{palmitoyl-CoA} + O_2 + \text{NADPH} + H^+ \rightarrow \text{palmitoleoyl-CoA} + \text{NADP}^+ + H_2O \tag{9.2}$$

Also involved, at least in such aerobic bacteria as *Micobacterium phlei*, are a flavin and ferrous ion. These cofactors are typical of mixed-function oxidases, but no oxygenated fatty acid has been shown to be an intermediate in this process. If one exists, it is apparently not released unchanged.

More information on the detailed mechanism has been obtained from other than bacterial sources. In the case of the phytoflagellate, *Euglena gracilis*, the desaturase is soluble and has been separated into three fractions: a flavin-containing NADPH dehydrogenase, a non-heme iron protein (ferredoxin), and the desaturase itself. These findings have led to proposal of the following as the electron transport chain in the desaturase reaction (Eq. 9.3):

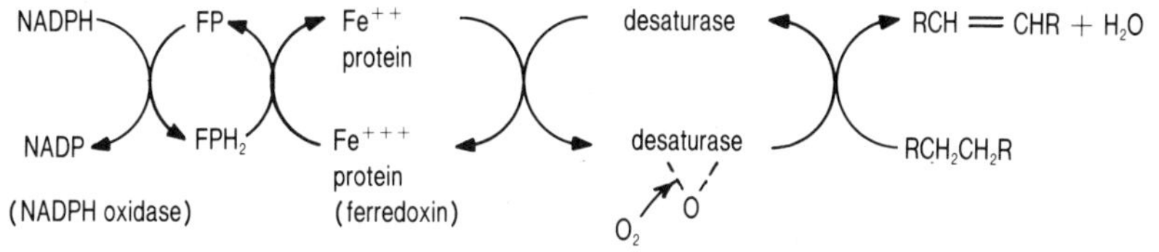

$$\begin{array}{c} \text{NADPH} \\ \text{NADP} \\ \text{(NADPH oxidase)} \end{array} \ce{><} \begin{array}{c} \text{FP} \\ \text{FPH}_2 \end{array} \ce{><} \begin{array}{c} \text{Fe}^{++} \text{ protein} \\ \text{Fe}^{+++} \text{ protein} \\ \text{(ferredoxin)} \end{array} \ce{->} \begin{array}{c} \text{desaturase} \\ \text{desaturase} \\ O_2 \end{array} \ce{><} \begin{array}{c} \text{RCH}=\text{CHR} + H_2O \\ \text{RCH}_2\text{CH}_2\text{R} \end{array} \tag{9.3}$$

In higher animals, the liver microsomal systems involved in desaturation have not been so readily dissected but appear to involve a similar mechanism. Although the desaturase itself is labile and not readily isolated, other components are recognizable as cytochrome b_5, a cytochrome b_5 reductase, and a small-molecular weight, phospholipid-containing protein. Oshino and Sato (1966) depicted the microsomal electron transport chain with various electron donors as shown in Eq. 9.4.

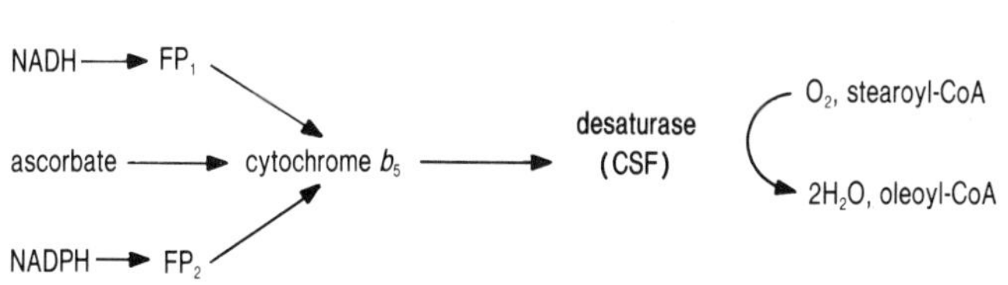

$$\begin{array}{c} \text{NADH} \rightarrow \text{FP}_1 \\ \text{ascorbate} \rightarrow \text{cytochrome } b_5 \rightarrow \text{desaturase (CSF)} \\ \text{NADPH} \rightarrow \text{FP}_2 \end{array} \begin{array}{c} O_2, \text{ stearoyl-CoA} \\ 2H_2O, \text{ oleoyl-CoA} \end{array} \tag{9.4}$$

They consider that different flavoproteins are concerned with electron transfer from different nucleotides but not from ascorbate and that a cyanide-sensitive factor (CSF) (the desaturase?) is involved in the actual transfer of electrons and protons from stearate to oxygen.

One of the most interesting aspects of the desaturase reaction is the stereochemistry of removal of hydrogen atoms at C_9 and C_{10}. The ingenious experiments of Schroepfer and Bloch (1965), with the desaturase from *Corynebacterium diphtheriae* and the four stereoisomeric 9- or 10-tritium-substituted stearates, together with [1-^{14}C]stearate, showed that only the 9- and 10-pro-*R* hydrogen atoms were removed. There was also an isotope effect at the 9-position, indicating that removal of the 9-pro-*R* hydrogen atom may be rate-limiting. Morris and coworkers (Morris, 1970), although agreeing with Schroepfer and Bloch on the stereochemistry of removal, do not distinguish between isotope effects in the 9- and 10-positions and suggest a "concerted" removal. In any event, the probability that a hydrocarbon chain can be held on the enzyme surface rigidly enough to permit stereospecific desaturation gives us some insight into the hydrophobic nature of the active site.

The question of whether membrane lipid fatty acyl groups may be desaturated *in situ* without prior conversion to a thiol ester in both plants and animals was first answered by Kates and his coworkers (Pugh and Kates, 1979) who showed that in rat liver (as had been shown in higher plants—see below), some desaturation of phospholipid fatty acids does occur. This finding led to the idea that there are two separate desaturation systems in the endoplasmic reticulum, one of which requires a thiol ester as substrate, while the other is specific for esterified fatty acids of phospholipids. The relative importance of the two systems has not yet been determined, but it is possible that an *in situ* desaturase would permit a more rapid response, requiring lower energy expenditure, to a need for such modification of membrane lipids. A rapid response of this sort may also be provided by a closely coupled system in which a fatty acid of a membrane lipid is transferred to the desaturase and the altered fatty acid transferred back to a membrane lipid without the intermediates ever leaving the membrane to equilibrate with a free fatty acid pool. At least one such reaction has been shown to occur in membrane lipid alteration although, in the case of desaturation, it might be difficult to distinguish between the two possible mechanisms.

9.3. The Aerobic System in Higher Plants

For several years it was thought that the desaturase system in plants represented a third mechanism. This was due to observations in several laboratories that in aerobic systems of this kind, stearic acid and stearoyl-CoA are not directly desaturated by chloroplasts, although oleic acid is produced from fatty acids or fatty acyl-CoAs of chain-length shorter than C_{16}. It was then found that the system has a specific thiol requirement: stearoyl-ACP, but not stearoyl-CoA, is readily desaturated by spinach chloroplasts. The higher plants apparently lack a stearoyl-CoA:ACP acyl transferase. Thus only stearoyl-ACP, synthesized by elongation of shorter-chain acyl-ACPs, is desaturated. This scheme, a modification of the proposal by Gurr (1971), is shown in Eq. 9.5, E representing elongation, D, desaturation, and T, transacylation.

DESATURATION OF FATTY ACIDS

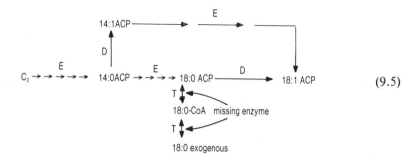

(9.5)

Thus, the mechanism of desaturation in higher plant systems is actually very similar to that of the bacterial or animal aerobic systems.

An interesting difference between the plant and animal systems concerns the further desaturation of oleic acid to linoleic and linolenic acids, which occurs in plants, but not in animals. The desaturating enzyme of *Chlorella* effects desaturation of oleic acid in acyl linkage in phosphatidylcholine. Thus there appear to be two desaturating enzymes in plants: one which attacks the fatty acid in thiol linkage and produces a double bond at the 9-, or in some cases, the 7-position from the carboxyl end, and another which attacks the fatty acid bound in acyl linkage to phosphatidyl-choline and desaturates in the $\omega 6$ and $\omega 3$ positions (i.e., from the methyl end).

9.4. Temperature Control in Desaturation

The control of desaturation in fatty acids is exerted at several different levels and at many enzyme sites. An apparently universal phenomenon appears to be the effect of the temperature of the tissue in which the lipid is situated. In poikilothermic organisms, for example, lipids tend to be more highly unsaturated, and thus to remain fluid, at lower temperatures. Moreover, in homeothermic animals, layers of fat exposed to lower temperatures tend to be more highly unsaturated (see Chapter 5). Despite the apparent universality of the phenomenon, however, its mechanism has remained obscure.

Fulco and his coworkers have addressed this question in the case of certain bacilli (Fulco, 1970; Fujii and Fulco, 1977). The site of desaturation appears to be important in temperature regulation in these bacteria. About half of the bacilli that were surveyed desaturated acids at the 8-, 9-, or 10-positions and appeared to do so with equal efficiency over a growth temperature range of 20–35°C. The remaining bacilli, such as *Bacillus subtilis* and *Bacillus megaterium*, desaturated fatty acids in the 5-position in a process that was under strict temperature control. Thus, in *B. megaterium* ATCC 14581, no unsaturated fatty acids were formed in cultures growing at temperatures above 30°C but, at temperatures below 30°C, the unsaturated fatty acid content of the membrane lipids increased as temperature decreased. In this organism the initial rate of the desaturation reaction was itself relatively insensitive to temperature changes. All of the significant temperature effects on desaturation could be ascribed either to changes in the stability of the desaturase or to alterations in the rate of synthesis of the desaturase. One control process directly responsive to temperature was the rate of inactivation of the desaturase enzyme. This, *in vivo*, followed strict first-order kinetics at all temperatures, and the enzyme half-life was determined solely and instantaneously by incubation temperature: 28 min at 20°C but less than 10 min at 35°C.

A second and much more complex control system mediated by temperature was that of desaturase "hyperinduction" and repression in cultures that were shifted from 30–35°C to below about 28°C. Cultures growing near 35°C contained neither unsaturated fatty acids nor desaturase enzyme. Within 5–10 min after a shift down to 20°C, however, desaturase synthesis began and continued at a high level for about one hour and then rapidly decelerated. This hyperinduction process (so called because the levels of desaturase enzyme initially formed in the cultures grown at lower temperatures far exceeded the levels found in cultures growing from inoculum at 20°C) was completely blocked by protein or RNA synthesis inhibitors added before or at the time of culture transfer from 35°C to 20°C. It thus appeared that these processes provided the bacterium with a way of rapidly increasing the unsaturated fatty acid content of its existing membrane lipids in response to a sudden shift down in temperature but that, once the adjustment in "old membrane" liquidity was complete, hyperinduction shut down and desaturase activity returned to a lower, steady-state level commensurate with new membrane lipid synthesis at the reduced temperature of growth. It was eventually determined that cultures of *B. megaterium* growing at 35°C lacked the messenger RNA responsible for desaturase synthesis and that, before desaturase synthesis could occur after a shift down to 20°C, the synthesis of specific messenger RNA (produced only at temperatures below 30°C) was required. Hyperinduction reflected unmodulated desaturase synthesis; the rapid decrease in desaturase synthesis that occurred about one hour after culture transfer from 35°C to 20°C, as well as the relatively low rate of desaturase synthesis in cultures growing from inoculum at 20°C, was explained by the action of a modulator protein which acted at the transcription level to repress desaturase synthesis. This modulator was absent in cultures growing at 35°C but was produced after transfer to temperatures below 30°C.

Although most species of bacilli appear to contain only one type of desaturase enzyme and thus produce monounsaturated fatty acids with double bonds either in the 5-position or in the 8-, 9-, or 10-positions, one species (*Bacillus licheniformis* ATCC 9259) contains both a temperature-insensitive $\Delta 10$-desaturase and a temperature-regulated $\Delta 5$-desaturase. During growth at 35°C, this bacillus produces only $\Delta 10$-monounsaturated fatty acids. When grown at 20°C, however, not only are $\Delta 10$- and $\Delta 5$-monounsaturated fatty acids synthesized but also a unique polyunsaturated fatty acid, 5,10-hexadecadienoic.

It appears that these tiny organisms, whose temperatures must closely follow changes in that of their environment, require a means for rapid formation of more fluid fatty acids for incorporation into membrane lipids. Failure to respond quickly to environmental temperature changes might well compromise the survival of the organism. Whether the mechanism found in bacilli is related in any way to those in higher organisms remains to be thoroughly investigated (there is some indication for existence of a similar type of regulation in crayfish).

9.5. The Formation of Polyunsaturated Fatty Acids in Animals

Present knowledge of the formation and metabolism of the polyunsaturated fatty acids originated largely from investigation of the "essential" members of the class. It seems difficult now to understand why the discovery that fat is necessary in the diet came so slowly. However, a number of highly respected scientists carried out carefully controlled experiments showing that rats could thrive on a diet that was believed to be free of fat. These early studies were apparently misinterpreted because it was not known that starch, used as a source of carbohydrate, clings tenaciously to traces of fatty acids. Thus it was not until 1929, when Burr and Burr (1930) and Evans and Burr (1927) substituted sucrose for starch in the diet of weanling rats, that a deficiency disease was discovered. As noted in these early experiments, the first deficiency symptoms were scaly paws and tails (leading eventually to caudal necrosis), growth retardation, severe dermatitis, fatty livers and kidney damage, high water consumption, and shortened longevity.

Additional symptoms noted by later workers included hair loss, reduced fertility, elevated respiratory quotient (reflecting conversion of carbohydrate to fat), high metabolic rate, and high skin permeability to water (in both directions). The water and heat loss probably resulted from this heightened permeability and led Thomasson to develop a more rapid means of inducing deficiency symptoms (and a more convenient assay for essential fatty acid (EFA) activity) by limiting water consumption (Thomasson, 1953).

Production of EFA deficiency is usually carried out by starting weanling rats on an artificial diet either fat-free or containing only fully saturated fat. It is difficult to produce deficiency symptoms in the adult rat, presumably because stored EFA meet the needs of the adult for protracted periods. However, if the adult rats are starved to approximately half their original weight and then placed on the deficient diet, symptoms of the deficiency appear. A chronic deficiency can be produced if animals are started on the deficient diet shortly after being weaned. In this condition no overt symptoms appear unless injury or other traumas induce resumption of rapid (albeit localized) growth. This deficiency disease appears to have been produced in all species investigated, including man. That the EFA requirement is associated with rapid growth has been confirmed in a number of studies. For example, Smedley-MacLean and Nunn (1941) carried out experiments on tumor transplantation in rats on a fat-free diet in which tumor growth was not reduced, presumably because the rats were not actually deficient in EFA. However, a transfer of EFA from other tissues to the tumor was shown. This ability of the tumor to cannibalize EFA was shown even more dramatically in fat-deficient but nonsymptomatic mice, who died of deficiency symptoms soon after tumor implantation. Some reduction in tumor growth rate was also noted. Thus, the tumor appears to have the ability to mobilize fat in order to obtain the EFAs that are present as components of the mixture of mobilized fatty acids. In fact, a substance causing extensive fat mobilization, probably a peptide, has been found in the blood of tumor-bearing animals and of human patients with cancer, as well as in extracts of tumors (Kitada et al., 1980).

The fatty acids necessary to prevent the deficiency symptoms proved to be linoleic acid (cis, cis-9,12-18:2), of plant origin, and arachidonic acid (all cis-5,8,11,14-20:4), formed in animals consuming linoleic acid. Two other acids that are related to linoleic in that they have a similar methyl-terminal ($\omega 6$) chain structure are also effective: all cis-8,11,14-20:3 and all cis-6,9,12-18:3 (γ-linolenic acid). α-Linolenic acid, all cis-9,12,15-18:3, present in certain seed oils, seems to be effective for growth promotion but does not prevent the skin symptoms. Thus, the fatty acids related structurally (at the methyl end) to linoleic acid, the $\omega 6$ family, appear to be the true EFA.

$$\overset{\omega 6}{CH_3-(CH_2)_4-}\overset{12}{CH=CH-CH_2-CH=}\overset{9}{CH-(CH_2)_7-COOH} \quad \text{Linoleic acid}$$

It has been found that this fatty acid is required as about 1–2% of the caloric intake of the diet of young animals. Many experiments have shown that linoleic and linolenic acids are not synthesized in the animal body, which seems to lack a type of desaturase, apparently present in all plants, that is highly specific with respect to structure of the methyl end of the chain.

Early determinations of the polyunsaturated fatty acids were carried out by the alkaline isomerization technique, which assays the number of double bonds in the fatty acid molecule but not its chain-length. By this method, metabolic studies were conducted in which it was found that feeding linoleic acid (18:2) to fat-deficient rats resulted in production of increased amounts of tetraene, that feeding 18:3 resulted in increased hexaene, and that continued subsistence on the fat-deficient diet led to an increase in triene in the tissues.

The nature of these transformations was revealed by a series of experiments with ^{14}C-labeled precursors and suspected intermediates (Mead, 1980). The conversion of diene to tetraene was found to proceed as shown in Fig. 9.2.

It can readily be seen why members of this series serve as EFA if arachidonic acid is indeed the ultimate essential substance. Arachidonic acid itself is subject to further elongation and desaturation to some extent, yielding successively 22:4 and 22:5. α-Linolenic acid was found to follow a similar pathway, leading ultimately to 22:6 and to other members of the $\omega 3$ family. The triene accumulating in fat deficiency proved to be 5,8,11-20:3, a member of the $\omega 9$ or oleic acid family, together with a small amount of 20:3$\omega 7$, derived from palmitoleic acid. The shorthand means of depicting the first three pathways is shown in Fig. 9.3.

A reasonable explanation for the accumulation of eicosatrienoic acid (20:3$\omega 9$) in the fat-deficient animal was forthcoming when it was noted that, with linoleic acid in barely adequate supply in the diet, ingestion of large amounts of oleic acid precipitates deficiency symptoms, suggesting involvement of a com-

linoleic acid $CH_3(CH_2)_4-CH=CH-CH_2-CH=CH-(CH_2)_7-COOH$

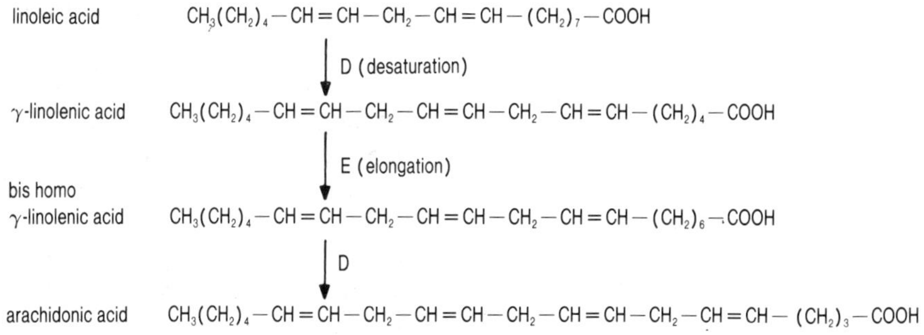

↓ D (desaturation)

γ-linolenic acid $CH_3(CH_2)_4-CH=CH-CH_2-CH=CH-CH_2-CH=CH-(CH_2)_4-COOH$

↓ E (elongation)

bis homo
γ-linolenic acid $CH_3(CH_2)_4-CH=CH-CH_2-CH=CH-CH_2-CH=CH-(CH_2)_6-COOH$

↓ D

arachidonic acid $CH_3(CH_2)_4-CH=CH-CH_2-CH=CH-CH_2-CH=CH-CH_2-CH=CH-(CH_2)_3-COOH$

FIGURE 9.2. Transformations of linoleic acid.

petition between members of the different families for one or more of the enzymes. This has since been confirmed in general, with the ω3 family taking precedence over the ω6 family, which has, in turn, much greater enzyme affinity than the ω9 family (see inhibition arrows in Fig. 9.3 and further discussion below). In the scarcity or absence of dietary linoleic (ω6) or α-linolenic (ω3) fatty acids, oleic (ω9) acid, ordinarily a weak competitor, is converted to the eicosatrienoic acid to an appreciable extent.

The enzymatic processes responsible for both the elongation and desaturation were found to reside in the microsomal fraction of liver homogenates. Nugteren (1962) and Stoffel (1971) independently studied systems including microsomes, NADPH, and an acyl-CoA. In the absence of oxygen and presence of

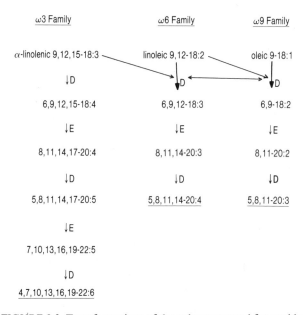

FIGURE 9.3. Transformations of the polyunsaturated fatty acids.

malonyl-CoA, elongation takes place (see Chapter 8). On the other hand, in the presence of oxygen, the desaturation reactions occur. Thus all the interconversions of the polyunsaturated fatty acids in the animal body take place in the endoplasmic reticulum of the cell, with essentially the same cofactors. For example, 18:2ω6 can yield 18:3ω6 aerobically and then 20:3ω6 anaerobically. Although Nugteren (1962) found the 18:2ω6 → 18:3ω6 conversion, in agreement with the postulated preferred scheme given in Fig. 9.3, Stoffel (1971) observed conversion of 18:2ω6 to 20:2ω6, indicating existence of an alternate pathway. This alternative (9,12-18:2 → 11,14-20:2) was originally believed to be a step in the major route from linoleic to arachidonic acid, largely because, in *in vitro* systems, 20:2ω6 could be isolated, while 18:3ω6 was found in small amounts, if at all. Sprecher and his coworkers (Ullman and Sprecher, 1971) then showed, however, that in all families of polyunsaturated fatty acids, when the double bond nearest the carboxyl group is at the 11-position, the next desaturation occurs at position 5. In the case of 11,14-20:2, the desaturation product is not 8,11,14-20:3, an intermediate in arachidonate synthesis, but 5,11,14-20:3, which appears to be a metabolic dead end. Similarly, 11-20:1 is converted to 5,11-20:2 and 11,14,17-20:3 is converted to 5,11,14,17-20:4. It would therefore appear that the route from linoleic to arachidonic acid involves initial desaturation to 6,9,12-18:3 rather than elongation to 11,14-20:2 and that a Δ8-desaturase may not exist.

The competitive interrelationships of these families of unsaturated fatty acids have been investigated by Brenner and his coworkers (1969) who concluded that the major discrimination occurs at the desaturase level—particularly at the Δ6-desaturase, which controls the rate-limiting step. The affinities of this desaturase appear to be: 18:3ω3 > 18:2ω6 > 18:1ω9. An optimum relationship between chain-length and unsaturation for desaturase affinity appears to be found in 20:3ω6. Inhibition within families also appears to occur: Brenner found that 22:5ω6 and 22:6ω3 inhibit earlier desaturation steps in their own families, thus providing some feedback inhibition for these reactions—possibly both at the desaturase level and at the acyl transferase (to form phospholipids). Work in Holman's laboratory (Mohrhauer *et al.*, 1967) has revealed some inhibition of the elongation steps as well. Saturated fatty acids (which would, of course, be present *in vivo*) inhibit the 18:2ω6 → 20:2ω6 elongation more strongly than the 18:2ω6 → 18:3ω6 desaturation, in further accord (see above) with the conclusion that desaturation is the major first step in the linoleic → arachidonic acid transformation. It has also been reported that some of the *trans* fatty acids, such as occur in hydrogenated oils, are specific inhibitors of the Δ6-desaturase (Mahfouz *et al.*, 1980).

It can readily be surmised that any attempt to predict tissue fatty acid composition from a knowledge of dietary fatty acid content, coupled with an understanding of competitive inhibition among and within the families, would be fraught with difficulties. Nevertheless, Sprecher and his coworkers (Bernert and Sprecher, 1975) carried out an extensive series of experiments aimed at determining the rate of each step in the transformations of several polyunsaturated fatty acids. Rates of desaturation, elongation, and incorporation into phospholipids were considered, using microsomal systems prepared from fat-deficient rats. Average rates for a number of linoleate transformations, expressed as nanomoles of product formed during a 3-min incubation of the substrate fatty acid (150 nmol of carboxy-labeled precursor) with 5 mg of microsomal protein and the required cofactors, are shown in Fig. 9.4.

From these conversion rates, together with other information, certain observations made in the whole animal are understandable. For example, the relatively rapid conversion of 6,9,12–18:3 to 8,11,14–20:3 is in agreement with the former being a very minor constituent of tissue lipids. The common observation that arachidonic acid (5,8,11,14–20:4) is the major end product of this family requires some additional explanation, since its further elongation to 7,10,13,16–22:4 is (at least under these conditions) more rapid than its formation by desaturation of 8,11,14–20:3. It is known, however, from studies with whole animals that the major fate of 7,10,13,16–22:4 is "retroconversion" to arachidonate, rather than incorporation into lipids. At least one of these transformations is therefore known to be reversible, by mechanisms possibly involving non-microsomal enzymes and/or other cofactors. Similar considerations may serve to explain why, in other families as well, certain fatty acids (e.g., 20:3ω9, 20:3ω7, and 22:6ω3) do not undergo extensive further elongation or desaturation.

The polyunsaturated fatty acid composition of tissue lipids is, of course, the net result of complex interrelationships of a number of major factors: the composition of the dietary fatty acids; the rates of oxidation of the fatty acids before incorporation into the lipids; the rates of desaturation and elongation, as outlined above; the relative rates of incorporation into lipids as opposed to further alteration; the retroconversion of the longest and most highly unsaturated members of the families; and competition among and within families for desaturation and elongation steps, as well as incorporation. Only a few of the many details governing the operation of this complex system are as yet understood, but these already serve to explain certain variations in composition revealed by analysis of tissue lipids.

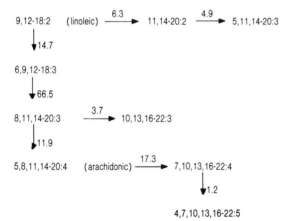

FIGURE 9.4. Rates of desaturation and elongation steps in *in vitro* conversion of linoleic to arachidonic acid (based on Bernert and Sprecher, 1975).

It would appear from what has been said so far that the structural requirements for EFA activity are somewhat rigid. However, Schlenk and Sand (1967) have shown that the family of homologous odd-chain fatty acids starting with 9,12-17:2 and leading to 5,8,11,14-19:4 have EFA activity. The 21:4 of the $\omega 7$ family (5,8,11,14-21:4) is also active. These fatty acids are found in fairly high amounts in mullet oil, although it is difficult to determine the nature of the ingested precursors of these substances.

In so complex an organism as the whole animal, attempts to determine the mechanism of action of the EFA have produced a confusing picture. It seems probable that they are involved in many processes, directly or indirectly, and a single initial site of action would appear to be difficult to determine.

Certainly, EFAs are involved in various transport phenomena; for example, in their absence a fatty liver develops. Without going into the detailed reasoning, we can logically suppose that this difficulty may reflect inadequate formation of the proper phospholipids needed for incorporation into lipoproteins necessary for transport of fat. Cholesterol also accumulates in the liver during this deficiency and it seems probable that transport of this substance in the form of fatty acyl esters is facilitated if the esters are polyunsaturated and thus more fluid. It is also probable that for the lipids of the transporting lipoproteins in general, the polyunsaturated fatty acyl groups aid in the maintenance of more fluid and thus more stable structures.

A second and probably related function is in the formation and maintenance of cellular membranes. This function is strongly suggested by many features of the nature of the EFA deficiency disease. In particular, the skin symptoms, such as dermatitis and water permeability, and the breakdown of liver mitochondrial function, as seen in decreased oxidative phosphorylation, point to a membrane problem. In simple organisms such as the mycoplasma, lack of unsaturated fatty acids in general results in a tipping of the critical transition temperature toward gel formation at usual temperatures in the cell membrane lipid bilayer and thence to membrane breakdown and cell death.

In some cells in culture, the unavailability of linoleic, or particularly arachidonic, acid results in a cessation of growth that is quickly restored when these fatty acids are made available.

The viscotropic properties of the membrane lipid bilayer are crucial in maintaining the osmotic properties of the membrane. They are also important in determining the conformation and thus the enzymatic and transport functions of the transmembrane proteins and in the regulation of cell surface receptors. For these functions, the unsaturated and polyunsaturated fatty acids and the ability of the cell to synthesize and incorporate them are of primary importance.

However, there is no evidence that the EFA are solely involved in these functions. As a matter of fact, the ability to "fluidize" a viscous membrane appears to be a function of the structure of the membrane phospholipid as well as of the melting point of the constituent fatty acids. For this fluidizing function there is no reason to believe that the $\omega 6$ or $\omega 3$ fatty acids have any special properties stemming from the location of their unsaturated centers. Therefore, these functions do not seem to be related to specific properties of the EFA but to those of unsaturated fatty acids in general.

An exceedingly important function of some of the EFA—20:4ω6, 20:3ω6, and 20:5ω3—is to serve as precursors to the prostaglandins (PG) and related substances. This subject is treated in detail in Chapter 10. However, there is good reason to believe that the lack of PG is not related to most of the symptoms usually associated with EFA deficiency. Administration of PG by various routes has not been notably successful in curing all symptoms of EFA deficiency, and even if this lack of effect is attributed to the short biological half-lives of the PG, there are other reasons to discount this function. First, the symptoms of EFA deficiency are not those usually associated with PG action. Second, the finding that columbinic acid ($t5,c9,c12$-18:3) acts as an efficient EFA but is not transformed into a PG and, indeed, even inhibits PG endoperoxide synthetase, the enzyme primarily concerned with PG formation, points to another function, possibly the intimate association of a membrane lipid containing an EFA with an ion-channel protein involved in water and electrolyte transport.

The mechanism of control of desaturation remains imperfectly understood. The work of Fulco and his coworkers on effects of ambient temperature on these processes in bacilli was considered in Section 4; there is also some evidence that similar, though not identical, regulation holds in higher invertebrates such as the crayfish.

In other cases it has been shown that desaturase activity responds to membrane fluidity regardless of temperature. This would seem to be a very logical means of control, since decreased fluidity (e.g., at lower temperature) could activate desaturases and result in increased proportions of unsaturated fatty acids and thus in increased fluidity. Increased fluidity should conversely result in suppression of desaturation and ultimately in restoration of "normal" viscosity to the membrane. There is, however, no general agreement that this is actually the means by which membrane fluidity is controlled, and it seems probable that a number of mechanisms may accomplish this end.

9.6. Hydrogenation

Finally, there is the reaction resulting in the opposite transformation of fatty acids—biohydrogenation.

Although there has been some study of this process in higher animals, there is no good evidence that they possess this capability. On the other hand, certain microorganisms, such as, notably, the rumen bacteria, have been known for a long time to be capable of hydrogenation reactions. In particular, Tove and his coworkers (Anonymous, 1980) showed that the mechanism of the reactions, as studied in *Butyrivibrio fibrisolvens,* involves isomerization of the *cis*-12 double bond in linoleate to *trans*-11, followed by hydrogenation of the resulting $c9,t11$-18:2 to $t11$-18:1. One hydrogen is derived from α-tocopherolquinol, with a proton from water furnishing the second, while the immediate electron donor may be the quinol, as illustrated in Eq. 9.6 (Rosenfield and Tove, 1971).

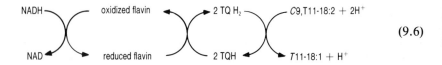

(9.6)

Control of this reaction by temperature or fluidity was also seen. With *Bacillus cereus* an increase in temperature from 20°C to 37°C results in induction of oleate hydrogenation and thus the formation of more saturated membrane lipids.

REFERENCES

Anonymous, 1980, Role of α-tocopherolquinol in biohydrogenation, *Nutr. Rev.* **38**:284.

Bernert, J. T., Jr., and Sprecher, H., 1975, Studies to determine the role rates of chain elongation and desaturation play in regulating the unsaturated fatty acid composition of rat liver lipids, *Biochim. Biophys. Acta* **298**:254.

Bloch, K., 1969, Enzymatic synthesis of monounsaturated fatty acids, *Accounts Chem. Res.* **2**:193.

Bloomfield, D. K., and Bloch, K., 1960, The formation of Δ9-unsaturated fatty acids, *J. Biol. Chem.* **235**:337.

Brenner, R. R., Peluffo, R. O., Nervi, A. M., and DeTomas, M. E., 1969, Competitive effect of α- and γ-linolenyl-CoA and arachidonyl-CoA in linoleyl-CoA desaturation to γ-linolenyl-CoA, *Biochim. Biophys. Acta* **176**:420.

Burr, G. O., and Burr, M. M., 1930, On the nature and role of the fatty acids essential in nutrition, *J. Biol. Chem.* **86**:587.

Evans, H. M., and Burr, G. O., 1927, A new dietary deficiency with highly purified diets, *Proc. Soc. Exp. Biol. Med.* **24**:740.

Fujii, D., and Fulco, A. J., 1977, Biosynthesis of unsaturated fatty acids by bacilli. Hyperinduction and modulation of desaturase synthesis, *J. Biol. Chem.* **252**:3660.

Fulco, A. J., 1970, The biosynthesis of unsaturated fatty acids by bacilli. II. Temperature-dependent biosynthesis of polyunsaturated fatty acids, *J. Biol. Chem.* **245**:2985.

Gurr, M. I., 1971, The biosynthesis of polyunsaturated fatty acids in plants, *Lipids* **6**:266.

Kitada, S., Hays, E. F., and Mead, J. F., 1980, A lipid mobilizing factor in serum of tumor-bearing mice, *Lipids* **15**:168.

Mahfouz, M. M., Johnson, S., and Holman, R. T., 1980, The effect of isomeric *trans*-18:1 acids on the desaturation of palmitic, linoleic and eicosa-8,11,14-trienoic acids by rat liver microsomes, *Lipids* **15**:100.

Mead, J. F., 1980, Nutrients with special functions: essential fatty acids, in: *Human Nutrition: A Comprehensive Treatise*, Vol. 3A (R. B. Alfin-Slater and D. Kritchevsky, eds.), Plenum Press, New York, pp. 213–238.

Mohrhauer, D., Christiansen, K., Gan, M. V., Deubig, M., and Holman, R. T., 1967, Chain elongation of linoleic acid and its inhibition by other fatty acids *in vitro*, *J. Biol. Chem.* **242**:4507.

Morris, L. J., 1970, Mechanisms and stereochemistry in fatty acid metabolism, *Biochem. J.* **118**:681.

Nugteren, D. H., 1962, Conversion *in vitro* of linoleic acid into γ-linolenic acid by rat liver enzymes, *Biochim. Biophys. Acta* **60**:656.

Oshino, N. Y., and Sato, R., 1966, Electron-transfer mechanism associated with fatty acid desaturation catalyzed by liver microsomes, *Biochim. Biophys. Acta* **128**:13.

Pugh, E. L., and Kates, M., 1979, Membrane-bound phospholipid desaturases, *Lipids* **14**:159.

Rosenfield, I. S., and Tove, S. B., 1971, Biohydrogenation of unsaturated fatty acids. VI. Source of hydrogen and stereospecificity of reduction, *J. Biol. Chem.* **246:**5025.

Schlenk, H., and Sand, D. M., 1967, A new group of essential fatty acids and their comparison with other polyenoic fatty acids, *Biochim. Biophys. Acta* **144:**305.

Schroepfer, G. J., Jr., and Bloch, K., 1965, The stereospecific conversion of stearic acid to oleic acid, *J. Biol. Chem.* **240:**54.

Smedley-MacLean, I., and Nunn, L. C. A., 1941, The relation of essential fatty acids to tumour formation in the albino rat, *Biochem. J.* **35:**983.

Stoffel, W., 1971, Enzymatische Untersuchungen zum Stoffwechsel der Polyenfettsäuren, *Wiss. Veröff. Deut. Ges. Ernähr.* **22:**12.

Thomasson, H. J., 1953, Biological standardization of essential fatty acids (a new method), *Internat. Rev. Vitamin-Res.* **25:**62.

Ullman, D., and Sprecher, H., 1971, An *in vitro* and *in vivo* study of the conversion of eicosa-11,14-dienoic acid to eicosa-5,11,14-trienoic acid and of the conversion of eicosa-11-enoic acid to eicosa-5,11-dienoic acid in the rat, *Biochim. Biophys. Acta* **248:**186.

SELECT BIBLIOGRAPHY

Holman, R. T. (ed.), 1971, *Progress in the Chemistry of Fats and Other Lipids,* Vol. 9, Pergamon Press, Oxford.

Mead, J. F., and Fulco, A. J., 1976, *The Unsaturated and Polyunsaturated Fatty Acids in Health and Disease,* Charles C. Thomas, Springfield, Ill.

10
PROSTAGLANDINS, THROMBOXANES, AND PROSTACYCLIN

One of the roles of three polyunsaturated fatty acids biosynthesized from linoleic and linolenic acids—all-*cis*-eicosa-8,11,14-trienoic, all-*cis*-eicosa-5,8,11,14-tetraenoic (arachidonic), and all-*cis*-eicosa-5,8,11,14,17-pentaenoic—is to act as precursors of prostaglandins and related substances, generally referred to as prostanoids. There is hardly an organ in the body that does not synthesize prostaglandins and whose function may not be modified by them. Prostaglandins are best looked upon as "local" hormones synthesized in many organs and cells in response to specific stimuli. They exert their effects mostly at the site of their synthesis, unlike the hormones produced by endocrine glands which are secreted into the blood and exert their effects in parts of the body distant from the site of synthesis. In this respect, the prostaglandins compare best with cyclic 3′,5′-adenosine monophosphate (cAMP), which is also a "local" regulator formed at specific loci as a second messenger in response to general hormones. In fact, many of the actions of the prostanoids are intimately linked to the regulation of levels of cAMP in cells. The many-sided effects of prostaglandins have been compared to those of the dollar which is used at many places by many people to buy a variety of goods.

Although our current knowledge about the structure, origin, action, and metabolism of prostaglandins dates back to 1960, the discovery of prostaglandins has a venerable background. Kurzrok and Lieb (1930) were the first to report that human seminal fluid either relaxed or contracted strips of human uterine muscle. Interestingly, they found that the uterine muscle from women who had had successful pregnancies responded with relaxation, but preparations from the womb of sterile women responded to the seminal fluid with contraction. Then M. W. Goldblatt (1933; 1935) reported that seminal fluid contained a substance, or substances, which lowered blood pressure and stimulated the contractions of uterine and intestinal smooth muscle. Von Euler (1934; 1935a,b) made the independent discovery of similar biological effects of an acidic component of lipid extracts of seminal fluid and seminal vesicles of sheep. Von Euler coined the term "prostaglandin" for the active principle of the extracts in the belief that the substance was synthesized in the prostate gland and was only stored in the seminal vesicle.

Although the term is a misnomer, it nevertheless became generally accepted even after Eliasson (1959) established the seminal vesicle as the primary source of these substances. The term prostaglandin is now commonly abbreviated to PG. It may be recorded as a curiosity that von Euler (1936) also coined the term "vesiglandin" to denote the biologically active principle in extracts of the semen and seminal vesicles of the rhesus monkey. Although such extracts had physicochemical properties similar to those of extracts of human semen and sheep seminal vesicles, and caused a sharp drop in the blood pressure of atropinized rabbits, they had only a weak action on the contraction of intestinal smooth muscle. Many years later, von Euler and Eliasson (1967, p. 143) suggested that the active substance in the extracts from the monkey may be related to prostaglandin A_1 which has a strong vasodepressor activity, but only about one-hundredth of the smooth-muscle-stimulating activity of prostaglandin E_1, which is one of the main constituents in extracts of sheep seminal vesicles.

Although much physiological work has been done in the years after the primary discovery of prostaglandins, and extracts of many tissues have been shown to have biological properties similar to those of extracts of the sheep seminal vesicles, no real progress was made in the study of prostaglandins until after 1960 when Bergström and Sjövall (1960a,b) reported the isolation in crystalline form of two structurally related prostaglandins, now known as PGE_1 and $PGF_{1\alpha}$. PGE_1 is a potent vasodepressor and smooth-muscle stimulant, whereas $PGF_{1\alpha}$ has only smooth-muscle-stimulating activity. The study of the biosynthesis of prostaglandins led to the discovery during the 1970s by Samuelsson and his colleagues that a precursor of prostaglandins, derived from arachidonic acid and known now as PGH_2, gives rise in blood platelets to a powerful platelet-aggregating and vasoconstricting compound, named thromboxane A_2 (TXA_2). The same precursor, PGH_2, was shown by Vane, Moncada, Needleman, and others to give rise in vascular endothelium and in the heart to a substance that counteracts the effects of TXA_2: it is a platelet-antiaggregating factor and a vasodilator and was named prostacyclin, or prostaglandin I_2 (PGI_2).

The early history of prostaglandin research has been admirably presented in a monograph by von Euler and Eliasson (1967). There have been very many reviews, chapters, and monographs written about prostaglandins; several of these will be referred to at appropriate places in the text. New journals devoted to prostaglandin research sprang up also: *Prostaglandins; Prostaglandins and Medicine; Prostaglandins and Related Lipids; Prostaglandins and the Gut; Prostaglandins, Leukotrienes and Medicine.* Also several volumes of *Advances in Prostaglandin and Thromboxane Research,* under the general editorship of B. Samuelsson and R. Paoletti, have been published by Raven Press, New York, since 1976. The structural identification of the prostaglandins, found only in minute amounts in organs and body fluids, stimulated also a wealth of new chemical syntheses not only of the prostaglandins themselves, but also of many analogues in attempts to produce inhibitors and substances of longer action than that of the natural compounds. Description of these syntheses is beyond the scope of this book; for surveys of the varied synthetic approaches the reader is referred to chapters by Garcia *et al.* (1977a,b) and to the monograph edited by Roberts and Newton (1982).

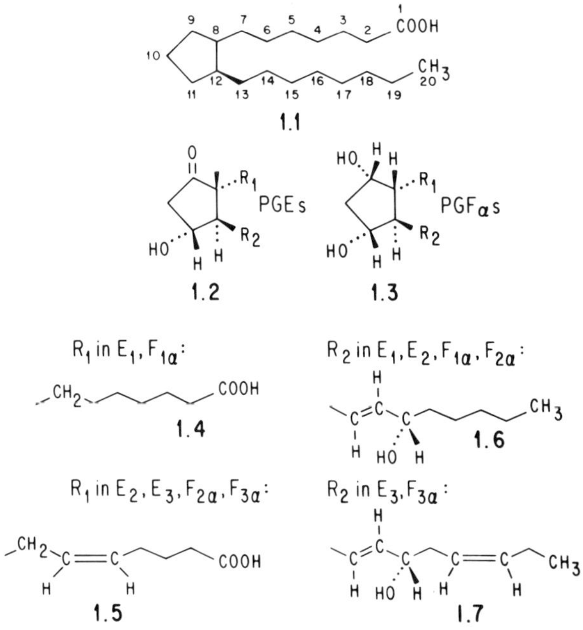

PROSTAGLANDINS, THROMBOXANES, AND PROSTACYCLIN 151

FIGURE 10.1. Reference prostanoic acid (**1.1**) and components of prostaglandins E and F.

Because of the thousands of publications on prostaglandins, only a few of the original articles on the biochemistry of prostaglandins can be cited here. In Sections 10.1–10.3 and 10.4 the chemistry and biochemistry of prostaglandins are described; Section 10.5 deals with thromboxanes and prostacyclin.

10.1. Chemistry of Prostaglandins

10.1.1. Structure of Prostaglandins

Bergström and his colleagues identified two groups of prostaglandins (PGs), the E group and F group; they are referred to as PGE and PGF. Both groups can be looked upon as modified forms of a C_{20} carboxylic acid which contains a cyclopentane ring and has been named prostanoic acid (**1.1** in Fig. 10.1).* Prostanoic acid, which does not occur naturally, is a disubstituted cyclopentane, one of the substituents being *n*-hexylcarboxylic acid (C_7) and the other an *n*-octyl side-chain.

There are three PGEs (PGE_{1-3}) and three PGFs ($PGF_{1\alpha-3\alpha}$). The subscript denotes the total number of double bonds in the two side-chains. In the E-series

---

*In several of the figures the formulas are numbered; thus formula **1.1** means formula 1 in Fig. 10.1, or **23.6** means formula 6 in Fig. 10.23. These numbered formulas are referred to in the text, in boldface type, to avoid reference to the compounds by their long systemic names.

the cyclopentane structure is replaced by cyclopentan-11-ol-9-one (**1.2**); in the F-series the ring ketone is reduced to a hydroxyl group (**1.3**), the two hydroxyl groups being in the syn (α) configuration. Thus, the PGFs are substituted pentanediols.

In both the E and F series the R_1 substituent on the ring is either of two groups. In E_1 and $F_{1\alpha}$, R_1 is *n*-hexylcarboxylic acid (**1.4**) as in prostanoic acid. In E_2, E_3, $F_{2\alpha}$, and $F_{3\alpha}$, R_1 is *cis*-hex-4-enylcarboxylic acid (**1.5**), i.e., the double bond in these PGs is between positions 5 and 6 according to the numbering of the positions in the PGs. There are also two forms of R_2, whether in the E or F series. The R_2 side-chain in E_1, E_2, $F_{1\alpha}$, and $F_{2\alpha}$ is identical: it is the 3-hydroxy-*trans*-oct-1-enyl residue (**1.6**), i.e., by the numbering of the positions in PGs the *trans* double bond is between positions 13 and 14 and the hydroxyl group is at C-15. In PGE_3 and $PGF_{3\alpha}$ the R_2 side-chain contains, in addition to the *trans* double bond, also a *cis* double bond in the ω3-position (**1.7**). The full formulas of the six primary prostaglandins are shown in Fig. 10.2.

The PGEs are readily convertible into PGFs by reduction of the 9-keto function with $NaBH_4$. The chemical reduction, lacking stereospecificity, gives two isomers: one identical with the natural F-series, known as the α-series, and the other in which the hydroxyl groups at positions 9 and 11 are in an anti or *trans* configuration. The latter is designated as the F_β series. Reduction of PGEs results predominantly in the formation of PGF_βs.

The PGEs have an interesting chemical property, which was successfully exploited in the identification of their chemical structure. Acetylation of PGEs with acetic anhydride–pyridine yields the 15-acetoxy derivative and results at the same time in the dehydration of the ring with the introduction of a double bond between positions 10 and 11. All PGEs are similarly dehydrated by weak acids or weak bases. The resulting compounds are known as PGAs and characteristically

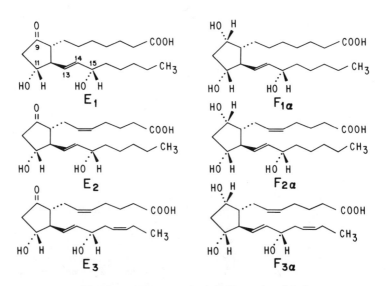

FIGURE 10.2. Structures of six primary prostaglandins.

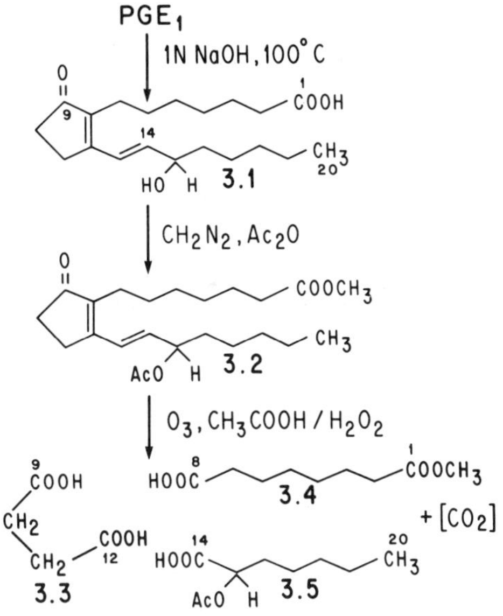

FIGURE 10.3. Degradation of prostaglandin B_1 methyl ester.

absorb UV light at 217–220 nm which is characteristic for substituted cyclopentenones. PGA_1 and PGA_2 are also known as naturally occurring metabolites of PGE_1 and PGE_2. Strong base (0.5–1.0 N NaOH at 100°C) causes not only a dehydration on the ring but also isomerization of the 10,11-double bond to the 8,12 position; the products are known as PGBs and, owing to the conjugated 8(12), 13-dien-9-one structure, have a characteristic UV absorption maximum at 278 nm (see structure **3.1** of Fig. 10.3). Both these dehydrations are clearly directed by the carbonyl function of C-9 and not by the homoallylic group at C-13, as they occur also in the catalytically reduced forms of PGEs, i.e., when both side-chains (R_1 and R_2) are saturated. They are not observed in the F-series which carries a hydroxyl instead of a carbonyl function at C-9. PGB_1 and PGB_2 occur naturally in human seminal fluid.

The systemic nomenclature of the prostaglandins is based on prostanoic acid as the reference compound. Thus PGE_1 is 11,15-dihydroxy-9-keto-13(*trans*)-prostenoic acid; PGE_2 is 11,15-dihydroxy-9-keto-5(*cis*)-13(*trans*)-prostadienoic acid, and PGE_3 is 11,15-dihydroxy-9-keto-5,17(*cis*)-13(*trans*)-prostatrienoic acid. The PGFs are similarly described as the 9,11,15-trihydroxyprostenoic acids, with the positions and geometry of the double bonds indicated exactly as in the corresponding PGEs, e.g., $PGF_{2\alpha}$ is 9,11,15-trihydroxy-5(*cis*)-13(*trans*)-prostadienoic acid.

10.1.2. Structural Identification of Prostaglandins

The structure of prostaglandins was deduced mostly by Bergström and his colleagues by a combination of spectroscopic techniques—UV, IR, NMR, and mass spectrometry—and chemical degradations and identification of the fragments by gas–liquid chromatography and mass spectrometry. Here we summarize, as examples, the identifications of the structures of PGE_1 and $PGF_{1\alpha}$, the first ones to be determined (Bergström et al., 1962, 1963). Samuelsson (1970) described in much detail the structural identification of all prostaglandins.

Physical methods showed PGE_1 to be a C_{20} carboxylic acid containing a cyclopentanone ring, two hydroxyl groups, and a *trans* double bond, and to have an elemental composition of $C_{20}H_{34}O_5$, with a molecular weight of 354 readily determined by mass spectrometry.* $PGF_{1\alpha}$ gave, on the other hand, a molecular weight of 356, commensurate with an elemental composition of $C_{20}H_{36}O_5$.

The main chemical features of PGE_1 and $PGF_{1\alpha}$ were deduced from the following observations. Both PGE_1 and $PGF_{1\alpha}$ absorbed 1 mol of H_2 when hydrogenated in the presence of PtO_2 in ethanol. At the same time the infrared absorptions of both compounds at 10.3 μm (971 cm^{-1}) disappeared, thus establishing the presence of one olefinic *trans* double bond in each. The existence of a cyclopentanone structure was similarly inferred from absorption in the infrared at 5.77 μm (1733 cm^{-1}). The correlation between PGE_1 and $PGF_{1\alpha}$ was deduced from the formation of the latter after reduction of PGE_1 with $NaBH_4$. The reduction of the ring carbonyl function produced not only the natural $PGF_{1\alpha}$, but also the isomeric $PGF_{1\beta}$, as referred to earlier (cf. p. 152). Acylation of the [^{14}C]methyl esters of PGE_1 and $PGF_{1\alpha}$ with *p*-nitrobenzoylchloride–pyridine indicated (from the ratio of ^{14}C to the absorption of the *p*-nitrobenzoyl group at 257 nm) that PGE_1 contained two acylable hydroxyl groups and $PGF_{1\alpha}$ contained three, in accord with NMR data which showed PGE_1 to contain two and $PGF_{1\alpha}$ three carbinol protons.

Further evidence regarding the structure of PGE_1 came from the chemical degradations of PGB_1, generated from PGE_1, and from the degradation of 13,14-dihydro-PGB_1, generated from the catalytically reduced 13,14-dihydro-PGE_1. A third degradation was also carried out on 15-acetoxy-PGA_1 (Bergström et al., 1963). In the first degradation (see Fig. 10.3) PGB_1 (**3.1**) was converted to the methyl ester and acetylated with acetic anhydride in pyridine. The acetoxy methyl ester (**3.2**) was then ozonized and the ozonide cleaved with peroxyacetic acid. The degradation products were separated by gas–liquid chromatography and identified by mass spectrometry as succinic acid (**3.3**), 2-acetoxyheptanoic acid (**3.5**), and monomethyl suberic acid (**3.4**). The identification of succinic acid, which must have come from the ring, demonstrated that the ring contained two vicinal methylene groups and also that the three carbon atoms carrying the side-chains and keto group must have been adjacent to one another. The finding of monomethyl suberic acid showed, on the other hand, that the carboxyl-carrying side-chain was attached to the carbon atom α to the keto group. This analysis

---

*With a modern high-resolution mass spectrometer it could be determined on 1 μg of material, or less, that the precise molecular weight of PGE_1 is 354.240609 with an accuracy of ± 2 parts per million.

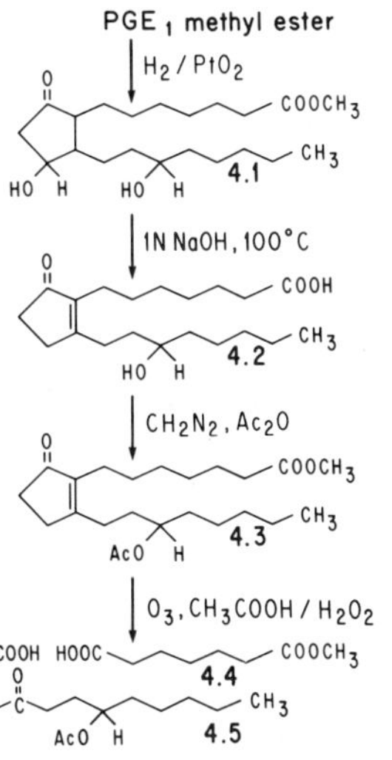

FIGURE 10.4. Degradation of 13,14-dihydroacetoxy-PGB$_1$ methyl ester, **4.3** (after Bergström et al., 1963).

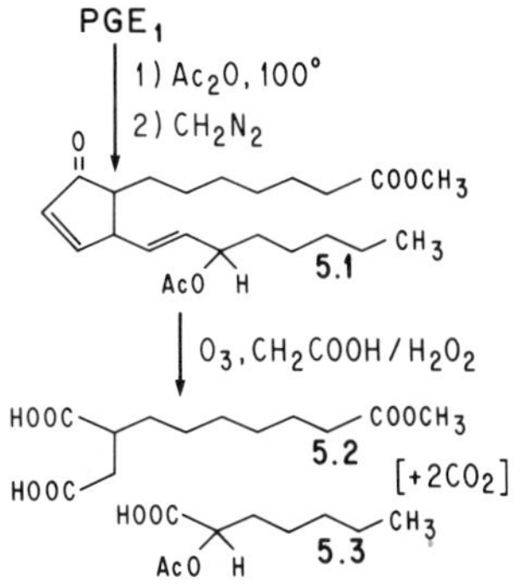

FIGURE 10.5. Degradation of acetoxy-PGA$_1$ methyl ester, **5.1** (after Bergström et al., 1963).

accounted for 19 of the 20 carbon atoms of PGB$_1$; the one missing carbon atom, probably lost as CO_2, must have been C-13.†

For the second degradation, PGE$_1$ methyl ester was hydrogenated in the presence of PtO$_2$ and then dehydrated and hydrolyzed with 1 N NaOH at 100°C to 13,14-dihydro-PGB$_1$ (Fig. 10.4). Bergström et al. (1963) originally called this product (structure **4.2**) PGE-237 on account of its UV absorption characteristics. The recovered free acid was methylated with CH_2N_2, acetylated with acetic anhydride, and the acetoxy methyl ester degraded by ozonolysis as before. This degradation gave two fragments: monomethyl suberate (**4.4**)—as did the degradation of acetoxy-PGB$_1$ methyl ester—and 4-keto-7-acetoxydodecanoic acid (**4.5**), and accounted for all 20 carbon atoms of PGE-237. The identification of monomethyl suberate emphasized once again that the carboxy side-chain of the dihydro-PGB$_1$ must have been attached to a carbon atom in a position α to the ring ketone. The larger fragment could be accounted for only by the assumption that the double bond in dihydro-PGB$_1$ was between positions 8 and 12.

For a third degradation (Fig. 10.5), the methyl ester of PGE$_1$ was acetylated with acetic anhydride at 100°C. The recovered and purified product had a UV absorption maximum at 220 nm characteristic for α,β-unsaturated cyclopenten-

---

†One of the primary products of the cleavage of the PGB$_1$ ozonide, from C-9 to C-13 inclusive, might have been α-ketoglutaric acid, which would have decarboxylated to succinic acid on being heated with the peroxyacetic acid.

ones.* The infrared spectrum of this derivative indicated the presence in the molecule of an ester carbonyl (5.77 μm, 1733 cm^{-1}), a conjugated carbonyl in a 5-membered ring (5.86 μm, 1706 cm^{-1}), a conjugate carbon-to-carbon double bond (6.30 μm, 1587 cm^{-1}), and an olefinic *trans* double bond (10.32 μm, 971 cm^{-1}). Ozonolysis of the methyl ester of this product gave two fragments, identified by mass spectrometry: 2-acetoxyheptanoic acid (**5.3**) and the monomethyl ester of octane-1,2,8-tricarboxylic acid (**5.2**). The latter was identified also by comparison with an authentic (*RS*)-octane-1,2,8-tricarboxylic acid sample.† In this degradation two carbon atoms, most probably C-10 and C-12, were lost. These observations established firmly the positions of the double bonds in acetoxy-PGA$_1$ methyl ester (**5.1**) between carbon atoms 10,11 and 13,14.

These three degradations, coupled with the spectroscopic data, established the structures of the side-chains, their attachments to the ring, and the position of the keto group in PGE$_1$. There remained the assignment of the position of the hydroxyl group in the five-membered ring. This assignment was arrived at indirectly. The position α to the keto group was excluded as neither PGE$_1$, PGF$_{1\alpha}$, nor PGF$_{1\beta}$ could be oxidized with periodate or lead tetraacetate. The tertiary carbon atoms (C-8, C-12) were also excluded as the carriers of the hydroxyl group by acylating experiments in which this hydroxyl group was shown to be readily acylable by various reagents, e.g., by *p*-nitro-, *p*-bromo-, and *p*-iodobenzoyl chlorides. The easy elimination of water by acid with the formation of the α,β-unsaturated cyclopentenone reinforced the conclusion that the ring hydroxyl was in a position β to the keto group.

The correlations between PGE$_1$, PGE$_2$, and PGE$_3$ were established by mass spectrometry which showed that PGE$_2$ had a m.w. 2 units less than that of PGE$_1$ and that PGE$_3$ had a m.w. 4 units less than that of PGE$_1$. Furthermore, selective hydrogenation of PGE$_3$ with Pd-catalyst gave PGE$_2$, and that of PGE$_2$ gave PGE$_1$. By full hydrogenation, all three compounds were reduced to the same product, identified as dihydro-PGE$_1$, 11-hydroxy-9-ketoprostanoic acid. The position of two of the double bonds in PGE$_3$ was established to be the same as in PGE$_2$ from mass spectral analysis and from chromic acid oxidation of the two compounds. Both PGE$_2$ and PGE$_3$ gave glutaric acid, which showed the presence of a double bond at position 5 in the carboxyl-carrying side-chain. The position of the third double bond in PGE$_3$ was deduced by NMR spectroscopy to be between carbon atoms 17 and 18 (Samuelsson, 1963).

The geometry of the additional double bonds in PGE$_2$ and PGE$_3$ was deduced indirectly. Änggård *et al.* (1965) and Änggård and Samuelsson (1965b) found that the soluble fraction of guinea-pig lung homogenates converted PGE$_2$ and PGE$_3$ into their 13,14-dihydro derivatives (13,14-dihydro-11,15-dihydroxy-9-keto-prost-5-enoic and 13,14-dihydro-11,15-dihydroxy-9-keto-prosta-5,17-dienoic acids). These enzymatically reduced products showed no infrared absorptions at

---

*Bergström *et al.* (1963) called this product PGE-220; it could be called 15-acetoxy-PGA$_1$ methyl ester, or methyl 15-acetoxy-9-keto-10,13-prostadienoate.

†From the knowledge of the stereochemistry of PGs it could now be inferred that the absolute configuration of the octane-tricarboxylic acid at C-2 must have been *R*.

10.3 μm (characteristic for *trans* olefins); hence the configuration of the double bonds at positions 5 and 17 was inferred to be *cis* (Z).

The absolute configuration of PGE$_1$ and, by inference, of PGF$_{1\alpha}$, was deduced from study by X-ray crystallography of the tri-*p*-bromobenzoate of the methyl ester of PGF$_{1\beta}$ (prepared by NaBH$_4$ reduction of PGE$_1$) and from the determination of the optical properties of 2-hydroxyheptanoic acid obtained by ozonolysis of PGE$_1$ and PGB$_1$. The X-ray crystallographic analysis showed that the two hydroxyl groups on the ring at C-9 and C-11 and also the two side-chains, attached at C-8 and C-12 in PGF$_{1\beta}$, were anti to one another, and also that the hydroxyl groups at C-11 and C-15 were on the same side of the molecule (Abrahamsson, 1963). By inference, the hydroxyl groups on the ring in PGF$_{1\alpha}$ must be syn to one another. The 2-hydroxyheptanoic acid, obtained either from the degradation of PGE$_1$ or PGB$_1$, had as the free acid in chloroform $[\alpha]_D^{20} = +6°$, but as the sodium salt in water had $[\alpha]_D^{20} = -13°$ (Nugteren *et al.*, 1966). These rotations are in the same sense as those of L-2-hydroxyhexanoic, L-2-hydroxyheptanoic, and L-2-hydroxyoctanoic acids. Thus, by the *RS*-convention the absolute configuration of the 2-hydroxyheptanoic acid derived from PGE$_1$ is *S*. From the crystallographic data, which established only the relative positions of the ligands, and from the optical measurements, the absolute configurations in PGE$_1$ and PGF$_{1\alpha}$ may be reconstructed as shown in Fig. 10.6.

In addition to the six primary prostaglandins, eight others have been found in human seminal fluid. These are: PGA$_1$, PGA$_2$, PGB$_1$, PGB$_2$, already referred to (pp. 152, 153) as derivable from the corresponding PGEs, and the 19-hydroxy derivatives of these four. The structures of the 19-hydroxy-PGs were deduced by methods similar to those which were used to determine the structures of other PGs: UV, IR, and mass spectrometry, and chemical degradation followed by identification of the fragments (cf. Samuelsson, 1970). The absolute configuration around C-19 in the hydroxylated compounds was deduced to be *R* by the degradation of 19-hydroxy-PGB$_1$. The methyl ester of 19-hydroxy-PGB$_1$ was acetylated to a mixture of the 15-acetoxy and 19-acetoxy derivatives. Oxidation of the latter with KMnO$_4$ in acetone gave, from the terminal four carbon atoms, D(−) or (*R*)-3-acetoxybutyric acid. The absolute configuration of 19-hydroxy-PGB$_1$ is shown in Fig. 10.6. It is not known whether the 19-hydroxy-PGBs arise by isomerization

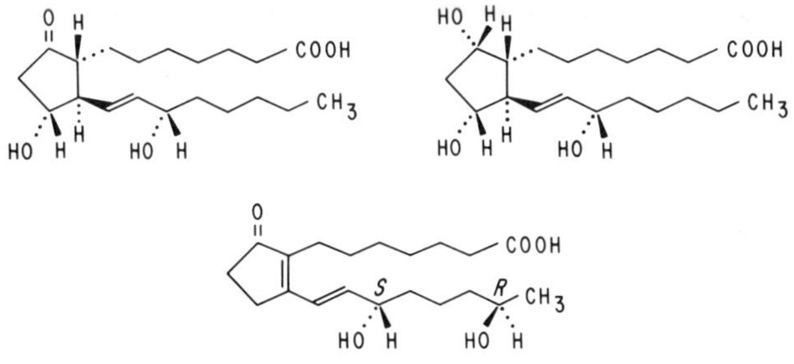

FIGURE 10.6. Absolute configurations of PGE$_1$ (top left), PGF$_{1\alpha}$ (top right), and of 19-hydroxy-PGB$_1$ (at bottom).

of 19-hydroxy-PGAs or whether the PGAs and PGBs can be hydroxylated at C-19 independently of one another.

10.1.3. *Prostaglandins of the Coral* Plexaura homomalla

Interesting structural analogues of prostaglandins were found in the marine soft coral, the gorgonian *Plexaura homomalla* (Esper) living in shallow waters off the coast of Florida and also in the Caribbean Sea. Weinheimer and Spraggins (1969) found that the air-dried cortex of this coral contains the 15-epi-(15R)-PGA$_2$ and its 15-acetoxy methyl ester (structure **7.1** of Fig. 10.7) to the extent of 0.2% and 1.3%, respectively. The structure of the (15R)-PGA$_2$ and its esters was established by methods similar to those outlined for the mammalian prostaglandins. These prostaglandins differ then from the mammalian compounds not only in being largely esterified, but also in their absolute configuration at C-15.

The absolute configuration of all mammalian PGs at C-15 is S. Subsequently, specimens of *P. homomalla* were found which contained the (15S)-15-acetoxy methyl ester of PGA$_2$ (**7.3**), and others which contained both the 15S and 15R isomers (Schneider *et al.*, 1972). *P. homomalla* (Var. S) was found in abundance in the Caribbean around the Cayman Islands and also near the Bahamas. This variant coral contains the (15S)-15-acetoxy methyl ester of PGA$_2$ in amounts corresponding to 1.5–2% of its wet weight and also about 0.06% of its wet weight as the (15S)-15-acetoxy methyl ester of PGE$_2$ (**7.2**). Besides, 5–15% of the total PGA$_2$ content of this coral was made up of the esters of 5-*trans*-PGA$_2$ (**7.4**). The abundance of the (15S)-15-acetoxy methyl esters of PGA$_2$ and PGE$_2$ in the coral made the esters valuable sources for large-scale production of PGE$_2$ and PGF$_{2\alpha}$ (Bundy *et al.*, 1972a,b; Schneider *et al.*, 1977). When specimens of *P. homomalla* are frozen in liquid N$_2$ or solid CO$_2$ immediately after harvesting, they contain exclusively the esterified forms of PGs. However, if they are kept in water, or homogenized and left to autolyze for 20–24 hr, the esters become hydrolyzed and the PGs can be extracted as the free acids. Apparently the coral contains esterases that can be isolated and used for the hydrolysis of the 15-acetoxy methyl esters (Prince *et al.*, 1973; Schneider *et al.*, 1977). The (15R)-PGA$_2$ and its esters are

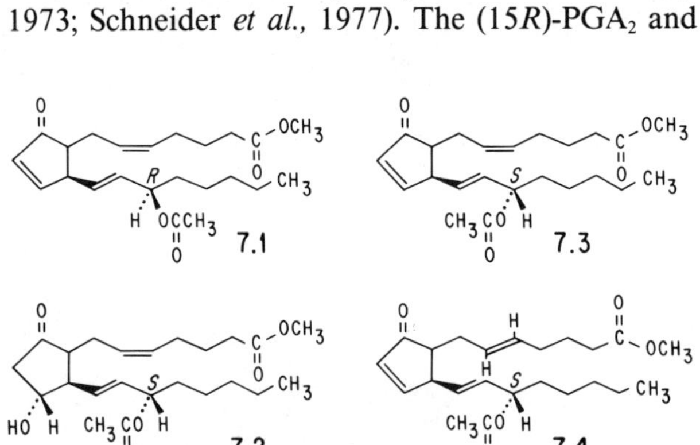

FIGURE 10.7. Structures of prostaglandins of *Plexaura homomalla* Var. *R* and Var. *S*.

devoid of the dramatic blood pressure-lowering effect of the mammalian PGA_2; their role in the physiology of the coral is not known. The structures of the PGs from the corals are shown in Fig. 10.7. The physical properties of most PGs have been summarized by Crabbé (1977).

10.1.4. Occurrence of Prostaglandins

Prostaglandins occur in many tissues and body fluids, although in minute amounts, about 0.3–0.5 µg/g tissue or less (see von Euler and Eliasson, 1967, pp. 41–59). The most common forms are PGE_2 and $PGF_{2\alpha}$, which probably reflect the greater abundance of arachidonic acid over the other two eicosapolyenoic precursors in tissue lipids. Human and sheep seminal fluids are the richest sources of PGs and contain five of the six primary PGs, $PGF_{3\alpha}$ being found only in bovine lung. The earliest analyses for PGs were based on their biological activity, most commonly by their contractile action on rabbit jejunal strips. Horton and Thompson (1964) examined 14 pooled samples of human semen and found wide variations ranging from 11 to 844 µg PGE_1 equivalents per ml, with a mean of 226 µg/ml. Bygdeman and Samuelsson (1966) developed a chemical and gas–liquid chromatographic method for the determination of the five primary prostaglandins in single specimens of human semen. First they separated by silicic acid chromatography as groups the PGEs and PGFs extracted from semen, then fractionated the PGEs on silver nitrate-impregnated thin-layer plates, converted them to PGBs with base, and measured their light absorption at 278 nm. All three PGBs derived from the three PGEs having similar extinction coefficients (ϵ = 26,000–28,600). The PGFs were first methylated with diazomethane and then were converted to the trimethylsilyl ethers and analyzed by gas–liquid chromatography. From the determinations on the semen of six men with normal fertility Bygdeman and Samuelsson obtained the following values in µg/ml: PGE_1, 24.5 ± 4.7; PGE_2, 23 ± 5.8; PGE_3, 5.5 ± 1.8; $PGF_{1\alpha}$, 3.6 ± 1.7, and $PGF_{2\alpha}$, 4.4 ± 2.0. These are much lower values than those obtained by Horton and Thompson (1964) by the biological assays. The difference is partly accounted for by the facts that PGE_2 is about five times more potent than PGE_1 in the biological assays and that human seminal fluid contains eight other prostaglandins, PGA_1, PGA_2, PGB_1, PGB_2, and their 19-hydroxylated derivatives. The concentrations of PGAs and PGBs in human semen are about the same as those of PGE_1 and PGE_2, but the concentrations of the 19-hydroxylated compounds are about four times higher (von Euler and Eliasson, 1967, p. 47). For the ram seminal fluid Bygdeman and Holmberg (1966) reported the following prostaglandin contents in µg/ml: PGE_1, 28; PGE_2, 3.2; PGE_3, 2; $PGF_{1\alpha}$, 5; and $PGF_{2\alpha}$, 2.3. The abundance of PGE_1 in ram seminal fluid relative to other PGs can be correlated with the very high content of the 20:3ω6 fatty acid in both the choline- and ethanolamine-containing phospholipids in the sheep seminal vesicle: 17.5 and 21.8 mol %, respectively. In contrast these phospholipids are rather poor in the 20:4ω6 fatty acid: 1.5 and 3.1 mol % (Samuelsson, 1970, p. 139). It is startling that the seminal fluid from horse, ox, dog, and rabbit contains no prostaglandins (von Euler and Eliasson, 1967, p. 50).

Quantitative determination of PGs in tissues and blood is fraught with difficulties on account of their rapid synthesis on stimulation of tissues during removal of organs and during homogenization, and even after the minor injury to cells at venipuncture. The methods of quantitative analysis of PGs by gas–liquid chromatography, gas chromatography–mass spectrometry, and radioimmunoassay have been reviewed by Samuelsson et al. (1975). Volume 5 of *Advances in Prostaglandin and Thromboxane Research* is devoted entirely to the various methods that have been developed for determination of PGs (Frölich, 1978). A critical evaluation of these methods can be found in an essay by Granström (1981b).

10.2. Biosynthesis of Prostaglandins

10.2.1. Precursors of Prostaglandins

Barely had the structures of the primary prostaglandins been deduced when the first reports on their biosynthesis began to appear. Van Dorp et al. (1964a), and Bergström et al. (1964a) reported simultaneously that [^{3}H]arachidonic acid, when incubated with homogenates of sheep seminal vesicles, gave PGE_2 in high yield. In subsequent publications in the same year by the same two groups of workers it was shown that all-*cis*-8,11,14-20:3 was the precursor of PGE_1, and all-*cis*-5,8,11,17-20:5 the precursor of PGE_3 (van Dorp et al., 1964b; Bergström et al., 1964b). Not only the seminal vesicle, but other organs too, notably the lung of various animals, can convert any one of the $\omega 6$ or $\omega 3$ all-*cis* eicosapolyenoic fatty acids into prostaglandins. Homogenates of the lung are particularly interesting as these convert the 20:3 acid not only into PGE_1 but also into $PGF_{1\alpha}$, whereas the sheep vesicular gland enzymes synthesize almost exclusively the E-series of prostaglandins. Experiments with the lung enzymes showed that the E- and F-series of PGs are synthesized independently of one another.

The prostaglandin synthetase system is associated with microsomes from which it can be solubilized with nonionic detergents and partially purified by precipitation with ammonium sulfate between 40% and 60% saturation. The prostaglandin synthetase system requires molecular oxygen, heme, and the boiled $100{,}000 \times g$ supernatant of the homogenates. Specific enzymes of prostaglandin synthesis and their cofactors are described in a subsequent section.

10.2.2. Source of Polyunsaturated Fatty Acids for Prostaglandin Synthesis

All the experiments in the study of prostaglandin synthesis were carried out with the free eicosapolyenoic fatty acids as substrates. However, these do not occur in tissues in the free form, or only in negligibly small amounts; they are usually found in esterified form at position 2 of phospholipids. It is, therefore, a legitimate question to ask whether the eicosapolyenoic acids may not be converted into prostaglandins on the phospholipids and stored in an esterified form and released on specific stimuli. Lands and Samuelsson (1968) carried out a revealing experiment to answer this question. They prepared a specimen of 1-palmitoyl-2-[^{14}C]eicosa-8,11,14-trienoyl-sn-glyceryl-3-phosphocholine and incubated it* with the 8,000 × g supernatant of homogenates of sheep seminal vesicles with or without the addition of the 100,000 × g supernatant of rat-liver homogenates. Incubation of the labeled substrate with the 8,000 × g supernatant of seminal-vesicle homogenate alone slowly produced free (unesterified) PGE_1, but addition of the liver supernatant resulted, in two experiments, in 8.4% and 12.9% conversion of the added substrate into free PGE_1 in 75 min. The 100,000 × g supernatant of the liver homogenate lacked the ability to form PGE_1, but released 12–19% of the [^{14}C]20:3 acid from the substrate. The residual phospholipids were separated and hydrolyzed with 0.5 N NaOH. If esterified PGE_1 had been formed, the alkaline hydrolysis would be expected not only to release it from the phospholipid, but also to convert it to PGB_1; however, none was found. Thus, it follows that the primary rate-limiting step in prostaglandin synthesis must be the release of the eicosapolyenoic fatty acids from lipids, triacylglycerols, or phospholipids that would result from activation of lipases, e.g., of phospholipase A_2. The two- to threefold increased release of PGE_1, PGE_2, and $PGF_{1\alpha}$ from rat epididymal fat pads during increased lipolysis, observed by Shaw and Ramwell (1968) and evoked by epinephrine, norepinephrine, acetylcholine, ACTH, and nerve stimulation, is in accord with such a conclusion. A similar mechanism, i.e., activation of phospholipase, may account for the release of prostaglandins in the contralateral somatosensory cortex in the central nervous system after stimulation of the radial nerve, or after direct electrical stimulation of the somatosensory cortex (for review see Bergström et al., 1968). Release of prostaglandins after nerve stimulation has been noted at several other sites also, e.g., from the spleen of dog after stimulation of the splenic nerve (Davies et al., 1968), or at nerve endings of the rat diaphragm after stimulation of the phrenic nerve, or after addition of epinephrine or norepinephrine to a diaphragm–phrenic-nerve preparation. Stimulation of the vagus nerve was observed also to be followed by release of PGE_1, PGE_2, and $PGF_{2\alpha}$ from the isolated rat stomach. Perfusion of the adrenals with acetylcholine, or ACTH, in the cat or rat also evokes the release of prostaglandins into the perfusion fluid. The presence of CA^{2+} in the perfusion fluid with acetylcholine is essential for the release of prostaglandins (Shaw and Ramwell, 1967), which suggests that Ca^{2+} may be needed for the activity of the phospholipase that releases the prostaglandin precursors from phospholipids.

Until recently phosphatidylcholine and phosphatidylethanolamine, known to carry polyunsaturated fatty acids at C-2 of the glycerol moiety, were favored as sources of prostaglandin precursors through the action of a phospholipase A_2. However, the long-known "inositol effect," i.e., the rapid turnover of phospho-

---

*Lands and Samuelsson (1968) chose this particular phospholipid because both the choline- and ethanolamine-containing phospholipids in the sheep seminal vesicles are particularly rich in 20:3ω6 fatty acid (17.5 and 21.8 mol % of all fatty acids), but are poor in the 20:4ω6 (1.5–3.1 mol %).

inositides in cells and tissues after a variety of stimuli (Hokin, 1968; Michell, 1975), called attention to phosphoinositides as a possible source of the polyunsaturated fatty acids for prostaglandin biosynthesis (Dawson and Irvine, 1978). It was envisioned that a phospholipase C could cleave phosphoinositol from such lipids, leaving a 1,2-diacylglycerol, which, after hydrolysis by diacylglycerol lipase, would provide the prostaglandin precursor. This idea became particularly attractive in view of prostaglandin and thromboxane release from stimulated platelets (see later, p. 192). Phosphoinositides of platelets contain almost exclusively stearic acid at C-1 (44.3–44.7%) and arachidonic acid at C-2 (41.2–44.1%; Marcus et al., 1969). Bell et al. (1979) found that 10^9 human platelets, stimulated with thrombin, released within 15 sec 1.2 nmol of 1-stearoyl-2-arachidonoylglycerol which was further hydrolyzed by a Ca^{2+}-activated diacylglycerol hydrolase attached to a particulate fraction of platelets. The activity of this lipase was sufficient to cause the release of 5–10 nmol of arachidonate from 10^9 platelets on thrombin stimulation. This diacylglycerol lipase, while equally active with 1-stearoyl- and 2-arachidonoylglycerol, hydrolyzes the 1-stearoyl-2-arachidonoylglycerol sequentially, releasing first the stearic acid and then the arachidonic acid (Prescott and Majerus, 1983). Agranoff et al. (1983) demonstrated that thrombin stimulation of human platelets prelabeled with ^{32}P specifically caused within 5 sec the cleavage of phosphatidylinositol 4,5-diphosphate, with the release of inositol triphosphate. However, this cleavage was followed by rapid resynthesis of phosphatidate and of phosphatidylinositol 4,5-diphosphate as judged by reincorporation of ^{32}P into these substances. A similar fast degradation of phosphatidylinositol 4,5-diphosphate occurs in horse platelets after addition of "platelet-activating factor" (1-O-alkyl-2-acetyl-sn-glyceryl-3-phosphocholine; cf. Chapter 18), which is known to cause release of arachidonic acid from membrane phospholipids and cause aggregation of platelets (Billah and Lapetina, 1983). It remains to be seen whether breakdown of phosphatidylinositols may provide the polyunsaturated fatty acids for prostaglandin synthesis in cells other than platelets.

10.2.3. Mechanism of Prostaglandin Biosynthesis

The mechanism of prostaglandin biosynthesis was elucidated mostly by Samuelsson and his colleagues and by van Dorp and colleagues. In the formation of all prostaglandins, from any one of the three eicosapolyenoic fatty acids, a new C—C bond is established between C-8 and C-12 (with the loss of the 8(9)-double bond in the fatty acids), a *trans* double bond is formed between C-13 and C-14, and three oxygen atoms are introduced at C-9, C-11, and C-15. In the synthesis of E_2 and $F_{2\alpha}$ and of E_3 and $F_{3\alpha}$ the original *cis* double bonds in the precursor fatty acids at positions 5(6) and 17(18), respectively, are retained.

The earliest experiments on the mechanism of prostaglandin biosynthesis aimed to determine the origin of the oxygen atoms introduced. When homogenates of sheep vesicular glands were incubated with all-*cis*-8,11,14-20:3 in an atmosphere of $^{18}O_2$ gas, the PGE_1 synthesized contained two atoms of ^{18}O associated with the two hydroxyl groups. This was concluded from the comparison of the

TABLE 10.1 Ions in the Mass Spectra of the Methyl Esters of [^{16}O]PGE$_1$ and [^{18}O]PGE$_1$[a]

Ion	[^{16}O]PGE$_1$	[^{18}O]PGE$_1$
	m/z	m/z
M$^+$	368	368,372
[M−H$_2$O]$^+$	350	350,352
[M−CH$_3$O]$^+$	337	337,341
[M−2H$_2$O]$^+$	332	332
[M−(CH$_3$O + H$_2$O)]$^+$	319	319,321
[M−C$_5$H$_{11}$]	297	297,301

[a]Data culled from the mass spectra published by Nugteren and van Dorp (1965). The presence of nonisotopic fragments in the spectrum of [^{18}O]PGE$_1$ methyl ester resulted from dilution of ^{18}O$_2$ with ^{16}O$_2$ in the incubations. Ryhage and Samuelsson (1965) converted their specimens of [^{16}O]PGE$_1$ and [^{18}O]PGE$_1$ to the dimethoxy ethyl esters for mass spectral analysis; their data gave the same information as those shown here.

mass spectra of PGE$_1$ synthesized in homogenates of sheep seminal vesicles in normal atmosphere and PGE$_1$ synthesized in an atmosphere of ^{18}O$_2$ (Nugteren and van Dorp, 1965; Ryhage and Samuelsson, 1965). As is shown in Table 10.1, in the mass spectrum of the methyl ester of PGE$_1$ synthesized in the ^{18}O$_2$ atmosphere, there were fragment ions four mass units higher ([M − 31]$^+$: m/z 341; [M − 71]$^+$: m/z 301) than in the mass spectrum of [^{16}O]PGE$_1$, indicating the presence of two atoms of ^{18}O. Elimination of one hydroxyl group as water gave a fragment ion two mass units higher (m/z 352) than the corresponding ion in the spectrum of [^{16}O]PGE$_1$ methyl ester. Elimination of two molecules of water led to the same ion (m/z 332) in the spectrum of [^{18}O]PGE$_1$ methyl ester as in the spectrum of [^{16}O]PGE$_1$ methyl ester. These observations clearly established the presence of ^{18}O in the two hydroxyl groups of PGE$_1$ synthesized in an atmosphere of ^{18}O$_2$.

The origin of the carbonyl oxygen in the ring of PGE$_1$ was at first uncertain because it exchanges with oxygen of water. The exchange is fortunately not very fast and can be stopped by reduction of the ketone with NaBH$_4$. Samuelsson (1965) demonstrated that the oxygen atom at C-9 of PGE$_1$ originates also from molecular oxygen and, moreover, that it originates from the same molecule of oxygen as the oxygen atom of the hydroxyl group at C-11. He incubated [2-^{14}C]eicosa-8,11,14-trienoic acid with the enzymes of sheep seminal vesicles in an atmosphere of ^{16}O$_2$ (46%) plus ^{16}O^{18}O (1%) plus ^{18}O^{18}O (53%) and terminated the reaction by the addition of NaBH$_4$ in ethanol, thereby converting the formed PGE$_1$ mainly into PGF$_{1\beta}$. This was isolated and was converted into the trimethoxy ethyl ester (structure **8.2** of Fig. 10.8). Oxidation of (**8.2**) with KMnO$_4$ and KIO$_4$ followed by esterification of the product with diazoethane gave the dimethoxy diethyl ester (**8.3**) in which the oxygen atoms attached to the ring were, presumably, those introduced during the biosynthesis. Mass spectral analysis of the derivative of this degradation product of PGE$_1$ showed that it consisted essentially of two molecular species: one contained no excess isotope and the other

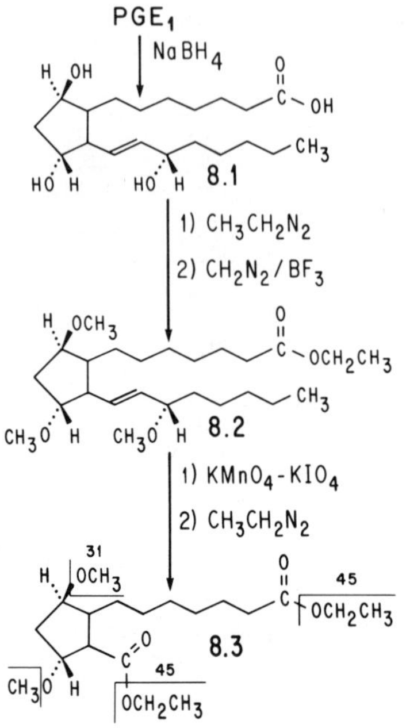

FIGURE 10.8. Degradation of PGE$_1$ through PGF$_{1\beta}$ for determination of origin of the ring-attached oxygen atoms by mass spectrometry of compound **8.3** (cf. Table 10.2; after Samuelsson, 1965).

contained two atoms of ^{18}O. There was an insignificant proportion of molecules which contained one atom of ^{16}O and one of ^{18}O (Table 10.2). These observations meant that not only were the two ring-attached oxygen atoms derived from molecular oxygen, but that they had to originate from the same molecule of oxygen. In other words these oxygen atoms were introduced into the molecule by a dioxygenase type of reaction and not by a monooxygenase reaction. This interesting finding, together with some further observations, led to the formulation of a very reasonable reaction mechanism for the biosynthesis of prostaglandins, with far-reaching consequences.

To explore the fate of the hydrogen atoms on the centers in 8,11,14-20:3 that

TABLE 10.2. Mass Spectral Data on Degradation Product of PGE$_1$ Biosynthesized in the Presence of ^{18}O$_2$ + ^{16}O$_2$[a]

Ion	Reference compound	Experimental sample
	m/z	m/z
M$^+$	358	358,362
[M$-$CH$_3$]$^+$	343	343,347
[M$-$(CH$_3$O + H)]$^+$	326	326,328
[M$-$CH$_3$CH$_2$O]$^+$	313	313,317
[M$-$(CH$_3$CH$_2$O + H + CH$_3$)]$^+$	297	297,301
[M$-$(CH$_3$CH$_2$OH + CH$_3$OH)]$^+$	280	280,282

[a]From data of Samuelsson (1965).

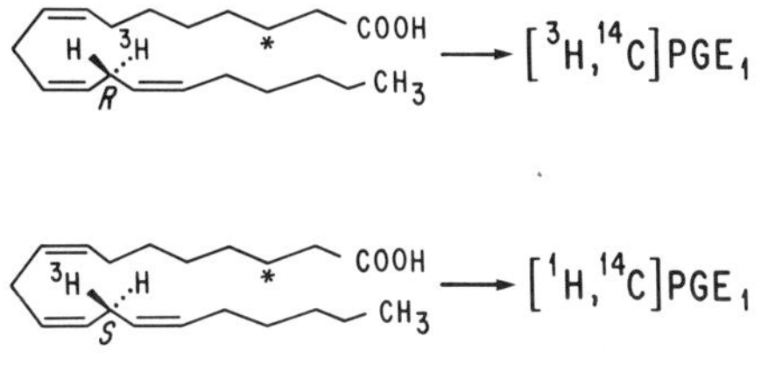

FIGURE 10.9. Conversions of [(13R)-13-^{3}H-3-^{14}C]- and [(13S)-13-^{3}H-3-^{14}C]eicosatrienoic acids into PGE$_1$s (after Hamberg and Samuelsson, 1967).

form the new carbon–carbon bond and C—OH bond in the cyclopentanolone ring of PGE$_1$, Klenberg and Samuelsson (1965) tested [8-^{3}H,3-^{14}C]-, [11-^{3}H,3-^{14}C]- and [12-^{3}H,3-^{14}C]eicosa-3,11,14-trienoic acids as precursors to PGE$_1$. The tritium at all three positions was retained in PGE$_1$ without loss. However, [9-^{3}H]eicosatrienoic acid gave—not unexpectedly—PGE$_1$ without tritium, but it was converted in homogenates of lung into [^{3}H]PGF$_{1\alpha}$, indicating that PGF$_{1\alpha}$ in the lung is not formed by the reduction of PGE$_1$. Homogenates of guinea-pig lung also synthesize PGE$_2$ and PGF$_{2\alpha}$ from 20:4ω6 but cannot convert PGE$_2$ into PGF$_{2\alpha}$ (Änggård and Samuelsson, 1965a). However, PG 9-ketoreductases and PG 9-dehydrogenases are known in other organs which interconvert PGEs to PGFs and PGFs to PGEs (Lee and Levine, 1974).

In critical experiments, Hamberg and Samuelsson (1967) tested eicosa-8,11,14-trienoic acid labeled stereospecifically with ^{3}H at position 13 and also with ^{14}C at C-3 as precursors of PGE$_1$. The [(13R)-13-^{3}H,3-^{14}C]eicosatrienoic acid gave [^{3}H,^{14}C]PGE$_1$ essentially without loss of tritium as judged by ^{3}H/^{14}C ratios in product and precursor. The [(13S)-13-^{3}H,3-^{14}C]-substrate yielded, however, PGE$_1$ free of tritium (Fig. 10.9). Examination of the residual substrate ([(13S)-13-^{3}H,^{14}C]eicosatrienoic acid), when the reaction had gone 75% to completion, showed a nearly threefold enrichment in ^{3}H (^{3}H/^{14}C ratio in starting substrate 1.36, and in residual substrate 3.87), indicating that the removal of the pro-S hydrogen atom from C-13 is stereospecific and a rate-limiting step in the synthesis of PGE$_1$. Based on all these observations, Samuelsson (1965) and Hamberg and Samuelsson (1967) proposed a free-radical mechanism for synthesis of prostaglandins from the eicosapolyenoic fatty acids (Fig. 10.10). If the hydrogen from

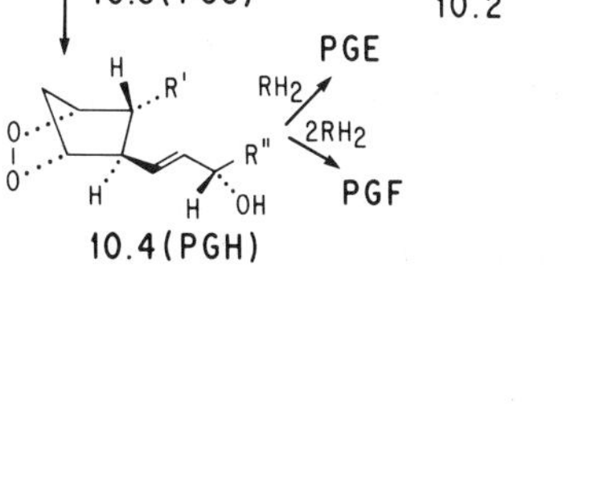

FIGURE 10.10. Mechanism of prostaglandin biosynthesis. The fishhook arrows are meant to indicate free-radical reactions (after Samuelsson, 1970).

C-13 of the acid were eliminated as a hydrogen atom, not as a proton, the reaction would create a free radical at C-13 (**10.1**). Attack of peroxy radical at C-11 now results in the addition of the peroxy radical and isomerization of the 11-*cis* double bond to 12-*trans* (**10.2**). In a second concerted sequence the 9,11-endoperoxide is formed with the addition of a second peroxy radical to C-15, and isomerization of the 12-*trans* double bond to 13-*trans* (**10.3**). Dismutation or reduction of the 9,11-endoperoxy-15-hydroperoxy intermediate (**10.3**) to the prostaglandin-15-hydroxy-9,11-endoperoxide (**10.4**) would furnish a common precursor to PGEs and PGFs: two-electron reduction of **10.4** would give PGE and four-electron reduction, PGF. This proposal became highly probable when it was found that prostaglandin $F_{1\alpha}$ that was synthesized with enzymes of guinea-pig lung from [9-^{3}H]eicosatrienoic acid retained the ^{3}H at C-9, whereas PGE_1, of course, lost it. The same mechanism can account for the syntheses of PGE_2, $PGF_{2\alpha}$ and of PGE_3, $PGF_{3\alpha}$, except that the precursors of these are arachidonic and eicosapentaenoic acids (see also discussion in Section 10.2.4.1).

The mechanism just outlined for biosynthesis of PGs, through the intermediacy of the 15-hydroperoxy- and 15-hydroxy-9,11-endoperoxides (**10.3** and **10.4**), remained only an attractive hypothesis until 1973–74, when Hamberg and Samuelsson (1973, 1974) and Nugteren and Hazelhof (1973) reported the detection and isolation of these substances from incubations of microsomal preparations of sheep seminal vesicles and of human platelets with [^{14}C]arachidonic acid. Hamberg and Samuelsson (1973) observed a burst of oxygen consumption by suspensions of sheep seminal-vesicle microsomes within a few seconds after the addition of arachidonic acid. The oxygen consumption reached a maximum in 10 sec and ceased after 30 sec. It was known that 1mM *p*-chloromercuribenzoate or *N*-ethylmaleimide in such microsomal preparations inhibited synthesis of PGE_2 from 20:4ω6. In spite of the presence of either of these reagents in the microsomal suspensions, arachidonic acid still evoked the burst of oxygen consumption. From brief (2-min) incubations of microsomes, and eventually of human platelets, with [^{14}C]arachidonic acid Samuelsson and his colleagues isolated two new substances that became known as PGG_2 and PGH_2 (see also Hamberg *et al.*, 1974a,b; 1975a,b). Nugteren and Hazelhof (1973), who discovered the prostaglandin endoperoxides independently of and almost simultaneously with Hamberg and Samuelsson (1973), called the PGHs synthesized from 20:3ω6 and 20:4ω6 prostaglandin R_1 and R_2, respectively, and they named the corresponding PGGs 15-hydroperoxyprostaglandin R_1 and R_2. The prostaglandin G and H terminology is used now universally.

Samuelsson and his colleagues proved that PGG_2 was 15-hydroperoxy-9,11-endoperoxyprosta-5,13-dienoic acid, and that PGH_2 was 15-hydroxy-9,11-endoperoxyprosta-5,13-dienoic acid, by the following reactions. PGG_2 gave a positive test for a hydroperoxy group with sodium ferrocyanide; its treatment with lead tetraacetate, followed by reduction with $SnCl_2$ or triphenylphosphine, gave 15-keto-$PGF_{2\alpha}$, which was known from studies on the metabolism of $PGF_{2\alpha}$ (cf. Section 10.3). Treatment of PGG_2 with $SnCl_2$ or triphenylphosphine gave $PGF_{2\alpha}$. PGH_2 gave no reaction for a hydroperoxy group, but, like PGG_2, could be reduced with either $SnCl_2$ or triphenylphosphine to $PGF_{2\alpha}$. Both PGG_2 and PGH_2

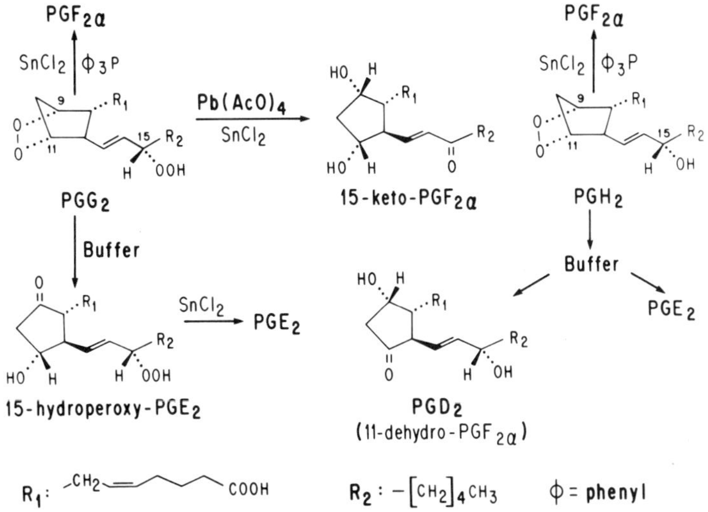

FIGURE 10.11. Chemical and spontaneous transformations of two prostaglandin endoperoxides, PGG_2 and PGH_2, derived from arachidonic acid (after Hamberg and Samuelsson, 1973; Hamberg et al., 1974b).

are unstable in aqueous buffered solutions: PGG_2 decomposes to 15-hydroperoxy-PGE_2, and PGH_2 decomposes mainly (80–85%) into PGE_2 and to a lesser extent into PGD_2. For these reasons PGG_2 and PGH_2 need to be handled in dry organic solvents for any chemical reactions. On silica gel thin-layer plates the methyl ester of PGH_2 decomposes to the methyl esters of PGE_2 and 9,15-dihydroxy-11-keto-prosta-5,13-dienoic acid, code-named PGD_2. The chemical and spontaneous transformations of PGG_2 and PGH_2 are shown in Fig. 10.11.

10.2.4. Enzymes of Prostaglandin Biosynthesis

The enzymes of prostaglandin biosynthesis have been most intensively studied in mammalian organs, particularly in microsomes of ovine and bovine seminal vesicles. However, biosynthesis of PGs is not confined to the seminal vesicles nor to mammals. Homogenates of organs of birds, fish, frog, toad, lobster, or homogenates of the whole animal among lower orders *(Annelida, Mollusca, Coelenterata)* synthesize PGEs and PGFs either from exogenous or endogenous polyenoic fatty acids. The gills of trout, carp, tench, lobster, and mussel contain lipids fairly rich in 20:4ω6 and 20:5ω3 fatty acids. Particularly high synthetic activities are present in mammalian lung and renal medullary tissue, in the gills of fish, lobster, and mussel, and in the frog's lung and toad bladder (Christ and van Dorp, 1972). The PG synthetase system of *Plexaura homomalla* (Var. *S*) is associated with microsomes, but absolutely requires 1 M NaCl for activity; it converts [^{3}H]-20:3ω6 and [^{3}H]20:4ω6 into unesterified (15*S*)-PGA_1 and (15*S*)-PGA_2 without detectable intermediacy of a PGE. The enzyme system of this coral must differ structurally from mammalian systems in fundamental ways, since potent inhibitors of mammalian PG synthetase systems, such as indomethacin or 5,8,11,14-eicosatetraynoic acid, do not inhibit it (Corey et al., 1973).

The association of the prostaglandin biosynthetic system with microsomes of the seminal vesicle or lung has been referred to earlier (cf. p. 160). The discovery of the prostaglandin endoperoxides resolved the enzymology of prostaglandin synthesis into two main sequences: (i) the oxygenase sequence leading to the formation of PGG and PGH, and (ii) the isomerization or reduction of the latter to the primary prostaglandins. Miyamoto et al. (1974) solubilized the prostaglandin biosynthetic system from microsomes of bovine seminal vesicles with ethylene glycol (20%) and the detergent Tween 20 (1%), and resolved the soluble proteins by chromatography on DEAE cellulose columns into two fractions, I and II. When fraction I was incubated with 20:3ω6 in the presence of hemoglobin and tryptophan, a labile compound was formed which, on reduction with $SnCl_2$, gave $PGF_{1\alpha}$, and when left at room temperature overnight decomposed into $PGF_{1\alpha}$, PGE_1, and PGD_1. These properties are characteristic of the behavior of PGH_1. When the product of fraction I was reincubated with fraction II and glutathione, it gave mostly PGE_1 (62.7%), a little $PGF_{1\alpha}$ (9.2%), PGD_1 (7.5%), and some unidentified products.

10.2.4.1. Prostaglandin Endoperoxide Synthetase and Prostaglandin Hydroperoxidase

The oxygenase component of the prostaglandin synthetase system, designated prostaglandin endoperoxide synthetase (EC 1.14.99.1), also referred to as fatty acid or prostaglandin cyclo-oxygenase (cf. Samuelsson et al., 1975), has been purified to homogeneity from microsomes of sheep and bovine seminal vesicles (Hemler et al., 1976; Miyamoto et al., 1976; Van der Ouderaa et al., 1977; Ogino et al., 1978). The enzyme from either source is a dimer of two apparently identical subunits (amino terminus: Ala) with M_r of about 70,000. The molecular weight of the native enzyme has been estimated at 126,000 from its Stoke's radius (53 Å) and sedimentation coefficient (7.4S). The sheep enzyme is a glycoprotein containing 12 mol mannose and 5 mol N-acetylglucosamine per mol. The glycoprotein of the bovine enzyme has not been reported.

The prostaglandin endoperoxide synthetase has two functions: (i) the bis-dioxygenase function that generates a PGG from either 20:3ω6 or 20:4ω6; and (ii) a peroxidase function that converts PGG to PGH. An intrinsic property of this synthetase is that it suffers an irreversible self-destruction after a brief period of catalysis (see later). Some investigators studied the enzyme with 20:3ω6, others with 20:4ω6 and with PGG_1 or PGG_2 as substrates; the results were analogous whichever substrate was used. For this reason we shall refer here mostly to the substrates as "the substrate fatty acids," or PGG or PGH, implying that these could be either 20:3ω6 or 20:4ω6 fatty acids or PGG_1, PGG_2, PGH_1, or PGH_2.

The protein purified from either sheep or bovine seminal vesicles is an inactive apoenzyme, devoid of heme or non-heme iron, but requiring heme for both its enzymatic activities, either in the form of free hematin or associated with certain hemoproteins, such as hemoglobin (Hb), myoglobin (Mb), or indoleamine 2,3-dioxygenase. The effect of the hemoproteins is indirect and results from the substrate fatty acid (arachidonic acid) causing dissociation of heme from these hemoproteins, followed by the avid binding of the released heme by the prostaglandin endoperoxide synthetase with an association constant of 0.6×10^8 M^{-1} (Ueno et al., 1982). Horseradish peroxidase and beef liver catalase, the hemes of which are not dissociated by arachidonic acid, cannot activate the endoperoxide synthetase; cytochrome c, in which the heme is covalently bound, is also ineffective. Apomyoglobin and apohemoglobin inhibit the enzyme activated by hematin, but the inactivation can be overcome by excess hematin. The apoenzyme of the PG endoperoxide synthetase, either from sheep or bovine seminal vesicles, can bind 1 or 2 mol of heme per 70,000-Da subunit to form the holoenzyme with an absorption maximum at 410–412 nm and extinction coefficients that have been reported in the ranges of 61 $mM^{-1} \cdot cm^{-1}$ and 120 $mM^{-1} \cdot cm^{-1}$ (Van der Ouderaa et al., 1979; Roth et al., 1981; Ueno et al., 1982).

The holoenzyme, reconstituted with hematin, catalyzes in the presence of O_2 the synthesis only of PGG from a substrate fatty acid, but in the presence of aromatic compounds (e.g., epinephrine, adrenochrome, guaiacol, phenol, tryptophan, hydroquinone, ascorbic acid, sulindac sulfide, tetramethylphenylenediamine, and several others), it rapidly converts PGG into PGH and thus displays a peroxidase function also (Ogino et al., 1977, 1978; Ohki et al., 1979; Egan et al., 1979). Low conversions at slow rates of PGG_1 into PGH_1 were noted only at high concentrations (5–10 μM) of hematin. With L-tryptophan, epinephrine, guaiacol, or phenol, and with several other compounds (α-lipoic acid, methional (3-methylthiopropionaldehyde), sulindac sulfide) added to the enzyme, the V_{max} for the conversion of PGG to PGH is much increased and is attained at a hematin concentration of 1 μM or less. Attempts at separating the bis-dioxygenase and peroxidase functions of the PG endoperoxide synthetase have not been successful; apparently the same protein catalyzes both types of reactions.

Ogino et al. (1978) made the interesting observation that manganese protoporphyrin IX, in place of hematin, can also activate the PG endoperoxide synthetase, but only to promote its first function—the bis-dioxygenation reaction—and not the peroxidase function. Thus, it is possible to discriminate between the two enzyme processes and some phenomena specifically associated with one or the other.

Over the last several years the mechanism of action of PG endoperoxide synthetase has been the subject of much exploration. The inference from the early experiments of Hamberg and Samuelsson (1967) was that the first rate-limiting step in the synthesis of a PG endoperoxide was the abstraction of the 13-pro-S hydrogen atom with the generation of the alkyl radical at C-13. Isomerization of the 11(12)-double bond to 12(13) would result in the creation of the alkyl radical at C-11 and the addition of the first O_2 molecule at that position. The oxygenation of a C_{20} polyenoic fatty acid at C-11 is apparently an intrinsic property of PG endoperoxide synthetase as it converts even eicosa-11,14-cis-dienoic acid specif-

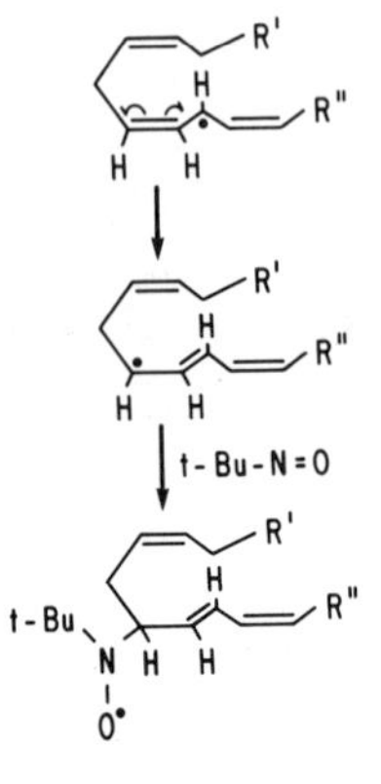

FIGURE 10.12. Trapping of free radical formed from arachidonic acid with 2-methyl-2-nitrosopropane (t-Bu$-$N$=$O) (after Mason et al., 1980).

ically to the 11-hydroperoxyeicosa[12-*trans*-14-*cis*-]dienoate (Hemler et al., 1978).*

Evidence for the alkyl radical at C-11 formed from arachidonic acid by PG endoperoxide synthetase was provided by Mason et al. (1980) who trapped this free radical with 2-methyl-2-nitrosopropane (t-Bu$-$N$=$O). They observed, concomitant with the uptake of O_2 on addition of 20:4 ω6 to a preparation of microsomes from sheep seminal vesicles, the appearance of a strong well-defined electron paramagnetic resonance (EPR) signal with parameters of α_N = 15.7 G and α_H = 2.5 G. These parameters were identical with those given by the radical chemically synthesized from arachidonic acid and 2-methyl-2-nitrosopropane (Fig. 10.12).

During catalysis by the PG endoperoxide synthetase some as yet unidentified hydroperoxide is probably formed and is essential for the activity of the enzyme. This is inferred from the fact that addition of glutathione (GSH) peroxidase and GSH to a reaction mixture during catalysis immediately stops the reaction. Reducing agents, such as α-naphthol, in micromolar concentrations, also inactivate the enzyme (Hemler and Lands, 1980). Hemler and Lands observed also that the few seconds' lag in the initiation of catalysis by the endoperoxide synthetase was progressively lengthened by increasing concentrations of NaCN without significantly affecting the maximum rate. 15-Hydroperoxyeicosatetraenoate (15-HPETE; cf. Chapter 11, Fig. 11.2), without being consumed or converted to some other product, completely abolished this inhibitory effect of NaCN. Second addition of fresh enzyme, to a mixture containing NaCN and in which the previous reaction stopped owing to self-destruction of the enzyme, restarts the reaction without delay. Apparently an activator peroxide, presumed to have been formed during the first reaction, can overcome the inhibiting effect of NaCN.

Hemler and Lands (1980) proposed a reaction mechanism for the PG endoperoxide synthetase that embodies all the experimental observations made with this enzyme. An essential feature of this mechanism is the activation of the

---

*The 12-*trans*-14-*cis* configuration for the 11-hydroperoxyeicosadienoate was not demonstrated.

enzyme by a hydroperoxide presumably generated from the substrate fatty acid during the first few seconds of enzyme catalysis (Eq. 10.1).

$$Enz(Fe^{III})_{inactive} \xrightleftharpoons{ROOH} Enz(Fe^{III})-ROOH \rightleftharpoons Enz(Fe^{II})-ROO\bullet_{active} \quad (10.1)$$

$$ROO\bullet + RH \rightarrow ROOH + R\bullet \quad (10.2)$$

Equation 10.1 postulates interaction of the ferriheme with the hydroperoxide to form an enzyme-bound peroxy radical. "This radical, with the help of orienting enzyme then propagates the reaction by abstracting the 13-S-hydrogen . . . " atom from the substrate fatty acid (Hemler and Lands, 1980, p. 6258). Isomerization of the alkyl radical at C-13 to the 12-$trans$-14-cis radical at C-11 leads to the addition of the first molecule of O_2 to that carbon atom. In Fig. 10.10 it may readily be seen that formation of the 9,11-endoperoxide and completion of the cyclopentane ring by the formation of the carbon-to-carbon bond between C-8 and C-12 force the isomerization of the 12(13)-double bond to 13(14) and the creation of yet a second radical at C-15 which now can react with the second molecule of O_2. Thus the primary product of endoperoxide synthetase would be the 9,11-endoperoxy-15-peroxy radical. The mechanism is then analogous to the mechanism of autoxidation of organic compounds. Indeed, autoxidation of arachidonic acid has been reported to yield stereoisomers of prostaglandins (cf. Chapter 6). If the above formulation of PGG synthesis were correct, a termination reaction for the reduction of the 15-peroxy radical would be required (Eq. 10.2). One might speculate that, since the pure holoenzyme of the endoperoxide synthetase requires no cofactors for the synthesis of PGG, the enzyme itself would provide the reducing RH component. The generated Enz$-$R$\bullet$ could react with another Enz$-$R$\bullet$ to form Enz$-$R$-$R$-$Enz dimer and might be responsible, at least in part, for the self-destruction referred to earlier.

The peroxidase activity of the PG endoperoxide synthetase is not specific for PGG as substrate, but is almost as effective with 15-hydroperoxy-PGE$_1$, with 15-hydroperoxyeicosa-8,11(cis),13($trans$)-trienoic and 15-hydroperoxyeicosa-5,8,11(cis),13($trans$)-tetraenoic acids as with PGG$_2$ as substrate. Guaiacol, phenol, tryptophan, and sulindac sulfide stimulate the reduction of these 15-hydroperoxy compounds to the 15-hydroxy compounds in a similar manner qualitatively and quantitatively. Peroxidase rates lower than those seen with substrates related to the PG-structures were noted with H_2O_2, t-butyl hydroperoxide, p-menthane hydroperoxide, and cumene hydroperoxide. In spite of its broad specificity, the second function of the PG endoperoxide synthetase may be referred to as "PG hydroperoxidase" (Ohki et al., 1979).

This peroxidase function of the prostaglandin endoperoxide synthetase aroused much interest because not only guaiacol and epinephrine (Ohki et al., 1979), phenol and sulindac sulfide (Egan et al., 1979), but also a number of other organic compounds become oxidized during the peroxidative phase of PG synthesis. Marnett et al. (1975) found that sheep vesicular gland microsomes oxygenated (i) 1,3-diphenylisobenzofuran (**13.1**) to o-dibenzoylbenzene (**13.2**), (ii)

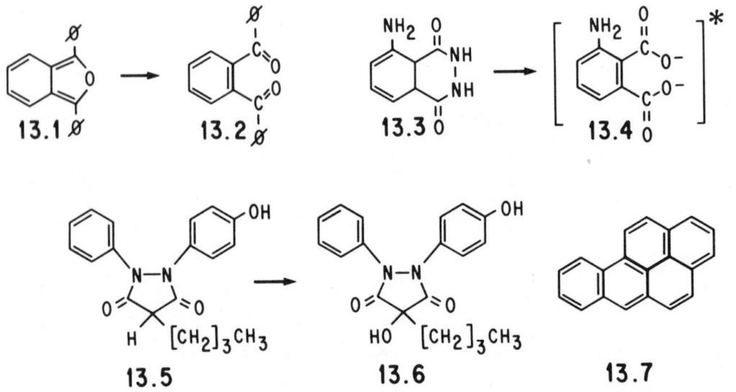

FIGURE 10.13. Oxidation of some aromatic compounds during peroxidase reaction catalyzed by prostaglandin endoperoxide synthetase (Marnett et al., 1975). ϕ = phenyl; **13.1**: diphenylisobenzofuran; **13.2**: dibenzoylbenzene; **13.3**: luminol; **13.4**: aminophthalate anion (excited); **13.5**: oxyphenbutazone; **13.6**: 4-hydroxyoxyphenbutazone; **13.7**: benzo(a)pyrene.

luminol (**13.3**) to the excited state of the aminophthalate ion (**13.4**), (iii) oxyphenbutazone (**13.5**) to 4-hydroxyphenbutazone (**13.6**), and (iv) benzo(a)pyrene to an unidentified product (Fig. 10.13). These oxygenations depended absolutely on the presence of arachidonic acid with the enzyme and were completely inhibited by indomethacin and by 2,3-dimercaptopropanol, known inhibitors of prostaglandin synthetase. However, the same co-oxygenations were also promoted by PGG_2 and by 15-hydroperoxy-5,8,11(cis),13(trans)-eicosatetraenoic acid, a product of the action of soybean lipoxygenase on arachidonic acid (cf. Chapter 11), but they were not inhibited by indomethacin. It has been known for some time that indomethacin and aspirin inhibit prostaglandin synthesis. At least some of the pharmacological actions of these drugs have been attributed to their inhibition of PG synthesis (Vane, 1971; Smith and Willis, 1971; Ferreira et al., 1971). It is now clear that aspirin and indomethacin inhibit only the oxygenase and not the peroxidase function of the PG endoperoxide synthetase. Aspirin inhibits by acetylating a serine residue within the polypeptide chain of the protein (Roth et al., 1975, 1980).

In the conversion of PGG to PGH the O—O bond in the 15-hydroperoxy group is cleaved. Three types of reactions could bring about such a transformation (Ohki et al., 1979): (i) oxygen transfer to an acceptor molecule (Eq. 10.3); (ii) a dismutation by the interaction of two PGG molecules (Eq. 10.4); and (iii) a peroxidase reaction (Eq. 10.5).

$$R-OOH + X \rightarrow R-OH + X-O \tag{10.3}$$

$$R-OOH + R-OOH \rightarrow 2R-OH + O_2 \tag{10.4}$$

$$R-OOH + H_2A \rightarrow R-OH + H_2O + A \tag{10.5}$$

The dismutation reaction can be excluded, as O_2 evolution during the transformation of PGG_1 into PGH_1 could not be demonstrated in a complete enzyme system containing purified prostaglandin endoperoxide synthetase, hematin, and tryptophan (Ohki et al., 1979). Among the various aromatic compounds listed earlier (cf. p. 169) as promoters of the PGG → PGH transformation, several are known to be hydrogen donors for other peroxidases. Epinephrine and guaiacol seem to fulfill this role in the conversion of PGG into PGH, as both these activators are dehydrogenated in stoichiometric amounts, together with the reduction of the hydroperoxide to the hydroxy compound.

The natural co-substrate in the PG peroxidase reaction is still a matter of speculation, for the PGG → PGH transformation can proceed also by a direct oxygen transfer to an acceptor molecule according to Eq. 10.3. Sulindac sulfide (for structure see Fig. 10.14), an anti-inflammatory drug, is a potent inhibitor of the dioxygenase function of PG endoperoxide synthetase (ID_{50} = 0.1 μM), but a potent stimulator of the PG peroxidase reaction at concentrations of 50–100 μM (Egan et al. 1979). Sulindac sulfide is oxidized to sulindac by PG peroxidase during conversion of 15-hydroperoxy-PGE_1 and 15-hydroperoxy-PGE_2 into PGE_1 and PGE_2, respectively (Egan et al., 1979; 1980). The molar ratios of PGE_1 or PGE_2 generated to sulindac range from about 0.7 to 1.0. Egan et al. (1980) demonstrated with the aid of [15-$^{18}O_2$]15-hydroperoxy-PGE_2 that one atom of ^{18}O was transferred quantitatively from the substrate hydroperoxide to sulindac sulfide with the formation of [^{18}O]sulindac, in accordance with Eq. 10.3 (Fig. 10.14). While it might be thought that the co-oxidations of the various compounds during the peroxidase phase of PG synthesis are also associated with oxygen transfer from a PGG, the oxidation of sulindac sulfide is, so far, the only reaction known in which PG peroxidase catalyzes such a direct oxygen transfer. The oxidation of diphenylisobenzofuran (DPBF) to dibenzoylbenzene (DBB) (cf. Fig. 10.13) by PG peroxidase results from a chain reaction, most likely free radical in nature, through the action of oxidizing radicals generated during the peroxidase reaction. There is no clear stoichiometric relation between DPBF oxidized and the amount of available substrate (PGG_2); at low concentrations of PGG_2 (<16 μM) the amount of DPBF oxidized far exceeds the amount of available substrate. Further,

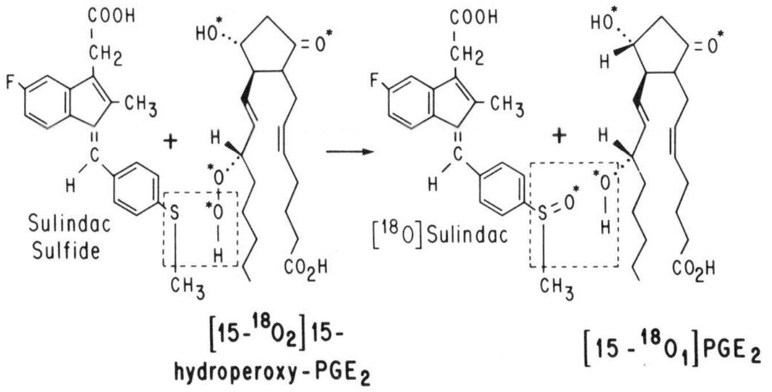

FIGURE 10.14. Transfer of ^{18}O (*O) from [15-$^{18}O_2$]15-hydroperoxy-PGE_2 to sulindac sulfide during peroxidase reaction catalyzed by prostaglandin endoperoxide synthetase. In the actual experiment the 15-hydroperoxy-PGE_2 contained ^{18}O also at C-11 and probably also at C-9, and ^{14}C at C-1 (Egan et al., 1980).

during the oxidation of DPBF one oxygen atom is introduced into DBB from atmospheric O_2 and none from the substrate, or water, as determined by experiments with $^{18}O_2$. The second oxygen atom in DBB is that originally contained in the furan ring (Marnett et al., 1979).

The role of tryptophan in activating the PG peroxidase is intriguing, as it is recovered from the reaction mixtures unchanged; it is neither oxidized nor dehydrogenated, as no ^3H could be detected in the water of incubations containing [G-^{3}H]Trp, or L-Leu-[*side-chain*-2,3,-^{3}H]Trp-Leu.* The fate of the O-atom removed from PGG in the peroxidase reaction promoted by tryptophan is not known. Tryptophan could, of course, be simply an allosteric effector of the enzyme. However, it probably participates in the enzyme catalysis in some manner, as it is raised to an excited state followed by light emission immediately on addition of a substrate hydroperoxide. Marnett et al. (1974) observed intense chemiluminescence on addition of 20:4ω6 and other polyunsaturated fatty acids to preparations of sheep vesicular gland microsomes and attributed the phenomenon to early steps of PG synthesis and to the possible participation of singlet oxygen ($^1O_2$) (Rahimtula and O'Brien, 1976) in these steps. However, the steps of PG biosynthesis associated with the chemiluminescence were not determined. The availability of the pure PG endoperoxide synthetase and the ability to differentiate between the bis-dioxygenase and peroxidase functions of the enzyme made it possible to ascribe the chemiluminescence entirely to the peroxidase reaction when this is activated by tryptophan and a few other indole derivatives. Thus, Yoshimoto et al. (1980) found that addition of arachidonic acid to an enzyme preparation activated with hematin and tryptophan immediately produced an intense chemiluminescence, which decayed with a half-life of about 1 min. On the other hand, in an enzyme preparation activated with manganese protoporphyrin IX, arachidonic acid did not elicit the chemiluminescence. Thus, the light emission cannot be attributed to the bis-dioxygenase reaction, since manganese protoporphyrin IX activates the apoenzyme to catalyze the conversion of 20:4ω6 only to PGG_2. "In contrast, the addition of PGG_2 as a substrate in the presence of hematin and tryptophan brought about an intense light emission . . . " (Yoshimoto et al., 1980, p. 10210). In conformity with the broad specificity of PG peroxidase, not only PGG_2 and PGG_1, but also 15-hydroperoxy-5,8,11(*cis*),13(*trans*)-eicosatetraenoic acid, cumene hydroperoxide, *t*-butyl hydroperoxide, *p*-menthane hydroperoxide, and hydrogen peroxide elicited the chemiluminescence in the presence of tryptophan and the hematin-activated enzyme. Heme compounds which do not activate the enzyme, such as horseradish peroxidase or cytochrome *c*, do not elicit the chemiluminescence. Chemiluminescence is also seen in the presence of *N*-acetyltryptophan and 3-indoleacetic and 3-indolebutyric acids, though only of about one-half the intensity seen in the presence of tryptophan. The most intense chemiluminescence, about three times as intense as seen with tryptophan, is elicited by skatole (3-methylindole) and 3-

---

*Tryptophan, even in a peptidic linkage, activates the PG endoperoxidase.

indolepropionic acid. Substitution of position 5 in the indole ring, as in 5-hydroxytryptophan and serotonin, abolishes the chemiluminescence, though not the ability of these compounds to activate the PG peroxidase reaction.

The emission spectrum from tryptophan during the observed chemiluminescence had maxima at 500, 550, and 600 nm, which are distinct from those attributed to singlet oxygen (520, 578, and 633 nm). 2,5-Dimethylfuran, a singlet-oxygen quencher, does not inhibit the tryptophan luminescence in the PG peroxidase reaction. The emission spectrum of the tryptophan luminescence in the peroxidase reaction is similar to that induced photochemically by irradiation of tryptophan at 365 nm. It is not known whether the mechanism of excitation of tryptophan in the PG peroxidase reaction is similar to that attributed to the photochemical reaction (cf. "Discussion" in: Yoshimoto et al., 1980). Nor is the connection between the excitation and enzyme catalysis clear, as the pH optimum for the PG peroxidase reaction is about 8, where the intensity of the tryptophan luminescence is at its lowest, corresponding to only 25% of the peak light emission seen at pH 6.0.

Chemiluminescence is not seen during activation of the peroxidase reaction either by epinephrine or by guaiacol (o-methoxyphenol), i.e., in a typical peroxidase reaction in which either of these compounds is dehydrogenated stoichiometrically with the amount of product (PGH) formed.

It was alluded to earlier that the nature of the physiological activator in vivo of the PG peroxidase was unknown; it could be tryptophan, or epinephrine. Another candidate might be uric acid. Ogino et al. (1979) found in the ultrafiltrate of the 105,000 $\times$ g supernatant of homogenates of bovine seminal vesicles an activator of the PG peroxidase, which they identified as uric acid. Uric acid is as effective as tryptophan in activating the peroxidase; its maximum effects are seen at concentrations of 2–4 mM. The effects of tryptophan and uric acid are not additive. The manganese protoporphyrin-assisted synthesis of PGG_1 from 20:3ω6 by the purified PG endoperoxide synthetase is not affected by uric acid. 2,8-Dihydroxyadenine, mimicking the lactim form of uric acid (Fig. 10.15), is as effective as uric acid; no other purine or pyrimidine derivative, nucleoside, or nucleotide can activate the PG peroxidase. The effect of uric acid is a new and previously unknown, possibly physiological, function of the substance.

A characteristic of the PG endoperoxide synthetase, whether attached to microsomes or reconstituted with hematin from the purified apoproteins, is that it suffers a self-catalyzed irreversible inactivation, first demonstrated by Smith and Lands (1972). Arachidonic acid, added either to fresh microsomes of seminal vesicles or to the purified PG endoperoxide synthetase activated with hematin,

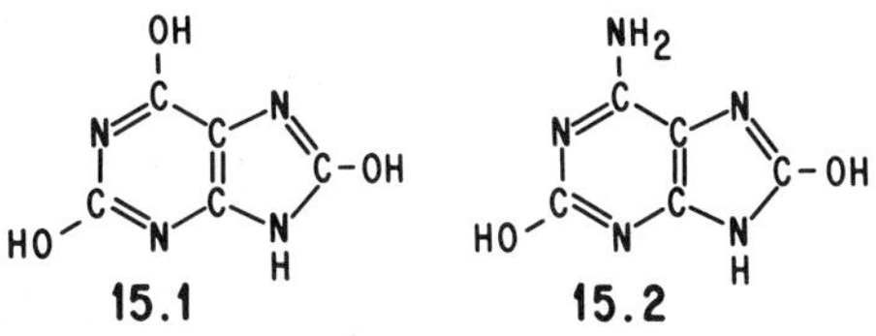

FIGURE 10.15. The lactim form of uric acid (15.1) and the structure of 2,8-dihydroxyadenine (15.2). Both are activators of prostaglandin peroxidase.

initiates PG synthesis, but the reaction proceeds linearly for only 15–30 s and comes to a stop after about 60 s, long before all substrate has been consumed (Egan *et al.,* 1976; Ogino *et al.,* 1978). Addition of further amounts of substrate does not restart the reaction. Fresh enzyme added to a stopped reaction mixture restarts the reaction, but the enzyme is rapidly inactivated as before. The same type of inactivation is observed with PGG_2 as substrate. Egan *et al.* (1976) ascribe this inactivation to the oxidation of the enzyme by oxygen-centered radicals generated during the reductive breakdown of the hydroperoxide on PGG_2. They noted that PGG_2 added to microsomal enzymes 5 s before addition of arachidonic acid had no effect on the arachidonic acid-dependent O_2 consumption, but a 2-min preincubation of the microsomes with PGG_2 reduced the rate of O_2 consumption on addition of arachidonic acid by 90%. However, the self-destruction cannot be ascribed solely to the peroxidase reaction, as the endoperoxide synthetase, activated with the manganese protoporphyrin, is also inactivated during catalysis despite the absence of hydroperoxidase activity. The self-destruction of the synthetase occurs both during the bis-dioxygenase and the peroxidase phases of the reaction catalyzed by this enzyme.

O'Brien and Rahimtula (1980) ascribe the self-catalyzed inactivation of the PG endoperoxide synthetase to the destruction of the heme on the enzyme by peroxide radicals. They observed that addition of arachidonic or 15-hydroperoxy-eicosa-5,8,11(*cis*),13(*trans*)-tetraenoic acid or PGG_2 to sheep vesicular gland microsomes resulted in the formation of a spectral complex absorbing at 435 nm which quickly decayed, but this decay was accompanied by a steadily deepening trough of absorption at 412 nm, the absorption band of the PG endoperoxide synthetase hemoprotein. PGH_2 did not produce these effects. The absorption at 435 nm is attributed to a higher oxidation state of the heme moiety of the enzyme, in analogy with the complex formed between cytochrome P-450 and hydroperoxides. Again, the destruction of heme on the synthetase cannot account fully for the inactivation of the enzyme since the rate of inactivation of the enzyme is much faster than the rate of the destruction of the heme as judged by the spectral disappearance of the Soret band of the hemoprotein. The self-destruction of this synthetase is still incompletely understood; it is regarded as a " . . . feature intrinsic to cyclooxygenase catalysis and probably caused by reaction intermediates" (Hemler and Lands, 1980, p. 6259). Whether the self-catalyzed inactivation of PG endoperoxide synthetase plays a significant role *in vivo* by limiting the amount of PGs synthesized in response to a stimulus is not known.

Many of the substances that activate the PG peroxidase reaction (phenol, guiacol, tryptophan, hydroquinone, methional (3-thiomethylpropionaldehyde), ascorbic acid) also partially protect the enzyme against the self-catalyzed inactivation. The protection is partial in the sense that the protectors prolong the life of the enzyme before inactivation, resulting in higher conversion of substrate into products than in the absence of the protectors. It is of interest that sulfhydryl-blocking agents, such as *N*-ethylmaleimide and *p*-chloromercuribenzoate, have no effect on the enzyme, but agents that reduce disulfide bonds, such as dithiothreitol, powerfully inhibit the enzyme, which suggests the existence of vital disulfide bridges in the molecule.

The reconstituted PG endoperoxide synthetase can be assayed in the presence of, e.g., phenol or tryptophan either by determination of PGH formed from a labeled substrate fatty acid, or by measurement of O_2 consumption with an O_2-electrode, 2 mol O_2 being consumed per mol of fatty acid used. A "coupled" assay was described by Roth et al. (1981). These authors found that flufenamic acid (3′-trifluoromethyldiphenylamine-2-carboxylic acid), an anti-inflammatory and analgesic drug, greatly stabilized the purified synthetase and that in its presence NADH was oxidized on addition of substrate (arachidonic acid) at a rate comparable to substrate consumption as measured by an O_2-electrode. The mechanism of this coupled assay is not known.

10.2.4.2. Prostaglandin Isomerases and Reductases

The enzymes that convert prostaglandin Hs to PGEs have been named prostaglandin endoperoxide:E isomerases (EC 5.3.99.3). Among these the best characterized is the enzyme solubilized and purified from microsomes of bovine seminal vesicles (Miyamoto et al., 1974; Ogino et al., 1977). This is a very labile enzyme with a half-life of about 30 min at pH 8.0 and 28°C. However, thiol compounds, such as glutathione, 2-mercaptoethanol, dithiothreitol, and cysteine substantially protect the enzyme against inactivation. Mercuric chloride and p-chloromercuribenzoate almost completely inactivate this isomerase even in the presence of 10- to 20-fold excess of glutathione. Other sulfhydryl-blocking agents, such as iodosobenzoate, N-ethylmaleimide, and iodoacetic acid, produce only partial inactivation. The enzyme catalyzes the isomerization of PGH_1 and PGH_2 to PGE_1 and PGE_2, with glutathione as the coenzyme. The exact role of glutathione in the reaction is not known, as it is not oxidized detectably during the isomerization. No other thiol, nor ascorbic acid, can substitute for glutathione to promote catalysis. The enzyme is also active with PGG_1 and PGG_2 as substrates, but the products of their isomerizations are 15-hydroperoxy-PGE_1 and 15-hydroperoxy-PGE_2, both of which can be reduced to PGE_1 and PGE_2, respectively, with $SnCl_2$ or sodium sulfite. The isomerization of the PGGs to the 15-hydroperoxy-PGEs raises the possibility that the PGEs could be derived also from the PGGs, through the sequential action of the endoperoxide isomerase on PGGs, followed by the prostaglandin peroxidase reaction. However, such sequence is thought unlikely, particularly as the endoperoxide PGE isomerase reacts with the PGGs at only one-half the rate displayed with PGHs.

The prostaglandin endoperoxide:E isomerase has also been solubilized with Triton X-100 and deoxycholate from microsomes of sheep seminal vesicles, but has not been purified. It has only been reported that it is a highly labile enzyme, that it is strongly inhibited by p-chloromercuribenzoate and N-ethylmaleimide, and that it requires glutathione as coenzyme, in analogy with the properties of the enzyme purified from bovine seminal vesicles (Nugteren and Christ-Hazelhof, 1980).

Prostaglandin endoperoxide:E isomerase has a wide distribution among the organs, but, in addition to the seminal vesicles, the kidney medulla and cortex are particularly rich sources of this enzyme. At both these sites the enzyme is associated with microsomes and is of exceptionally high activity ($V_{max} \sim 30$ nmol/min·mg), although its precise tubular localization is not known. Although the renal cortex is rich in the PG endoperoxide:E isomerase, it is very poor in the endoperoxide synthetase ($V_{max} \sim 0.2$ nmol/min·mg), which is much higher in the renal medulla with a V_{max} of 2.0 nmol/min·mg (Sheng et al., 1982).

Prostaglandins D_1 and D_2 have been referred to earlier (cf. p. 167) as spontaneous degradation products of PGH_1 and PGH_2 in buffered solution. They were originally thought to be biologically inactive, or of no significance. However, PGDs turned out to have important biological actions. Smith et al. (1974b) demonstrated that PGD_2 was about twice as active as PGE_1 in inhibiting ADP- or collagen-induced aggregation of human platelets. Needleman et al. (1976a) recorded that microsomes of sheep seminal vesicles converted eicosapentaenoic acid (20:5ω3) into PGH_3, but—in contrast to PGH_2, which causes indomethacin-treated platelets to aggregate (cf. Section 10.5)—PGH_3 inhibited aggregation. It turned out that PGH_3 added to platelet-rich plasma was converted by components of the plasma into PGD_3, which was the antiaggregating factor (Whitaker et al., 1980). PGD_2 is also highly active in causing smooth-muscle contractions, e.g., on strips of gerbil colon or rat stomach, on rabbit aorta or tracheal and bronchiolar smooth muscle (Hamberg et al., 1975a; Wasserman et al., 1977). PGDs assumed further importance when it was discovered that preparations of some organs, e.g., homogenates of rat brain, strips of the rat gastrointestinal tract or peritoneal macrophages, synthesized far more PGD_2 than any other prostaglandin from arachidonic acid (Abdel-Halim et al., 1977; Knapp et al., 1978; Roberts et al., 1978a; cf. also Granström, 1981a). Prostaglandin D_2 is also a marker of malignant melanoma cells (Fitzpatrick and Stringfellow, 1979).

Prostaglandin Ds, 9,15-dihydroxy-11-ketoprostenoic acids, must be classed now, after their late discovery, among the primary prostaglandins. Thus, the total number of primary prostaglandins is nine and not six, as was related in the earlier part of this chapter (cf. p. 152). The PGDs (PGD_1, PGD_2, and PGD_3) have the same structures as the correspondingly numbered PGEs, except that the keto group is on the ring in position 11 instead of position 9. They are distinguished from PGEs not only by their slightly different chromatographic behavior, but also by their reaction with $NaBH_4$. Whereas reduction of PGEs with $NaBH_4$ gives a mixture of PGF_αs and PGF_βs, reduction of PGDs with $NaBH_4$ gives exclusively PGF_αs.

The first prostaglandin endoperoxide:D isomerases were recognized in animal sera, specifically associated with albumin. During the "spontaneous" decomposition of PGHs in buffer the main product is PGE and a little PGD; the usual ratio of PGD:PGE in such spontaneous decomposition is 0.20–0.25. Serum albumins of cow, sheep, pig, guinea pig, and to a lesser extent of man promote the isomerization of PGH_1 and PGH_2 into PGD_1 and PGD_2, respectively. In the presence of bovine or guinea-pig serum albumin (3–30 mg/ml), PGH_2 is converted mostly into PGD_2 and to a lesser extent into PGE_2 with PGD_2:PGE_2 ratios of 4.4–4.6 (Hamberg and Fredholm, 1976; Christ-Hazelhof et al., 1976). Eicosanoic acid (20:0) at 10^{-3} M (23 mol 20:0/mol bovine serum albumin (BSA)) severely inhibited the isomerization, indicating that the fatty acid binding sites of albumin need to be free for albumin to be able to catalyze the isomerization. Boiling of a solution of BSA for 20 min abolished the ability of the protein to promote the isomerization of PGH_2 to PGD_2. Whether the isomerization activity was truly the property of serum albumin or of some other agent attached to it could not be judged from the reported experiments. The situation could have been thought to be akin to that of lecithin–cholesterol acyltransferase (LCAT), which is difficult to dissociate from plasma high-density lipoproteins (HDL). However, this uncertainty regarding the possible catalytic activity of serum albumin was elegantly resolved by Watanabe *et al.* (1982) who studied the mechanism of the biosynthesis of PGD_2 in human platelet-rich plasma. They clearly demonstrated that PGD_2 arose from PGH_2 synthesized by the platelets and released into the plasma and that PGH_2 was there isomerized to D_2. They identified the PGH_2:PGD_2 isomerase with serum albumin by sodium dodecyl sulfate/polyacrylamide gel electrophoresis, by electrofocusing, and by immunoelectrophoresis. They carried out appropriate kinetic experiments and showed that the amount of PGD_2 formed from PGH_2 was proportional to the amount of protein, that the pH optimum of the reaction was 9.0, that the isomerase reaction required no cofactors and that it was abolished by boiling the protein. The K_m value for PGH_2 was 6.2 μM, and the enzyme was saturated at 50 μM concentration of PGH_2. This isomerase function of serum albumin is one of its two catalytic activities known so far.

Christ-Hazelhof *et al.* (1976) purified from sheep lung two isozymes of glutathione-S-transferase, SL_1 and SL_2, which, with glutathione as cofactor, converted PGH_2 into PGD_2 and $PGF_{2\alpha}$. SL_1 and SL_2 have an M_r of 23,500. These transferases are similar to the known four glutathione-S-transferases (A,B,C and E) of rat liver, which also consist of two subunits and have an M_r of 45,000. These liver transferases also catalyze the conversions of PGH_2 into $PGF_{2\alpha}$, PGE_2, and PGD_2. The lung glutathione-S-transferase is similar to transferase B of rat liver, also known as ligandin. These glutathione-S-transferases thus appear to function both as prostaglandin endoperoxide:E isomerases and F reductases.

A prostaglandin endoperoxide:D isomerase has been purified from rat-brain homogenates (Shimizu *et al.*, 1979a). In contrast to the prostaglandin endoperoxide:E isomerase, which in brain—as in seminal vesicles—is attached to microsomes, the PGD isomerase is present in the cytosol of brain homogenates. The enzyme is a single polypeptide with M_r of 80,000–85,000. Its K_m value for PGH_2 is 8 μM, and its pH optimum is 8.0, although at that pH it is rapidly inactivated. Glutathione, dithiothreitol, and 2-mercaptoethanol stabilize the enzyme, though they are not required for catalysis. This enzyme is also active with PGG_2, which it converts to 15-hydroperoxy-PGD_2. The latter is readily converted to PGD_2 by purified PG endoperoxide synthetase (peroxidase) in the presence of heme and tryptophan.

Prostaglandin endoperoxide:$F_{2\alpha}$ reductase has so far been identified only in microsomes of cow and guinea-pig endometrium (Wlodawer *et al.*, 1976). These microsomes efficiently convert PGH_2 into $PGF_{2\alpha}$ without any added cofactors. The same or similar reductase was not found in guinea-pig lung, liver, or kidney microsomes; an odd situation, as the synthesis of $PGF_{2\alpha}$ from 20:4ω6, independent of the synthesis of PGE_2, was first demonstrated in homogenates of guinea-pig lung.

10.3. Metabolism of Prostaglandins

Prostaglandins are rapidly metabolized; their metabolic transformations have been extensively studied in several species, including man (male and female) mostly with the aid of isotopically labeled compounds. The reader is mostly referred to reviews by Samuelsson (1970), Samuelsson *et al.* (1971, 1975, 1978), Samuelsson (1978), Lands (1979), Granström (1981a), and Flower (1981) for references to the very large number of publications on this topic; only a few of the original papers will be cited here.

The simplest transformations of prostaglandins of the E-series are their dehydration to PGAs and isomerization of the latter to PGBs and the hydroxylations of both PGAs and PGBs at C-19, as evidenced by their presence in human seminal fluid (cf. pp. 157, 159). However, the presence of a prostaglandin-E dehydratase or of a prostaglandin-A isomerase or of 19-hydroxylase in male accessory organs has not been reported. On the other hand, human serum contains three prostaglandin-metabolizing enzymes: (i) a dehydratase that converts PGEs into PGAs; (ii) a PGA → PGC isomerase (PGC_1 = 15-hydroxy-9-ketoprosta-11(12),13-dienoic acid); and (iii) a PGC → PGB isomerase. The latter is apparently associated with crystalline human serum albumin and also with rabbit serum albumin. The rabbit serum is very active in respect of the two isomerases, but contains no PGE dehydratase (Polet and Levine, 1975). Hydroxylation of PGA, at C-19 and C-20, has been demonstrated with the microsomes, boiled 105,000 × *g* supernatant of guinea-pig liver and NADPH, and also with the microsomal fraction of human liver. The ω- and ω1-hydroxylations of the various PGs are catalyzed by typical monooxygenase systems of microsomes with NADPH as the coenzyme in analogy with the ω-oxidation of fatty acids (cf. Chapter 7).

Intravenously administered PGs rapidly disappear from the circulation and their metabolites are excreted in the urine within a few hours. In an early study, examination by radioautography of whole body sections of female mice injected with radioactive PGE_1 and $PGF_{2\alpha}$ showed accumulation of radioactivity in liver, kidneys, myometrium, but not in endometrium, and an unexplained and striking accumulation in connective tissue. Those early pictures may have foreshadowed the discovery of specific PG receptors.

In man, male or female, intravenously injected [^{3}H]PGE_2 or [^{3}H]$PGF_{2\alpha}$ disappears so rapidly from the circulation that in 1.5 min only about 3% of the dose is identifiable as the unchanged compounds and by 5 min they vanish. One and one-half minute after such injections about 40% of the injected PGs are found in the circulation in the form of their 15-keto-13,14-dihydro derivatives, which are eliminated from the blood with a half-life of 8–9.5 min. These initial transformations of PGEs and PGFs into biologically inactive products are attributed to metabolism mainly by the lungs. Indeed, the same rapid disappearance of PGEs and PGFs from the circulation is seen during perfusion of isolated lungs with these PGs. Only PGAs seem to escape transformations by the lung; for this reason PGAs have been thought to be circulating hormones. Enzymes inactivating PGAs exist in other organs.

From the examination of the metabolites in the blood and urine of human subjects and animals, and from the transformations of PGs in extracts of organs, a composite picture of the metabolism of PGs has emerged. Generally, two phases of PG metabolism can be distinguished: (i) a rapid phase that modifies the C-13 to C-15 region of the molecules, and (ii) a slower phase that results in the β-oxidation of the carboxyl-bearing side-chain and ω-oxidation coupled with β-oxidation from the ω-end of the side-chain attached to C-12. A third type of reaction involves specifically PGEs and PGDs by the reduction of the respective 9- and 11-keto groups on the ring, thereby leading to metabolites similar to those derived from $PGF_\alpha s$.

The first enzymatic attack on PGs is on the 15-hydroxy group by 15-hydroxyprostaglandin dehydrogenases (PGDHs), which convert the PGs into their 15-keto derivatives. There are several PGDHs found in many organs with different substrate specificities. The most widely studied PGDH is that found most abundantly in the lungs which is specific for PGEs and PGFs and only poorly active with PGAs and totally inactive with PGBs and PGDs. Two types have been recognized: the type I enzyme needs NAD^+ and the type II uses $NADP^+$ as coenzyme (Lee and Levine, 1975); both are found in many organs besides the lungs (liver, kidney, adrenals, intestine, spleen). The reaction catalyzed by the 15-hydroxy-PGDHs is generally thought to be irreversible. However, a PGDH purified from human placenta, and which is most active with PGE_1, PGE_2, and $PGF_{2\alpha}$, also catalyzes the reversible reduction of the 15-keto-PGs with NADH (Braithwaite and Jarabak, 1975). Another reductase that specifically reduces the 15-keto group in 13,14-dihydro-15-keto-PGEs or PGFs must also exist, but its organ distribution and coenzyme requirement have not been established. The existence of this reductase has been inferred from the appearance of 13,14-dihydro-11α, 15S-dihydroxyprostenoic acid devoid of isotope at C-15 after incubation of [15-^{2}H]PGE_2 with the 100,000 $\times$ g supernatant of guinea-pig liver homogenate (cf. Fig. 10.16).

The soluble fraction of rabbit kidney papilla contains a 15-hydroxy PGDH which is specific for PGAs but—surprisingly—does not require NAD^+ as cofactor. This enzyme is inactive with PGEs and PGFs. Extracts of kidney cortex, however, dehydrogenate E,F, and A PGs with NAD^+ as coenzyme (Oien et al., 1976).

An $NADP^+$-linked PGDH specific for PGD_2 has been identified in the cytosol of pig brain. This enzyme is inactive with prostaglandins A_1, A_2, B_1, and B_2 and is also very poorly active with PGD_3, E_2, or $F_{2\alpha}$. Another NAD^+-linked PGDH active with PGB_2, E_2, and $F_{2\alpha}$ has also been found in pig brain (Watanabe et al., 1980).

The action of PGDHs seems closely linked to that of a cytosolic Δ^{13}-reductase which, probably with NADPH, reduces the 13(14)-double bond of the 15-keto-PGs to form the 13,14-dihydro-15-keto-PGs. It is not known whether multiple

FIGURE 10.16. The first metabolic transformations of prostaglandins catalyzed by 16-hydroxyprostaglandin dehydrogenase and Δ^{13}-reductase. X and Y represent the substituents at positions 9 and 11 in the PGs; in PGF$_a$s X = Y: HO—; in PGEs X: O=, Y: HO—; in PGDs X: HO—, Y: O=. The 15-hydroxy-13,14-dihydro-PGs are products of reduction of the 15-keto-13,14-dihydro-PGs and not of the reduction of the 13,14-double bond in the PGs.

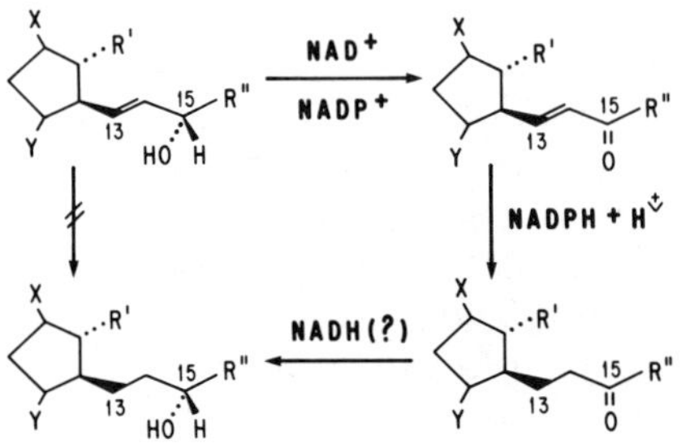

forms of this reductase exist that might be specific for individual 15-keto-PGs. A Δ^{13}-reductase from brain that reduced 15-keto-PGD$_2$ with NADPH was purified together with the PGDH specific for PGD$_2$ (Watanabe et al., 1980).

The effect of these initial transformations of PGs, summarized in Fig. 10.16, is largely to inactivate them. Thus, for example, 15-keto-PGE$_1$ and 13,14-dihydro-15-keto-PGE$_1$ have lost virtually all of the vasodepressor and smooth-muscle-stimulating activities of PGE$_1$. Interestingly, rereduction of the 15-keto to the 15-hydroxy function in 13,14-dihydro-15-keto-PGE$_1$ not only restores, but even augments, the vasodepressor activity of the resulting 13,14-dihydro-PGE$_1$ as compared to PGE$_1$ although it only partially restores the smooth-muscle-stimulating activity of PGE$_1$ (cf. Fig. 2 in: Flower, 1981). PGE$_1$ is also an inhibitor of platelet aggregation, but 15-keto-PGE$_1$ is even more active in this respect, being equipotent with prostacyclin, PGI$_2$ (Wong et al., 1979a; quoted by Granström, 1981a: Ref. 20).

ω-Oxidation produces ω-hydroxy- and ω-carboxy-PGs, and β-oxidation of the carboxyl-bearing side-chain and of the ω-carboxy-PGs may result in loss of two, four, or six carbon atoms. The β-oxidations usually involve not the primary PGs but the substances that have already undergone modifications in the C-13 to C-15 region, though primary PGs (E$_1$, E$_2$, F$_{1\alpha}$, F$_{2\alpha}$) and PGA$_2$ and 15-keto-PGE$_2$ can all be oxidized by liver mitochondria. The β-oxidations may well be carried out by the β-oxidation system of fatty acids as they absolutely require carnitine and ATP, and are "sparked" by the addition of malate (Johnson et al., 1972).

These reactions may be further complicated by the interconversions, probably mostly in the liver, of PGEs and PGF$_a$s, and the conversion of PGD$_2$ into PGF$_{2\alpha}$. These diverse reactions and their combinations lead to a multiplicity of metabolites of PGs.

The simplest transformations can be seen in incubations of PGs with soluble extracts of organs, such as, e.g., of PGE$_2$ in liver extract illustrated in Fig. 10.17. In the whole organism, where the PGs are exposed to the multiple enzyme systems of whole cells, they usually suffer much greater transformations. In man, male and female, intravenous injection of PGE$_1$ or PGE$_2$ results in the excretion in the urine of one main metabolite, 7α-hydroxy-5,11-diketotetranorprosta-1,16-dioic acid (Fig. 10.18). When tracer doses of radiolabeled PGE$_1$ or PGE$_2$ are administered to man, compound **18.1** is excreted in amounts equal to 15–16% of the dose. By quantitative determination in the urine of this common metabolite

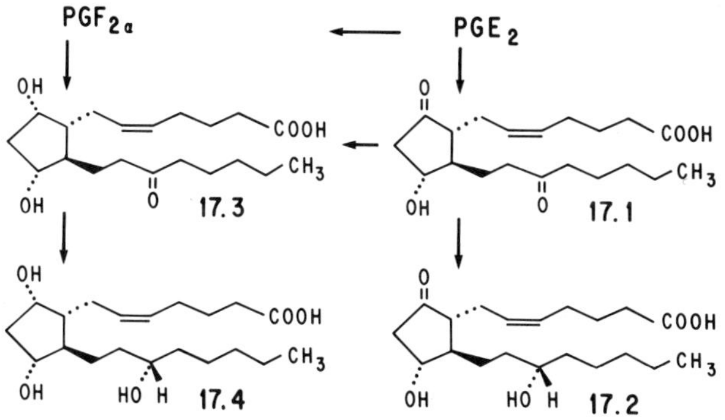

FIGURE 10.17. Metabolism of PGE$_2$ in the 100,000 × g supernatant of guinea-pig liver homogenate. Neither compound **17.2** nor **17.4** is formed directly from PGE$_2$ or PGF$_{2\alpha}$, but from the 13,14-dihydro-15-keto derivatives, **17.1** and **17.3**, respectively (after Hamberg and Israelsson, 1970; Hamberg and Samuelsson, 1972.)

of PGE$_1$ and PGE$_2$ it is possible to arrive at an estimate of the secretion of these two PGs in the body (Hamberg and Samuelsson, 1971). In two female subjects values of 23 and 48 μg, and in three males 109, 184, and 226 μg per 24 hr were found.

There are apparently species differences in the metabolism of PGE$_1$ and PGE$_2$. In the guinea pig the main urinary metabolite of these PGs is the 5β,7α-dihydroxy-11-ketotetranorprostanoic acid (Fig. 10.18). The basal daily excretion of metabolite **18.2** in the male guinea pig is 1.97 + 0.44 μg/kg body weight (calculated from data of Hamberg and Samuelsson, 1972), and it is almost completely abolished by the daily administration of 50 mg of indomethacin. The effect of

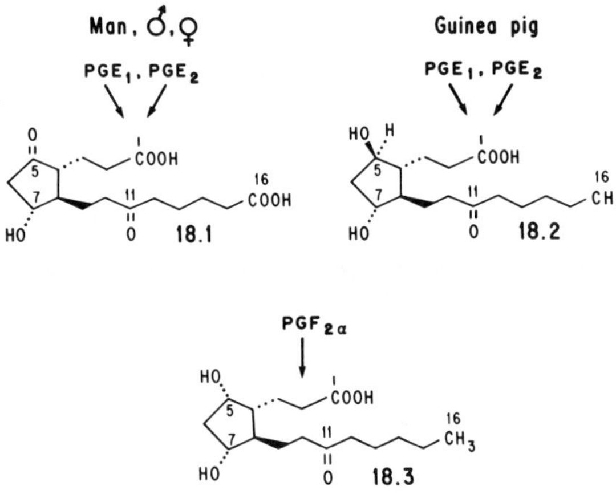

FIGURE 10.18. Major urinary metabolites of PGE$_1$ and PGE$_2$ in man, male or female: 7α-hydroxy-5,11-diketotetranorprosta-1,16-dioic acid (**18.1**); and in the guinea pig: 5β,7α-dihydroxy-11-ketotetranorprostanoic acid (**18.2**). One of the major urinary metabolites of PGF$_{2\alpha}$ in the guinea pig and monkey is the 5α,7α-dihydroxy-11-ketotetranorprostanoic acid (**18.3**).

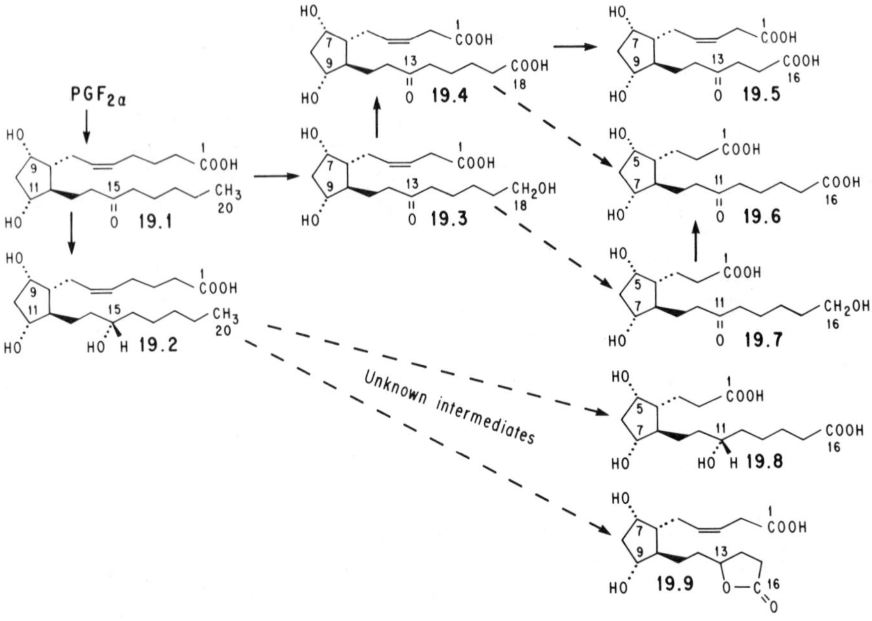

FIGURE 10.19. Metabolism of PGF$_{2\alpha}$ in women. The main urinary metabolite is 5,7-dihydroxy-11-ketotetranorprosta-1,16-dioic acid (**19.6**). Compounds **19.6, 19.7,** and **19.8** can also exist as their γ-lactones formed between the 1-carboxyl and 5-hydroxyl group, and **19.9** can be hydrolyzed to the 13-hydroxy-1,16-dioic acid (after Granström and Samuelsson, 1971a,b; Granström, 1972).

indomethacin is clearly on the synthesis of the PGs and not on their degradation since tritio-**18.2** is excreted in the urine in the usual amounts (about 28% of the dose) after administration of [^{3}H]PGE$_2$ to indomethacin-treated guinea pigs.

The multiplicity of products that are formed during the metabolism of prostaglandins is well illustrated by the metabolism of PGF$_{2\alpha}$ in women (Fig. 10.19) and of PGD$_2$ in the monkey. In women the main urinary metabolite of PGF$_{2\alpha}$ is the 5,7-dihydroxy-11-ketotetranorprosta-1,16-dioic acid (**19.6**), but in addition at least eight others have been recognized, which arise by the combination of the various reactions described earlier in this section. The tetranor acids, such as compounds **19.6, 19.7,** and **19.8,** probably arise by two successive β-oxidations of the original carboxyl-bearing side-chain. Their formation raises the question of the changes in the remaining double bond (the original 5(6)-*cis*-double bond in PGF$_{2\alpha}$) after the first β-oxidation. It is not known whether this double bond is first hydrated to form a β-hydroxycarboxylic acid and then dehydrogenated to the β-keto acid before cleavage, or whether it is first reduced to be followed by the known reactions of β-oxidation of the fatty acids. In truth, it is not known with certainty whether the β-oxidation of the PGs proceeds via their coenzyme A derivative, though both Hamberg (1968) and Johnson *et al.* (1972), who studied the mitochondrial oxidation of PGs, added coenzyme A to their incubations of mitochondria with PGs.

The most diversified *in vivo* metabolism among the various PGs was observed for PGD_2 in a cynomolgus monkey. Ellis *et al.* (1979) isolated from the urine of the monkey infused with [^{3}H]PGD_2 no less than 20 metabolites. Eleven of these were $PGF_{2\alpha}$ metabolites, i.e., with the cyclopentane-1,3-diol structures; the others retained the characteristics of PGD_2 and contained the cyclopentan-1-ol-3-one structure. The main urinary metabolite of PGD_2 was the C_{18} dinor-$PGF_{2\alpha}$. Among the $F_{2\alpha}$-metabolites, in addition to $PGF_{2\alpha}$ itself, six were identical with structures found in the urine of women who were given $PGF_{2\alpha}$ intravenously: compounds **18.3, 19.3, 19.4, 19.5, 19.6,** and **19.7**. The metabolites with the cyclopentan-1-ol-3-one structure were mostly the counterparts of the metabolites with the cyclopentane-1,3-diol structure. Ellis *et al.* (1979) called attention to the fact that quantitative determination of $PGF_{2\alpha}$ metabolites in the urine "may not be a specific indicator of prostaglandin $F_{2\alpha}$ biosynthesis." Similarly, since PGD_2 is converted into $PGF_{2\alpha}$, $PGF_{2\alpha}$ may contribute to the biological effects of PGD_2.

The structure of PGD_2 suggests that this compound, apart from being subject to multiple enzymatic attacks *in vivo*, might also be fairly unstable and prone to dehydrations. The instability is suggested (i) by the proximity of the C-10 methylene group to the ketone at C-11, in analogy with the structure of PGE_2, and (ii) by the presence at C-12 of an allylic hydrogen atom which is also α to a ketone. Indeed, Fukushima *et al.* (1982) reported the dehydration of PGD_2 in Tris-Cl buffer at pH 7.2 at 37°C to the 9-deoxy-Δ^9-PGD_2 (15-hydroxy-11-keto-5,9,13-prostatrienoic acid) and proposed, following tradition, PGJ_2 as the symbol for this substance (**20.1** in Fig. 10.20).

PGJ_2 is much weaker than PGD_2 in preventing ADP-induced aggregation of platelets and has barely detectable smooth-muscle-stimulating activity. However, PGJ_2 is a potent inhibitor of cell growth, about three times more effective than PGD_2. The IC_{50} value of PGJ_2 for the inhibition of the growth of cultured L1210 leukemia cells is 1.8 μM as compared to the IC_{50} value of 5.2 μM for PGD_2 (Fukushima *et al.*, 1982).

Serum albumin also catalyzes dehydrations and isomerizations of PGD_2 in a dose-dependent fashion and as a first-order reaction (Fitzpatrick and Wynalda, 1983). The structures of the dehydration and isomerization products of PGD_2 are

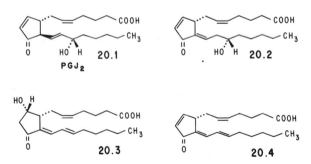

FIGURE 10.20. Dehydration and isomerization products of PGD_2, 9,15-dihydroxy-11-ketoprosta-5,13-dieonic acid. Compounds **20.2–20.4** are the products of catalysis by albumin on PGD_2. The formation of PGJ_2 (**20.1**) was observed in aqueous buffer.

shown in Fig. 10.20. Three products of the catalytic action of albumin on PGD_2 have been identified by mass spectrometry and NMR spectrometry: (i) 9-deoxy-11-keto-15S-hydroxy-5,9,12-prostatrienoic acid (**20.2**); (ii) 15-deoxy-9α-hydroxy-11-keto-5,12,14-prostatrienoic acid (**20.3**); and (iii) 9,15-dideoxy-11-keto-5,9,12,14-prostatetraenoic acid (**20.4**).

10.4. Mode of Action of Prostaglandins

Prostaglandins have many regulatory functions, but they differ from the classical hormones synthesized in specific organs since PGs are synthesized in almost every organ and cell, and probably exert their effects at or near their site of synthesis. They are very short-lived and even those that might reach the circulation are rapidly inactivated by a single passage through the lungs with the possible exceptions of PGAs and prostacyclin (PGI_2; see Section 10.5) though the natural concentrations of these in blood are unknown. Generally, PGs are regarded as local hormones, falling in the same class as histamine and serotonin, and for which the term "autactic" hormones (from the Greek words autos = self and akos = cure, remedy; hence meaning roughly "self-curing" or "self-medicating") has been coined. Such substances might also be called "cellular" hormones, since they exert their effects on some intracellular component at the site of their synthesis (cf. Zor and Lamprecht, 1977). Thus, the existence of high-affinity receptors specific for PGEs and PGFs on the plasma membranes of a large variety of cells (for review, cf. Samuelsson *et al.*, 1978) may seem to be redundant. However, these specific receptors could be regarded as having evolved to promote intercellular communication within the same or neighboring tissues (see also below). Prostaglandin receptors on endoplasmic reticulum, on mitochondrial membranes, and in the cytosol have also been identified.

The ubiquitous nature of PGs, their diverse effects in different organs, their rapid inactivation, and the different effects evoked by PGs with relatively minor differences in their structures, e.g., PGEs vs. PGFs, make it difficult if not impossible to ascribe a single mode of action to the various PGs, or to study their effects by methods other than pharmacological techniques. The vast amount of research carried out during the last 20 years on the effects of PGs has been done mostly by the use of exogenous PGs injected into experimental animals or added to isolated organ systems. It is hard to decide whether the exogenous PGs truly mimic the actions of the endogenously synthesized substances. However, the results of such studies have not infrequently been corroborated by the use of PG endoperoxide synthetase inhibitors, such as aspirin, indomethacin, eicosa-5,8,11,14-tetraynoic acid, the acetylenic analogue of arachidonic acid, or some analogues of PG endoperoxides, e.g., 9,11-azido-15-hydroxyprostadienoic acid and others.

In one of the earliest studies on the biochemical actions of prostaglandins it was found that PGE_1 at a very low concentration (0.1 μg/ml) and to a lesser extent PGE_2 (but not $PGF_{2\alpha}$) counteracted in rat epididymal fat pads the lipolytic effects of catecholamines, glucagon, ACTH, and thyroid-stimulating hormone (TSH) and the activation of the hormone-sensitive triacylglycerol lipase by these hormones (Steinberg et al., 1964; Steinberg and Vaughan, 1967). Since the activation of the adipose tissue lipase by epinephrine, norepinephrine, and the other hormones is mediated through 3′,5′-cyclic-AMP (cAMP)-dependent phosphorylation of the hormone-sensitive triacylglycerol lipase, the countereffect of PGEs was attributed to inhibition of adenylate cyclase, particularly as the increased lipolysis elicited by cAMP or N^6-2′-O-dibutyryl cAMP added to intact rat epididymal fat pads was not counteracted by PGEs. The idea that PGEs acted by inhibiting adenylate cyclase was further reinforced by the observations that PGEs also inhibited the increased water permeability and electrolyte exchange in toad bladder elicited by vasopressin. The effect of vasopressin is known to be mediated by cAMP.

PGE_1 also blocked in dogs the vasopressor effects of epinephrine and norepinephrine, and also largely suppressed the rise in plasma free fatty acids, but not the rise in glucose, after the injection of epinephrine.

Although the simple concept that PGEs generally acted by inhibiting adenylate cyclase was subsequently shattered and had to be revised, the observations on the effects of PGEs on hormone-stimulated lipolysis established the first link between PGs and cyclic nucleotides.

The revision of the concept that PGEs acted by inhibiting adenylate cyclase became necessary when some paradoxical observations came to light. Injection of relatively large doses of PGE_1 into dogs (50 μg/kg) seemingly confirmed the observations made on hormone-stimulated epididymal fat pads in that such injections caused a drop of free fatty acid levels, commensurate with the inhibition of lipolysis, in the venous blood of normal animals or animals given epinephrine. However, continuous infusion of 0.1–0.2 μg of PGE per kg in man and unanesthetized dogs caused a rise in blood free fatty acids (bound to serum albumin), indicating an increased lipolysis. This rise could be prevented by agents blocking β-adrenergic receptors. Accordingly, it seemed that PGE in vivo had a dual action: one, seen at low concentrations, mediated through the sympathetic nervous system, very likely causing release of epinephrine or norepinephrine, which enhance lipolysis, and the other, apparent at high doses, suppressing adenylate cyclase and thus inhibiting lipolysis.

The real surprise and puzzle came when it was found that the cAMP content of epididymal fat pads incubated with PGE_1 actually rose substantially and that this rise was not accompanied by an increased lipolysis, whereas cAMP, or even theophyllin, added to fat pads increased lipolysis. The puzzle was solved when adipocytes were separated from the other components of fat pads (described as stromatovascular components) and it was found that the rise in cAMP was confined entirely to the stromatovascular elements (Butcher and Baird, 1968). The only type of cells in which prostaglandins (E_1, E_2, and to lesser extent $F_{1\alpha}$ and A_1) lower levels of cAMP are the hormonally stimulated white and brown adipocytes (Hittelman and Butcher, 1973). The basal levels of cAMP in isolated adipocytes are not affected by PGs, and the mode of action of PGs on the cAMP contents of the hormonally stimulated adipocytes is obscure, as PGs do not affect in any way the phosphodiesterase that hydrolyses cAMP.

Shaw and Ramwell (1968) observed that hormone-stimulated lipolysis in the rat epididymal fat pad is accompanied by an increased formation of PGs (as referred to earlier, p. 161) and proposed that the PGs could be part of a feedback loop regulating the levels of cAMP and thus the extent of lipolysis also. The increased PG synthesis during hormone-stimulated lipolysis could be the result of greater availability of substrate or specific stimulation by cAMP. Stimulation of PG biosynthesis by cAMP in cultured mammalian cells, thyroid cells, Graafian follicles, adrenal glands, and adipocytes has been demonstrated. Cyclic-AMP causes a selective release of 20:4ω6 and 20:3ω6 from mouse 3T3 fibroblasts and of 20:4ω6 from pig thyroid slices, probably by activation of a triacylglycerol lipase (cf. Samuelsson et al., 1978).

In all tissues and cells other than the hormone-stimulated adipocytes (spleen, lung, diaphragm, uterus, fetal bone, platelets, the endocrine glands, corpus luteum, leukocytes, lymphocytes, neuroblastoma cells) PGEs increase the cAMP content by stimulation of adenylate cyclase (cf. Gorman, 1975).

If endogenous PGEs have actions similar to those of exogenous PGEs, one might understand the evolution of specific PGE receptors on plasma membranes. One might envisage a biochemical loop in which PGs are synthesized intracellularly on endoplasmic reticulum and then released from the cell and recaptured by the receptors, thereby achieving the activation of adenylate cyclase which is embedded in the plasma membrane.

PGEs may exert their effects by mechanisms other than regulating cAMP levels in cells as they also stimulate the Mg^{2+}-dependent Na^+/K^+-activated ATPase and may have profound effects on intracellular ions. Mobilization of Ca^{2+} ions has also been suggested as one of the primary actions of these hormones. PGE_1 is known to increase permeability of mitochondrial membranes and to stimulate efflux of Ca^{2+} from rat-liver mitochondria (cf. Zor and Lamprecht, 1977).

The action of PGFs is usually opposite to that of PGEs. Thus $PGF_{2\alpha}$ contracts the smooth muscle of blood vessels and bronchioles, whereas PGEs relax these structures. $PGF_{2\alpha}$ was found to contract bovine and canine vein strips and to cause two- to fivefold increases in the cyclic guanosine monophosphate (cGMP) content of the strips in spite of the fact that the basal (control) levels of cGMP in the strips varied five- to tenfold (Dunham et al., 1974). The $PGF_{2\alpha}$-elicited vasoconstriction is also associated with a rise of cGMP/cAMP ratio without any consistent change in cAMP levels. However, the vasodilator PGE_2 causes a rise in the cAMP content of the veins and the rise occurs before a measurable relaxation of the vessels. The bronchoconstriction and the contraction of uterine muscle elicited by $PGF_{2\alpha}$ are also known to be associated with a rise of cGMP content of the tissues. In other tissues and cells (liver, brain, cultured 3T3 fibroblasts) $PGF_{2\alpha}$ causes also a rise in cGMP content.

Thus out of these varied observations arose an attractive hypothesis postulating that PGEs and PGFs, in addition to having opposing effects on cyclic nucleotides, represent opposing arms of a bidirectional controlling mechanism in a variety of biological processes. Ca^{2+} ions may also be intimately linked in this "dualism" theory inasmuch as these ions inhibit adenylate cyclase, but stimulate guanylate cyclase (cf. Goldberg et al., 1973; Goldberg, 1976; Harris et al., 1979).

10.4.1. Biological Effects of Prostaglandins

Most of the biological effects of PGs have been inferred from the application of the exogenous compounds, often in pharmacological doses, and it is difficult to decide whether the effects observed are simply the magnified physiological effects of endogenously produced PGs *in vivo* or pathological effects of pharmacologically potent agents. Here we summarize briefly the most widely studied effects of PGs, not mentioned previously; with the exception of a few specific cases, we exclude the discussion of thromboxanes and prostacyclin as the roles of these prostanoids will be dealt with in the next sections.

10.4.1.1. Vascular Effects

Prostaglandins of the E and A series administered intravenously lower the arterial pressure in all animals including man by causing general venous vasodilation which is accompanied by an increased cardiac output and heart rate. However, the vasodilatory effect depends on the route of administration. PGE_2 and PGA_1 given intravenously dilate the cerebral basilar artery, but when applied topically to the same blood vessel they produce vasospasm. Similarly PGE_2 applied topically to other cerebral blood vessels acts as a vasoconstrictor and is thought to be the mediator of migraine. There is no explanation of these diametrically opposite effects. $PGF_{1\alpha}$ and $PGF_{2\alpha}$ are vasoconstrictors in all species studied and in all areas of the vasculature irrespective of the mode of administration (for reviews, see Horton, 1969; Wolfe and Coceani, 1979; White, 1982).

10.4.1.2. Effects on Reproductive Systems

The follicle-stimulating hormone (FSH) and the luteinizing hormone (LH) of the pituitary gland cause a rise of PG content in Graafian follicles, but only in preovulatory follicles, i.e., in those "destined" to maturation. PGE_2 injected *in vivo* or added to Graafian follicles cultured *in vitro* mimics completely the effects of LH whose functions are (i) to induce maturation of the oocyte to ovum in preovulatory follicles, (ii) to cause rupture of the follicles, and (iii) to promote the conversion of the granulosa cells (the innermost cell layer surrounding the oocyte and lining the inside of the Graafian follicle) to luteal cells. The effect of LH on the maturation of the ovum is apparently not mediated through PGs, as indomethacin fails to antagonize this effect of LH. However, indomethacin prevents the rupture of the follicles in rats and rabbits and the conversion of granulosa cells into luteal cells, but PGE_2 can overcome the inhibitory effect of indomethacin on these two functions of LH, which thus seem to be regulated through PGs. It is a conjecture that PGE_1 and PGE_2 in the follicles activate the plasminogen activator which converts plasminogen to the protease plasmin which in turn weakens the theca (the fibrous capsule of the follicle) and thus leads to the rupture of the follicle and transfer of the ovum into the Fallopian tube (cf. Zor and Lamprecht, 1977; Behrman, 1979).

The prostaglandins in human semen (mostly PGE_1 and PGE_2) may have a fundamental role in fertility. First, the PG content of the human semen is sufficiently high to cause some dilation of the cervix after coitus and promote the passage of spermatozoa into the uterus. Second, PGs are absorbed from the vagina, and the Fallopian tubes are particularly sensitive to PGEs in the postovulatory phase. PGE_1 and PGE_2 contract the proximal part (near the uterus) of the tubes, but relax their distal portions. This action of the PGEs permits the passage of the sperm into the Fallopian tubes, but prevents the transfer of the ovum to the uterus until after fertilization and development of the blastocyst.

The corpus luteum formed in the ovary after ovulation has a brief life in the nonpregnant animal subject to normal estrus or menstrual cycles, whereas after fertilization of the ovum it persists longer and provides the progesterone needed for normal implantation of the fertilized ovum in the uterus and for the maintenance of pregnancy. The transient existence of the corpus luteum after ovulation in a nonpregnant animal is ascribed to the transfer from the uterus to the ovary of a luteolytic substance, which causes the regression of the corpus luteum. This luteolytic substance has been identified as $PGF_{2\alpha}$, but how it is transferred from the uterus to the ovary is uncertain. In the sheep the ovarian artery is tightly wound around the uteroovarian vein, and thus $PGF_{2\alpha}$ produced in the uterus could enter this vein, from which it could be transferred by a countercurrent distribution into the ovarian artery. However, such anatomical connections in other species are not known and it is uncertain how the luteolytic substance reaches the ovaries from the uterus. It is not known whether lymph vessels may provide the connection.

The mechanism of luteolysis induced by $PGF_{2\alpha}$ is also uncertain. Since $PGF_{2\alpha}$ is a vasoconstrictor, attempts were made to correlate decreased ovarian and luteal blood flow with the regression of corpus luteum, but such correlation could not be established. In the rat the effect of $PGF_{2\alpha}$ on the corpus luteum manifests itself in a rapid inhibition of LH-activated adenylate cyclase and a slower repression of LH receptors. Behrman (1979) suggests that these two effects would tend to isolate the luteal cells from the "trophic influences of gonadotropins necessary for continued function of the luteal cell culminating in luteolysis."

During early pregnancy, the amniotic fluid contains traces of PGE_1 and PGE_2, but at the end and at the onset of labor $PGF_{1\alpha}$ and $PGF_{2\alpha}$ predominate. $PGF_{2\alpha}$ has oxytocic properties and promotes labor. $PGF_{2\alpha}$ is a powerful contractor of the pregnant uterus and very small doses, given either intravenously or introduced directly into the uterine cavity, will induce labor or abortion (Karim, 1971a,b).

The function of the PGs in the male genitalia is uncertain; however, PGEs sensitize the seminal vesicles and vas deferens to contractions elicited by catecholamines; thus they may promote ejaculation. Males who secrete less than 50 µg of PGs per ml of semen are of low fertility. (For reviews, see Bergström et al., 1968; Horton, 1969; Zor and Lamprecht, 1977; Behrman, 1979).

10.4.1.3. Prostaglandins and the Gastrointestinal Tract

Prostaglandins are released in the gastrointestinal tract on stimulation of the vagus nerve. All the PGs contract *in vitro* the longitudinal smooth muscle of stomach and intestine, but PGEs relax the circular muscles of stomach and intestine, whereas $PGF_{2\alpha}$ contracts these. Intravenously given PGs greatly increase intestinal motility with accompanying diarrhea.

In the stomach the prostaglandins, together with prostacyclin (cf. Section 10.5.2), which is the major prostanoid produced in the stomach (Moncada and Vane, 1979), have special functions. Prostaglandins, prostacyclin, and some of their stable, long-acting synthetic analogues suppress acid secretion and can prevent experimental ulcers produced by parenterally administered salicylates (Kauffman and Grossman, 1978; Kauffman et al., 1979). In addition, PGE_2 (50–75 µg/kg), $PGF_{2\beta}$ (300–500 µg/kg), and the synthetic 16,16-dimethyl-PGE_2 (1 µg/kg) given orally or subcutaneously to anesthetized rats prevent the necrosis of gastric mucosa caused by absolute ethanol, 0.75 N HCl, 0.2 N NaOH, or boiling water! The mechanism of this remarkable "cytoprotective" effect is not known (Nezamis et al., 1977).

10.4.1.4. Prostaglandins and the Nervous System

Release of prostaglandins on nerve stimulation in end-organs has been mentioned earlier. Stimulation of a peripheral sensory nerve results in the release of PGs in the contralateral somatosensory cerebral cortex. Electrical stimulation of cerebral cortex also causes release of PGs in the contralateral cortex, but section of the corpus callosum abolishes this effect. Even bathing the exposed cerebral cortex with isotonic solutions causes release of PGs. These have mostly been identified as PGE_2 and $PGF_{2\alpha}$ (cf. Ramwell et al., 1965, 1966; Ramwell and Shaw, 1966). However, the recent discovery of PGD_2 as the most prominent product in brain homogenates may call for the reexamination of the cerebral PGs. Prostaglandin $H_2:D_2$ isomerase (PGD_2 synthetase) is present in the spinal cord, in all parts of the brain, and at particularly high levels in the anterior hypothalamic–preoptic area. Cultured neuroblastoma cells (but not gliomas) also synthesize PGD_2. PGD_2 added to neuroblastoma cells causes a very rapid rise in cAMP content of the cells. The implication is that PGD_2 might be a neuronal regulator (Shimizu et al., 1979b).

Prostaglandins are now recognized as central thermoregulators. The main agent is probably PGE_2, a normal constituent of the anterior hypothalamic–preoptic region which is considered to be the main temperature-regulating center. Injection of PGE_1 or PGE_2 into this area produces fever. Pyrogen-induced fever is associated with the appearance of PGE_2 in the cerebrospinal fluid where normally only $PGF_{2\alpha}$ (<100 pg/ml) is found. PGE_2-induced fever, unlike pyrogen-induced fever, does not respond to antipyretics. Intracranial bleeding (subarachnoid hemorrhage, stroke) is associated with fever; this may result from exposure of the anterior hypothalamic–preoptic area to massive amounts of prostaglandins, as under such conditions levels of $PGF_{2\alpha}$ and PGE_2 are much increased (200–3000 pg/ml) in the cerebrospinal fluid. After hemorrhage one would expect also substantial formation of thromboxane A_2 (cf. Section 10.5.1) as a result of platelet aggregation and blood clotting. Wolfe and Coceani (1979) quote unpublished observations by F. Coceani that thromboxane B_2 (the hydrolytic product of thromboxane A_2; cf. Section 10.5.1) is a normal constituent (800–3000 pg/ml) of the cerebrospinal fluid, at least in the cat.

Prostaglandins E_1, E_2, or E_3 introduced into the lateral cerebral ventricle in the cat produce sleep, stupor, and catatonic state that may persist for 24 hr; $PGF_{2\alpha}$ is ineffective in this respect. Whether the profound effects of PGEs seen in experimental animals have any relation to human disease is not known, though in epilepsy PGE_2 is found in varying amounts in the cerebrospinal fluid. Whether prostaglandins, by their effects on cAMP and cGMP levels, may modulate actions of neurotransmitters is a matter of debate (cf. Wolfe and Coceani, 1979; White, 1982).

10.4.1.5. Prostaglandins and Renal Function

The enzymes of PG synthesis in the kidneys were discussed earlier (see pp. 167, 177) and have been reviewed by Morrison and Needleman (1979). The complex physiology of renal function and its hormonal regulation by the renin–angiotensin–aldosterone axis is beyond the scope of this book. Prevailing hypotheses postulate that renal kinins and angiotensin (either by direct action or indirectly through the induced vasoconstriction) may activate phospholipase A_2 and thus release precursors for PG synthesis. The prostanoids are assumed to interact with the juxtaglomerular cells that regulate renin release (cf. Dunn, 1979; Smith and Dunn, 1981). Grossly, PGE_1 and PGA_1 infused into renal artery increase urine volume and excretion of Na^+, K^+, and Cl^- ions, whereas PGE_2 causes Na^+ retention and may be a factor in hypertension (Greenberg, 1982).

10.5. Thromboxane and Prostacyclin

Thromboxane(s) and prostacyclin will be discussed in parallel because— although they differ in their structures—they have a common precursor, PGH_2, and because their biological functions are anticomplementary. Thromboxane aggregates platelets and is a vasoconstrictor, whereas prostacyclin is an antiaggregating factor and a vasodilator.

10.5.1. Thromboxanes

Aggregation of platelets by arachidonic acid has been known since 1973 (Ingerman et al., 1973; Silver et al., 1973; Vorgaftig and Zirinis, 1973). It has also been known that the aggregation is accompanied by release of ADP and serotonin from the platelets. Examination of the metabolism of arachidonic acid by platelets revealed that the platelets contained not only a hitherto unrecognized lipoxygenase, but also the prostaglandin endoperoxide synthetase (fatty acid cyclo-oxygenase), and that PGG_2 and PGH_2, first identified in incubations of microsomes from sheep seminal vesicles, played an important role in the aggregation of platelets (Smith et al., 1974a; Hamberg and Samuelsson, 1974).

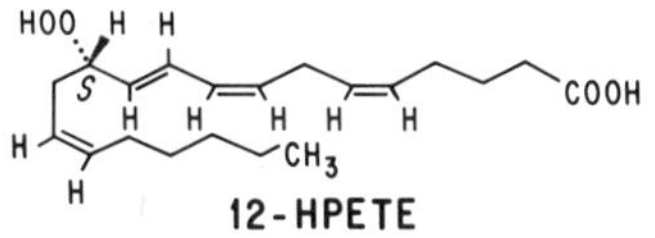

FIGURE 10.21. 12S-Hydroperoxyeicosatetraenoic (12-HPETE) and 12S-hydroxyeicosatetraenoic acids (12-HETE), products derived from arachidonic acid by action of platelet lipoxygenase.

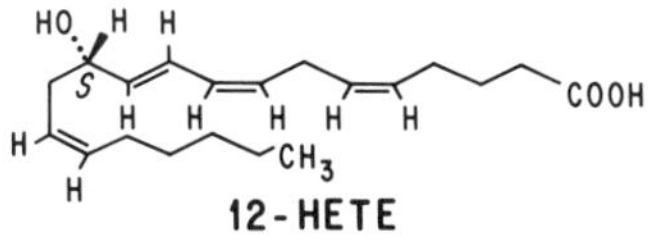

The lipoxygenase of platelets differs from the previously recognized lipoxygenases that attack eicosapolyenoic fatty acids either at C-5 or at the $\omega 6$ position (cf. Chapter 11, Section 1.1.1). The platelet lipoxygenase attacks 20:4ω6 at C-12 with the isomerization of the 11Z(cis)-double bond to 10E(trans). The product formed from 20:4ω6 in platelets was identified as 12S-hydroxyeicosa-5,8Z,10E,14Z-tetraenoic acid (Fig. 10.21; 12-HETE) formed through the intermediacy of 12S-hydroperoxyeicosatetraenoic acid (Fig. 10.21; 12-HPETE) (Hamberg and Samuelsson, 1974). The biological function of HETE is not known, but HPETE is a potent inhibitor of thromboxane synthetase and may limit the formation of thromboxane in platelets. In psoriasis (a scaly skin disease) the concentration of HETE and arachidonic acid was found to be 30 times higher in the areas of lesions in the epidermal skin than in unaffected areas (cf. Discussion in: Samuelsson, 1978).

The details of transformations of arachidonic acid in platelets and the role of its metabolites in platelet aggregation were elucidated mostly by Hamberg and Samuelsson and their colleagues (Hamberg *et al.*, 1974a,b; Hamberg *et al.*, 1975a,b). Platelet aggregation is a complex process and there are several agents that may initiate it (for review, see Barnhart, 1978). The most powerful aggregating agents are thrombin, collagen, especially collagen IV (a component of the basement membrane of the subendothelial layer of blood vessels) and collagen III (found in the media of blood vessels), epinephrine, ADP, and arachidonic acid. Platelet aggregation is generally considered to be biphasic, consisting of an initial stickiness of the platelets and adhesion to one another and a second secretory phase, which completes the aggregation and during which contents of dense granules in the central zone of the platelets, ADP, serotonin, fibrinogen, epinephrine, and Ca^{2+} ions are extruded (secreted) through channel systems open to the surface of the platelets. These two phases of platelet aggregation can be clearly distinguished after application of relatively mild aggregating agents, such as epinephrine. Thrombin, collagen, and arachidonic acid, on the other hand, usually elicit a single massive wave of aggregation which is also accompanied by the secretion of the contents of the dense granules.

The interesting fact that emerged through the study of arachidonic acid metabolism in platelets is that platelet aggregation initiated by whatever means can be accounted for by a common biochemical process: the transformation of arachidonic acid (exogenous or liberated from endogenous lipids) through the prostaglandin endoperoxide intermediates, PGG_2 and PGH_2, into a very short-

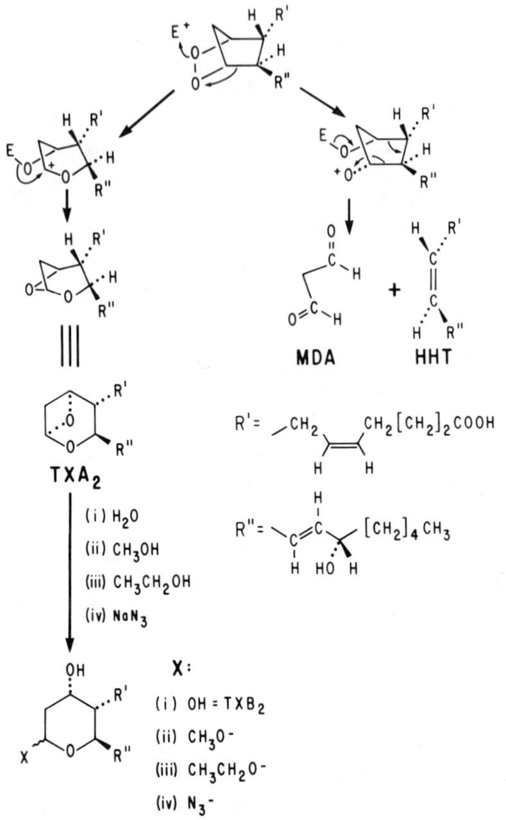

FIGURE 10.22. Theoretical mechanisms of thromboxane A_2 (TXA$_2$) and thromboxane B_2 (TXB$_2$) formation from PGH$_2$, and cleavage of PGH$_2$ into malonic dialdehyde (MDA) and 12S-hydroxyheptadeca-5Z,8,11E-trienoic acid (HHT) (after Diczfalusy *et al.*, 1977; and Fried and Barton, 1977).

lived ($t_{1/2} \sim 32$ sec) substance, containing an oxane and oxetane ring, which became known as thromboxane A_2 (TXA$_2$) (cf. Fig. 10.22). TXA$_2$ could never be isolated on account of its extreme instability, but could be trapped by nucleophilic reagents. In the earliest experiments involving incubations of platelets with arachidonic acid for several minutes only a hemiacetal hydrolytic product of TXA$_2$, 8-(1-hydroxy-3-oxopropyl)-9,12S-dihydroxy-5,10-heptadecadienoic acid, was isolated. This substance, devoid of biological activity, was subsequently named thromboxane B_2 (TXB$_2$; Fig. 10.22). However, in short incubations of 30 sec or less, when the reactions were stopped either by the addition of large volumes of methanol, ethanol, or a solution of NaN$_3$, besides TXB$_2$, the methoxy, ethoxy, or azido derivatives of TXB$_2$ were isolated (Fig. 10.22). In incubations of longer duration (5 min) only the hemiacetal form of TXB$_2$ was found.

Because the designation of the structure of TXB$_2$ by systemic chemical nomenclature is awkward and because the numbering of carbon atoms in TXB$_2$ by this nomenclature is totally different from that in the parent PGH$_2$, Roberts *et al.* (1978b) proposed for thromboxanes a system similar to that of prostanoic acid used for prostaglandins. According to this proposal a basic C_{20}-substituted tetrahydropyran is called thrombanoic acid with retention of the numbering of the carbon atoms used, for example, in PGH_2 (cf. Fig. 10.27). By this system TXB_2 is called $9\alpha,11,15(S)$-trihydroxythromba-5,13-dienoic acid. The thrombanoic acid nomenclature is particularly useful in denoting metabolites of TXB_2 (see later).

Further evidence, linking the formation of TXA_2 to the generation of PGG_2 and PGH_2, was obtained from the use of platelets treated with aspirin or indomethacin. These two drugs specifically inhibit the PG endoperoxide synthetase (cf. Section 10.2.4.1). First, platelets treated with aspirin or indomethacin fail to aggregate after addition of thrombin, collagen, or arachidonic acid. However, indomethacin-treated platelets rapidly aggregate on addition of preformed PGG_2 or PGH_2. The aggregating factor is not identical with either of the PG endoperoxides, but is derived from them. When indomethacin-treated platelets were incubated for 30 sec with PGG_2 (75 ng) and 0.1 ml of the incubation mixture was transferred to another preparation of indomethacin-treated platelets, rapid aggregation was induced. The amount of PGG_2 + PGH_2 transferred to the second preparation (6.4 ng) was too small to induce aggregation. Hence it followed that a potent aggregating factor must have been derived from the PG endoperoxides. The half-life of this factor, whether generated from arachidonic acid or PGG_2, was very similar, about 32 sec.

The derivation of TXB_2 via TXA_2 from arachidonic acid, PGG_2, or PGH_2 was also demonstrated by the use of isotopic labeling. Thus the hydrogen atoms (^2H) at positions 5,6,8,9,11,12,14, and 15 in arachidonic acid or PGG_2 appeared without change in TXB_2. Also, when synthesis of TXB_2 from arachidonic acid was conducted in an atmosphere of $^{18}O_2$ gas, TXB_2 became labeled with ^{18}O in the hydroxyl group at C-15 (thrombanoic acid numbering), in the non-hemiacetal hydroxyl group at C-9, and in the ether oxygen of the oxane ring between C-11 and C-12 (cf. Fig. 10.22). Hamberg et al. (1975b) also provided evidence that at least one component of the so-called rabbit aorta-contracting substance, observed by Piper and Vane (1969) to be released from the lung of sensitized guinea pigs after perfusion with antigen, was TXA_2.

The prostaglandin endoperoxide synthetase and thromboxane synthetase are associated with platelet microsomes derived from the intermediate zone of microtubules of the platelets (Needleman et al., 1976b; Hammarström and Falardeau, 1977). These enzymes can be solubilized from the microsomes with detergent and resolved by chromatography on DEAE cellulose columns into two fractions (Hammarström and Falardeau, 1977). Fraction I, which is not adsorbed onto the column, contains the PG endoperoxide synthetase and converts arachidonic acid into PGG_2 and PGH_2. The second fraction (actually fraction III of Hammarström and Falardeau), eluted from the column with 0.2 M potassium phosphate/0.1% Triton X-100, pH 7.4, cannot convert arachidonic acid into PGG_2 or PGH_2, but converts PGH_2 into TXB_2. This second enzyme was named thromboxane synthetase; it could also be called prostaglandin H_2:thromboxane isomerase.

An interesting facet of thromboxane formation in platelets is that it is accompanied by the simultaneous and stoichiometric cleavage of some of the PGH_2 into malonic dialdehyde and 12S-hydroxyheptadeca-5Z,8,10E-trienoic acid (MDA and HHT; Fig. 10.22). Diczfalusy *et al.* (1977) showed that the partially purified thromboxane synthetase catalyzed the cleavage of PGH_1 into malonic dialdehyde and 12S-hydroxyheptadeca-8,10E-dienoic acid (HHD), and the transformations of PGH_2 into TXB_2 and malonic dialdehyde and HHT. By the use of four inhibitors of thomboxane synthetase, Diczfalusy *et al.* (1977) demonstrated that inhibition of thromboxane synthesis and of HHT formation from PGH_2 were exactly parallel and concluded that the same enzyme catalyzes thromboxane synthesis and HHT formation. The functional significance of the cleavage of PGH_2 into malonic dialdehyde and HHT is not understood. However, malonic dialdehyde is a highly reactive substance and could modify proteins either by intra- or intermolecular cross-linking through lysine residues. Fogelman and his colleagues (cf. Haberland *et al.*, 1982) found, e.g., that malonic dialdehyde modified low-density lipoproteins and changed their cellular uptake.

The hypothetical mechanism of thromboxane and HHT formation from PGH_2 shown in Fig. 10.22 embodies the ideas of Diczfalusy *et al.* (1977) and of Fried and Barton (1977). The hypothesis assumes that an electrophilic site of thromboxane synthetase attacks the oxygen atom attached to C-9 of PGH_2, thereby breaking the peroxide bond. Migration of the bond between C-11 and C-12 to the electron-deficient oxygen results in the insertion of that oxygen between C-11 and C-12 and creates an electron-deficient center at C-11. The highly strained TXA_2 is then formed by the attachment of the enzyme-bound oxygen to C-11 and liberation of the enzyme E^+. A slightly different rearrangement of the same initial enzyme-bound intermediate leads to the cleavage of malonic dialdehyde (C-9−C-11) and generation of HHT.

The importance of prostaglandin endoperoxide synthesis in normal hemostasis was demonstrated by Malmsten *et al.* (1975) who studied the platelets of a man who had prolonged bleeding time and a tendency for easy bruising. The platelets of this individual failed to aggregate in response to either collagen or arachidonic acid, but responded normally to PGG_2. When washed platelets of this individual were incubated with [^{14}C]arachidonic acid, only traces of TXB_2 and HHT were formed; on the other hand the formation of 12-HETE (see Fig. 10.21) was enhanced. The platelets of this individual clearly were deficient in the PG endoperoxide synthetase, but had a normal complement of the platelet-specific lipoxygenase and also of thromboxane synthetase. The report by Malmsten *et al.* (1975) left some tantalizing questions: if the PG endoperoxide synthetase catalyzes not only the formation of PGG_2 but also the latter's conversion to PGH_2 and if TXA_2 is derived from PGH_2, how was PGG_2 converted to PGH_2 in the platelets of this individual? Were other tissues or cells of this individual also deficient in PG endoperoxide synthetase or are there multiple alleles for this enzyme, finding expression in some cells, but not in others? Can PGG_2 be converted directly into TXA_2 without PGH_2 as an intermediate, or has PGG_2 itself a direct aggregatory potency?

10.5.2. Prostacyclin (PGI_2)

The discovery of prostacyclin has a curious history. There can be very little doubt that prostacyclin was first obtained by Pace-Asciak and Wolf (1971) from

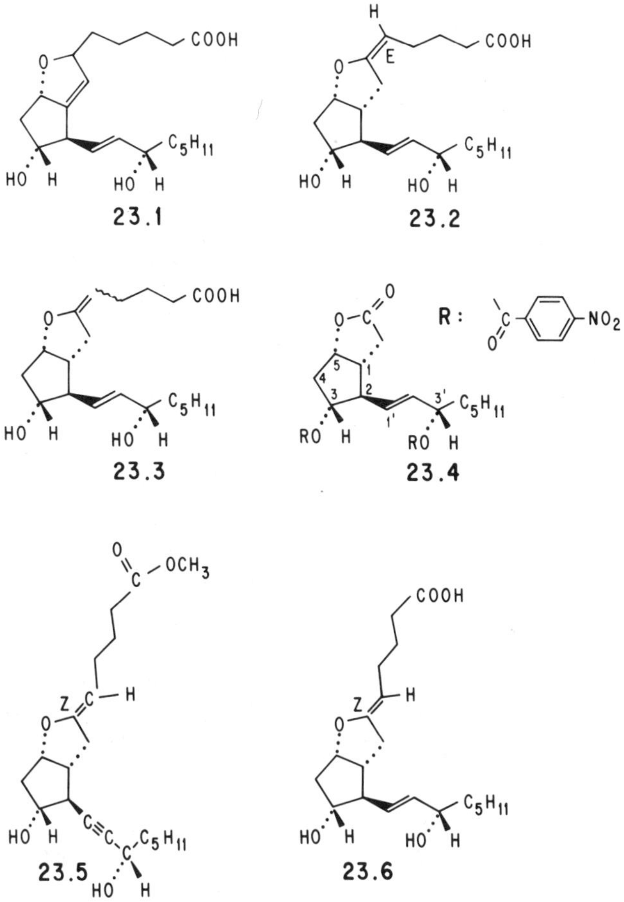

FIGURE 10.23. Structures proposed at various times for prostacyclin (PGX = PGI$_2$). Structures **23.1** and **23.2** are those of Pace-Asciak and Wolf (1971); structure **23.3** is that first reported by Johnson *et al.* (1976) for the synthetic PGI$_2$; **23.4** is the degradation product obtained from a mixture of biosynthetic PGX and compound **23.3**. Compound **23.5** is the 13,14-dehydro (acetylenic) analogue of PGI$_2$ (Fried and Barton, 1977); **23.6** is the accepted structure of PGI$_2$.

incubations of homogenates of the fundus of the rat stomach with [^{3}H]arachidonic acid. In addition to PGE$_2$ and PGF$_{2\alpha}$, they isolated two compounds which they correctly recognized to be bicyclic and, on the basis of spectroscopic evidence, assigned to them the 6(9)-oxy-7,13-prostadienoic and the 6(9)-oxy-5,13-prostadienoic acid structures (Fig. 10.23, **23.1** and **23.2**). Subsequently, Pace-Asciak (1972) found that epinephrine and dopamine stimulated the synthesis not only of PGE$_2$, but also of "prostacyclin" in homogenates of rat stomach, and in 1976 he also isolated from incubations of homogenates of rat-stomach fundus, with arachidonic acid, PGG$_2$, or PGH$_2$ as substrates, 6-ketoprostaglandin F$_{1\alpha}$ (Fig. 10.24), now recognized to be the hydrolytic product of prostacyclin. Pace-Asciak (1976) also recognized the hemiketalic (or lactol) form of 6-ketoprostaglandin F$_{1\alpha}$ and also found that when this substance was biosyn-

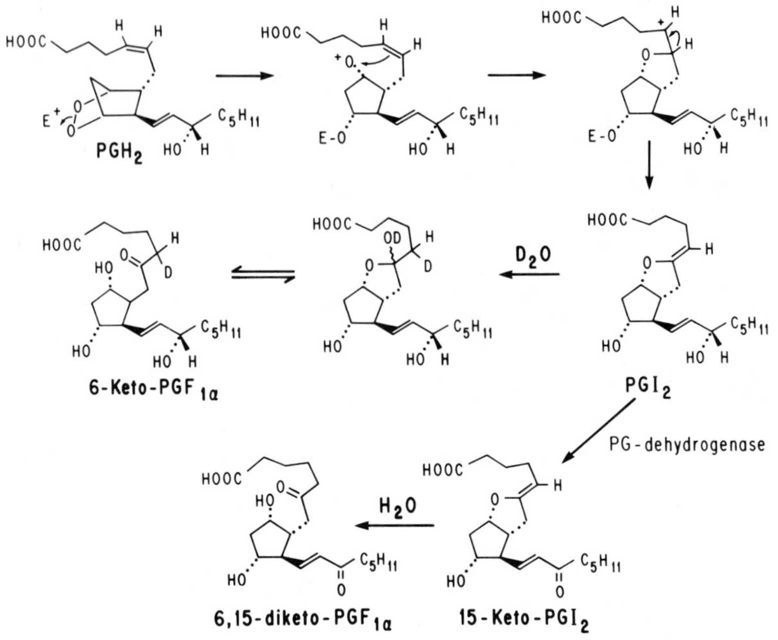

FIGURE 10.24. Suggested mechanism of prostacyclin (PGI$_2$) synthesis from PGH$_2$ after Fried and Barton (1977). The nonenzymatic hydrolysis in D$_2$O of PGI$_2$ to 6-keto-PGF$_{1\alpha}$ is shown to indicate uptake of deuterium at C-5. The known enzymatic dehydrogenation of PGI$_2$ to 15-keto-PGI$_2$ and the latter's hydrolysis to 6,15-diketo-PGF$_{1\alpha}$ are also shown (cf. Gorman, 1979).

thesized from [5,6,8,9,11,12,14,15-^{2}H$_8$]-PGG$_2$ it lost one atom of deuterium. The "prostacyclin" isolated by Pace-Asciak and Wolf (1971) aroused little interest on account of its weak biological activity in a gerbil colon test.

The discovery of thromboxane stimulated a new interest in the metabolism of arachidonic acid, PGG$_2$, and PGH$_2$ not only in platelets but also in blood vessels and led to the "rediscovery" of prostacyclin as an extremely important and potent biological agent. Needleman and his colleagues (Needleman et al., 1977) observed a paradoxical effect of arachidonic acid and PGH$_2$ on bovine coronary arteries in that they caused dilation instead of the expected contraction (PGE$_2$ and PGF$_{2\alpha}$ cause contraction and PGE$_1$ causes relaxation of coronary arteries). They concluded that coronary arteries contain little, if any, PGH$_2$:PGE$_2$ isomerase and that instead these arteries contain an enzyme that converts PGH$_2$ into a novel potent PG-like substance which has an effect opposite to that of PGE$_2$ or TXA$_2$. Moncada et al. (1976) found that microsomes from rabbit or pig aortas and rings of arteries transformed PGG$_2$ and PGH$_2$ into an unstable compound, which they labeled PGX, that inhibited platelet aggregation and relaxed isolated blood vessels. PGX was 30 times more potent than PGE$_1$ as an antiaggregatory agent and 5 to 20 times more potent than PGD$_2$. The antiaggregatory activity of PGX disappeared in 20 min at 22°C in aqueous solution, or in 10–15 sec after boiling, but PGX was stable in dry acetone at −20°C. A little later, Moncada et al. (1977) demonstrated that human arteries and veins also generate PGX from either arachidonic acid, PGG$_2$, or PGH$_2$.

Generally Johnson et al. (1976) are credited with the identification of the chemical structure of PGX as the 9-deoxy-6,9α-epoxy-Δ^5-PGF$_{1\alpha}$ (an enol ether) and they named it prostacyclin, later to become known as PGI$_2$. In their first publication Johnson et al. (1976) did not define the geometry of the Δ^5-double bond in the enol ether (cf. Fig. 10.23). However, they refer to "The availability of enol-ether . . . by chemical synthesis" from the unpublished work of F. H. Lincoln and R. A. Johnson (Ref. 18 in Johnson et al.,1976), which prompted them to compare the properties of PGX with those of the synthetic compound. Johnson et al. (1976) found that during the biosynthesis of PGX with aortic microsomes from [5,6,8,9,11,12,14,15-^{3}H$_8$]PGH$_2$ one atom of tritium was lost from the [^{3}H$_8$]PGH$_2$, presumably from position 6 (see later for mechanism of PGX biosynthesis and Fig. 10.23). Thus such PGX may be designated as [^{3}H$_7$]PGX. The main proof of the identity of PGX with the synthetic enol ether was provided by the biological activity of the synthetic substance and by the chemical degradation of a mixture of the methyl ester of [^{3}H$_7$]PGX and the synthetic methyl ester of the 9-deoxy-6,9α-epoxy-Δ^5-PGF$_{1\alpha}$. The mixture was first converted to the 11,15-*bis*-*p*-nitrobenzoate and then oxidized with *N*-methylmorpholine oxide and osmium tetroxide. The crude product was then cleaved with sodium periodate to the 3,3′-*p*-nitrobenzoate of 3α,5α-dihydroxy-2β[3′*S*-hydroxy-1′-*trans*-octenyl]-cyclopentane-1α-acetic acid-γ-lactone (cf. Fig. 23.4), which was available and a known intermediate in the chemical synthesis of PGs. The crystalline degradation product retained about 75% of the radioactivity of the starting material, or 85% of the theoretical amount (assuming that the [^{3}H$_7$]PGX contained one atom of ^{3}H at C-6 which would have been lost in the degradation). Johnson et al. (1976) demonstrated unequivocally that 6-keto-PGF$_{1\alpha}$ was the hydrolytic degradation product of PGX. When the hydrolysis was carried out in D$_2$O, 6-keto-PGF$_{1\alpha}$ acquired one atom of D at position 5 (cf. Fig. 10.24). It is worth noting that while the free acid of prostacyclin is unstable, its Na salt is not.

Fried and Barton (1977) proposed a reaction mechanism for the biosynthesis of prostacyclin from PGH$_2$. Their basic assumption is similar to that made for the mechanism of TXA$_2$ synthesis except that the enzyme electrophile attacks the oxygen attached to C-11 with the cleavage of the peroxide bond. The electron-deficient 9α-oxygen atom can now accept electrons from the 5(6)-double bond with ring closure and creation of electron deficiency at C-5, which is stabilized by the elimination of the proton from C-6 (Fig. 10.24). Examination of atomic models shows that the conformation of the carboxyl-bearing side-chain of PGH$_2$ on the enzyme would determine whether, in the final product, the geometry around the 5(6)-double bond would be *E* or *Z*. In Fig. 10.24 we show a plausible conformation leading to the *Z* geometry.

Fried and Barton (1977) also synthesized the 13,14-dehydro (13,14-acetylenic) analogue of prostacyclin with the *Z* geometry around the 5(6)-double bond (**23.5**). This analogue mimicked qualitatively and quantitatively the antiaggregatory and vasodilating effects of natural prostacyclin and, therefore, Fried and Barton postulated that the geometry in natural prostacyclin around the 5(6)-double bond is in all probability also *Z* rather than *E* as formulated originally by Pace-Asciak and Wolf (1971). In a subsequent publication Johnson et al. (1977) established the 5*Z* geometry in prostacyclin (**23.6**) from its NMR spectrum. It is pertinent to mention that the chemical degradation of PGX to compound **23.4** used by Johnson *et al.* (1976) could not distinguish between the Z and E isomers.

Figure 10.24 summarizes the hypothetical mechanism of the biosynthesis of PGI_2, its enzymatic dehydrogenation, and nonenzymatic hydrolysis to 6-keto-$PGF_{1\alpha}$.

Prostacyclin is not exclusively the product of blood vessels but is also the major prostanoid made in gastric mucosa; it reduces acid secretion induced by pentagastrin (Moncada and Vane, 1979).

10.5.3. Interaction of Thromboxane A_2 and Prostacyclin

There are three excellent reviews discussing the "anticomplementary" relation between thromboxane A_2 (TXA_2) and prostacyclin (PGI_2) and their role in homeostasis between platelets and blood vessels: these are by Moncada and Vane (1979), Sivakoff *et al.* (1979), and Gorman (1979). These reviews contain also references to the many original articles that could not be cited here.

The discovery of PGI_2 as platelet antiaggregatory factor elaborated by vascular endothelium dispelled an ancient idea that vascular endothelium was inert towards platelets. On the contrary, the endothelium actively keeps platelets from adhering and initiating thrombus formation. Endothelial cells can elaborate PGI_2 not only from endogenous substrates but also from PG endoperoxides (mainly PGH_2) released from platelets. Thus, there may be even some "cooperation" between platelets and the vascular endothelium (Moncada and Vane, 1979). It is well known that platelets adhere to surfaces of blood-vessel walls where the endothelium has been damaged or to those which have been denuded of the endothelium (see Barnhart, 1978). Such adhesions can be explained by a local deficiency of PGI_2 synthesis and by the contact of platelets with the subendothelial collagen which is one of the powerful aggregating agents. The predilection for thrombus formation in arteries over an ulcerated atheromatous lesion is also understandable, as the endothelium over such lesions is destroyed and moreover such lesions contain lipid peroxides which are powerful inhibitors of prostacyclin synthetase. Contact of platelets with subendothelial and even medial collagen would favor platelet aggregation that cannot now be counteracted by prostacyclin. The anticomplementary relation between thromboxane and prostacyclin is illustrated by the model shown in Fig. 10.25 (Gorman, 1979).

The opposing effects of TXA_2 and PGI_2 can be also understood in biochemical terms and can be accounted for by modulations of adenylate cyclase and the cAMP content of cells. It has been known for some time that PGE_1 inhibits platelet aggregation and that it causes an increase in the cAMP content of platelets. PGD_2 is also an antiaggregatory agent and likewise raises the cAMP levels in platelets. PGI_2 (at 0.14 μM) also can increase six- to eightfold the cAMP content of platelets, and PGH_2 (at 2.8 μM) substantially counteracts the effect of PGI_2, although PGH_2 by itself does not lower the basal level of cAMP in platelets. According to the hypothesis proposed by Gorman (1979), the effects of cAMP, PGH_2, and TXA_2 may be intimately linked to sequestration and mobilization of

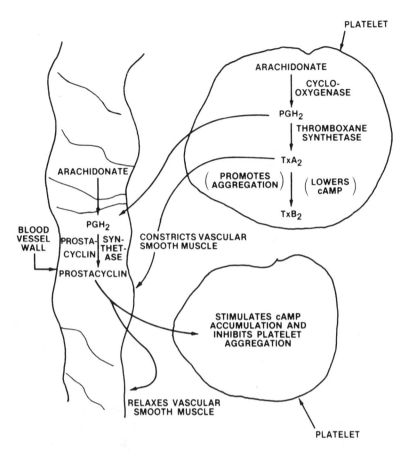

FIGURE 10.25. Model of human platelet homeostasis according to Gorman (1979). By permission of the author and the copyright holder, the Federation of American Societies for Experimental Biology.

Ca^{2+} in the platelets, as depicted in Fig. 10.26. As is shown, cAMP may have multiple sites of action: it may inhibit release of arachidonic acid from phospholipids, synthesis of PGH_2, and mobilization of Ca^{2+} from storage sites. On the other hand, TXA_2 promotes mobilization of Ca^{2+} ions which in turn stimulate the platelet release reaction and inhibit adenylate cyclase. PGI_2, by stimulation of adenylate cyclase and by raising the cAMP content of platelets, has the opposite effects.

The aggregation of platelets elicited by collagen or 20:4ω6 is accompanied not only by a decrease in cAMP content, but also by an increase in cGMP paralleling the rise of PGG_2 and PGH_2 in platelets. Aspirin and indomethacin prevent both aggregation and the rise in cGMP. However, PGG_2 induces platelet aggregation and the rise in cGMP even after treatment of platelets by aspirin or indomethacin. PGG_2 and 20:4ω6 activate the soluble form of guanylate cyclase obtained from disrupted platelets, but $PGF_{2\alpha}$, PGE_1, and PGE_2 have no effect on the enzyme (Goldberg, 1976).

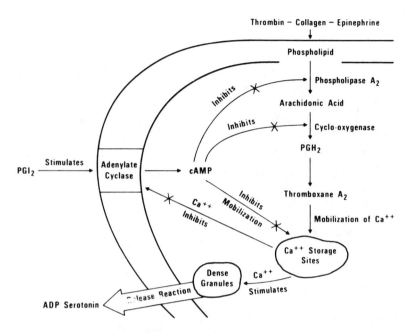

FIGURE 10.26. Postulated relationship between Ca^{2+}, TXA_2, PGI_2, and adenylate cyclase in human platelets according to Gorman (1979). By permission of the author and the copyright holder, the Federation of American Societies for Experimental Biology.

It needs to be emphasized that the actual aggregating agent of platelets is ADP liberated in the release reaction. The action of collagen, thrombin, and epinephrine, the natural aggregating agents, is merely to activate in some unknown manner the phospholipase(s) that releases arachidonic acid which then initiates the cascade of events culminating in the platelet release reaction. Platelets of patients with the Hermansky–Pudlak syndrome (albinism with hemorrhagic diathesis; see Witkop *et al.,* 1978) are grossly deficient in storage granules and ADP and are almost refractory to PGG_2, but respond normally to the aggregating effects of added ADP. In Glanzmann's thrombasthenia (see Ratnoff, 1978), the platelets, owing to some abnormality of the plasma membrane, do not respond by aggregation to thrombin, collagen, epinephrine, ADP, or PGG_2, although they synthesize normally TXA_2 and TXB_2 from added arachidonic acid (Samuelsson, 1978).

The homeostatic mechanism dependent on the balance between TXA_2 and PGI_2 is probably of great physiological significance. The platelets may normally be under a continuous influence of PGI_2 as this agent may be a truly circulating hormone continuously released by the lungs which do not rapidly inactivate it (see Moncada and Vane, 1979). Likewise, PGI_2 is the major prostanoid produced in the blood vessels of the heart, but not by the myocardium; its production is stimulated there by bradykinin and can be detected in coronary venous effluents (see Sivakoff *et al.,* 1979).

The possibilities of pharmacological interventions to turn the balance in favor of PGI_2 over TXA_2 have been discussed by Moncada and Vane (1979). Ideally, to diminish the risk of thrombotic events, one would wish to suppress selectively TXA_2 synthesis. However, a therapeutically useful inhibitor of TXA_2 synthetase is not yet available. As an alternative it is possible to exploit inhibitors of phosphodiesterase which slow down the hydrolysis of cAMP. Indeed, the mechanism of action of dipyridamol as an antithrombotic agent and coronary vasodilator is attributed to inhibition of phosphodiesterase and enhancement of the effects of circulating PGI_2. Theophyllin, another phosphodiesterase inhibitor, has similarly an antiaggregatory effect on platelets *in vivo*. The use of aspirin in analgesic and anti-inflammatory doses, 1.5 g per day, as an inhibitor of prostaglandin endoperoxide synthetase is not recommended because it equally inhibits the enzyme in platelets and vascular endothelium, thereby neutralizing the whole PGI_2-TXA_2 system. However, small doses of aspirin might be beneficial as even one tablet per day (5 grains $\simeq$ 325 mg) inhibits platelet endoperoxide synthetase by 89%. The inhibition of this enzyme by aspirin is irreversible and as platelets do not have a nucleus they cannot resynthesize the enzyme. Hence the effect of aspirin on platelets is long-lasting and will wear off only as new platelets come into the circulation from the bone marrow. In contrast, the endoperoxide synthetase in vascular endothelium is said not to be as sensitive to aspirin as that in platelets, and, furthermore, endothelial cells, being nucleated, contain the protein-synthesizing machinery and can regenerate the enzyme.

The remarkable effects of PGI_2 on platelet aggregation and blood vessels raised high hopes that this substance might become a useful therapeutic agent either in diseases associated with platelet destruction (thrombotic thrombocytopenic purpura; hemolytic uremic syndrome) or vascular occlusions. These hopes so far have not been fulfilled; in fact most trials have failed. One problem is that PGI_2 to be effective has to be given by continuous intravenous infusion, but the maximum that a conscious patient can tolerate is no more than 8–16 ng/kg·min, although in anesthetized patients as much as 100 ng/kg·min have been safely used. Besides, infusion of PGI_2 has many unpleasant side effects: headache, abdominal cramps, nausea, pallor and sweating, restlessness, and collapse with bradycardia (Lewis and Dollery, 1983).

10.5.4. Metabolic Transformations of Thromboxanes and Prostacyclin

The metabolic transformations of TXA_2, TXB_2, and PGI_2 *in vivo* are of much interest, first because identification of some major metabolite either in blood or urine could shed light on the balance of these substances under physiological and pathological conditions, and second because of the possibility that some metabolites might have new biological activities.

Since TXA_2 is rapidly hydrolyzed to the inactive TXB_2, most of the studies have concentrated on the latter substance, and because the simple trauma of venipuncture causes a very large artifactual release of TXB_2 into blood samples, urinary metabolites of TXB_2 were sought. All the investigators used $[^3H_8]TXB_2$ (generated with platelets or lung microsomes from $[5,6,8,9,11,12,14,15-^3H_8]$-

PGH$_2$) administered either by intravenous infusion or subcutaneous injection. After intravenous injection into a female cynomolgus monkey [^{3}H]TXB$_2$ is lost from the blood relatively slowly, with a half-life of about 10 min, and the radioactivity remaining in the blood is associated with unchanged TXB$_2$ (Kindahl, 1977). Intravenous infusion of [^{3}H]TXB$_2$ into monkey or man at rates of 50–500 μg/hr causes no changes in blood pressure or pulse rate and 74–80% of the ^{3}H is recovered in the urine within 8–13 hr (Roberts *et al.*, 1977a,b). The major urinary metabolite of TXB$_2$ in man, monkey, and guinea pig is dinor-TXB$_2$, i.e., the product of one-step β-oxidation from the carboxyl end, but its proportion can vary between 17% and 32% of the excretion products (Kindahl, 1977; Roberts *et al.*, 1977a,b). Svensson (1979) developed a quantitative method for the determination of dinor-TXB$_2$ in the urine and found in guinea pigs a rather varying basal excretion rate of 65 ± 36 ng/kg in 24 hr. This value corresponds to a daily synthesis of 543 ± 300 ng/kg, and is presumably derived from TXA$_2$; this value may, however, be much too low. Maclouf *et al.* (1980) found that substantial amounts of TXA$_2$ formed from PGH$_2$ in platelet-rich plasma become covalently linked to

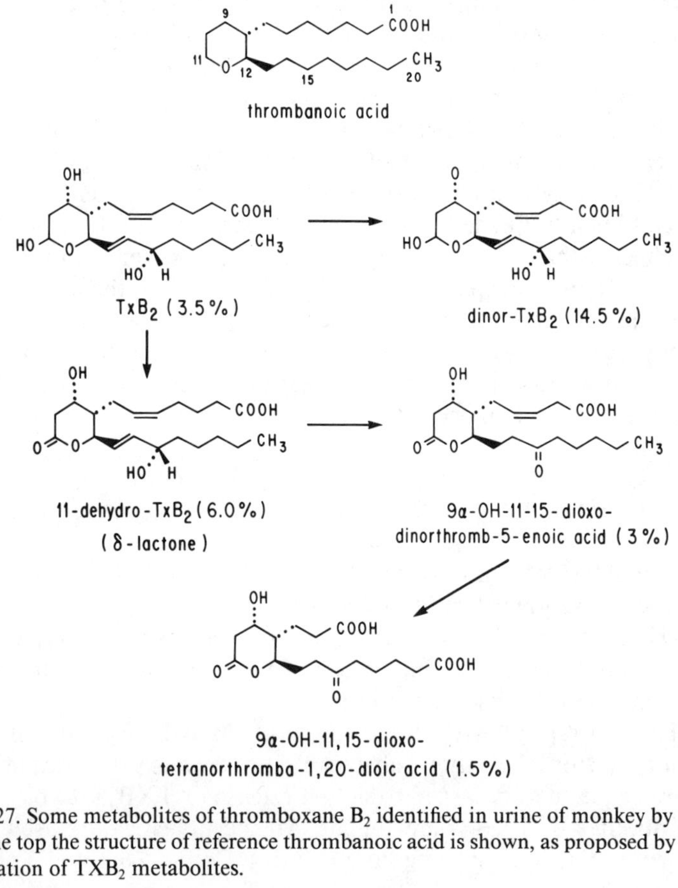

FIGURE 10.27. Some metabolites of thromboxane B$_2$ identified in urine of monkey by Roberts *et al.* (1978b). At the top the structure of reference thrombanoic acid is shown, as proposed by these authors for easier notation of TXB$_2$ metabolites.

serum albumin. If this finding is applicable to *in vivo* conditions, much of the TXA_2 may not be converted to TXB_2 and could be available for general metabolism. The fate of the albumin-bound TXA_2 or its chemical form is not known.

Detailed fractionation of the urinary metabolites of $[^3H_8]TXB_2$ intravenously infused into a male cynomolgus monkey revealed, in addition to dinor-TXB_2, several substances, all of which were derived probably from a common parent, 11-dehydro-TXB_2 (Roberts *et al.*, 1978b). The systemic description of 11-dehydro-TXB_2 is the δ-lactone of 3-hydroxy-4-(1′,4′-dihydroxynona-2′-enyl)-6-undecendioic acid, or by the thrombanoic acid nomenclature it is 9α,15(S)-dihydroxy-11-oxothromba-5,13-dienoic acid. This substance is apparently further metabolized by reactions common to other prostaglandins: dehydrogenation at C-15, reduction of the 13(14)-double bond, one- and two-step β-oxidation of the carboxyl-bearing side-chain, and ω-oxidation to a dicarboxylic acid (Fig. 10.27).

The metabolic transformations of PGI_2 have been studied *in vivo*, in perfusions of organs and with isolated enzymes with either PGI_2 or its nonenzymatic

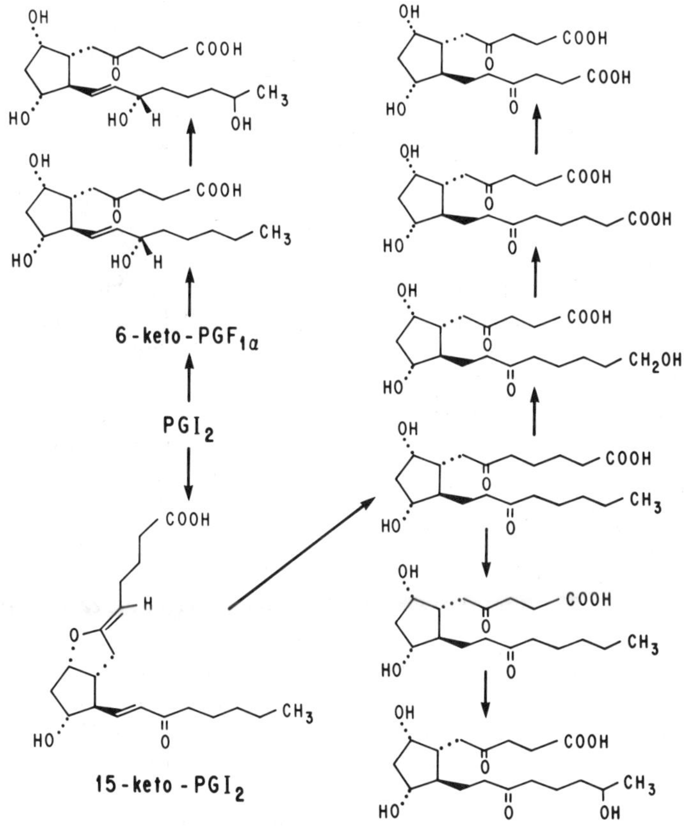

FIGURE 10.28. Metabolic transformations of prostacyclin (PGI_2) in the monkey and the rat (after Sun *et al.*, 1979). The precursor product relationships indicated by arrows, while logical, are purely conjectural. For structures of PGI_2 and 6-keto-$PGF_{1\alpha}$ see Figs. 10.23 and 10.24.

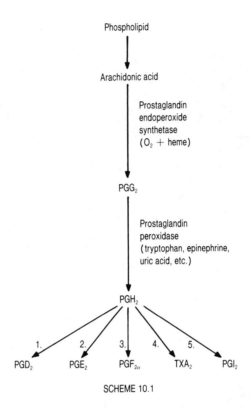

SCHEME 10.1

hydrolysis product, 6-ketoprostaglandin $F_{1\alpha}$, as substrates (Sun *et al.*, 1979; Wong *et al.*, 1979b). After a single intravenous injection of [^{3}H]PGI$_2$ to rats, about one-third of the dose is excreted in the urine within 24 hr, most of it during the first 2 hr. However, nearly one-half of the dose gradually appears in the feces over three days when the cumulative excretion in urine and feces amounts to about 80% of the dose. Fifteen minutes after intravenous administration of [11-^{3}H]-PGI$_2$ to rats the ^{3}H is almost equally distributed between PGI$_2$ and 6-keto-PGF$_{1\alpha}$, the largest amounts being found in the liver, small intestine, and kidneys. The organ distribution coupled with the excretion pattern strongly suggest the existence of an enterohepatic circulation for PGI$_2$ and its metabolites (Sun *et al.*, 1979). The metabolism of PGI$_2$ has well-recognizable features which differ from those of the metabolism of the primary prostaglandins only in that PGI$_2$ yields also the nonenzymatic hydrolysis product of 6-keto-PGF$_{1\alpha}$, which is a poor substrate for the 15-hydroxyprostaglandin dehydrogenase (PGDH) and the Δ^{13}-reductase. PGI$_2$ itself is a good substrate for PGDH and the reductase (cf. Granström, 1981c). Characteristically, nearly all the metabolites of PGI$_2$, after nonenzymatic cleavage of the enol-ether structure to the 9,11-dihydroxy-6-keto compound, suffer a one-step β-oxidation on the carboxyl-bearing side-chain. About one-half of the metabolites are products of action by the PGDH and Δ^{13}-reductase and some undergo also ω-oxidation and another one-step β-oxidation from the ω-end. In about one-third of the metabolites the C-15 allylic alcohol structure is maintained; these metabolites arise presumably from 6-keto-$PGF_{1\alpha}$ which, as mentioned, is a poor substrate for PGDH (cf. Fig. 10.28).

10.6. Correlation of Prostanoids

Scheme 10.1 shows the relationship of the prostanoids derived from arachidonic acid as an example. The immediate precursor to all the prostanoids is the 15-hydroxy-9,11-endoperoxyprosta-5,13-dienoic acid, PGH_2. The prostaglandin endoperoxide synthetase–peroxidase pair is ubiquitous, present in all organs and cells. Reactions 1 and 2 from PGH_2 are catalyzed by prostaglandin D and E isomerases, respectively; reaction 3 is the product of prostaglandin F reductases. The isomerases and reductases are widely distributed and are found in many organs. Reaction 4, catalyzed by thromboxane synthetase, is largely confined to blood platelets, but it also occurs in the lungs as part of anaphylactic phenomena. Reaction 5 is catalyzed by the prostacyclin synthetase present in all vascular endothelium and gastric mucosa, and possibly also in the lungs.

References

Abdel-Halim, M. S., Hawberg, M., Sjöquist, B., and Änggård, E., 1977, Identification of prostaglandin D_2 as a major prostaglandin in homogenates of rat brain, *Prostaglandins* **14**:633.
Abrahamsson, S., 1963, A direct determination of the molecular structure of prostaglandin F_{2-1}, *Acta Cryst.* **16**:409.
Agranoff, B. W., Murthy, P., and Seguin, E., 1983, Thrombin-induced phosphodiesteratic cleavage of phosphatidylinositol biphosphate in human platelets, *J. Biol. Chem.* **258**:2076.
Änggård, E., and Samuelsson, B., 1965a, Biosynthesis of prostaglandins from arachidonic acid in guinea pig lung. Prostaglandins and related factors 38, *J. Biol. Chem.* **240**:3518.
Änggård, E., and Samuelsson, B., 1965b, The metabolism of prostaglandin E_2 in guinea pig lung, *Biochemistry* **4**:1864.
Änggård, E., Gréen, K., and Samuelsson, B., 1965, Synthesis of tritium-labeled prostaglandin E_2 and studies on its metabolism in guinea pig lung, *J. Biol. Chem.* **240**:1932.
Barnhart, M. I., 1978, Platelet responses in health and disease, *Mol. Cell. Biochem.* **22**:113.
Behrman, H. R., 1979, Prostaglandins in hypothalamo-pituitary and ovarian function, *Annu. Rev. Physiol.* **41**:685.
Bell, R. L., Kennerly, D. A., Stanford, N., and Majerus, P. W., 1979, Diglyceride lipase: A pathway for arachidonate release from human platelets, *Proc. Nat. Acad. Sci. U.S.A.* **76**:3238.
Bergström, S., and Sjövall, J., 1960a, The isolation of prostaglandin F from sheep prostate glands, *Acta Chem. Scand.* **14**:1693.
Bergström, S., and Sjövall, J., 1960b, The isolation of prostaglandin E from sheep prostate glands, *Acta Chem. Scand.* **14**:1701.
Bergström, S., Krabisch, L., Samuelsson, B., and Sjövall, J., 1962, Preparation of prostaglandin F from prostaglandin E, *Acta Chem. Scand.* **16**:969.
Bergström, S., Ryhage, R., Samuelsson, B., and Sjövall, J., 1963, Prostaglandins and related factors 15. The structures of prostaglandin E_1, $F_{1\alpha}$, and $F_{1\beta}$, *J. Biol. Chem.* **238**:3555.

Bergström, S., Danielsson, H., and Samuelsson, B., 1964a, The enzymatic formation of prostaglandin E_2 from arachidonic acid. Prostaglandins and related factors 32, *Biochim. Biophys. Acta* **90**:207.

Bergström, S., Danielsson, H., Klenberg, D., and Samuelsson, B., 1964b, The enzymatic conversion of essential fatty acids into prostaglandins. Prostaglandins and related factors 34, *J. Biol. Chem.* **239**:PC4006.

Bergström, S., Carlson, L. A., and Weeks, J. R., 1968, The prostaglandins: A family of biologically active lipids, *Pharmacol. Rev.* **20**:1.

Billah, M. M., and Lapetina, E. G., 1983, Platelet-activating factor stimulates metabolism of phosphoinositides in horse platelets: Possible relationship to Ca^{2+} mobilization during stimulation, *Proc. Nat. Acad. Sci. U.S.A.* **80**:965.

Braithwaite, S. S., and Jarabak, J., 1975, Studies on a 15-hydroxyprostaglandin dehydrogenase from human placenta, *J. Biol. Chem.* **250**:2315.

Bundy, G. L., Daniels, E. G., Lincoln, F. H., and Pike, J. E., 1972a, Isolation of a new naturally occurring prostaglandin, 5-*trans*-PGA$_2$. Synthesis of 5-*trans*-PGE$_2$ and 5-*trans*-PGF$_{2\alpha}$, *J. Am. Chem. Soc.* **94**:2124.

Bundy, G. L., Schneider, W. P., Lincoln, F. H., and Pike, J. E., 1972b, The synthesis of prostaglandins E_2 and $F_{2\alpha}$ from (15*R*)- and (15*S*)-PGA$_2$, *J. Am. Chem. Soc.* **94**:2123.

Butcher, R. W., and Baird, C. E., 1968, Effects of prostaglandins on adenosine 3′,5′-monophosphate levels in fat and other tissues, *J. Biol. Chem.* **243**:1713.

Bygdeman, M., and Holmberg, O., 1966, Isolation and identification of prostaglandins from ram seminal plasma, *Acta Chem. Scand.* **20**:2308.

Bygdeman, M., and Samuelsson, B., 1966, Analyses of prostaglandins in human semen. Prostaglandins and related factors 44, *Clin. Chim. Acta* **13**:465.

Christ, E. J., and van Dorp, D. A., 1972, Comparative aspects of prostaglandin biosynthesis in animal tissues, *Biochim. Biophys. Acta* **270**:537.

Christ-Hazelhof, E., Nugteren, D. H., and van Dorp, D. A., 1976, Conversion of prostaglandin endoperoxides by glutathione-*S*-transferases and serum albumins, *Biochim. Biophys. Acta* **450**:450.

Corey, E. J., Washburn, W. N., and Chen, J. C., 1973, Studies on the prostaglandin A_2 synthetase complex from *Plexaura homomalla*, *J. Am. Chem. Soc.* **95**:2054.

Crabbé, P., 1977, Appendix [Physical properties of prostaglandins], in: *Prostaglandin Research* (P. Crabbé, ed.), Academic Press, New York, pp. 315–319.

Davies, B. N., Horton, E. W., and Withrington, P. G., 1968, The occurrence of prostaglandin E_2 in splenic venous blood of the dog following splenic nerve stimulation, *Brit. J. Pharmacol. Chemother.* **32**:127.

Dawson, R. M. C., and Irvine, R. F., 1978, Possible role of lysosomal phospholipases in inducing tissue prostaglandin synthesis, in: *Advances in Prostaglandin and Thromboxane Research*, Vol. 3 (C. Galli, G. Galli, and G. Porcellati, eds.), Raven Press, New York, pp. 47–54.

Diczfalusy, U., Falardeau, P., and Hammarström, S., 1977, Conversion of prostaglandin endoperoxides to C_{17}-hydroxy acids catalyzed by human platelet thromboxane synthase, *FEBS Lett.* **84**:271.

Dunham, E. W., Haddox, M. K., and Goldberg, N. D., 1974, Alteration of vein cyclic 3′:5′ nucleotide concentrations during changes in contractility, *Proc. Nat. Acad. Sci. U.S.A.* **71**:815.

Dunn, M. J., 1979, Renal prostaglandins: Influences on excretion of sodium and water, the renin–angiotensin system, renal blood flow and hypertension, in: *Hormonal Function and the Kidney. Contemporary Issues in Nephrology*, Vol. 4 (B. M. Brenner and J. H. Stein, eds.), Churchill Livingstone, New York, pp. 89–114.

Egan, R. W., Paxton, J., and Kuehl, F. A., Jr., 1976, Mechanism for irreversible self-deactivation of prostaglandin synthetase, *J. Biol. Chem.* **251**:7329.

Egan, R. W., Gale, P. H., and Kuehl, F. A., Jr., 1979, Reduction of hydroperoxides in the prostaglandin biosynthetic pathway by a microsomal peroxidase, *J. Biol. Chem.* **254**:3295.

Egan, R. W., Gale, P. H., VandenHeuvel, W. J. A., Baptista, E. M., and Kuehl, F. A., Jr., 1980, Mechanism of oxygen transfer by prostaglandin hydroperoxidase, *J. Biol. Chem.* **255**:323.

Eliasson, R., 1959, Studies on prostaglandin. Occurrence, formation and biological actions, *Acta Physiol. Scand.* **46**(Suppl. 158):1.

Ellis, C. K., Smigel, M. D., Oates, J. A., Oelz, O., and Sweetman, B. J., 1979, Metabolism of prostaglandin D_2 in the monkey, *J. Biol. Chem.* **254**:4152.

Ferreira, S. H., Moncada, S., and Vane, J. R., 1971, Indomethacin and aspirin abolish prostaglandin release from the spleen, *Nature New Biology,* **231**:237.
Fitzpatrick, F. A., and Stringfellow, D. A., 1979, Prostaglandin D_2 formation in malignant melanoma cells correlates inversely with cellular metastatic potential, *Proc. Nat. Acad. Sci. U.S.A.* **76**:1765.
Fitzpatrick, F., and Wynalda, M. A., 1983, Albumin-catalyzed metabolism of prostaglandin D_2. Identification of products formed *in vitro, J. Biol. Chem.* **258**:11713.
Flower, R. J., 1981, Basic biochemistry of prostaglandin metabolism with especial reference to PGDH, in: *The Prostaglandin System* (F. Berti and G. P. Velo, eds.), Plenum Press, New York, pp. 49–66.
Fried, J., and Barton, J., 1977, Synthesis of 13,14-dehydroprostacyclin methyl ester: A potent inhibitor of platelet aggregation, *Proc. Nat. Acad. Sci. U.S.A.* **74**:2199.
Frölich, J. C. (ed.), 1978, *Methods in Prostaglandin Research. Advances in Prostaglandin and Thromboxane Research,* Vol. 5, Raven Press, New York.
Fukushima, M., Kato, T., Ota, K., Arai, Y., Narumiya, S., and Hayaishi, O., 1982, 9-Deoxy-Δ^9-prostaglandin D_2, a prostaglandin D_2 derivative with potent antineoplastic and weak smooth muscle-contracting activities, *Biochem. Biophys. Res. Commun.* **109**:626.
Garcia, G. A., Maldonado, L. A., and Crabbé, P., 1977a, Total syntheses [of prostaglandins], in: *Prostaglandin Research* (P. Crabbé, ed.), Academic Press, New York, pp. 121–221.
Garcia, G. A., Maldonado, L. A., and Crabbé, P., 1977b, Syntheses of modified prostaglandins, in: *Prostaglandin Research* (P. Crabbé, ed.), Academic Press, New York, pp. 223–313.
Goldberg, N. D., 1976, Regulation of guanylate cyclase, in: Meeting Report; Cyclic Nucleotides as Regulators of Proliferation and Differentiated Cell Function, Internat. Titisee Conference, *J. Cyclic Nucleotide Res.* **2**:189.
Goldberg, N. D., O'Dea, R. F., and Haddox, M. K., 1973, Cyclic GMP, *Adv. Cyclic Nucleotide Res.* **3**:155.
Goldblatt, M. W., 1933, A depressor substance in seminal fluid, *J. Soc. Chem. Ind.* (London), **52**:1056.
Goldblatt, M. W., 1935, Properties of human seminal plasma, *J. Physiol.* **84**:208.
Gorman, R., 1975, Prostaglandin endoperoxides: possible new regulators of cyclic nucleotide metabolism, *J. Cyclic Nucleotide Res.* **1**:1.
Gorman, R. R., 1979, Modulation of human platelet function by prostacyclin, *Fed. Proc.* **38**:83.
Granström, E., 1972, On the metabolism of prostaglandin $F_{2\alpha}$ in female subjects. Structures of two metabolites in blood, *Eur. J. Biochem.* **27**:462.
Granström, E., 1981a, Biosynthesis of prostaglandins and thromboxanes, in: *The Prostaglandin System* (F. Berti and G. P. Velo, eds.), Plenum Press, New York, pp. 15–25.
Granström, E., 1981b, Assay methods for prostaglandins and thromboxanes: Gas chromatographic-mass spectrometric methods and radioimmunoassay, in: *The Prostaglandin System* (F. Berti and G. P. Velo, eds.), Plenum Press, New York, pp. 67–72.
Granström, E., 1981c, Metabolism of prostaglandins and thromboxanes, in: *The Prostaglandin System* (F. Berti and G. P. Velo, eds.), Plenum Press, New York, pp. 39–48.
Granström, E., and Samuelsson, B., 1971a, On the metabolism of prostaglandin $F_{2\alpha}$ in female subjects, *J. Biol. Chem.* **246**:5254.
Granström, E., and Samuelsson, B., 1971b, On the metabolism of prostaglandin $F_{2\alpha}$ in female subjects. II. Structures of six metabolites, *J. Biol. Chem.* **246**:7470.
Greenberg, S., 1982, Prostaglandins and vascular smooth muscle hypertension, in: *Prostaglandins: Organ- and Tissue-Specific Actions* (S. Greenberg, P. J. Kadowitz, and T. F. Burks, eds.), Marcel Dekker, New York, pp. 25–88.
Haberland, M., Fogelman, A. M., and Edwards, P. A., 1982, Specificity of receptor-mediated recognition of malondialdehyde-treated low density lipoproteins, *Proc. Nat. Acad. Sci. U.S.A.* **79**:1712.
Hamberg, M., 1968, Metabolism of prostaglandins in rat liver mitochondria, *Eur. J. Biochem.* **6**:135.
Hamberg, M., and Fredholm, B. B., 1976, Isomerization of prostaglandin H_2 into prostaglandin D_2 in the presence of serum albumin, *Biochim. Biophys. Acta* **431**:189.
Hamberg, M., and Israelsson, U., 1970, Metabolism of prostaglandin E_2 in guinea pig liver. I. Identification of seven metabolites, *J. Biol. Chem.* **245**:5107.
Hamberg, M., and Samuelsson, B., 1967, On the mechanism of the biosynthesis of prostaglandin E_1 and $F_{1\alpha}$, *J. Biol. Chem.* **242**:5336.

Hamberg, M., and Samuelsson, B., 1971, On the metabolism of prostaglandins E_1 and E_2 in man, *J. Biol. Chem.* **246**:6713.
Hamberg, M., and Samuelsson, B., 1972, On the metabolism of prostaglandins E_1 and E_2 in the guinea pig, *J. Biol. Chem.* **247**:3495.
Hamberg, M., and Samuelsson, B., 1973, Detection and isolation of an endoperoxide intermediate in prostaglandin biosynthesis, *Proc. Nat. Acad. Sci. U.S.A.* **70**:899.
Hamberg, M., and Samuelsson, B., 1974, Prostaglandin endoperoxides. Novel transformations of arachidonic acid in human platelets, *Proc. Nat. Acad. Sci. U.S.A.* **71**:3400.
Hamberg, M., Svensson, J., and Samuelsson, B., 1974a, Prostaglandin endoperoxides. A new concept concerning the mode of action and release of prostaglandins, *Proc. Nat. Acad. Sci. U.S.A.* **71**:3824.
Hamberg, M., Svensson, J., Wakabayashi, J., and Samuelsson, B., 1974b, Isolation and structure of two prostaglandin endoperoxides that cause platelet aggregation, *Proc. Nat. Acad. Sci. U.S.A.* **71**:345.
Hamberg, M., Hedqvist, P., Strandberg, K., Svensson, J., and Samuelsson, B., 1975a, Prostaglandin endoperoxides IV. Effects on smooth muscle, *Life Sci.* **16**:451.
Hamberg, M., Svensson, J., and Samuelsson, B., 1975b, Thromboxanes: A new group of biologically active compounds derived from prostaglandin endoperoxides, *Proc. Nat. Acad. Sci. U.S.A.* **72**:2994.
Hammarström, S., and Falardeau, P., 1977, Resolution of prostaglandin endoperoxide synthase and thromboxane synthase of human platelets, *Proc. Nat. Acad. Sci. U.S.A.* **74**:3691.
Harris, R. H., Ramwell, P. W., and Gilmer, P. J., 1979, Cellular mechanisms of prostaglandin action, *Annu. Rev. Physiol.* **41**:653.
Hemler, M. E., and Lands, W. E. M., 1980, Evidence for a peroxide-initiated free radical mechanism of prostaglandin biosynthesis, *J. Biol. Chem.* **255**:6253.
Hemler, M., Lands, W. E. M., and Smith, W., 1976, Purification of the cyclooxygenase that forms prostaglandins. Demonstration of two forms of iron in the holoenzyme, *J. Biol. Chem.* **251**:5575.
Hemler, M. E., Crawford, C. G., and Lands, W. E. M., 1978, Lipoxygenation activity of purified prostaglandin-forming cyclooxygenase, *Biochemistry* **17**:1772.
Hittelman, K. J., and Butcher, R. W., 1973, Cyclic AMP and mechanism of action of the prostaglandins, in: *Prostaglandins: Pharmacological and Therapeutic Advances* (M. F. Cuthbert, ed.), Heinemann Medical Books, London, pp. 151–165.
Hokin, L. E., 1968, Dynamic aspects of phospholipids during protein secretion, *Internat. Rev. Cytol.* **23**:187.
Horton, E. W., 1969, Hypotheses on physiological roles of prostaglandins, *Physiol. Rev.* **49**:122.
Horton, E. W., and Thompson, C. J., 1964, Thin-layer chromatography and bioassay of prostaglandins in extracts of semen and tissues of the male reproductive tract, *Brit. J. Pharmacol. Chemother.* **22**:183.
Ingerman, C., Smith, J. B., Kocsis, J. J., and Silver, M. J., 1973, Arachidonic acid induces platelet aggregation and platelet prostaglandin formation, *Fed. Proc.* **32**:219.
Johnson, M., Davison, P., and Ramwell, P. W., 1972, Carnitine-dependent oxidation of prostaglandins, *J. Biol. Chem.* **247**:5656.
Johnson, R. A., Morton, D. R., Kinner, J. H., Gorman, R. R., McGuire, J. C., Sun, F. F., Whittaker, N., Bunting, S., Salmon, J., Moncada, S., and Vane, J. R., 1976, The chemical structure of prostaglandin X (prostacyclin), *Prostaglandins* **12**:915.
Johnson, R. A., Lincoln, F. H., Thompson, J. L., Nidy, E. G., Mizsak, S. A., and Axen, U., 1977, Synthesis and stereochemistry of prostacyclin and synthesis of 6-ketoprostaglandin $F_{1\alpha}$, *J. Am. Chem. Soc.* **99**:4182.
Karim, S., 1971a, Action of prostaglandin in the pregnant woman, *Ann. N.Y. Acad. Sci.* **180**:483.
Karim. S. M. M., 1971b, Prostaglandins as abortifacients, *New Engl. J. Med.* **285**:1534.
Kauffman, G. L., Jr., and Grossman, M. I., 1978, Prostaglandin and cimetidine inhibit the formation of ulcers produced by parenteral salicylates, *Gastroenterology* **75**:1099.
Kauffman, G. L., Jr., Whittle, B. J. R., Aures, D., Vane, J. R., and Grossman, M. I., 1979, Effects of prostacyclin and a stable analogue, 6_β-PGI_1, on gastric acid secretion, mucosal blood flow, and blood pressure in conscious dogs, *Gastroenterology* **77**:1301.
Kindahl, H., 1977, Metabolism of thromboxane B_2 in the cynomolgus monkey, *Prostaglandins* **13**:619.

Klenberg, D., and Samuelsson, B., 1965, The biosynthesis of prostaglandin E_1 studied with specifically ^{3}H-labelled 8,11,14-eicosatrienoic acids, *Acta Chem. Scand.* **19**:534.

Knapp, H. R., Oelz, O., Sweetman, B. J., and Oates, J. A., 1978, Synthesis and metabolism of prostaglandins E_2, $F_{2\alpha}$, and D_2 by the rat gastrointestinal tract. Stimulation by a hypertonic environment in vitro, *Prostaglandins* **15**:751.

Kurzrok, R., and Lieb, C. C., 1930, Biochemical studies of human semen II. The action of semen on the human uterus, *Proc. Soc. Exp. Biol. Med.* **28**:268.

Lands, W. E. M., 1979, The biosynthesis and metabolism of prostaglandins, *Annu. Rev. Physiol.* **41**:633.

Lands, W. E. M., and Samuelsson, B., 1968, Phospholipid precursors of prostaglandins, *Biochim. Biophys. Acta* **164**:426.

Lee, S.-C., and Levine, L., 1974, Prostaglandin metabolism. I. Cytoplasmic reduced nicotinamide adenine dinucleotide phosphate-dependent and microsomal reduced nicotinamide adenine dinucleotide-dependent prostaglandin E 9-ketoreductase activities in monkey and pigeon tissue, *J. Biol. Chem.* **249**:1369.

Lee, S.-C., and Levine, L., 1975, Prostaglandin metabolism. II. Identification of two 15-hydroxyprostaglandin dehydrogenase types, *J. Biol. Chem.* **250**:548.

Lewis, P. J., and Dollery, C. T., 1983, Clinical pharmacology and potential of prostacyclin, *Brit. Med. Bull.* **39**:281.

Maclouf, J., Kindahl, H., Granström, E., and Samuelsson, B., 1980, Thromboxane A_2 and prostaglandin H_2 form covalently linked derivatives with human serum albumin, in: *Advances in Prostaglandin and Thromboxane Research,* Volume 6 (B. Samuelsson, P. W. Ramwell, and R. Paoletti, eds.), Raven Press, New York, pp. 283–286.

Malmsten, C., Hamberg, M., Svensson, J., and Samuelsson, B., 1975, Physiological role of an endoperoxide in human platelets: Hemostatic defect due to platelet cyclo-oxygenase deficiency, *Proc. Nat. Acad. Sci. U.S.A.* **72**:1446.

Marcus, A. J., Ullman, H. L., and Safier, L. B., 1969, Lipid composition of subcellular particles of human blood platelets, *J. Lipid Res.* **10**:108.

Marnett, L. J., Wlodawer, P., and Samuelsson, B., 1974, Light emission during the action of prostaglandin synthetase, *Biochem. Biophys. Res. Commun.* **60**:1286.

Marnett, L. J., Wlodawer, P., and Samuelsson, B., 1975, Co-oxygenation of organic substrates by the prostaglandin synthetase of sheep vesicular gland, *J. Biol. Chem.* **250**:8510.

Marnett, L. J., Bienkowski, M. J., and Pagels, W. R., 1979, Oxygen 18 investigation of the prostaglandin synthetase-dependent co-oxidation of diphenylisobenzofuran, *J. Biol. Chem.* **254**:5077.

Mason, R. P., Kalyanaraman, B., Tainer, B. E., and Eling, T. E., 1980, A carbon-centered free radical intermediate in the prostaglandin synthetase oxidation of arachidonic acid. Spin trapping and oxygen uptake studies, *J. Biol. Chem.* **255**:5019.

Mitchell, R. H., 1975, Inositol phospholipids and cell surface receptor function, *Biochim. Biophys. Acta* **415**:81.

Miyamoto, T., Yamamoto, S., and Hayaishi, O., 1974, Prostaglandin synthetase system—Resolution into oxygenase and isomerase components, *Proc. Nat. Acad. Sci. U.S.A.* **71**:3645.

Miyamoto, T., Ogino, N., Yamamoto, S., and Hayaishi, O., 1976, Purification of prostaglandin endoperoxide synthetase from bovine vesicular gland microsomes, *J. Biol. Chem.* **251**:2629.

Moncada, S., Gryglewski, R., Bunting, S., and Vane, J. R., 1976, An enzyme isolated from arteries transforms prostaglandin endoperoxides to an unstable substance that inhibits platelet aggregation, *Nature* **263**:663.

Moncada, S., Higgs, E. A., and Vane, J. R., 1977, Human arterial and venous tissues generate prostacyclin (prostaglandin X), a potent inhibitor of platelet aggregation, *Lancet* **1**:18.

Moncada, S., and Vane, J. R., 1979, The role of prostacyclin in vascular tissue, *Fed. Proc.* **38**:66.

Morrison, A. R., and Needleman, P., 1979, Biochemistry and pharmacology of renal prostaglandins, in: *Hormonal Function and the Kidney. Contemporary Issues in Nephrology,* Volume 4 (B. M. Brenner and J. H. Stein, eds.), Churchill Livingstone, New York, pp. 68–88.

Needleman, P., Minkes, M., and Raz, A., 1976a, Thromboxanes: Selective biosynthesis and distinct biological properties, *Science* **193**:163.

Needleman, P., Moncada, S., Bunting, S., Vane, J. R., Hamberg, M., and Samuelsson, B., 1976b, Identification of an enzyme in platelet microsomes which generates thromboxane A_2 from prostaglandin endoperoxides, *Nature* **261**:558.

Needleman, P., Kulkarni, P. S., and Raz, A., 1977, Coronary tone modulation: Formation and actions of prostaglandins, endoperoxides, and thromboxanes, *Science* **195**:409.

Nezamis, R. J. E., Lancaster, C., and Hauchar, A. J., 1977, Gastric cytoprotective property of prostaglandins, *Gastroenterology* **72**:1121.

Nugteren, D. H., and Christ-Hazelhof, E., 1980, Chemical and enzymic conversion of the prostaglandin endoperoxide H_2, *Adv. Prostaglandin Thromboxane Res.* **6**:129.

Nugteren, D. H., and Hazelhof, E., 1973, Isolation and properties of intermediates in prostaglandin biosynthesis, *Biochim. Biophys. Acta* **326**:448.

Nugteren, D. H., and van Dorp, D. A., 1965, The participation of molecular oxygen in the biosynthesis of prostaglandins, *Biochim. Biophys. Acta* **98**:654.

Nugteren, D. H., van Dorp, D. A., Bergström, S., Hamberg, M., and Samuelsson, B., 1966, Absolute configuration of the prostaglandins, *Nature* **212**:38.

O'Brien, P. J., and Rahimtula, A. D., 1980, Mechanism of oxygen activation involved in the prostaglandin synthetase mechanism, *Adv. Prostaglandin Thromboxane Res.* **6**:145.

Ogino, N., Miyamoto, T., Yamamoto, S., and Hayaishi, O., 1977, Prostaglandin endoperoxide isomerase from bovine vesicular gland microsomes, a glutathione-requiring enzyme, *J. Biol. Chem.* **252**:890.

Ogino, N., Ohki, S., Yamamoto, S., and Hayaishi, O., 1978, Prostaglandin endoperoxide synthetase from bovine vesicular gland microsomes. Inactivation and activation by heme and other metalloporphyrins, *J. Biol. Chem.* **253**:5061.

Ogino, N., Yamamoto, S., Hayaishi, O., and Tokuyama, T., 1979, Isolation of an activator for prostaglandin hydroperoxidase from bovine vesicular gland cytosol and its identification as uric acid, *Biochem. Biophys. Res. Commun.* **87**:184.

Ohki, S., Ogino, N., Yamamoto, S., and Hayaishi, O., 1979, Prostaglandin hydroperoxidase, an integral part of prostaglandin endoperoxide synthetase from bovine vesicular gland microsomes, *J. Biol. Chem.* **254**:829.

Oien, H. G., Ham, E. A., Zanetti, M. E., Ulm, E. H., and Kuehl, F. A., Jr., 1976, A 15-hydroxyprostaglandin dehydrogenase specific for prostaglandin A in rabbit kidney, *Proc. Nat. Acad. Sci. U.S.A.* **73**:1107. (Correction: 1976, *Proc. Nat. Acad. Sci. U.S.A.* **73**:2528.)

Pace-Asciak, C., 1972, Prostaglandin synthetase activity in rat stomach fundus. Activation by L-norepinephrine and related compounds, *Biochim. Biophys. Acta* **280**:161.

Pace-Asciak, C., 1976, Isolation, structure, and biosynthesis of 6-ketoprostaglandin $F_{1\alpha}$ in rat stomach, *J. Am. Chem. Soc.* **98**:2348.

Pace-Asciak, C., and Wolf, L. S., 1971, A novel prostaglandin derivative formed from arachidonic acid by rat stomach homogenates, *Biochemistry* **10**:3657.

Piper, P. J., and Vane, J. R., 1969, Release of additional factors in anaphylaxis and its antagonism by anti-inflammatory drugs, *Nature* **223**:29.

Polet, H., and Levine, L., 1975, Metabolism of prostaglandins E, A, and C in serum, *J. Biol. Chem.* **250**:351.

Prescott, S. M., and Majerus, P. N., 1983, Characterization of 1,2-diacylglycerol hydrolysis in human platelets. Demonstration of an arachidonoyl-monoacylglycerol intermediate, *J. Biol. Chem.* **258**:764.

Prince, A., Alvarez, F. S., and Young, J., 1973, Preparation of prostaglandins A_2, E_2 and 11-epi-E_2 from their 15 acetates methyl esters using an endogenous enzyme system present in Plexaura homomalla (Esper), *Prostaglandins* **3**:531.

Rahimtula, A., and O'Brien, P. J., 1976, The possible involvement of singlet oxygen in prostaglandin biosynthesis, *Biochem. Biophys. Res. Commun.* **70**:893.

Ramwell, P. W., and Shaw, J. E., 1966, Spontaneous and evoked release of prostaglandins from cerebral cortex of anesthetized cats, *Am. J. Physiol.* **211**:125.

Ramwell, P. W., Shaw, J. E., and Nucharski, J., 1965, Prostaglandin: Release from the rat phrenic nerve–diaphragm preparation, *Science* **149**:1390.

Ramwell, P. W., Shaw, J. E., Douglas, W. W., and Poisner, A. M., 1966, Efflux of prostaglandin from adrenal glands stimulated with acetycholine, *Nature* **210**:273.

Ratnoff, O. D., 1978, Hereditary disorders of hemostasis, in: *Metabolic Basis of Inherited Disease* (J. B. Stanbury, J. B. Wyngaarden, and D. S. Fredrickson, eds.), 4th ed., McGraw-Hill, New York, pp. 1755-1791.
Roberts, L. J., II, Sweetman, B. J., Morgan, J. L., Payne, N. A., and Oates, J. A., 1977a, Identification of the major urinary metabolite of thromboxane B_2 in the monkey, *Prostaglandins* **13**:631.
Roberts, L. J., II, Sweetman, B. J., Payne, N. A., and Oates, J. A., 1977b, Metabolism of thromboxane B_2 in man, *J. Biol. Chem.* **252**:7415.
Roberts, L. J., II, Lewis, R. A., Lawson, J. A., Sweetman, B. J., Austen, K. F., and Oates, J. A., 1978a, Arachidonic acid metabolism by rat mast cells, *Prostaglandins* **15**:717.
Roberts, L. J., II, Sweetman, B. J., and Oates, J. A., 1978b, Metabolism of thromboxane B_2 in the monkey, *J. Biol. Chem.* **253**:5305.
Roberts, S. M., and Newton, R. F. (eds.), 1982, *Prostaglandins and Thromboxanes*, Butterworths Monographs in Chemistry, Butterworth Scientific, London, pp. 37-136.
Roth, G. J., Stanford, N., and Majerus, P. W., 1975, Acetylation of prostaglandin synthetase by aspirin, *Proc. Nat. Acad. Sci. U.S.A.* **72**:3073.
Roth, G. J., Siok, C. J., and Ozois, J., 1980, Structural characteristics of prostaglandin synthetase from sheep vesicular gland, *J. Biol. Chem.* **255**:1301.
Roth, G. J., Machuga, E. T., and Strittmatter, P., 1981, The heme-binding properties of prostaglandin synthetase from sheep vesicular glands, *J. Biol. Chem.* **256**:10018.
Ryhage, R., and Samuelsson, B., 1965, The origin of oxygen incorporated during the biosynthesis of prostaglandin E_1, *Biochem. Biophys. Res. Commun.* **19**:279.
Samuelsson, B., 1963, Prostaglandins and related factors. 17. The structure of prostaglandin E_3, *J. Am. Chem. Soc.* **85**:1878.
Samuelsson, B., 1965, On the incorporation of oxygen in the conversion of 8,11,14-eicosatrienoic acid to prostaglandin E_1, *J. Am. Chem. Soc.* **87**:3011.
Samuelsson, B., 1970, Structures, biosynthesis and metabolism of prostaglandins, in: *Lipid Metabolism* (S. J. Wakil, ed.), Academic Press, New York, pp. 107-153.
Samuelsson, B., 1978, Prostaglandins and thromboxanes, *Recent Prog. Hormone Res.* **34**:239.
Samuelsson, B., Granström, E., Gréen, K., and Hamberg, M., 1971, Metabolism of prostaglandins, *Ann. N.Y. Acad. Sci.* **180**:138.
Samuelsson, B., Granström, E., Green, K., Hamberg, M., and Hammarström, S., 1975, Prostaglandins, *Annu. Rev. Biochem.* **44**:669.
Samuelsson, B., Goldyne, M., Granström, E., Hamberg, M., Hammarström, S., and Malmsten, C., 1978, Prostaglandins and thromboxanes, *Annu. Rev. Biochem.* **47**:997.
Schneider, W. P., Hamilton, R. D., and Rhuland, L. E., 1972, Occurrence of esters of (15S)-prostaglandin A_2 and E_2 in coral, *J. Am. Chem. Soc.* **94**:2122.
Schneider, W. P., Bundy, G. L., Lincoln, F. H., Daniels, E. G., and Pike, J. E., 1977, Isolation and chemical conversions of prostaglandins from *Plexaura homomalla:* Preparation of prostaglandin E_2, prostaglandin $F_{2\alpha}$, and their 5,6-trans isomers, *J. Am. Chem. Soc.* **99**:1222.
Shaw, J. E., and Ramwell, P. W., 1967, Prostaglandin release from the adrenal gland, in: *Nobel Symposium 2. Prostaglandins* (S. Bergström and B. Samuelsson, eds.), Interscience, New York, pp. 291-299.
Shaw, J. E., and Ramwell, P. W., 1968, Release of prostaglandin from rat epididymal fat pad on nervous and hormonal stimulation, *J. Biol. Chem.* **243**:1498.
Sheng, W. Y., Wyche, A., Lysz, T., and Needleman, P., 1982, Prostaglandin E_2 isomerase is dissociated from prostaglandin endoperoxide synthetase in the renal cortex, *J. Biol. Chem.* **257**:14632.
Shimizu, T., Yamamoto, S., and Hayaishi, O., 1979a, Purification and properties of prostaglandin D synthetase from rat brain, *J. Biol. Chem.* **254**:5222.
Shimizu, T., Mizuno, N., Amano, T., and Hayaishi, O., 1979b, Prostaglandin D_2, a neuromodulator, *Proc. Nat. Acad. Sci. U.S.A.* **76**:6231.
Silver, M. J., Smith, J. B., Ingerman, C., and Kocsis, J. J., 1973, Arachidonic acid-induced human platelet aggregation and prostaglandin formation, *Prostaglandins* **4**:863.
Sivakoff, M., Pure, E., Hsueh, W., and Needleman, P., 1979, Prostaglandins and the heart, *Fed. Proc.* **38**:78.
Smith, M. C., and Dunn, M. J., 1981, Renal kallikrein, kinins and prostaglandins in hypertension, in:

Hypertension. Contemporary Issues in Nephrology, Volume 8 (B. M. Brenner and J. H. Stein, eds.), Churchill Livingston, New York, pp. 168–202.
Smith, W. L., and Lands, W. E. M., 1972, Oxygenation of polyunsaturated fatty acids during prostaglandin biosynthesis by sheep vesicular gland, *Biochemistry* **11:**3276.
Smith, J. B., and Willis, A. L., 1971, Aspirin selectively inhibits prostaglandin production in human platelets, *Nature New Biology* **231:**235.
Smith, J. B., Ingerman, C., Kocsis, J. J., and Silver, M. J., 1974a, Formation of an intermediate in prostaglandin biosynthesis and its association with the platelet release reaction, *J. Clin. Invest.* **53:**1468.
Smith, J. B., Silver, M. J., Ingerman, C. M., and Kocsis, J. J., 1974b, Prostaglandin D_2 inhibits the aggregation of human platelets, *Thromb. Res.* **5:**291.
Steinberg, D., and Vaughan, M., 1967, In vitro and in vivo effects of prostaglandins on free fatty acid metabolism, in: *Nobel Symposium 2. Prostaglandins* (S. Bergström and B. Samuelsson, eds.), Interscience Publishers, New York, pp. 109–121.
Steinberg, D., Vaughan, M., Nestel, P. J., Strand, O., and Bergström, S., 1964, Effects of prostaglandins on hormone-induced mobilization of free fatty acids, *J. Clin. Invest.* **43:**1533.
Sun, F. F., Taylor, B. M., McGuire, J. C., Wong, P. Y-K., Malik, K. U., and McGiff, J. C., 1979, Metabolic disposition of prostacyclin, in: *Prostacyclin* (J. R. Vane and S. Bergström, eds.), Raven Press, New York, pp. 119–131.
Svensson, J., 1979, Structure and quantitative determination of the major urinary metabolite of thromboxane B_2 in the guinea pig, *Prostaglandins* **17:**351.
Ueno, R., Shimizu, T., Kondo, K., and Hayaishi, O., 1982, Activation mechanism of prostaglandin endoperoxide synthetase by hemoproteins, *J. Biol. Chem.* **257:**5584.
Van der Ouderaa, F. J., Buytenhek, M., Nugteren, D. H., and van Dorp, D. A., 1977, Purification and characterization of prostaglandin endoperoxide synthetase from sheep vesicular glands, *Biochim. Biophys. Acta* **487:**315.
Van der Ouderaa, F. J., Buytenhek, M., Slikkerveer, F. J., and van Dorp, D. A., 1979, On the haemoprotein character of prostaglandin endoperoxide synthetase, *Biochim. Biophys. Acta* **572:**29.
van Dorp, D. A., Beerthuis, R. K., Nugteren, D. H., and Vonkeman, H., 1964a, The biosynthesis of prostaglandins, *Biochim. Biophys. Acta* **90:**204.
van Dorp, D. A., Beerthuis, R. K., Nugteren, D. H., and Vonkeman, H., 1964b, Enzymatic conversion of all-*cis*-polyunsaturated fatty acids into prostaglandins, *Nature* **203:**839.
Vane, J. R., 1971, Inhibition of prostaglandin synthesis as a mechanism of action for aspirin-like drugs, *Nature New Biology* **231:**232.
von Euler, U. S., 1934, The pharmacological effects of the secretion and extracts of male accessory glands, *Naunyn-Schmiedeberg's Arch. Exptl. Pathol. Pharmacol.* **175:**78.
von Euler, U. S., 1935a, Über die spezifische blutdrucksenkende Substanz des menschlichen Prostata und Samenblasensekretes, *Klin. Wchnschr.* **14:**1182.
von Euler, U. S., 1935b, A depressor substance in the vesicular gland, *J. Physiol.* (London) **84:**21P.
von Euler, U. S., 1936, On the specific vasodilating and plain muscle stimulating substances from accessory genital glands in man and certain animals (prostaglandin and vesiglandin), *J. Physiol.* (London) **88:**213.
von Euler, U. S., and Eliasson, R., 1967, *Prostaglandins,* Academic Press, New York.
Vorgaftig, B. B., and Zirinis, P., 1973, Platelet aggregation induced by arachidonic acid is accompanied by release of potential inflammatory mediators distinct from PGE_2 and $PGF_{2\alpha}$, *Nature New Biology* **244:**114.
Wasserman, M. A., DuCharm, D. W., Griffin, R. L., DeGraaf, G. L., and Robinson, F. G., 1977, Bronchopulmonary and cardiovascular effects of prostaglandin D_2 in the dog, *Prostaglandins* **13:**255.
Watanabe, K., Shimizu, T., Iguchi, S., Wakatsuka, H., Hayashi, M., and Hayaishi, O., 1980, An NADP-linked prostaglandin D dehydrogenase in swine brain, *J. Biol. Chem.* **255:**1779. (Correction: 1980, *J. Biol. Chem.* **255:**5992.)
Watanabe, T., Narumiya, S., Shimizu, T., and Hayaishi, O., 1982, Characterization of the biosynthetic pathway of prostaglandin D_2 in human platelet-rich plasma, *J. Biol. Chem.* **257:**14847.
Weinheimer, A. J., and Spraggins, R. L., 1969, The occurrence of two new prostaglandin derivatives

(15-epi-PGA$_2$ and its acetate, methyl ester) in the gorgonian *Plexaura homomalla, Tetrahedron Lett.* No. 59, p. 5185.

Whitaker, M. O., Needleman, P., Wyche, A., Fitzpatrick, F. A., and Sprecher, H., 1980, PGD$_3$ is the mediator of the antiaggregatory effects of the trienoic endoperoxide PGH$_3$, *Adv. Prostaglandin Thromboxane Res.* **6**:301.

White, R. P., 1982, Prostaglandins and cerebral vasospasm, in: *Prostaglandins: Organ- and Tissue-specific Actions* (S. Greenberg, P. J. Kadowitz, and T. F. Burks, eds.), Marcel Dekker, New York, pp. 341–365.

Witkop, C. J., Jr., Quevedo, W. C., Jr., and Fitzpatrick, T. B., 1978, Albinism, in: *Metabolic Basis of Inherited Disease,* 4th ed. (J. B. Stanbury, J. B. Wyngaarden, and D. S. Fredrickson, eds.), McGraw-Hill, New York, pp. 295–303.

Wlodawer, P., Kindahl, H., and Hamberg, M., 1976, Biosynthesis of prostaglandin F$_{2\alpha}$ from arachidonic acid and prostaglandin endoperoxides in the uterus, *Biochim. Biophys. Acta* **431**:603.

Wolfe, L. S., and Coceani, F., 1979, The role of prostaglandins in the central nervous system, *Annu. Rev. Physiol.* **41**:669.

Wong, P., Malik, K., Sun, F., Lee, W., and McGiff, J., 1979a, Hepatic metabolism of PGI$_2$ in the rabbit: Formation of a potent inhibitor of platelet aggregation. Abstracts IV International Prostaglandin Conference, Washington, D.C., May 1979. (Cited by Granström, 1981a.)

Wong, P. Y-K., Sun, F. F., Malik, K. U., Cagen, L., and McGiff, J. C., 1979b, Metabolism of prostacyclin in the isolated kidney and lung of the rabbit, in: *Prostacyclin* (J. R. Vane and S. Bergström, eds.), Raven Press, New York, pp. 133–145.

Yoshimoto, T., Yamamoto, S., Sugioka, K., Nakano, M., Takyu, C., Yamagishi, A., and Inaba, H., 1980, Studies on the tryptophan-dependent light emission by prostaglandin hydroperoxidase reaction, *J. Biol. Chem.* **255**:10199.

Zor, U., and Lamprecht, S. A., 1977, Mechanism of prostaglandin action in endocrine glands, in: *Biochemical Action of Hormones,* Volume IV (G. Litwack, ed.), Academic Press, New York, pp. 85–133.

11
EICOSANOIDS: LEUKOTRIENES AND SLOW-REACTING SUBSTANCES OF ANAPHYLAXIS

11.1. Introduction

The role of prostaglandins in many physiological and pathological processes, including inflammations, has been extensively studied. Prostaglandins and unsaturated hydroxy acids have been noted to affect some functions of polymorphonuclear leukocytes, e.g., chemotaxis and phagocytosis (for review, see Willoughby et al., 1973). To obtain detailed information about the possible role of polyunsaturated fatty acids in the life of leukocytes, Samuelsson and his colleagues began about 1976 to investigate the transformations of arachidonic (5,8,11,14-cis-eicosatetraenoic; 5,8,11,14-cis-20:4) and dihomo-γ-linolenic (8,11,14-cis-eicosatrienoic; 8,11,14-cis-20:3) acids by polymorphonuclear leukocytes, in continuation of their work on prostaglandins. The experiments were at first carried out either with rabbit leukocytes, collected from the peritoneal cavity of the animals injected intraperitoneally with suspensions of glycogen, or with human leukocytes isolated from freshly drawn blood, and subsequently also with murine mastocytoma cells and with cultured peritoneal macrophages. These investigations led to the discovery of a number of new C_{20} hydroxy-polyenoic fatty acids and the leukotrienes (A, B, C, D, and E). Three of these, leukotrienes C, D, and E, were identified as the long-sought "slow-reacting substances of anaphylaxis" (SRS-A, or SRS), which accompany the release of histamine in anaphylactic shock and in other allergic reactions.

The term "leukotriene" was proposed by Samuelsson et al. (1979) to denote compounds like SRS and noncyclic C_{20}-polyenoic carboxylic acids that contain one or two oxygen substituents and three conjugated double bonds. While it is not stated in this proposal, the "leuko" part of the term was, no doubt, inspired by the fact that the first representatives of this class of compounds were obtained from leukocytes.

A historical parallel exists between the discovery of SRS and the prostaglandins in that the original observations with respect to both were made in the 1930s, and some 40 years passed before they were identified. SRS was discovered by Feldberg and Kellaway in 1938 after the perfusion of the lungs and liver of dogs and monkeys with fluids containing cobra venom, and was found again in 1940 by Kellaway and Trethewie after perfusion of sensitized guinea-pig lungs with antigen. A further parallel between SRS and the prostaglandins, as it turned out, is that eicosapolyenoic fatty acids, bound probably to phospholipids, are the precursors of both.

11.2. Hydroperoxy and Hydroxy Acids Derived from Eicosapolyenoic Fatty Acids

11.2.1. Reactions of Polyenoic Fatty Acids Catalyzed by Lipoxygenases

Since the precursors of the hydroxy acids and of the leukotrienes, derived from eicosapolyenoic fatty acids by the action of leukocytes and other types of cells, are probably the products of reactions catalyzed by lipoxygenases, the properties of such enzymes will be first described. Lipoxygenases have been recognized in many plants, but only the enzymes of soybean and of the potato tuber have been well characterized.

Both types of enzyme produce hydroperoxy acids from polyunsaturated fatty acids containing the *cis,cis*-1,4-pentadiene system (Fig. 11.1a,b). The soybean contains three lipoxygenases, L-1, L-2, and L-3. Of these L-1 is the best characterized and was the first one obtained in pure and crystalline form by Theorell *et al.* (1947). The soybean lipoxygenase L-1 has an M_r of about 100,000 (reported values: 100,000–108,000) and a pH optimum of 9.0. It attacks eicosapolyenoic fatty acids, such as dihomo-γ-linolenic, arachidonic, and 5,8,11,14,17-*cis*-20:5 acids, specifically at the 14-double bond, introducing the O_2 at the ω6 (C-15) carbon atom, with the isomerization of the *cis*-14-double bond to the *trans*-13 with well-defined stereochemistry. Hamberg and Samuelsson (1967) have shown that the product obtained from 8,11,14-*cis*-20:3 by the action of L-1 was the 15*S*-hydroperoxy-8,11-*cis*-13-*trans*-20:3. Moreover, they demonstrated, with the aid of [(13*R*)-13-^{3}H$_1$-3-^{14}C]- and [(13*S*)-13-^{3}H$_1$-3-^{14}C]-8,11,14-*cis*-20:3 that, in the isomerization of the double bond, the 13-pro-*S* hydrogen atom was stereospecifically eliminated (cf. Fig. 11.2). In other words, the hydrogen atom was eliminated from the methylene group of the *cis,cis*-pentadiene system from the face of the

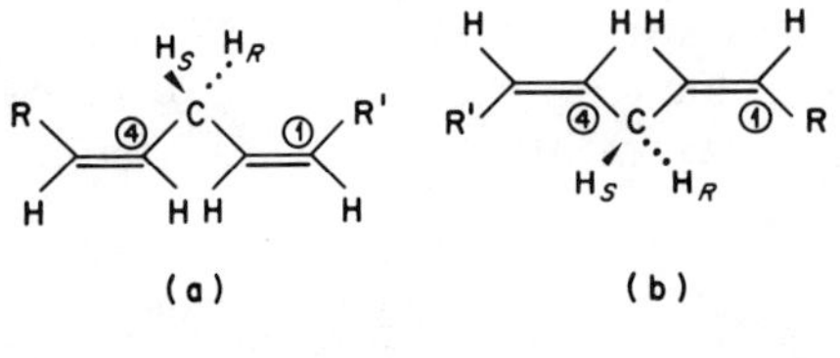

FIGURE 11.1. Structural requirements in substrates for lipoxygenases. The prochirality of the hydrogen atoms on the methylene carbon atom in (a) and (b) is assigned on the assumption that R has priority over R' by the sequence rule.

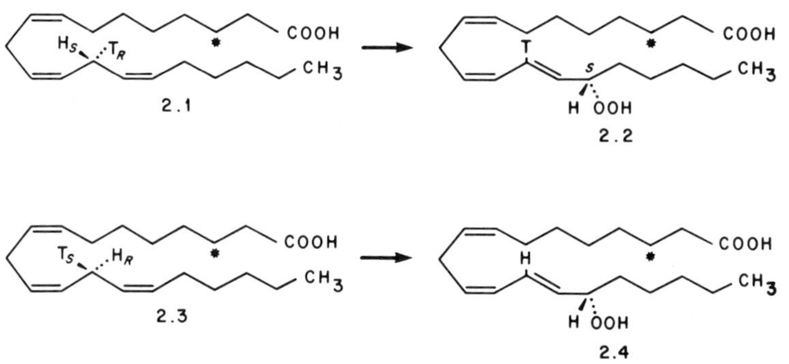

FIGURE 11.2. Conversion of dihomo-γ-linolenic acids, labeled at C-3 with $^{14}C(*)$ and stereospecifically with tritium at C-13, to 15S-hydroperoxy-8,11-cis-13-trans-20:3, by soybean lipoxygenase. Note the stereospecific elimination of the 13-pro-S-hydrogen atom. From Hamberg and Samuelsson (1967).

molecule opposite to that onto which the O_2 was added. They also found that in the conversion of [(13S)-13-$^3H_1$-3-^{14}C]-8,11,14-cis-20:3 to the 15S-hydroperoxy-8,11-cis-13-trans-20:3 there was an appreciable ^{3}H-isotope effect: in the experiments in which the substrate was not used completely, the residual substrate became substantially enriched with ^{3}H, as judged by ^{3}H/^{14}C ratios higher in the residual than in the starting substrate. This observation suggests that removal of the 13-pro-S hydrogen atom might be the first rate-limiting step in the reaction, in analogy with the phenomenon described for the initial step in prostaglandin biosynthesis from [(13S)-13-$^3H_1$-1-^{14}C]5,8,11,14,-cis-20:4 (cf. Chapter 10, p.165).

In addition to L-1, often referred to as the "Theorell" enzyme, two other lipoxygenases, L-2 and L-3, were identified in extracts of soybean by Christopher et al. (1970, 1972). L-2 and L-3 have the same molecular weight as L-1, but differ from it in their pH optima and their isoelectric points. While the pH optimum of L-1 is about 9.0, that of L-2 is 6.5, the same as that of the potato lipoxygenase (see below). L-3 has a broad pH activity curve with maxima between pH 6.0 and 7.0, but it is inactive at pH 9.0.* The three soybean lipoxygenases have also different isoelectric points. The pIs of L-1, L-2, and L-3 are 5.68, 6.25, and 6.15, respectively. In terms of their pH optima, L-2 and L-3 resemble the lipoxygenase from potato tuber. The positional selectivity of L-2 and L-3 at their pH optima for introduction of O_2 into a polyenoic fatty acid is not as strict as that of L-1 at pH 9. L-2 converts, e.g., linoleic acid equally into 13-hydroperoxy-9-cis-11-trans-18:2 and into 9-hydroperoxy-10-trans-12-cis-18:2; L-3 on the other hand produces more (65%) of the 9-hydroperoxy than of the 13-hydroperoxy derivative (Axelrod et al., 1981).

Although the molecular weights of L-1, L-2, and L-3 are indistinguishable from one another, it is uncertain whether these enzymes consist of a single polypeptide chain or of dissociable subunits. Stevens et al. (1970), who probably purified the soybean L-1 enzyme, reported that their homogeneous enzyme had an

---

*The delayed discovery of these two additional lipoxygenases in soybean may be attributed to the general prescription of pH 9.0 for assays of lipoxygenase activity.

M_r of 108,000 ± 2,000 (determined by sedimentation equilibrium in the ultracentrifuge), and that in 6.0 M guanidine + 0.5% 2-mercaptoethanol it partially dissociated into subunits of M_r 58,000. In sedimentation equilibrium experiments in the presence of 0.5% sodium dodecyl sulfate, the sedimentation coefficient of the native enzyme was reduced from 5.2S to 2.8S. These data suggest that at least the L-1 enzyme consists of two subunits. Christopher et al. (1972) reported, on the other hand, that the L-2 and L-3 enzymes were not dissociated into subunits by sodium dodecyl sulfate/polyacrylamide gel electrophoresis.

All three isozymes of soybean lipoxygenase contain 1 g-atom of iron per mol, probably as Fe^{3+} (Pistorius and Axelrod, 1974). The Fe^{3+}-chelator, Tiron (4,5-dihydroxy-*m*-benzenedisulfonic acid) removes the iron from these enzymes slowly, but an Fe^{2+}-chelator, such as *o*-phenanthroline, can remove the iron only after reduction of the enzymes with 2-mercaptoethanol.

The lipoxygenase from potato tubers has a pH optimum of 5.5–6.0 and is almost inactive at pH 9.0. This enzyme, in contrast to L-1 of soybean, attacks polyunsaturated fatty acids at the double bond nearest to the carboxyl group (Galliard and Phillips, 1971). Thus potato lipoxygenase converts linoleic (9,12-*cis*-18:2) acid almost exclusively (95%) to the 9*S*-hydroperoxy-10-*trans*-12-*cis*-18:2 acid; the 13-hydroperoxy isomer is only a minor (5%) product (Fig. 11.3).

The remarkable difference between the lipoxygenases from the two sources is that the L-1 enzyme from soybean at pH 9 attacks almost exclusively the ω6 position of the polyunsaturated fatty acids, as in linoleic, arachidonic, or 5,8,11,14,17-*cis*-20:5 acids, but the potato enzyme "looks" only for the first double bond of the *cis,cis*-pentadiene system nearest to the carboxyl group, irrespective of the number of methylene groups intervening between the carboxyl and the

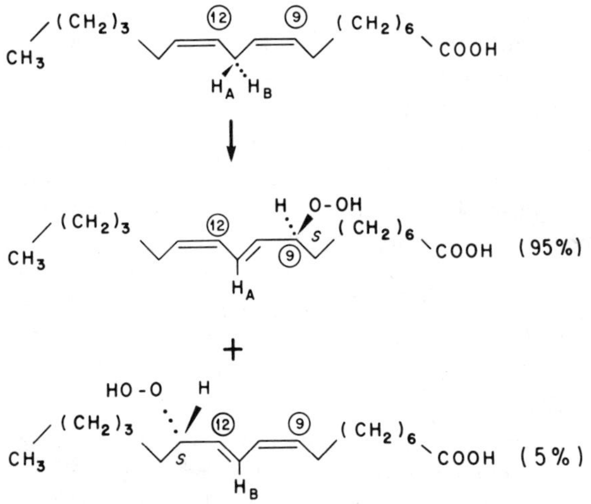

FIGURE 11.3. Conversion of linoleic acid (9,12-*cis*-18:2) by potato lipoxygenase to the 9*S*-hydroperoxy-10-*trans*-12-*cis*-18:2 as the major product and to the 13-hydroperoxy-9-*cis*-11-*trans*-18:2 (Galliard and Phillips, 1971). The stereochemistry of elimination of hydrogen atoms from C-11 is hypothetical and is based on the assumption that it occurs from the face of the molecule opposite to that onto which the hydroperoxy group is added.

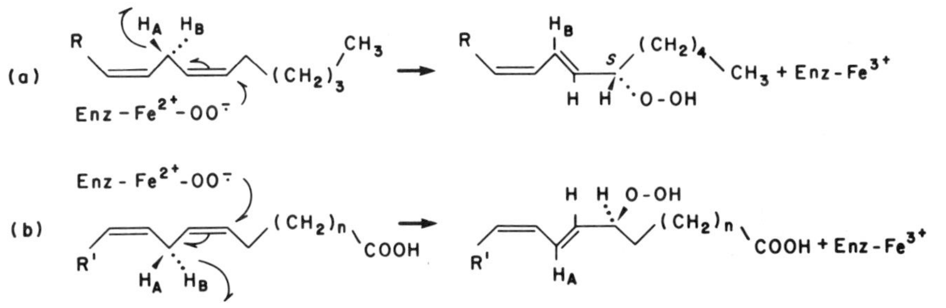

FIGURE 11.4. Specificities and possible mechanisms of reactions catalyzed by lipoxygenase from soybean (a), and from potato (b). The elimination of H_B from the methylene carbon atom in (b) is based on the assumption that the stereospecificity of the potato enzyme is analogous to that of the reaction catalyzed by the soya enzyme in that the hydrogen atom is eliminated from the face of the molecule opposite to that onto which the hydroperoxy group is added. The arrows with fishhooks (instead of full arrows) are meant to suggest a free-radical reaction sequence. In formulas shown in (b) the n subscript to (CH_2) could be 2 (as in arachidonic acid, 5,8,11,14-20:4), 5 (as in dihomo-γ-linolenic acid, 8,11,14-20:3), or 6 (as in linoleic acid, 9,12-18:2). The R-residue in (a) represents the carboxyl end, and the R'-residue in (b) the hydrocarbon tail of the molecules (cf. Walsh, 1979).

first double bond of the pentadiene system. Thus Corey, Albright et al. (1980) found that the potato lipoxygenase converted arachidonic acid exclusively into the 5S-hydroperoxy-6-*trans*-8,11,14-*cis*-20:4 acid (5-HPETE), although in the relatively poor yield of 15%.* The reactions catalyzed by the soybean and potato lipoxygenases are summarized in Fig. 11.4.

The high degree of selectivity of the L-1 lipoxygenase for the hydroperoxidation at the ω6 position in polyunsaturated fatty acids is seen only at pH 9. At lower pH values (6.5–7.5) this enzyme also catalyzes a second hydroperoxidation on polyenoic fatty acids containing at least the all-*cis*-ω6,9,12-octatriene moiety provided that after the first hydroperoxidation a *cis,cis*-1,4-pentadiene system still remains intact in the molecule. On the second hydroperoxidation the O_2 is introduced at the ω13 position with isomerization of the *cis*-ω12(13)-double bond to *trans*-ω11(12). Thus lipoxygenase L-1 converts arachidonic acid at pH 6.5–7.5 into the 8,15-dihydroperoxy-5-*cis*-9-*trans*-11-*cis*-13-*trans*-20:4 acid (= ω6,13-dihydroperoxy-ω7-*trans*-ω9-*cis*-ω11-*trans*-ω15-*cis*-20:4), and γ-linolenic acid (all-*cis*-ω6,9,12-18:3) to the 6,13-dihydroperoxy-7-*trans*-9-*cis*-11-*trans*-18:3 (= ω6,13-dihydroperoxy-ω7-*trans*-ω9-*cis*-ω11-*trans*-18:3). However, α-linolenic acid (ω3,6,9-18:3) is attacked only at the ω6 position; a second hydroperoxidation fails because after the isomerization of the *cis*-ω6-double bond to *trans*-ω7 there is no *cis,cis*-1,4-pentadiene system left in the ω6 hydroperoxy-18:3 acid.

The double hydroperoxidation of the polyenoic fatty acids containing the all-*cis*-ω6,9,12-octatriene moiety is biphasic: the first hydroperoxidation at the ω6 position is rapid, but the second at the ω13 position is seemingly slow. This difference is accounted for, in the case of arachidonic acid, by a more than two-

---

*The poor yield obtained by Corey, Albright et al. (1980) can be accounted for by their experimental conditions: they incubated the potato enzyme with arachidonic acid at pH 9.0, the optimum for the soybean enzyme, but not for the potato enzyme.

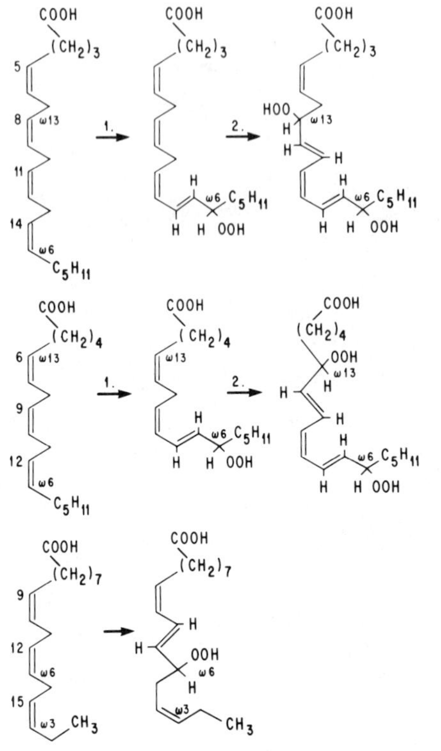

FIGURE 11.5. Hydroperoxidation of fatty acids containing all-cis-ω6,9,12-octatriene systems by soybean lipoxygenase L-1 at pH 6.8–7.5. The numbers at the left of formulas mark positions counted from the carboxyl end, and those at the right of formulas or above some positions mark positions counted from the ω-end. Note the two-step hydroperoxidation of arachidonic and γ-linolenic acids, but a single peroxidation of α-linolenic acid (after Bild et al., 1977).

hundredfold difference between the K_m value of arachidonic acid for the first oxygenation, 8.6×10^{-5} M, and the K_m value for the monohydroperoxy product in the second oxygenation, 1.8×10^{-2} M (Bild et al., 1977). The hydroperoxidation of three acids with the all-cis-ω6,9,12-octatriene system is shown in Fig. 11.5.

Polymorphonuclear leukocytes from man and the rabbit appear to contain two lipoxygenases: one has the same topical selectivity and stereochemistry as the lipoxygenase of potato tubers, the other attacks arachidonic acid at C-15 (see later). Human blood platelets, in addition to the prostaglandin cyclo-oxygenase, contain another type of lipoxygenase that introduces O_2 into arachidonic acid at C-12 with the isomerization of the 11-cis double bond to 10-trans and forms the 12S-hydroxy-5,8-cis-10-trans-14-cis-20:4 (Hamberg et al., 1974).

The mechanism of action of these lipoxygenases is probably very similar, except for their positional specificity. Conceivably, the first step could be the abstraction of a hydrogen atom from the methylene carbon atom of the cis,cis-pentadiene system, thus creating a pentadienyl free radical (cf. Walsh, 1979).

11.2.2. Hydroxy Acids and Leukotrienes Derived from Polyenoic Fatty Acids

Borgeat and coworkers first studied the transformations of dihomo-γ-linolenic and arachidonic acids by leukocytes (Borgeat et al., 1976; Borgeat and Samuelsson, 1979a,b,c). Dihomo-γ-linolenic acid (**6.1** in Fig. 11.6) incubated with rabbit leukocytes gave the 8*S*-hydroxyeicosa-9-*trans*-11,14-*cis*-20:3 acid (**6.2b**). Its absolute configuration was established to be 8*S*, as its ozonolysis yielded not only *n*-hexanoic acid (**6.3**), but also L-2-hydroxyazelaic acid (**6.4a,b**). It is important to recognize that the carboxyl group adjacent to the hydroxyl group of this hydroxyazelaic acid was C-9 of **6.2b**. The conjugated *trans/cis* double bond system was established by UV and IR spectroscopy. It is probable that the precursor of **6.2b** was the 8*S*-hydroperoxyeicosatrienoic acid, **6.2a**, the product of a lipoxygenase type of enzyme. The same hydroperoxy acid can be obtained by the action of potato lipoxygenase on dihomo-γ-linolenic acid.

The transformations of arachidonic acid, studied in both rabbit and human polymorphonuclear leukocytes, led to the recognition of several hydroxy acids and some of the leukotrienes, derived probably from one common precursor, 5*S*-hydroperoxy-6-*trans*-8,11,14-*cis*-eicosatetraenoic acid (5-HPETE), generated by a lipoxygenase, in analogy with the reaction of dihomo-γ-linolenic acid, although the hydroperoxy acids have not been isolated so far from incubations of leukocytes with any of the polyenoic fatty acids. The synthesis of the various derivatives of arachidonic acid is much stimulated by a Ca^{2+}-mobilizing ionophore, A 23187 (Eli Lilly). The effect of the ionophore is thought to be mediated by an increase in cytoplasmic Ca^{2+} and the stimulation of a phospholipase that releases arachidonic acid from phospholipids. However, the ionophore stimulates synthesis of the hydroxy acids and leukotrienes not only from endogenous but also from exogenous arachidonic acid; thus it might also activate the synthetic enzymes directly (or through Ca^{2+}).

A 23187, isolated from cultures of *Streptomyces chartreusensis*, is an antibiotic and a divalent cation ionophore. Its structure, deduced by Chaney *et al.* (1974), is shown in Fig. 11.7. It simultaneously uncouples oxidative phosphoryl-

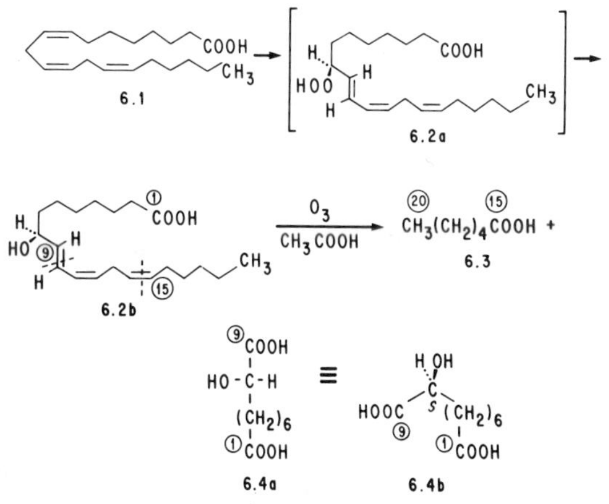

FIGURE 11.6. Conversion of dihomo-γ-linolenic acid (8,11,14-*cis*-20:3) by leukocytes into 8*S*-hydroxy-9-*trans*-11,14-*cis*-20:3 and method of deducing the absolute configuration at C-8. The hydroperoxy acid shown in brackets was not isolated. The dotted lines across the two double bonds in **6.2b** indicate the cleavage by ozonolysis. (After Borgeat *et al.*, (1976).)

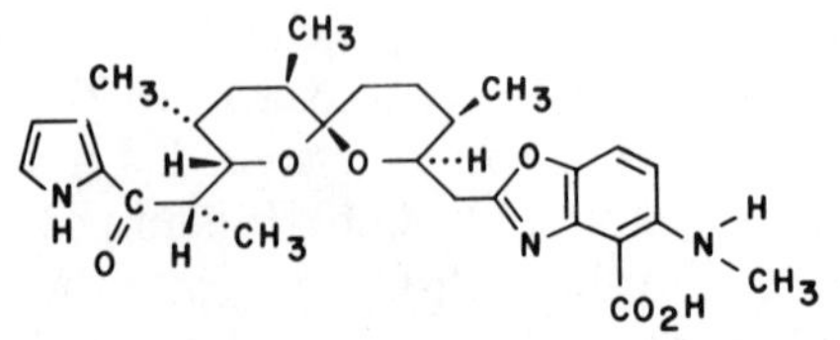

FIGURE 11.7. Structure of Ca^{2+} ionophore A 23187 (Chaney et al., 1974).

ation and inhibits hydrolysis of ATP by rat liver mitochondria. It acts as a carrier transporting Ca^{2+} and Mg^{2+}, but not K^+, from buffered aqueous medium at pH 7.4 into a bulk organic phase (Reed and Lardy, 1972).

The first derivative of arachidonic acid identified in incubations of polymorphonuclear leukocytes was 5S-hydroxy-6-*trans*-8,11,14-*cis*-eicosatetraenoic acid (**8.3** in Fig. 11.8; 5-HETE), and it was assumed that this resulted from the reduction of 5-HPETE (**8.2**). 5-HPETE is easily reduced with $NaBH_4$ to 5-HETE (Corey, Albright et al., 1980). The second derivative of arachidonic acid found in incubations of leukocytes was the 5S,12R-dihydroxyeicosatetraenoic acid containing three conjugated double bonds (Fig. 11.9). This substance became known as leukotriene B_4 (LTB_4; the subscript denotes the number of double bonds in the molecule). The absolute configuration of these acids, but not the geometry of their

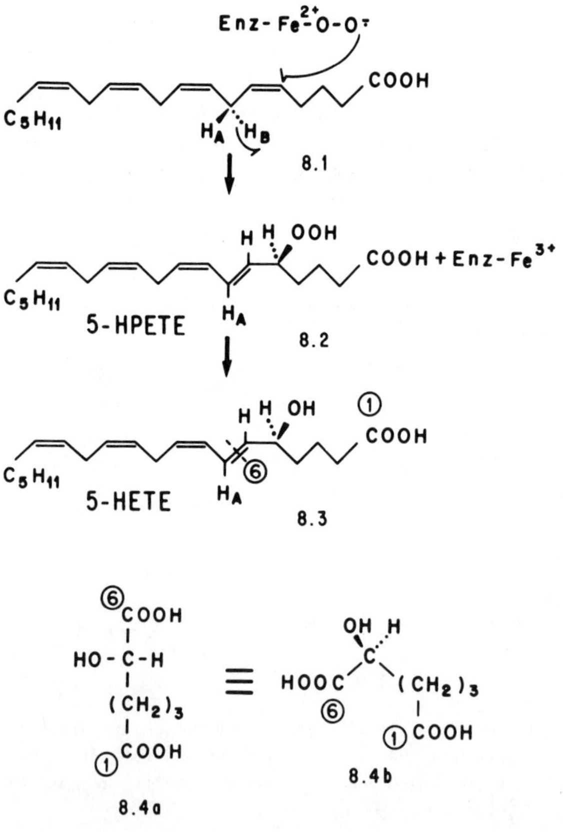

FIGURE 11.8. Suggested mechanism of formation of 5-HETE (5S-hydroxy-6-*trans*-8,11,14-*cis*-20:4) from arachidonic acid and determination of its absolute configuration. Dotted line in 5-HETE indicates cleavage by ozonolysis. The stereospecific elimination of H_B from **8.1** is indicated on the assumption that the hydrogen atom is abstracted from the face of the molecule opposite to that onto which the O_2 is added. (After Borgeat et al. (1976).)

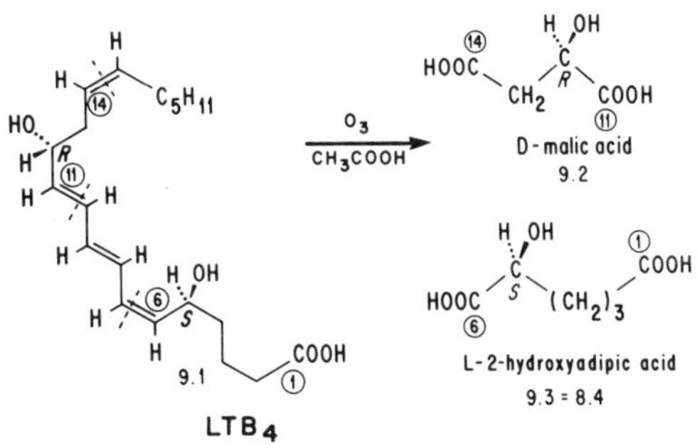

FIGURE 11.9. Structure of leukotriene B$_4$ (LTB$_4$) and its ozonolysis products. The structure shown for LTB$_4$ is that established by Corey et al. (1980b).

double bonds, was established from the analysis of their ozonolysis products: 5-HETE (cf. Figs. 11.8 and 11.9) gave L-2-hydroxyadipic acid (**8.4**), and LTB$_4$ (**9.1**) gave D-malic acid (**9.2**) in addition to L-2-hydroxyadipic acid.

Although the absolute configurations of the asymmetric centers in 5-HETE and in LTB$_4$ were correctly deduced from their ozonolysis products, the geometry of the double bonds in LTB$_4$ remained uncertain for some time. The structure shown in Fig. 11.9 is that established by Corey et al. (1980b) by chemical synthesis and comparison with the natural compound. LTB$_4$ has a UV absorption spectrum characteristic of a substituted conjugated triene with absorption maxima at 260, 270.5, and 281 nm (ϵ = 38,000, 50,000, and 39,000). LTB$_4$ was the first leukotriene isolated, whose structure was first tentatively formulated by Borgeat and Samuelsson (1979a).

That the precursor of 5-HETE (**8.3**) and of LTB$_4$ (**9.1**) was the 5-hydroperoxy acid (**8.2**), or some other unstable oxygenated substance derived from the latter, became highly probable from the experiments of Borgeat and Samuelsson (1979c), who carried out very short (30–45 sec) incubations of polymorphonuclear leukocytes with arachidonic acid in an atmosphere of $^{18}O_2$. They found that the oxygen at C-5 of both 5-HETE and LTB$_4$ was derived from the atmospheric $^{18}O_2$, but that the oxygen at C-12 of LTB$_4$ was not. When such short-term incubations were terminated by acid, in addition to the previously identified products two further pairs of hydroxy acids were formed, 5,6-dihydroxy- and 5,12-dihydroxyeicosatetraenoic acids, each existing as pairs of diastereomers, containing three conjugated double bonds and ^{18}O at C-5 (Fig. 11.10). When methanol or ethanol was used to terminate the reaction, instead of the dihydroxy acids, the 5-hydroxy-6-O-alkyl and the 5-hydroxy-12-O-alkyl (methoxy or ethoxy) derivatives were formed. The four dihydroxy acids were clearly nonenzymatic stabilization products of a carbonium ion derived from an unstable intermediate, which was postulated, and subsequently proved, to be the 5,6-epoxide formed enzymatically, probably from the 5-hydroperoxyeicosatetraenoic acid (cf. Figs. 11.10 and 11.11). The correct structure of this, at first hypothetical, intermediate was established by ingenious guess and chemical synthesis to be that of 5S,6S-5(6)-oxido-7,9-*trans*-11,14-*cis*-eicosatetraenoic acid (Corey et al., 1979; 1980a), which displayed the same types of instability as the natural compound ultimately isolated from very large incubations of [^{14}C]arachidonic acid with mixtures of human neutrophil and eosinophil granulocytes (Rådmark et al., 1980). This substance was

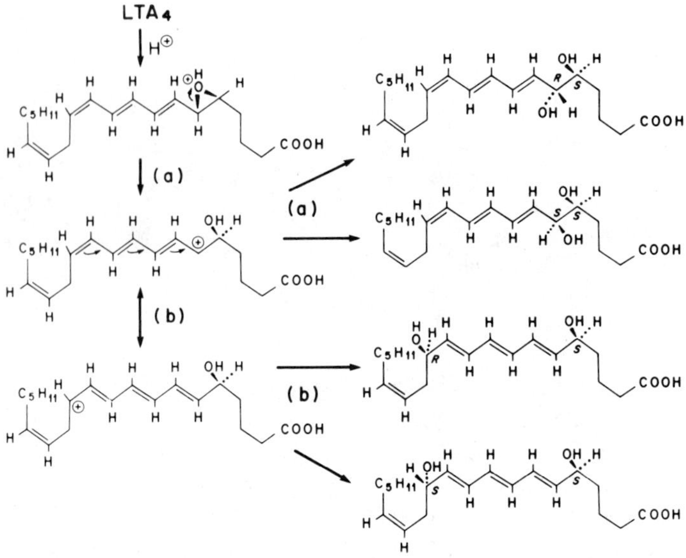

FIGURE 11.10. Acid-catalyzed solvolysis of leukotriene A$_4$ (LTA$_4$) to the diastereomeric 5,6-dihydroxy- and 5,12-dihydroxyeicosatetraenoic acids.

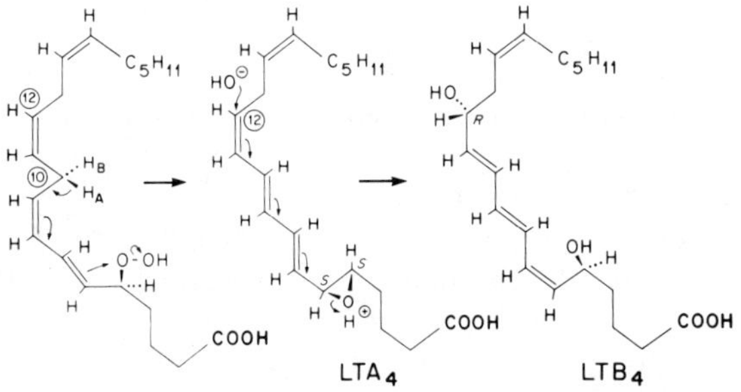

FIGURE 11.11. Possible reaction mechanisms of the enzymatic formation of LTA$_4$ and LTB$_4$. Note that in LTA$_4$ the two hydrogen atoms on the oxide ring are anti to one another. The stereospecific elimination of the pro-R H$_A$ from C-10 of 5-HPETE would have been hypothetically predicted; it has been tested experimentally also.

named leukotriene A$_4$ (LTA$_4$); its structure is shown in Fig. 11.11, together with the derivation of LTB$_4$ from it. The methyl ester of LTA$_4$ has a UV spectrum similar to that of LTB$_4$ but with a bathochromic shift of λ_{max} to 279 nm and shoulders at about 270 and 292 nm.

The hypothetical mechanisms shown in Fig. 11.11 for the formation of LTA$_4$ from 5-HPETE and of LTB$_4$ from LTA$_4$ are our interpretations of views expressed by Samuelsson, and by Corey and their colleagues. Corey *et al.* (1980b) considered all three possible geometric isomers for the conjugated triene system in LTB$_4$: (i) 6-*cis*-8,10-*trans;* (ii) 6-*trans*-8-*cis*-10-*trans;* and (iii) 6,8-*trans*-10-*cis*. On the assumption that LTB_4 was derived enzymatically from LTA_4, they concluded from the analysis "of the transition states for cation formation" that the energetically most favored conformation of LTA_4 for the formation of LTB_4 should be "that affording the 6-*cis*-8-*trans*-10-*trans* triene," as shown in Fig. 11.11. As shown in that figure, the transformation of LTA_4 to LTB_4 could be initiated either by a nucleophilic attack at C-12 of LTA_4, or by protonation of the 5,6-oxide ring. In contrast, the acid-catalyzed solvolysis of LTA_4 to the diastereomeric 5,12-dihydroxy-20:4 acids would kinetically favor the formation of the 6,8,10-*trans*-triene system (Corey *et al.*, 1980b), as shown in Fig. 11.10.

In the mechanism proposed for the formation of LTA_4 from 5-hydroperoxyeicosatetraenoic acid (5-HPETE) at the left of Fig. 11.11, elimination of H_A at C-10 is shown. By the RS-convention H_A at C-10 of 5-HPETE is the pro-R hydrogen atom. The stereospecific elimination of that hydrogen atom would have been predictable on the assumption that all addition and elimination reactions in this instance proceeded in a *trans* manner. Maas *et al.* (1982) and Hammarström (1983) examined this problem experimentally. Maas *et al.* used the 20:4ω6 fatty

acid labeled stereospecifically with ^3H at C-10 and concluded that the pro-R hydrogen atom at C-10 was lost and the pro-S hydrogen atom was retained during formation of LTA_4 from (5S)-5-HPETE. Hammarström (1983), on the other hand, used the 20:5ω3 acid labeled stereospecifically with ^3H at C-10 and concluded that the pro-S hydrogen atom was lost from C-10 and the pro-R hydrogen atom was retained during the synthesis of LTA_5 and LTC_5 from the precursor fatty acid (LTA_5 is an analogue of LTA_4; for LTA_5 and LTC_5 see Section 11.1.2.2. and Fig. 11.20). These seemingly contradictory results by Maas *et al.* and by Hammarström may be puzzling to those not fully conversant with the rules of the RS-convention and its extension to notations at prochiral centers. Inspection of the formulas of 20:4ω6 and 20:5ω3 shows that of the two nonidentical ligands attached to C-10 in the 20:4ω6 acid the carboxyl-bearing group (C-1 to C-9) has priority over the terminal alkenyl group (C-11 to C-20). However, in the 20:5ω3 acid, on account of the 17(18)-double bond, the ligand comprising C-11 to C-18 has priority over the ligand consisting of the segment C-9 to C-2. Thus, although H_A in both acids is similarly oriented spatially, in the 20:4ω6 acid it must be designated as a pro-R hydrogen atom and in the 20:5ω3 acid it must be designated as a pro-S hydrogen atom. It follows that the observations by Maas *et al.* (1982) and by Hammarström (1983) are in perfect harmony.

Although it seemed at first that leukocytes contained only a potato-type lipoxygenase, more recent evidence indicates that at least some elements of the white-cell population can also attack the ω6-position of arachidonic acid. Lundberg *et al.* (1981) examined the metabolism of 15-HPETE (generated by soybean

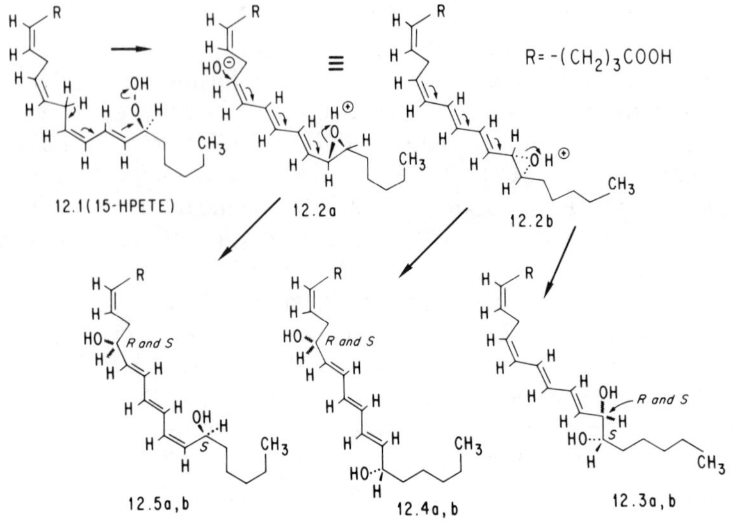

FIGURE 11.12. Hypothetical mechanisms for the conversion of 15-HPETE into 8,15-LTB$_4$ and 14,15-LTB$_4$ through 14,15-LTA$_4$.

lipoxygenase) by human leukocytes (cf. **12.1** in Fig. 11.12). The exact composition of the leukocyte preparations used was not reported, but it is probable that all elements of the white-cell population were present. From the incubation of 100 ml of cells (30 × 10^6 cells/ml) with 80 μM 15-HPETE for 30 min (without added ionophore A 23187) two pairs of diastereomeric dihydroxy-20:4 acids were obtained. Both pairs had UV spectra characteristic of conjugated trienes. The first pair (compounds I and II) had a UV spectrum with λ_{max} at 270 nm and shoulders at 281 and 261 nm (and another just perceptible shoulder absorption at about 252 nm). These two compounds were identified by their chromatographic behavior and by their mass spectra to be the diastereomers of 14,15-dihydroxy-5,8,10,12-20:4 acid **(12.3a,b)**. The UV spectrum of the second pair (compounds III and IV) had λ_{max} at 269 nm and shoulder absorptions at 280, 259, and about 249 nm. The two were distinct chromatographically but the mass spectra of the trimethylsilyl derivatives of their methyl esters were almost identical and were compatible with the derivatives of 8,15-dihydroxy-20:4 acids **(12.4 and 12.5)**. The isolation of the same two pairs of dihydroxy acids from incubations of human leukocytes with arachidonic acid (without added ionophore A 23187) was also reported by Jubiz *et al.* (1981).

Although neither the stereochemistry nor the geometry of the double bonds in these new substances has been determined, these authors assumed that they probably arose through the intermediacy of the 14(15)-oxide (14,15-LTA$_4$; **12.2**) in analogy with the formation of LTB$_4$ from LTA$_4$. For this reason they recommended the use of the leukotriene nomenclature for these dihydroxy acids also, labeling them 8,15-LTB$_4$ and 14,15-LTB$_4$.

These new leukotrienes are reminiscent of the diastereomeric 5,6- and 5,12-

dihydroxy-20:4 acids formed by acid-catalyzed solvolysis of LTA$_4$. Neither Lundberg et al. (1981) nor Jubiz et al. (1981) reported the relative amounts of the four dihydroxy acids formed in their experiments and therefore it is not possible to decide whether any one was an enzymatic product; they may have been only solvolytic products of a 14,15-oxide of 20:4. However, the absolute configuration and the geometry of the double bonds in 15-HPETE are known, i.e., 15S-hydroperoxy-5,8,11-cis-13-trans-20:4 (12.1). Hence it is safe to assume that the absolute configuration of the 14,15-oxide formed from 15-HPETE (if it is formed) should be 14S,15S (12.2a,b) and the position and the geometry of the double bonds in it should be 5,11-cis-10,12-trans. By whatever mechanism the 14,15- and 8,15-dihydroxy compounds are formed from 14,15-LTA$_4$, their absolute configuration at C-15 should be S. The diastereomers of the two sets of dihydroxy acids, on the other hand, are presumably 14R and 14S, and 8R and 8S, respectively. The double bonds in the 14,15-dihydroxy compounds should remain 5,11-cis-10,12-trans, but the geometry of the double bonds in the 8,15-dihydroxy compounds is unpredictable as this would depend on whether they are formed enzymatically or by solvolysis. In Fig. 11.12 two conformations are shown for 14,15-LTA$_4$. One of these (12.2a) would favor the formation of 8,15-LTB$_4$s with the 5-cis-9,11-trans-13-cis geometry in a reaction analogous to that observed in the enzymatic transformation of 5,6-LTA$_4$ to 5,12-LTB$_4$. The other conformation (12.2b) leads to the diastereomeric 8,15-LTB$_4$s with the 5-cis-9,11,13-trans-geometry.

Shortly after the publications by Lundberg et al. (1981) and Jubiz et al. (1981) on the synthesis of the diastereomeric 8,15-LTB$_4$s and 14,15-LTB$_4$s, Maas et al. (1981) reported the synthesis of two pairs of diastereomeric 8,15-LTB$_4$s (compounds I–IV) and two 14,15-LTB$_4$s (compounds V and VI) from arachidonic acid by porcine polymorphonuclear leukocytes and eosinophils. The absolute configurations of the four diastereomeric 8,15-dihydroxy-20:4 acids (8,15-LTB$_4$s) were shown to be 8R,15S (compounds I and III) and 8S,15S (compounds II and IV). Compounds I, IV and II, III are probably also geometric isomers as they differ in their chromatographic behavior and UV spectra: II and III have their λ_{max} at 268.5 nm, and I and IV have their λ_{max} at 269 nm. These two pairs differ from one another in other respects also. Experiments with $^{18}O_2$ and $H_2^{18}O$ showed that in I and IV the oxygen atom at C-15 was derived from molecular oxygen, and the one at C-8 from water, whereas in II and III the oxygen atoms at both of those positions came from molecular oxygen. The geometry of the double bonds in I and IV was probably 5-cis-9,11,13-trans as these compounds were identical with the solvolytic products of chemically synthesized 14S,15S-14,15-oxidoeicosa-5,8-cis-10,12-trans-20:4 (14,15-LTA$_4$; cf. Fig. 11.12). The geometry of the conjugated triene was not assigned either in II and III or in V and VI. The absolute configurations of the asymmetric centers in the two 14,15-dihydroxy compounds (V and VI) were not determined but the relative positions of the hydroxy groups were deduced to be *erythro*. These two compounds were assumed to be geometric isomers; V has a UV λ_{max} at 272.5 nm and VI at 270 nm. Added interest in these two *erythro*-14,15-LTB$_4$s (our terminology) is that the oxygen atoms at both C-14 and C-15 come from molecular oxygen. Maas et al. (1981) speculated that these compounds might be derived from 14,15-LTA$_4$ by the action of reactive oxygen metabolites (superoxide anion, singlet oxygen) which are known to be generated in leukocytes after various stimuli as part of the bacteriocidal properties of these cells. These authors suggested that the *erythro-* configuration about the 14,15-diol unit in compounds V and VI "would be compatible with a reactive oxygen-mediated opening of the *trans* epoxide ring, with inversion of configuration at C-14." In such an event the absolute configuration in these substances might be 14R,15S, but the reason for their geometric isomerism, and its nature, are uncertain. It is probable that at least some of the compounds found by Maas *et al.* (1981) are identical with those described by Lundberg *et al.* (1981) and Jubiz *et al.* (1981), although a direct comparison with the products noted by the latter authors is not possible as they have not determined the absolute configurations of the pairs of 8,15-LTB$_4$s or 14,15-LTB$_4$s, nor have they examined the sources of the oxygen atoms at C-5 and C-14. The hypothetical scheme shown in Fig. 11.12 predicts the possible formation of two pairs of diastereomeric 8,15-LTB$_4$s and two diastereomers of 14,15-LTB$_4$s, the number found by Maas *et al.* (1981).

The hydroperoxidation of arachidonic acid at C-15 (ω6) by leukocytes is well supported by the experiments of Narumiya *et al.* (1981) who reported the formation of 15-HPETE from arachidonic acid in the 100,000 $\times$ g supernatant of sonically disrupted rabbit peritoneal polymorphonuclear leukocytes. They also purified the "arachidonic acid 15-lipoxygenase" from such supernatants. The enzyme has an M_r of about 61,000 and a pH optimum of 6.5, quite different from that of the L-1 lipoxygenase of soybean. *o*-Phenanthroline had no effect on the enzyme before or after treatment with cysteine. It is not known whether this lipoxygenase contains iron or not. *p*-Chloromercuribenzoate and iodoacetamide at 1 mM concentration caused 70% and 30% inhibition, respectively. Although the enzyme was found in the soluble supernatant fraction of the sonically disrupted cells, its exact subcellular location is not known. When the cells were disrupted by freezing and thawing, or by brief homogenization with a Polytron, the

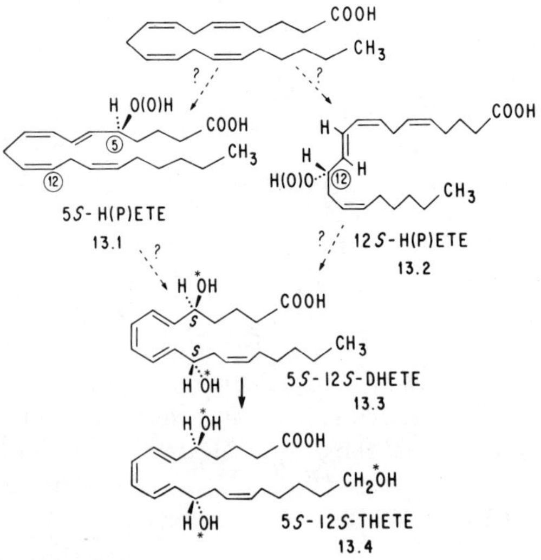

FIGURE 11.13. Possible routes of the origin of 5S,12S-dihydroxy- and 5S,12S,20-trihydroxyeicosatetraenoic acids (Lindgren *et al.*, 1981). The oxygen atoms marked with an asterisk are derived from atmospheric O_2.

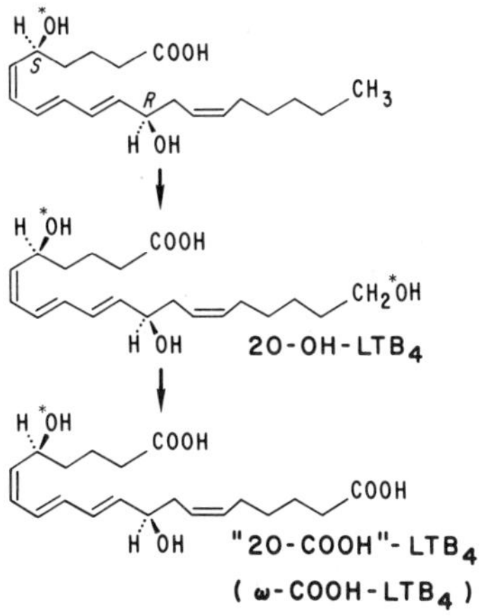

FIGURE 11.14. Products of ω-oxidation of LTB$_4$ (Hansson et al., 1981). The oxygen atoms marked with an asterisk are derived from atmospheric O$_2$.

enzyme was associated with a pellet sedimentable at low speed. The synthesis of 15-HPETE by leukocytes also gives added credence to the postulated synthesis of 14,15-LTA$_4$ by these cells.

In addition to the oxygenated derivatives of arachidonic acid so far described, Lindgren *et al.* (1981) and Hansson *et al.* (1981) identified four more oxygenated derivatives of arachidonic acid, two being derived directly from arachidonic acid and the other two from the metabolism of LTB$_4$. Lindgren *et al.* (1981) carried out large-scale and relatively long (10-min) incubations of polymorphonuclear leukocytes (mainly neutrophils contaminated with platelets) suspended in phosphate-buffered saline, pH 7.4, containing 0.87 mM CaCl$_2$, 5 μM of the ionophore A 23187, and 150 μM arachidonic acid either in air or in an atmosphere of ^{18}O$_2$ + argon. From the extracts of such incubations they isolated two novel hydroxy-20:4 acids. One was identified as 5S,12S-dihydroxyeicosatetraenoic acid (5S,12S-DHETE) and the other as 5S,12S-5,12,20-trihydroxyeicosatetraenoic acid (5S,12S,20-THETE). The probable structures of these are shown in Fig. 11.13. The remarkable feature of these two acids is that the new oxygen atoms in both are derived entirely from O$_2$, as shown by the incorporation of ^{18}O$_2$. One possible origin of **13.3** could be the 12S-hydroperoxy-5,8-*cis*-9-*trans*-14-*cis*-20:4 acid (12S-HPETE; **13.2**), known to be formed by platelets (cf. Section 11.1.1.1.), followed by the action of the potato-type lipoxygenase of the leukocytes and reduction to the hydroxy acids. The other possible origin of this acid could be the 5S-hydroperoxy-6-*trans*-8,11,14-*cis*-20:4 acid (5S-HPETE; **13.1**), which is then further oxygenated at C-12 by the lipoxygenase of platelets. The 5,12,20-trihydroxy-20:4 compound **(13.4)** is obviously derived by ω-oxidation of the 5S,12S-DHETE.

Hansson *et al.* (1981) found, in incubations of human leukocytes, similar to those of Lindgren *et al.* (1981), that arachidonic acid gave rise also to 20-hydroxy-LTB$_4$ and 20-carboxy-LTB$_4$ (Fig. 11.14).* When the incubations were carried out

---

*This designation by the authors is incorrect, strictly speaking, as the carboxyl group is not a substituent on C-20; "ω-carboxy-LTB$_4$" would be correct.

in an atmosphere of $^{18}O_2$ + argon, the 20-hydroxy-LTB_4 contained ^{18}O at C-5 and C-20 only. These two compounds are clearly formed by the ω-oxidation of LTB_4 and may represent initial steps in its degradation in analogy with the metabolism of prostaglandins (see Chapter 10, p. 180).

The importance of the studies on the hydroxy acids derived from eicosapolyenoic fatty acids is not only that they shed light on their metabolism, but also that they led to the discovery of LTA_4, which, as will be described, turned out to be the immediate precursor of the slow-reacting substances of anaphylaxis.

11.2.3. Discovery and Identification of the Slow-Reacting Substance(s) of Anaphylaxis as Product(s) of Arachidonic Acid

11.2.3.1. Historical Background

As mentioned briefly in the introduction to this chapter, the original observations leading to the recognition of the "slow-reacting substance of anaphylaxis" (SRS-A, or SRS) date back to 1938 and 1940. The active principle obtained by Feldberg and Kellaway (1938) from the perfusates of the lungs and liver of dogs and monkeys treated with cobra venom, and the one obtained by Kellaway and Trethewie (1940) by perfusing sensitized guinea-pig lungs with antigen had similar pharmacological properties, distinct from those of histamine, and caused, with a delayed onset, prolonged contraction of strips of isolated guinea-pig ileum, followed by slow relaxation. Concrete proof that SRS-A is not identical with acetylcholine, bradykinin, or histamine was provided by Brocklehurst (1960), who carried out the assays for SRS-A on strips of guinea-pig ileum in Tyrode solution containing 10^{-6} M atropine and 10^{-6} M mepyramine (= pyrilamine, an antihistaminic drug). This assay system remains up to the present one of the standard methods of assaying for the SRS. Brocklehurst demonstrated that the SRS-A was soluble in 75% ethanol, and that it was released not only from the lungs of several species in response to challenge by antigen after sensitization, but also from blood vessels, although its major source was the sensitized lung. His was probably also the first demonstration of release of SRS-A from the lung tissue of two human asthmatics sensitive to birch pollen. Brocklehurst showed further that, in the anaphylactic reaction, the release of histamine was rapid, "explosive," whereas the formation of SRS-A was prolonged, its maximum being reached about 1.5 min after that of histamine.

A little before Brocklehurst's clear differentiation of SRS-A from other smooth-muscle stimulating agents, Chakravarty *et al.* found in 1959 that histamine and an agent similar in pharmacological action to that of SRS-A were also released from the isolated cat's paw perfused with a compound coded 48/80 (48/80 is a condensation product of *p*-methoxyphenylethylmethylamine with formaldehyde). This was the first indication that a substance, or substances, with a pharmacological action similar to that of SRS-A may be released in reactions evoked by agents other than antigens and antibodies, and from tissues other than the lungs. The SRS produced in response to compound 48/80 was soluble in ethanol, diethyl ether, and methyl ethyl ketone (2-butanone): hence probably a lipid. Further corroboration of the evidence that SRS might be related to lipids was provided several years later by Orange et al. (1973), who found, in analogy with the observations of Brocklehurst, that lung tissue, obtained from patients sensitive to ragweed pollen, produced SRS-A on addition of a ragweed extract to the perfusion fluid of the lungs. They also provided evidence that the SRS-A had lipid characteristics, as it was adsorbed by Amberlite XAD-2 resin (a nonionic lipophilic resin) from which it could be eluted with a mixture of ethanol:NH_4^+:H_2O (6:3:1; v/v), and that it was an acidic lipid of low molecular weight; in its acidic form it could be quantitatively transferred from an ethereal solution into aqueous sodium hydroxide.

The last scenes, before the events of the identification of the chemical structures of SRS-As, were set between 1974 and 1977. Bach and Brashler (1974) found that intraperitoneal injection of the ionophore A 23187 to rats, or its addition to incubations of peritoneal macrophages, caused the release of SRS in addition to histamine. This ionophore was to play an important role subsequently in the production of SRS-A on a scale large enough for chemical analysis. Lewis et al. (1974) and Wasserman et al. (1975) showed that SRS-A could also be generated from macrophages of rat peritoneum, in animals sensitized actively or passively, by challenge with the antigen, and also from fragments of human lung sensitized with IgE in response to the appropriate antigen. Wasserman et al. (1975) also reported that human basophilic leukemia cells release SRS-A on treatment with ionophore A 23187, and that aryl sulfatase B, purified from human eosinophilic leukocytes, inactivated the SRS-A. A similar claim with respect to the effect of aryl sulfatases on SRS-A was also made by Orange et al. (1974). It was inferred that SRS-A had not only the characteristics of a lipid, but that it also contained a sulfur atom.

This was a serendipitously correct conclusion, but for the wrong reason. As will be described in the next section, leukotrienes C, D, and E, identified as the SRS-As, contain a sulfur atom, but in a thioether linkage. There is no evidence that pure aryl sulfatases could cleave a thioether bond. It is probable that the preparations of Orange et al. (1974) and Wasserman et al. (1975) were contaminated by peptidases which can inactivate SRS-A. Such peptidases were resolved by Sok et al. (1981) from aryl sulfatases which do not inactivate SRS-A.

The epilogue to a monograph, containing the accounts of a symposium on SRS-A and leukotrienes, gives in chronological order references to the development of the subject between 1974 and 1980 (Piper, 1981). The successes achieved in latter years can be attributed in no small measure to the introduction of high-pressure (or high-performance) liquid chromatography (HPLC), which has generally become the last step in the purification not only of the various hydroxylated derivatives of polyenoic fatty acids, but also of the SRS-As.

11.2.3.2. The Chemical Identification of the Slow-Reacting Substances of Anaphylaxis as Derivatives of Arachidonic Acid

The first experiments suggesting that SRS-A might be a derivative of arachidonic acid were presented in 1977. Bach et al. (1977) noted that eicosa-5,8,11,14-

tetraynoic acid, the acetylenic analogue of arachidonic acid, and also indomethacin severely inhibited the formation of SRS by rat peritoneal macrophages in response to incubation with cysteine and the ionophore A 23187. They postulated that SRS-A might be synthesized from a prostaglandin endoperoxide derived from arachidonic acid. Their surmise was correct only to the extent that arachidonic acid is indeed a precursor of SRS-A. More convincing evidence as to the precursor role of arachidonic acid in the synthesis of SRS was presented by Jakschik *et al.* (1977) who found that arachidonic acid (0.04–0.2 mg/ml) added to incubations of rat basophilic leukemia cells, stimulated by ionophore A 23187, caused about a three- to fourfold increase in the yield of SRS. Furthermore, they demonstrated that [^{14}C]arachidonic acid was incorporated into two radiochromatographically pure substances, each of which had SRS activity. Theirs was probably the first demonstration that more than one substance may have SRS activity. Jakschik *et al.* (1977) concluded that SRS is a previously undescribed metabolite of arachidonic acid "formed through the lipoxygenase pathway."

The first indication of the chemical nature of a substance with the pharmacological properties of SRS-A was reported by Murphy *et al.* in 1979. They found that mouse mastocytoma cells, grown as an ascites tumor and harvested from the peritoneal cavity, made an SRS when incubated with L-cysteine (0.01 M) and the ionophore A 23187 (10 µg/ml) in much larger amounts than any other cell-type tested before. The pure SRS made by the mastocytoma cells is said to be about 250 times more potent than histamine in causing contractions of strips of guinea-pig ileum; its action is rapidly reversed by an SRS-antagonist, FPL 55712 (Augstein *et al.*, 1973; Sheard, 1981; see Fig. 11.15).

This SRS has a UV absorption spectrum with λ_{max} at 280 nm ($\epsilon = 40,000$) and shoulders at 260, 270, and 292 nm (Murphy *et al.*, 1979), similar to that of LTC$_4$ (Fig. 11.16), characteristic for this metabolite of arachidonic acid with the conjugated triene chromophore.

When [^{3}H$_8$]arachidonic acid and [^{35}S]cysteine were included in the incubations of the mastocytoma cells, the purified SRS contained both isotopes. Experiments in which [3-^{3}H]Cys + [^{35}S]Cys or [3-^{3}H]Cys + [U-^{14}C]Cys were used gave SRS preparations with about equal incorporation of the differently labeled cysteines. Amino-group determination with fluorescamine indicated one free NH$_2$-group per molecule.

The structure provisionally assigned to this SRS, based on the following additional information, was that of 5-hydroxy-6-(*S*-cysteinyl)-7-*trans*-9,11,14-*cis*-eicosatetraenoic acid. "Oxidative" ozonolysis (i.e., cleavage of the ozonide with performic or peracetic acid) of the SRS gave pentane-1,5-dioic acid, probably representing C-1 to C-5 of the SRS. "Reductive" ozonolysis (i.e., treatment of the

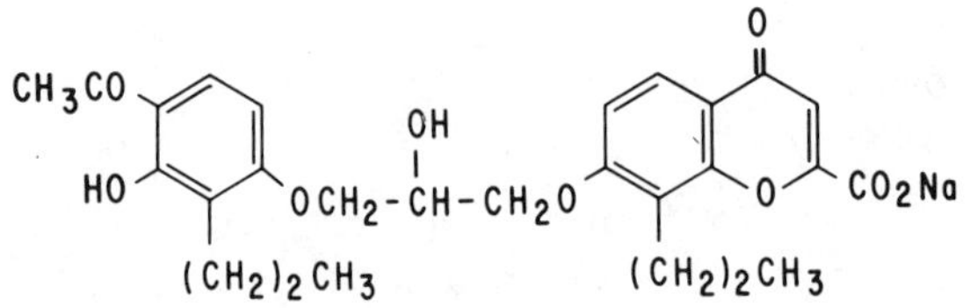

FIGURE 11.15. Structure of SRS-A antagonist FPL 55712 (Augstein *et al.*, 1973).

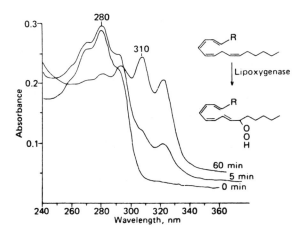

FIGURE 11.16. Ultraviolet spectrum of LTC-1 (LTC$_4$) before and after treatment with soybean lipoxygenase (Murphy et al., 1979). With the permission of the authors.

ozonide with LiAlH$_4$ at low temperature) yielded 1-hexanol, derived most probably from the terminal six carbon atoms of the compound, and suggesting the presence of a double bond in the ω6 position. Desulfurization and reduction of the SRS with Raney-nickel in refluxing ethanol, followed by methylation of the products, yielded methyl arachidate (20:0) as a minor product, and methyl 5-hydroxyarachidate as the major product; the observation also indicated that the sulfur atom in the SRS was present in a thioether linkage.

If the 11,14-*cis* double bonds of arachidonic acid were still contained in the SRS, soybean lipoxygenase L-1 was expected to introduce oxygen at the C-15 (ω6) position (cf. Section 11.1.1.1 on lipoxygenases) with the isomerization of the 14-*cis* double bond to 13-*trans*, creating a conjugated tetraene system in the SRS, and resulting in a bathochromic shift in its UV spectrum. Indeed, treatment of 2 μg of the SRS at 20°C in Tyrode buffer with 10 μg of soybean lipoxygenase for 60 min caused a shift of the λ_{max} of the SRS from 280 to 310 nm, with similar shifts of the shoulder absorptions (cf. Fig. 11.16). This SRS was named leukotriene C-1 (LTC-1), and it was suggested that its immediate precursor was leukotriene A$_4$ (cf. Fig. 11.11), although at that time neither the existence nor the correct structure of LTA$_4$ was known.

We have already described the finally determined structure of LTA$_4$ and indicated that the deduction of its stereochemistry was not the result of the usual course of events, i.e., the identification of the structure by the analysis of the natural product, followed by proof by chemical synthesis, but by the synthesis of a structure deduced by logical application of chemical theory to the hypothesis that the—at first elusive—LTA existed and that its most probable precursor was the 5S-hydroperoxy-6-*trans*-8,11,14-*cis*-20:4 acid (cf. Corey et al., 1979; 1980a,b). This same course of events was followed in the definition of the structure of the first SRS-A identified.

Within a few months of the provisional ascription of the structure of SRS (LTC-1, now usually referred to as LTC$_4$) produced by the mouse mastocytoma cells, Hammarström *et al.* (1979) reexamined its structure on a specimen larger than that at first available. From an incubation of 10^{10} mastocytoma cells they obtained about 0.2 µmol of LTC_4 (estimate based on UV absorption at 280 nm with $\epsilon = 40,000$). This time the amino acid analysis revealed the presence of γ-glutamylcysteinylglycine, i.e., glutathione, in the molecule. The report by Hammarström *et al.* (1979) also contained a preliminary account of the stereospecific synthesis of LTA_4 with the structure 5*S*,6*S*-5(6)-oxido-7,9-*trans*-11,14-*cis*-20:4 acid (cf. Fig. 11.11). When the methyl ester of this optically active LTA_4 ($[\alpha]_D^{25}$-21.9° in cyclohexane; Lewis *et al.*, 1980a) was treated with glutathione in the presence of triethylamine in methanol, it gave the methyl ester of LTC_4 from which the free acid, labeled LTC-1, was obtained by mild alkaline hydrolysis (for LTC-2 see below). Its absolute configuration was that of 5*S*-hydroxy-6*R*-(S-glutathionyl)-7,9-*trans*-11,14-*cis*-20:4 acid, clearly the product of an S_N2 type of reaction on LTA_4 (Corey *et al.*, 1980a). Reaction of the methyl ester of LTA_4 with cysteinylglycine under the same conditions, followed by hydrolysis of the ester, gave the 5*S*-hydroxy-6*R*-(S-cysteinylglycyl)-7,9-*trans*-11,14-*cis*-20:4, named leukotriene D, LTD_4 (Corey *et al.*, 1980a). In a second method, described briefly by Corey *et al.* (1980a), and a little later in greater detail by Lewis *et al.* (1980a), the methyl ester of LTA_4 was treated with *N*-trifluoroacetylglutathione methyl ester, or with *N*-trifluoroacetylcysteinylglycine methyl ester in a concentrated solution in methanol in the presence of triethylamine under argon at 23°C for 4 hr. From the resulting reactions the trimethyl ester of *N*-trifluoroacetyl-LTC_4 and the dimethyl ester of *N*-trifluoroacetyl-LTD_4, respectively, were obtained in 75–80% yield. The free acids were obtained from the *N*-trifluoroacetyl esters by mild alkaline hydrolysis (0.05 M K_2CO_3 in 1,2-dimethoxyethane/water). The dimethyl ester of the *N*-trifluoroacetyl-LTD_4 was incompletely hydrolyzed and gave, besides the free LTD_4, the *N*-trifluoroacetyl-LTD_4. From the chromatographic purification of the synthetic LTC_4 not only LTC-1 (corresponding to the natural product), but also the 11-*trans* isomer of LTC_4, were obtained as a small contaminant. The latter was labeled LTC-2. Lewis *et al.* (1980a) noted that while LTC_4 (LTC-1) was reasonably stable on storage and subsequent handling, LTD_4 was sensitive to exposure to air, thereby losing biological activity. The *N*-trifluoroacetyl esters of LTC_4, LTD_4 and LTE_4, and the free acids derived from them, are shown in Fig. 11.17 (for LTE see later). By the synthesis of LTD_4, the chemists stole a march on the biochemists, for LTD_4 was not yet known to be a natural product.

Very soon after the report by Corey *et al.* (1980a) of the first chemical syntheses of LTC_4 and LTD_4, LTD_4 was recognized as a natural product from the experiments of Örning *et al.* (1980). These workers incubated rat basophilic leukemia cells (RBL-1) with 10 mM cysteine and 10 µg/ml of ionophore A 23187 and isolated from the incubations an SRS that had chromatographic properties different from those of LTC_4. This new SRS had the same UV absorption spectrum as LTC_4 and was altered in the same way by treatment with soybean lipoxygenase as the spectrum of LTC_4 (cf. Fig. 11.16). Its reduction with Raney-nickel gave 5-hydroxyeicosanoic acid, and amino acid analysis showed that it contained only cysteine and glycine, but no glutamic acid. One-step Edman degradation gave glycine. Örning *et al.* (1980) also found that γ-glutamyl transpeptidase (EC

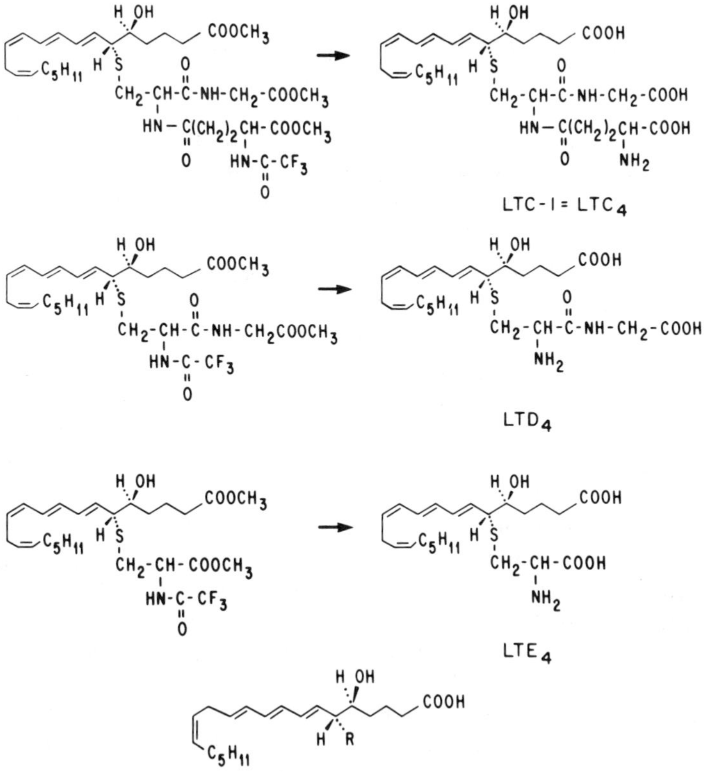

(a) R = S-glutathionyl = LTC-2 (11-*trans*-LTC_4)

(b) R = S-cysteinylglycyl = 11-*trans*-LTD_4

(c) R = S-cysteinyl = 11-*trans*-LTE_4

FIGURE 11.17. *N*-Trifluoroacetyl intermediates in the synthesis of LTC_4, LTD_4, and LTE_4 (Corey *et al.*, 1980a; Lewis *et al.*, 1980a) and the free acids (SRS-As) derived from them.

2.3.2.2) converted LTC_4 into a substance identical with that isolated from the leukemic cells. Hence they concluded that this new SRS was in all probability identical with LTD_4 synthesized by Corey *et al.* (1980a) and Lewis *et al.* (1980a), i.e., 5*S*-hydroxy-6*R*-(S-cysteinylglycyl)-7,9-*trans*-11,14-*cis*-eicosatetraenoic acid. In biological tests on guinea-pig ileum LTD_4 appeared to be about seven times more potent than LTC_4; its action, like that of LTC_4, was rapidly abolished by the anti-SRS compound FPL 55712 referred to earlier. Morris *et al.* (1980a,b) have also independently identified LTD_4 as the 5-hydroxy-6-(cysteinylglycyl)-7,9,11,14-20:4 acid as a product of RBL-1 cells stimulated by the ionophore A 23187 and released into the perfusion fluid of sensitized guinea-pig lungs in response to challenge with antigen. They recognized that the substance contained a conjugated triene system, though they did not define its stereochemistry nor the geometry of its double bonds.

The proposal that LTD_4 arose from LTC_4 by the action of γ-glutamyl transpeptidase (Örning et al., 1980) was further supported by the observations of Örning and Hammarström (1980) that rat basophilic leukemia cells, which produce LTD_4 exclusively when stimulated with the ionophore A 23187, accumulated LTC_4 in the presence of 20 mM serine–borate complex, a potent inhibitor of γ-glutamyl transpeptidase. 5,8,11-Eicosatriynoic acid, a lipoxygenase inhibitor, prevented the synthesis of LTC_4 in mouse mastocytoma cells (Örning and Hammarström, 1980).*

Lewis et al. (1980a) examined the nature of the SRS-As induced immunologically in rat peritoneal cavity and in human lung fragments. In the rat peritoneal cavity both LTC_4 and LTD_4 were produced in nearly equal amounts in response to the anaphylactic reaction. In human lung, however, 87–97% of the SRS-A activity is attributable to LTD_4.

The production of SRS-A in mouse peritoneal cavity (probably by macrophages), in response to an antigen challenge after passive or active immunization of the animals, has been referred to earlier in the section on the historical background to the discovery of SRS-A. Rouzer et al. (1980) demonstrated that cultured mouse peritoneal macrophages that had ingested unopsonized zymosan particles† also released an SRS. In contrast to the findings of Lewis et al. (1980a) with the SRS-As produced *in vivo* in rat peritoneal cavity in response to an immune reaction, Rouzer et al. (1980) identified the SRS released by the cultured macrophages in response to phagocytosis as LTC. They preincubated the macrophages overnight with [^{3}H]arachidonic acid, [^{35}S]cysteine, and [^{14}C]glutamic acid and found that the purified SRS contained all three isotopes and had an SRS activity, measured by the guinea-pig ileum assay, 500–1,000 times more potent than that of histamine.‡ It is noteworthy that the phospholipids of peritoneal macrophages are rich in arachidonic acid (Scott et al., 1980), up to 50% of which may be released in the form of oxygenated metabolites (mostly prostaglandins) in response to a phagocytic stimulus (Bonney et al., 1978).

The experiments of Örning and Hammarström (1980) on rat basophilic leukemia cells and with the serine–borate complex strongly suggested that the "proto-SRS" is LTC_4 formed by the enzymatic condensation of LTA_4 with glutathione (GSH). This interpretation was well supported by the work of Rouzer et al. (1981), who exposed primary cultures of mouse peritoneal macrophages for several hours to 200 μM buthionine sulfoximine (BSO; Fig. 11.18), a specific inhibitor of γ-glutamylcysteine synthetase that catalyzes the first step of GSH synthesis (Griffith and Meister, 1979). This agent caused an exponential drop in the GSH content of the cells, with a half-life of 2.8 hr, leading to complete depletion

---

*5,8,11-Eicosatriynoic acid is also an inhibitor of platelet lipoxygenase, but its ID_{50} for inhibition of LTC_4 synthesis is only 5.2 μM as compared to the ID_{50} of 24 μM needed for the inhibition of the synthesis of 12-hydroxy-5,8-*cis*-10-*trans*-14-*cis*-20:4 from arachidonic acid in platelets.

†Zymosan, a fine gray powder, is a crude yeast-cell-wall preparation almost insoluble in water; it is an anticomplementary agent. Unopsonized means not treated with serum or complement.

‡This is a far higher potency than reported by anyone else for LTC. It is possible that the preparations of Rouzer et al. (1980) also contained LTD, which is more potent than LTC.

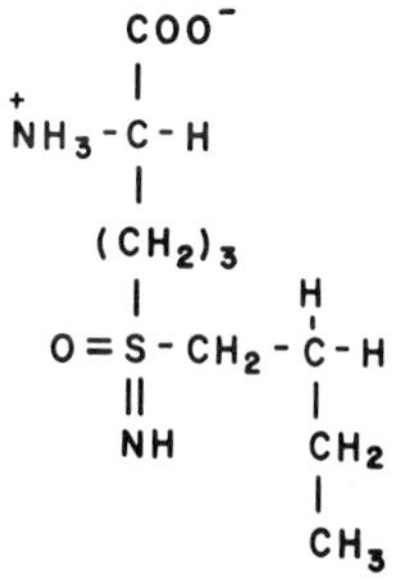

FIGURE 11.18. Buthionine sulfoximine = S-n-butylhomocysteine sulfoximine (Griffith and Meister, 1979).

$$\begin{array}{c} COO^- \\ | \\ \overset{+}{N}H_3-C-H \\ | \\ (CH_2)_3 \\ | \\ O=S-CH_2-\overset{H}{\underset{|}{C}}-H \\ \| \quad\quad | \\ NH \quad\quad CH_2 \\ \quad\quad\quad | \\ \quad\quad\quad CH_3 \end{array}$$

in 16 hr. This depletion in GSH content was completely mirrored by decreased formation of LTC_4 on phagocytic stimulus by unopsonized zymosan. BSO *per se* did not inhibit LTC_4 synthesis and had no effect on the release of 20:4 acid from phospholipids after zymosan challenge. In fact, the production of hydroxyeicosatetraenoic acids (HETEs) increased, so that the sum of LTC_4 and HETEs at any time after exposure to BSO was the same as observed with control cells. Also BSO had no effect on either PGE_2 or 6-keto- $PGF_{1\alpha}$ synthesis by the macrophages.

In addition to LTC_4 and LTD_4, a third substance, LTE_4, has been identified as a naturally occurring SRS-A. Lewis *et al.* (1980a,b) found that in two of five experiments, in which SRS-A was produced immunologically in rat peritoneal cavity, in addition to LTC_4 and LTD_4 a third major fraction with SRS-A activity could be separated by high-pressure liquid chromatography. This substance was identified as 5S-hydroxy-6R-(S-cysteinyl)-7,9-*trans*-11,14-*cis*-20:4 acid, and was labeled LTE_4. It was also synthesized chemically by reactions analogous to those described earlier for the synthesis of LTC_4 and LTD_4: the methyl ester of LTA_4 was reacted with the methyl ester of N-trifluoroacetyl cysteine in a minimum amount of methanol in the presence of triethylamine and hydroquinone. The resulting N-trifluoroacetyl-LTE_4 dimethyl ester was hydrolyzed quantitatively to LTE_4 with 0.13 M potassium carbonate in methanol/water containing hydroquinone (cf. Fig. 11.17). (When the deprotection is carried out in the absence of hydroquinone, LTE_4 is accompanied by some of the 11-*trans* isomer, paralleling observations made in the course of the syntheses of LTC_4 and LTD_4.) LTE_4 has a UV spectrum with absorption maxima in methanol at 280 nm (ϵ = 40,000), 270 nm (ϵ = 31,000), and 290 nm (ϵ = 31,000). The absorption maxima of 11-*trans*-LTE_4 are at 268 (ϵ = 31,000), 278 (ϵ = 40,000), and 288 nm (ϵ = 31,000). The biosynthetic relationship of the leukotrienes is shown schematically in Fig. 11.19. The reaction catalyzed by γ-glutamyl transpeptidase is reversible; at high concentrations of GSH (>0.5–1.0 mM) the LTC → LTD transformation is inhibited and the LTD → LTC reaction is promoted (Hammarström, 1981c).

Arachidonic acid is not the only eicosapolyenoic fatty acid from which a slow-reacting substance may be synthesized. Thus Hammarström (1980, 1981a) found that, in incubation of mouse mastocytoma cells with the ionophore A 23187, addition of 5,8,11,14,17-*cis*-20:5 or of 5,8,11-*cis*-20:3 acids resulted in formation of analogues of LTC_4, in addition to LTC_4 itself. The substance derived from the 20:5 acid was identified as the 5S-hydroxy-6R-(S-glutathionyl)-7,9-

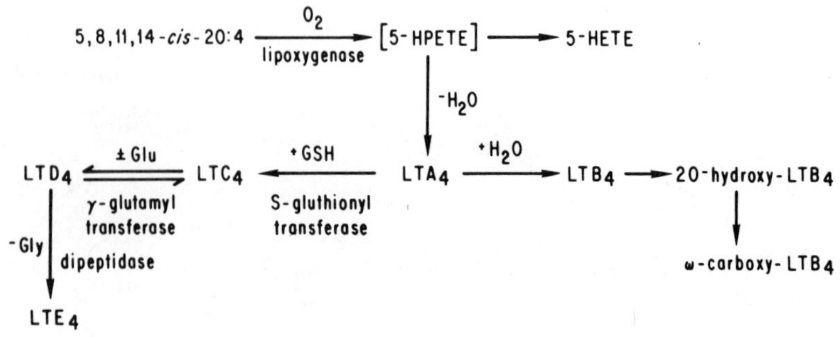

FIGURE 11.19. Relations of leukotrienes derived from arachidonic acid.

trans-11,14,17-cis-20:5 compound and was named leukotriene C_5 (LTC$_5$). In assays on strips of guinea-pig ileum its potency was about one-fourth that of LTC$_4$. The 5,8,11-cis-20:3 acid gave rise to two substances: the 5S-hydroxy-6R-(S-glutathionyl)-7,9-trans-11-cis-20:3 (LTC$_3$) and also a lesser amount of the 11-trans isomer of LTC$_3$ (about one-fifth of LTC$_3$) (Fig. 11.20). The UV spectrum of

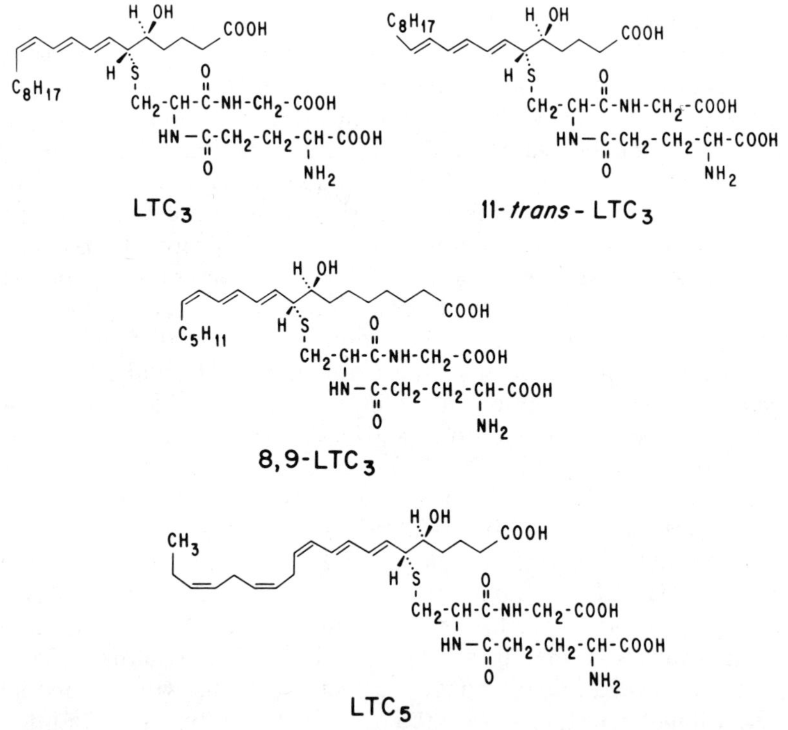

FIGURE 11.20. Structures of leukotrienes C_3 and C_5. γ-Glutamyl transpeptidase and a renal dipeptidase convert these substances into the corresponding leukotrienes D and E by removal of glutamic acid and glycine, respectively (Hammarström, 1980; 1981a,b; Bernström and Hammarström, 1981).

LTC$_3$ is very similar to that of LTC$_4$ except that its absorption maxima are shifted by 1nm to shorter wavelengths (λ_{max} 279 nm, shoulders at 269 and 291 nm). The UV spectrum of the 11-*trans*-LTC$_3$ showed a further hypsochromic shift to λ_{max} 277 nm. Soybean lipoxygenase (L-1) had no effect whatever on the UV spectrum of LTC$_3$, in conformity with its complete topical specificity at pH 9 for a double bond at the ω6-position in C$_{20}$ polyenoic fatty acids containing the 11,14-*cis*-pentadiene structure (cf. Section 11.1.1.1 on lipoxygenases). LTC$_3$ and 11-*trans*-LTC$_3$, when treated with γ-glutamyl transpeptidase, yielded LTD$_3$ and 11-*trans*-LTD$_3$, i.e., the 6-cysteinylglycyl derivatives of the 5-hydroxyeicosatrienoic acid. The contractile effects of LTC$_3$ and LTD$_3$ in the standard guinea-pig-ileum assays were identical with those of LTC$_4$ and LTD$_4$, but 11-*trans*-LTC$_3$ was about one-tenth as potent as LTC$_3$ or LTC$_4$. LTD$_3$ and LTD$_5$, like LTD$_4$, when treated with an enzyme (?dipeptidase) from porcine kidney, yield the corresponding LTEs (Bernström and Hammarström, 1981).

Hammarström (1981b) reported further that even dihomo-γ-linolenic acid can give rise to a leukotriene, isomeric with the LTC$_3$ formed from 5,8,11-*cis*-20:3. The addition of dihomo-γ-linolenic acid to mouse mastocytoma cells stimulated with the ionophore A 23187 caused the reduction of the yield of LTC$_4$ by about 80% and the appearance of a new leukotriene eluted from a reversed-phase HPLC column just before LTC$_4$. This new leukotriene was shown to have a hydroxyl group at C-8 and a glutathione substituent α to three conjugated double bonds. The UV absorption spectrum of this new leukotriene is almost identical with that of LTC$_4$ except for a 1-nm shift of its absorptions to shorter wavelengths, with λ_{max} at 279 nm. On the assumption that this isomeric LTC$_3$ was formed from dihomo-γ-linolenic acid through the 8-hydroperoxyeicosatrienoic acid and its transformation to the 8,9-epoxide, followed by the conjugation of the latter with glutathione, the 8-hydroxy-9-(S-glutathionyl)-10,12-*trans*-14-*cis*-eicosatrienoic structure was assigned to it, and it was designated as 8,9-LTC$_3$. Although the absolute configuration of 8,9-LTC$_3$ has not been established, it is presumed to be 8S,9R, in analogy with the known absolute configuration of LTC$_4$. Treatment of 8,9-LTC$_3$ with γ-glutamyl transpeptidase and a kidney dipeptidase produced two new substances which had the same chromatographic relations as the LTD$_4$ and LTE$_4$ produced by action of the same enzymes from LTC$_4$. Hence these two substances were assumed to be the 9-(S-cysteinylglycyl) and 9-(S-cysteinyl) derivatives of the 8-hydroxyeicosatrienoic acid. The biological activity of the 8,9-LTC$_3$ is not known. The structures of these new leukotrienes are shown in Fig. 11.20.

The synthesis of leukotrienes C$_3$ and C$_5$ under artificial conditions, when 5,8,11-*cis*-20:3 (ω9) or 5,8,11,14,17-*cis*-20:5 (ω3) acids are offered as substrates to either murine mastocytoma or to rat basophilic leukemia cells (RBL-1), raises the question of why the synthesis of LTC$_3$ or LTC$_5$ had not been observed either in response to stimulation of cells by the ionophore A 23187 or in an anaphylactic reaction on lung tissues.

This question is the more intriguing as ω9 and ω3 eicosapolyenoic fatty acids are present, albeit in small amounts, as constituents of phospholipids in normal animal cells. There are two main possibilities to explain this puzzle: (i) the phospholipase (as yet not identified) liberating the polyunsaturated fatty acid(s) from phospholipids has low specificity for the $\omega 9$ and $\omega 3$ polyunsaturated fatty acids, and/or (ii) the K_m values of the $\omega 9$ and $\omega 3$ fatty acids for the C-5 lipoxygenase of animal cells which generates the 5-hydroperoxy acids and for the enzyme converting the hydroperoxy acids to the leukotriene As are much higher than the K_m value of the $\omega 6$ acid. The problem is related also to the relative paucity of E_1, $F_{1\alpha}$ and E_3, $F_{3\alpha}$ prostaglandins.

The experiments of Murphy et al. (1981) shed some light on some of these questions. These authors grew murine mastocytoma cells as an ascites tumor in mice fed a normal chow diet and also in mice brought up on a diet in which the sole source of fat (5%, w/w) was menhaden fish oil rich in $\omega 3$ fatty acids. The lipids of the tumor cells grown in mice kept on the standard diet contained 9.2 mol % of the 20:4 $\omega 6$ and only 0.5 mol % of the 20:5 $\omega 3$ fatty acids, whereas the lipids of the tumor cells harvested from hosts brought up on the fish-oil diet contained only 3.9 mol % of the $\omega 6$ fatty acid and 4.5 mol % of the $\omega 3$ fatty acid. The response of the tumor cells from the two sources to stimulation by the ionophore A 23187 (20 μM) was dramatically different. The cells harvested from animals fed the standard diet produced in 20 min 200 $\pm$ 40 pmol of LTC_4 and 270 $\pm$ 30 pmol of LTB_4 per 10^7 cells. The tumor cells from the mice fed with fish oil on the other hand produced only 15 pmol of LTC_4 and 6.3 pmol of LTB_4 per 10^7 cells, but they also synthesized 1.5 pmol of LTC_5 and 9.6 pmol of LTB_5 per 10^7 cells. These values are remarkable in that the change in fatty acid composition of the tumor lipids produced a disproportionate decrease in the synthesis of LTC_4 and LTB_4 and that, while the ratio of LTC_4/LTB_4 from the cells rich in the $\omega 3$ fatty acid was about 2.4, the ratio of LTC_5/LTB_5 was only about 0.16. The data do suggest that the $\omega 3$ fatty acid has a lower affinity for the C-5 lipoxygenase than the $\omega 6$ fatty acid and also that the LTA_5 (the presumed precursor of LTC_5 and LTB_5) has a particularly low affinity for the glutathione transferase. It would be most interesting to know what might be the composition of SRS-As released by tissues or cells of Eskimos who consume a high-fish diet.

11.2.3.3. Biochemical Relations of Slow-Reacting Substances of Anaphylaxis and Leukotrienes C, D and E

Thoughts were entertained, after the elucidation of the structure of LTC_4 synthesized by the murine mastocytoma cells in response to the ionophore A 23187, that this particular slow-reacting substance may not be identical with the slow-reacting substance of anaphylaxis SRS-A (cf. Morris et al., 1981, p. 23). The evidence is now very clear that LTC_4 is indeed the primary SRS-A or "proto-SRS-A," as we have dubbed it earlier. Whether the SRS-A secreted will be LTC, LTD, LTE, or their mixture seems to depend on the enzymatic complement and the glutathione and cysteine contents of particular tissues or cells. Tissues rich in GSH and in γ-glutamyl transpeptidase, such as the liver, will degrade LTC_4 into LTD_4 only to a limited extent, as in the presence of high concentrations of GSH the conversion of LTD_4 into LTC_4 is favored (cf. Fig. 11.19). In the lung, relatively poor in GSH, but rich in γ-glutamyl transpeptidase and containing also dipeptidase(s), mainly LTD_4 and some LTE_4 may appear in response to an anaphylactic reaction. The correctness of the biochemical relations among the substances with SRS-A activity depicted in Fig. 11.19, LTC → LTD → LTE, has recently been reemphasized by the experiments of Örning et al. (1981) who found that not only murine mastocytoma cells, as recorded earlier (p. 236), but also rat basophilic leukemia cells (RBL-1) stimulated by the ionophore A 23187 will synthesize from added 5,8,11-cis-20:3 and 5,8,11,14,17-cis-20:5 acids LTC_3 and LTC_5, respectively, in addition to LTC_4 from endogenous arachidonic acid. These LTCs are degraded stepwise by the RBL-1 cells to the corresponding leukotrienes D and E. The sequential degradation of an LTC is well illustrated by the reactions of the RBL-1 cells on [^{3}H] LTC_3: after a 10-min incubation only bare traces of the labeled substrate remain, 60% of it being converted to LTD_3 and 40% into LTE_3, but after 33 min 30% of it is found in LTD_3 and 70% in LTE_3. "Acivicin" [L-(αS),(5S)-α-amino-3-chloro-4,5-dihydro-5-isoxazoleacetic acid], a potent irreversible inhibitor of γ-glutamyl transpeptidase, prevents the degradation of leukotrienes C to D. High concentrations of cysteine (10 mM), on the other hand, stimulate the leukotriene C → D conversion, possibly by acting as acceptor of glutamine, but inhibit the D → E transformation. Brief reviews on the biochemistry of the leukotrienes and SRS-As and their biological effects are available (Samuelsson, 1981; Piper et al., 1981). Samuelsson's Nobel Lecture of 1982 on leukotrienes was published in 1983.

11.2.3.4. Metabolism of Leukotrienes and Slow-Reacting Substances of Anaphylaxis

Apart from the ω-oxidation of LTB_4 and the sequential transformation of LTC to LTD and LTE in various isolated cell types (mastocytoma, rat basophilic leukemia cells, peritoneal macrophages) or in homogenates of organs under artificial conditions, information is available so far about the in vivo metabolism only of LTC_3.

[^{3}H] LTC_3 injected subcutaneously into guinea pigs is rapidly metabolized and is excreted quantitatively in 4–5 days in the form of highly polar unidentified metabolite(s) in the feces (60%) and urine (40%), 80% being excreted during the first 24 hr (Hammarström, 1981c). The large elimination in the feces suggests excretion with the bile. Indeed, examination by radioautography of whole-body sections and by chromatography of extracts of organs of mice injected intravenously or intramuscularly with [^{3}H] LTC_3 showed that within 5 min after the injection most of the ^{3}H was in the liver and bile, with appreciable amounts also in the kidney cortex, pleural fluid, salivary glands, and connective tissue fasciae around muscles. The ^{3}H in the lungs appeared in patches at first and accumulated there gradually, reaching a maximum at 60 min when most of the ^{3}H found there was in LTE_3 and only very little in LTC_3 or LTD_3. In the liver, bile, and intestine, LTC_3 and LTD_3 are the main leukotrienes retained. It is of interest that although intravenously injected [^{3}H] LTC_3 is taken up rapidly by the organs, nevertheless leukotrienes C_3, D_3, and E_3 are circulating in the blood even 4 hr after the injection and are acccompanied by large amounts of some unidentified highly polar ^{3}H-containing substance(s). The liver, lungs, kidneys, and intestine also convert the [^{3}H] LTC$_3$ into such polar substance(s). The only organ in which ^{3}H from LTC$_3$ was not seen was the brain, although the pituitary (lobe unspecified) contained some (Appelgren and Hammarström, 1982). No doubt, with the availability of leukotrienes C$_4$, D$_4$, and E$_4$ in large amounts from chemical syntheses, information on the metabolism of these substances will be forthcoming.

11.3. Biological Activity of Hydroxyeicosatetraenoic Acids and Leukotrienes

11.3.1. Leukotriene B and Monohydroxyeicosatetraenoic Acids

The biological activity of the HETEs, specifically that of LTB$_4$, is directed towards leukocytes: they are chemotactic and chemokinetic agents for these cells. Thus these substances stimulate both the migration of leukocytes along a concentration gradient of the attractants and also their motility in the absence of a chemotactic gradient, activities succinctly defined by Snyderman and Goetzl (1981) as "directed cellular migration" and "nondirectional cellular motility." Among the HETEs, LTB$_4$ is probably the only one of physiological importance as it is one of the most potent chemotactic and chemokinetic agents for polymorphonuclear leukocytes. Its activity is comparable to that of the C5a polypeptide (the fifth component of a cleavage product of serum complement; 74 amino acid residues) and of the synthetic tripeptide N-formyl-methionyl-leucyl-phenylalanine (f-Met-Leu-Phe) (Ford-Hutchinson et al., 1980). LTB$_4$ also stimulates in vitro the movements of other elements of white-cell population, such as eosinophils, monocytes, and macrophages, but does not aggregate platelets. LTB$_4$ also causes a degranulation and release of lysosomal enzymes, lysozyme, and β-glucuronidase from polymorphonuclear leukocytes treated with cytochalasin B,* but its effect in this respect is only about one-half of that elicited by f-Met-Leu-Phe (Hafstrom et al., 1981). This degranulation and release of lysosomal enzymes by LTB$_4$ or f-Met-Leu-Phe occurs only in the presence of cytochalasin B and is enhanced by extracellular Ca^{2+}. Preincubation of the cells with LTB$_4$ (10^{-7}M) rapidly ($t_{1/2}$ = 15 sec) inactivates the response to cytochalasin B (Showell et al., 1982).

Intravenous injection of 1 μg of LTB$_4$ into rabbits produces profound leukopenia lasting for only 3 min without affecting the monocyte or platelet counts. This leukopenia can be attributed to a transient adherence of polymorphonuclear leukocytes to the endothelium of postcapillary venules (Dahlén et al., 1981). Extravascular LTB$_4$ induces leukocyte accumulation and this effect is much enhanced at the site of injection by PGE$_2$ (Bray et al., 1981). LTB$_4$, in contrast to the actions of LTC$_4$, LTD$_4$, and LTE$_4$, has no effect on contractions of guinea-pig ileum, but it contracts guinea-pig lung parenchymal strips at 7.5×10^{-12}–7.5×10^{-11}M. However, pretreatment of the ileal strips with LTB$_4$ enhances the subse-

---

*Cytochalasin B, $C_{29}H_{37}NO_5$, is one of six similar fungal metabolites. Its most remarkable effect is the extrusion of nuclei from cells, e.g., from fibroblasts. For structure see Merck Index, 9th edition, 1976, entry number 2787, p. 365.

quent response of the preparations to other agonists, such as acetylcholine and histamine.

The 5S,12R-stereochemistry and also the 6-*cis*-8,10-*trans* geometry in the conjugated triene system of LTB$_4$ are vital for its activity; other isomers of 5,12-dihydroxy-20:4 acids that are formed by solvolysis of LTA$_4$ have but weak potencies, if any (cf. Showell *et al.,* 1982).

It is not known whether leukocytes have specific receptors for LTB$_4$. It is known that leukocytes possess specific receptors for N-formyl-methionyl peptides, but LTB$_4$ does not bind to these. Nevertheless, LTB$_4$ may exert its effects in leukocytes by mechanisms similar to those of f-Met-Leu-Phe. Both agents cause a rapid, apparently specific and saturable uptake of extracellular Ca^{2+} and Na$^+$ by rabbit polymorphonuclear leukocytes, although f-Met-Leu-Phe is more potent than LTB$_4$; the tripeptide at 10^{-9}M causes about twice as great influx of Ca^{2+} as 3.3×10^{-8}M LTB$_4$ or about 2.5 times as much uptake of Na$^+$ than 6.5×10^{-8}M LTB$_4$. When leukocytes are preloaded with ^{45}Ca^{2+} for a time sufficiently long to attain a steady-state level of the ions in the cells and are then exposed to LTB$_4$ in a medium still containing 0.5 mM Ca^{2+} + ^{45}Ca^{2+}, there is still a rapid but transient rise in cellular exchangeable Ca^{2+}. On the other hand, in the absence of extracellular Ca^{2+}, LTB$_4$ causes a rapid transient and dose-dependent efflux of Ca^{2+}, but in a few minutes a new steady-state level of intracellular ions is reestablished (Sha'afi *et al.,* 1981b; Molski *et al.,* 1981).

In spite of the seeming similarity between the actions of f-Met-Leu-Phe and LTB$_4$, there are some further differences between these agents apart from their not sharing common cell receptors. Preincubation of polymorphonuclear leukocytes for a few minutes with high concentrations of chemotactic agents diminishes their subsequent response to the same or other stimuli, a phenomenon known as deactivation. Thus preincubation of rabbit neutrophilic granulocytes for 5 min with 10^{-7}M f-Met-Leu-Phe, or with 2.6×10^{-7}M LTB$_4$, followed by removal of these agents, abolishes the specific saturable uptake of Ca^{2+} on subsequent exposure of the cells to these substances. The preincubation with the tripeptide abolishes the subsequent response to itself and to LTB$_4$, but LTB$_4$ produces only a homologous deactivation, the response to f-Met-Leu-Phe being totally unaffected by pretreatment with LTB$_4$. The deactivation phenomenon applies also to cells in steady-state conditions: when cells are preloaded with ^{45}Ca^{2+}, deactivated cells respond with smaller release of Ca^{2+} than do non-deactivated cells. LTB$_4$ has, again, a much smaller effect than the formyl-tripeptide (Sha'afi *et al.,* 1981a). Perhaps these differences may be accounted for by the fact that LTB$_4$ does not bind to the tripeptide-specific receptors. Notwithstanding these differences, Sha'afi *et al.* (1981a) conclude that a causal relationship exists between elevated intracellular Ca^{2+} levels, or increased levels of exchangeable Ca^{2+}, previously sequestered in cells, and initiation of neutrophil responsiveness to chemotactic and chemokinetic factors.

In contrast to LTB$_4$, the monohydroxy-20:4 acids (5-, 12-, or 15-HETE) have but weak biological activities and thus their role in cellular regulation is uncertain (for reviews cf. Snyderman and Goetzl, 1981; Ford-Hutchinson, 1981). Neither 15-HETE nor 15-HPETE have any chemotactic effect, nor do they have an effect on Ca^{2+} uptake by leukocytes. The strict relationship between LTB_4 structure and function is further emphasized by the ineffectiveness of 5S,12S-dihydroxy-[6,8,10-*trans*-14-*cis*?]-20:4 acid either as a chemotactic or Ca^{2+}-mobilizing agent (Sha'afi et al., 1981b).

11.3.2. Biological Activity of the Slow-Reacting Substances of Anaphylaxis: Leukotrienes C, D and E

Traditionally the SRS-As are tested *in vitro* mostly on strips of guinea-pig ileum suspended in oxygenated Tyrode's buffer containing 10^{-6}M atropine and 10^{-6}M mepyramine (= pyrilamine), a method originally devised by Brocklehurst (1960). In such assays the SRS-A activity is expressed in units, one unit being equivalent to the contraction elicited by a solution of histamine, 5 ng/ml. With the availability of methods for the purification of SRS-As from natural sources and of the pure synthetic compounds, it has been possible to define quantitatively the specific activities of the SRSs relative to the activity of histamine. Generally the potencies of the SRS-As are in the order LTD > LTC > LTE. Thus Lewis et al. (1980a) found that the specific activities of synthetic LTC_4 and LTD_4 were 1.93 ± 0.13 and 6.10 ± 1.15 units/pmol, respectively, values that correlated well with the specific activities of 1.69 ± 0.43 and 7.14 ± 0.51 units/pmol of LTC_4 and LTD_4, respectively, generated immunologically in rat peritoneal cavity. The synthetic LTC-2 (the 11-*trans* isomer of LTC_4) and the *N*-trifluoroacetyl-LTD_4 (cf. Fig. 11.17) are also active, but have specific activities of only 0.91 and 1.0 unit/pmol respectively.

The SRS-As are also assayed on parenchymal strips of the guinea-pig lung (strips cut from the subpleural portions of the organ containing the smooth muscle of peripheral airways) and on tracheal spirals. In such assays the tension developed in isometric contractions of the preparations is measured. The activity of the SRS-As in such assays is usually expressed as the concentration that produces one-half the maximum contraction elicited by 100 μM histamine; this is referred to as the EC_{50} value, the 50% effective concentration, of an SRS-A. In such assays the EC_{50} value of LTC_4 is 3 nM and that of LTD_4 is 20 fM (Lewis et al., 1980a). Thus these two leukotrienes are, respectively, about 200 and 20,000 times more active than histamine on the smooth muscle on the peripheral airways. LTC_4 and LTD_4 are equally active, on a molar basis, in inducing contractions of guinea-pig tracheal spirals and are 30 to 100 times more potent than histamine in this respect (Drazen et al., 1980).

When applied topically in a superfusion fluid to the everted cheek pouch of anesthetized golden hamsters, these two leukotrienes cause rapid and dramatic contractions of arterioles, similar to those evoked by angiotensin (Dahlén et al., 1981). The LTC_4- and LTD_4-induced arteriolar contractions last for a few minutes and are followed by leakage of plasma into the extravascular spaces. Angiotensin, in contrast, does not increase vascular permeability. LTC_4, which is not as potent as LTD_4 in causing bronchiolar or intestinal smooth muscle contractions, is about five times more potent than LTD_4 and about 5,000 times more potent than histamine in increasing vascular permeability.

These two leukotrienes injected subcutaneously into guinea pigs or rats also cause first arteriolar contractions followed by increased vascular permeability; however, in rabbits the increased vascular permeability could not be observed (Ueno et al., 1981). The existence of specific receptors for LTC_4 and LTD_4 on bronchiolar smooth-muscle cells has been implied (Drazen et al., 1980) although their existence has not been directly demonstrated.

The *in vitro* effects of the SRS-As and the *in vivo* effects after their topical application are reflected also in the *in vivo* effects after systemic administration or administration in aerosols. Smedegård et al. (1982) reported such experiments on anesthetized and artificially respired monkeys *(Macaca irus)*. Injection of LTC_4 (0.50–30 nmol) into the right atrium of the heart causes a prompt rise in peripheral arterial pressure, a rise in pressure in the pulmonary artery and in left and right atria lasting for a few seconds, followed by long hypotension. The initial rise in these pressures reflects increased resistance in the pulmonary and systemic vascular bed. This may be a corollary of the vasoconstrictions noted on topical application of LTC_4 and LTD_4. The hypotension, in spite of persistent increased systemic vascular resistance, is a consequence of greatly reduced cardiac output (from 733 ml/min to 449 ml/min in 4 min) without change in heart rate. Increased hematocrit values (from 42 to 46 in 15 min) can be attributed to plasma leakage, also noted after topical applications. The injection of 30 nmol of LTC_4 or of histamine into the right atrium causes also an increased trans-pulmonary pressure (= difference between tracheal and intrapleural pressure) of about equal magnitude, but the histamine-induced rise peaks within 10–15 sec and ends after 1 min, whereas the effect of LTC_4 reaches its maximum 2 min after the injection and is sustained for 10–15 min.

LTC_4 (20 nmol) administered in an aerosol causes a gradual and sustained rise in trans-pulmonary pressure (maximum at about 7 min; 163% above normal), and severe pulmonary hypertension accompanied by a 40% decrease in arterial P_{O_2}. These effects, attributable to bronchiolar contraction, may persist as long as 45 min, whereas the similar effects of histamine, produced by 50–300 times larger doses, vanish in 5–10 min. The differences between the potencies of LTC, LTD, and histamine may explain the relative ineffectiveness of antihistaminic agents in alleviating symptoms of asthma.

The biochemical mechanism of the action of the SRS-As (LTC_4, LTD_4, LTE_4) on the bronchiolar smooth muscles (or on intestinal smooth muscle) is as yet uncertain. It seems likely that their effects are not direct, but result from triggering the synthesis of prostaglandins. Mathé et al. (1977) injected into the pulmonary artery of perfused guinea-pig lungs partially purified SRS produced by the perfusion of cat's paw with the compound 48/80 (cf. p. 232). The chemical identity of the SRSs in the preparation of Mathé et al. (1977) is not known. However, these workers noted (i) that the perfused guinea-pig lungs (not sensitized or sensitized with ovalbumin) released prostaglandins E_2, $F_{2\alpha}$, and 15-keto-13,14-dihydroprostaglandin in response to the SRS; (ii) that indomethacin (a known inhibitor of prostaglandin cyclo-oxygenase, cf. Chapter 10) inhibited the action of the SRS; and (iii) that passively sensitized human lung released prostaglandins when incubated with the SRS or with antigen. These early observations, made before the structure of SRSs was known, were fully corroborated by Piper and Samhoun

(1981) who noted that the SRS-A released by antigen challenge from sensitized guinea-pig lungs, the SRS synthesized by rat basophilic leukemic cells (RBL-1) in response to stimulation by ionophore A23817, and synthetic LTC_4 and LTD_4, when injected into the pulmonary artery of perfused guinea-pig lungs, caused the release of thromboxane A_2 and prostaglandins. These effects, as well as the contractile effects of the SRS-As on parenchymal strips of the guinea-pig lung, were much, though not completely, inhibited by indomethacin. The residual activities were abolished by the anti-SRS compound FPL 55712 (cf. p. 234). The effects of low doses of LTC_4 ($1-2 \times 10^{-11}$ mol) on the guinea-pig lung parenchymal strips are completely abolished by indomethacin. The available evidence suggests that the activity of the SRSs on the smooth muscle of peripheral airways has two components: (i) direct contractile effect (the lesser one), and (ii) the stimulation of the synthesis of thromboxane A_2 and of prostaglandins which amplify their direct action.

The knowledge of the biochemistry of the SRS-As is far from complete, which is not surprising as their structural identity was revealed only during 1979-80. There is nothing known yet about the nature of the enzymes, or the cell compartments of these involved in their biosynthesis—only intelligent guesses. We have to invoke (i) an immunological receptor system (IgE–receptor complex?) on the cells, which, on binding an antigen, activates a phospholipase, or some other esterase, for the release of arachidonic acid from phospholipids or from other structures; (ii) the intervention of a lipoxygenase and of a dehydratase which converts the 5-hydroperoxy-20:4 acid to the 5,6-epoxide; and (iii) a glutathione transferase for the generation of the "proto-SRS," LTC_4.

The mechanism of action of the SRSs is equally intriguing, as the SRSs trigger the activation of another (?) phospholipase that releases further amounts of arachidonic acid and activates another oxygenase, the prostaglandin cyclo-oxygenase. All these phenomena may be intimately linked with Ca^{2+} metabolism, as is the biosynthesis of the leukotrienes.

A radioimmunoassay is also available for the determination of the SRS-As (Levine et al., 1981). The antigen, prepared by conjugating bovine serum albumin with LTD_4 through its C-1 carboxyl group, elicited antibodies in rabbits which have comparable affinities for leukotrienes C_4, D_4, and E_4 and their 11-trans-isomers. Thus the inhibition of the binding of 11-trans-[^{3}H]LTC_4 to the antiserum by other unlabeled leukotrienes can be used for the assays. The average association constant for the specific immunoglobulins is 2.8×10^9 M^{-1} at 37°C. The antibodies do not recognize minor changes in the sulfidopeptide region of the SRS-As; thus the homocysteinyl analogue of LTD_4 [5S-hydroxy-6R-(S-homocysteinylglycine) derivative] is as effective as LTD_4 itself, and deamino-LTD_4 is nearly three times as potent as LTD_4 in inhibiting binding of 11-trans-[^{3}H]LTC_4. On the other hand, the antibodies recognize the conjugated triene system in the lipid domain as well as the spatial orientation and the position of the glutathione or cysteinylglycine substituents. Binding of 11-trans-[^{3}H]LTD_4 to the antibodies is not inhibited by arachidonic acid, glutathione, cystinylbisglycine, or 5-HETE; LTB_4 is only about 1/1,000th as active as LTC_4, LTD_4 or LTE_4. The radioimmunoassay permits the approximate determination of total SRS-As, but individual members can be determined only after their separation by HPLC.

11.3.3. Structure–Function Relations among Leukotrienes

As is the case with LTB_4, the biological activity of the SRS-As (LTC, LTD, and LTE) critically depends on their structure. In assays on strips of guinea-pig ileum the activities of the SRS-As are in the order $LTD_4 = LTD_3 > LTC_4 = LTC_3 > LTE_4$. The equal activity of LTD_3 (cf. Fig. 11.20) to that of LTD_4, and of LTC_3 to that of LTC_4, clearly suggests that the 14(15)-*cis* ($\omega6$) double bond is not essential for high potency. However, the additional $\omega3$-double bond in LTC_5 is undesirable, as the potency of LTC_5 is only about one-fourth that of LTC_4. The geometry of the double bond at C-11 also appears to be critical, as the 11-*trans*-LTC_3 was reported to be only one-tenth as active as LTC_3 or LTC_4.

Far more critical for activity are the stereochemistry and the nature of the polar substituents at positions 5 and 6. Lewis *et al.* (1981) examined 23 structural analogues of LTC_4 and LTD_4 and compared their activities to that of LTD_4 both on parenchymal strips of guinea-pig lung and strips of guinea-pig ileum.

No improvements on the natural compounds could be found. On the contrary, every modification produced impairment, if not complete loss, of activity. Thus the 6-epimers (6S) of LTD_4 and LTC_4, and the 5-epimer (5R) of LTD had only 0.4–0.5% of the activity of the natural compound. The free carboxyl group of LTD at C-1 can be converted to the amide without impairing its activity when assayed on pulmonary parenchymal strips, although such modification reduces its activity on ileal strips by 75%. The free carboxyl group of glycine in LTD_4 is more critical for activity than a free carboxyl group at C-1, as its conversion to the amide reduces the activity to 10% of that of LTD_4. The bis-amide of LTD_4 (amide at C-1 and at the carboxyl group of glycine) is totally inactive. Analogues in which the glycyl residue is replaced by D- or L-alanine, proline, glutamic acid, or valine have only 0.8–3% of the activities of LTD_4. Even the conservative replacement of L-Cys by L-homocysteine gives an analogue with only 20–26% activity of the natural compound, and replacement of L-Cys with D-Cys reduces the activity of the compound by 90% or more.

The very strict relation of structure to function, revealed by the studies of Lewis *et al.* (1981), supports further the idea of the existence of specific cell receptors for the SRS-As.

REFERENCES

Appelgren, L.-E., and Hammarström, S., 1982, Distribution and metabolism of ^{3}H-labeled leukotriene C_3 in the mouse, *J. Biol. Chem.* **257**:531.

Augstein, J., Farmer, J. B., Lee, T. B., Sheard, P., and Tattersall, M. L., 1973, Selective inhibitor of slow reacting substance of anaphylaxis, *Nature New Biology* **245**:215.

Axelrod, B., Cheesbrough, T. M., and Laakso, S., 1981, Lipoxygenase from soybeans, in: *Methods in Enzymology*, Vol. 71 (J. M. Loewenstein, ed.), Academic Press, New York, pp. 441–451.

Bach, M. K., and Brashler, J. R., 1974, In vivo and in vitro production of slow reacting substance in the rat upon treatment with calcium ionophores, *J. Immunol.* **113**:2040.

Bach, M. K., Brashler, J. R., and Gorman, R., 1977, On the structure of slow reacting substance of anaphylaxis: evidence of biosynthesis from arachidonic acid, *Prostaglandins* **14**:21.

Bernström, K., and Hammarström, S., 1981, Metabolism of leukotriene D by porcine kidney, *J. Biol. Chem.* **256**:9579.

Bild, G. S., Ramadoss, C. S., and Axelrod, B., 1977, Multiple dioxygenation by lipoxygenase of lipids containing all-*cis*-1,4,7-octatriene moieties, *Arch. Biochem. Biophys.* **184**:36.

Bonney, R. J., Wightman, P. D., Davies, P., Sadowski, S. J., Kuehl, F. A., Jr., and Humes, J. L., 1978, Regulation of prostaglandin synthesis and selective release of lysosomal hydrolases by mouse peritoneal macrophages, *Biochem. J.* **176**:433.

Borgeat, P., and Samuelsson, B., 1979a, Transformation of arachidonic acid by rabbit polymorphonuclear leukocytes. Formation of a novel dihydroxyeicosatetraenoic acid, *J. Biol. Chem.* **254**:2643.

Borgeat, P., and Samuelsson, B., 1979b, Arachidonic acid metabolism in polymorphonuclear leukocytes: Effects of ionophore A 23187, *Proc. Nat. Acad. Sci. U.S.A.* **76**:2148.

Borgeat, P., and Samuelsson, B., 1979c, Arachidonic acid metabolism in polymorphonuclear leukocytes: Unstable intermediate in formation of dihydroxy acids, *Proc. Nat. Acad. Sci. U.S.A.* **76**:3213.

Borgeat, P., Hamberg, M., and Samuelsson, B., 1976, Transformation of arachidonic acid and homo-γ-linolenic acid by rabbit polymorphonuclear leukocytes, *J. Biol. Chem.* **251**:7816. (Correction: 1977, *J. Biol. Chem.* **252**:8772.)

Bray, M. A., Ford-Hutchinson, A. W., and Smith, M. J. H., 1981, In vivo properties of leukotriene B_4, *Brit. J. Pharmacol.* **74**:788P.

Brocklehurst, W. E., 1960, The release of histamine and formation of a slow-reacting substance (SRS-A) during anaphylactic shock, *J. Physiol.* **151**:416.

Chakravarty, N., Högberg, C., and Uvnäs, B., 1959, Mechanism of the release by compound 48/80 of histamine and of a lipid-soluble smooth muscle stimulating principle ("SRS"), *Acta Physiol. Scand.* **45**:255.

Chaney, M. O., Demarco, P. V., Jones, N. D., and Occolowitz, J. L., 1974, The structure of A 23187, a divalent cation ionophore, *J. Am. Chem. Soc.* **96**:1932.

Christopher, J. P., Pistorius, E. K., and Axelrod, B., 1970, Isolation of an isozyme of soybean lipoxygenase, *Biochim. Biophys. Acta* **198**:12.

Christopher, J. P., Pistorius, E. K., and Axelrod, B., 1972, Isolation of a third isozyme of soybean lipoxygenase, *Biochim. Biophys. Acta* **284**:54.

Corey, E. J., Albright, J. O., Barton, A. E., and Hashimoto, S-i., 1980, Chemical and enzymic synthesis of 5-HPETE, a key biological precursor of slow-reacting substance of anaphylaxis (SRS), and 5-HETE, *J. Am. Chem. Soc.* **102**:1435.

Corey, E. J., Arai, Y., and Mioskowski, C., 1979, Total synthesis of ($\pm$)-5,6-oxido-7,9-*trans*-11,14-*cis*-eicosapentaenoic acid, a possible precursor of SRS-A, *J. Am. Chem. Soc.* **101**:6748.

Corey, E. J., Clark, D. A., Goto, G., Marfat, A., Mioskowski, C., Samuelsson, B., and Hammarström, S., 1980a, Stereospecific total synthesis of a "slow reacting substance" of anaphylaxis leukotriene C-1, *J. Am. Chem. Soc.* **102**:1436. (Correction: 1980, *J. Am. Chem. Soc.* **102**:3663.)

Corey, E. J., Marfat, A., Goto, G., and Brion, F., 1980b, Leukotriene B. Total synthesis and assignment of stereochemistry, *J. Am. Chem. Soc.* **102**:7984.

Dahlén, S.-E., Björk, J., Hedqvist, P., Arfors, K.-E., Hammarström, S., Lindgren, J.-Å., and Samuelsson, B., 1981, Leukotrienes promote plasma leakage and leukocyte adhesion in postcapillary venules: In vivo effects with relevance to the acute inflammatory process, *Proc. Nat. Acad. Sci. U.S.A.* **78**:3887.

Drazen, J. M., Austen, K. F., Lewis, R. A., Clark, D. A., Goto, G., Marfat, A., and Corey, E. J., 1980, Comparative airway and vascular activities of leukotrienes C-1 [LTC_4] and D [LTD_4] in vivo and in vitro, *Proc. Nat. Acad. Sci. U.S.A.* **77**:4354.

Feldberg, W., and Kellaway, C. H., 1938, Liberation of histamine and formation of lyso[le]cithin-like substances by cobra venom, *J. Physiol.* **94**:187.

Ford-Hutchinson, A. W., 1981, Leukotriene B_4 and neutrophil function: a review, *J. Roy. Soc. Med.* **74**:831.

Ford-Hutchinson, A. W., Bray, M. A., Doig, M. V., Shipley, M. E., and Smith, M. J. H., 1980, Leukotriene B, a potent chemokinetic and aggregating substance released from polymorphonuclear leukocytes, *Nature* **286**:264.

Galliard, T., and Phillips, D. R., 1971, Lipoxygenase from potato tubers. Partial purification and properties of an enzyme that specifically oxygenates the 9-position of linoleic acid, *Biochem. J.* **124**:431.

Griffith, O. W., and Meister, A., 1979, Potent and specific inhibition of glutathione synthesis by buthionine sulfoximine (*S-n*-butylhomocysteine sulfoximine), *J. Biol. Chem.* **254**:7558.

Hafstrom, I., Palmblad, J., Malmsten, C. L., Rådmark, O., and Samuelsson, B., 1981, Leukotriene B_4—A stereospecific stimulator for release of lysosomal enzymes from neutrophils, *FEBS Lett.* **130**:146.

Hamberg, M., and Samuelsson, B., 1967, On the specificity of the oxygenation of unsaturated fatty acids catalyzed by soybean lipoxidase, *J. Biol. Chem.* **242**:5329.

Hamberg, M., Svensson, J., and Samuelsson, B., 1974, Prostaglandin endoperoxides. A new concept concerning the mode of action and release of prostaglandins, *Proc. Nat. Acad. Sci. U.S.A.* **71**:3824.

Hammarström, S., 1980, Leukotriene C_5: a slow reacting substance derived from eicosapentaenoic acid, *J. Biol. Chem.* **255**:7093.

Hammarström, S., 1981a, Conversion of 5,8,11-eicosatrienoic acid to leukotrienes C_3 and D_3, *J. Biol. Chem.* **256**:2275.

Hammarström, S., 1981b, Conversion of dihomo-γ-linolenic acid to an isomer of leukotriene C_3, oxygenated at C-8, *J. Biol. Chem.* **256**:7712.

Hammarström, S., 1981c, Metabolism of leukotriene C_3 in the guinea pig, *J. Biol. Chem.* **256**:9573.

Hammarström, S., 1983, Stereospecific elimination of hydrogen at C-10 in eicosapentaenoic acid during the conversion to leukotriene C_5, *J. Biol. Chem.* **258**:1427.

Hammarström, S., Murphy, R. C., Samuelsson, B., Clark, D. A., Mioskowski, C., and Corey, E. J., 1979, Structure of leukotriene C. Identification of the amino acid part, *Biochem. Biophys. Res. Commun.* **91**:1266.

Hansson, G., Lindgren, J. Å., Dahlén, S-E., Hedqvist, P., and Samuelsson, B., 1981, Identification and biological activity of novel ω-oxidized metabolites of leukotriene B_4 from human leukocytes, *FEBS Lett.* **130**:107.

Jakschik, B. A., Falkenhein, S., and Parker, C. W., 1977, Precursor role of arachidonic acid in release of slow reacting substance from rat basophilic leukemia cells, *Proc. Nat. Acad. Sci. U.S.A.* **74**:4577.

Jubiz, W., Rådmark, O., Lindgren, J. Å., Malmsten, C., and Samuelsson, B., 1981, Novel leukotrienes: Products formed by initial oxygenation of arachidonic acid at C-15, *Biochem. Biophys. Res. Commun.* **99**:976.

Kellaway, C. H., and Trethewie, E. R., 1940, The liberation of a slow-reacting smooth muscle stimulating substance in anaphylaxis, *Quart. J. Exp. Physiol.* **30**:121.

Levine, L., Morgan, R. A., Lewis, R. A., Austen, K. F., Clark, D. A., Marfat, A., and Corey, E. J., 1981, Radioimmunoassay of the leukotrienes of slow reacting substance of anaphylaxis, *Proc. Nat. Acad. Sci. U.S.A.* **78**:7692.

Lewis, R. A., Wasserman, S. I., Goetzl, E. J., and Austen, K. F., 1974, Formation of slow-reacting substance of anaphylaxis in human lung tissue and cells before release, *J. Exp. Med.* **140**:1133.

Lewis, R. A., Austen, K. F., Drazen, J. M., Clark, D. A., Marfat, A., and Corey, E. J., 1980a, Slow reacting substances of anaphylaxis: Identification of leukotrienes C-1 and D from human and rat sources, *Proc. Nat. Acad. Sci. U.S.A.* **77**:3710.

Lewis, [R.] A., Drazen, J. M., Austen, K. F., Clark, D. A., and Corey, E. J., 1980b, Identification of the C(6)-S-conjugate of leukotriene A with cysteine as a naturally occurring slow reacting substance of anaphylaxis (SRS-A), *Biochem. Biophys. Res. Commun.* **96**:271.

Lewis, R. A., Drazen, J. M., Austen, K. F., Toda, M., Brion, F., Marfat, A., and Corey, E. J., 1981, Contractile activities of structural analogs of leukotrienes C and D: Role of the polar substituents, *Proc. Nat. Acad. Sci. U.S.A.* **78**:4579.

Lindgren, J. Å., Hansson, G., and Samuelsson, B., 1981, Formation of novel hydroxylated eicosatetraenoic acids in preparations of human polymorphonuclear leukocytes, *FEBS Lett.* **128**:329.

Lundberg, U., Rådmark, O., Malmsten, C., and Samuelsson, B., 1981, Transformation of 15-hydroperoxy-5,9[sic],11,13-eicosatetraenoic acid into novel leukotriene, *FEBS Lett.* **126**:127.

Maas, R. L., Brash, A. R., and Oates, J. A., 1981, A second pathway of leukotriene biosynthesis in porcine leukocytes, *Proc. Nat. Acad. Sci. U.S.A.* **78**:5523.

Maas, R. L., Ingram, C. D., Taber, D. F., Oates, J. A., and Brash, A. R., 1982, Stereospecific removal of the D_R hydrogen atom at the 10-carbon of arachidonic acid in the biosynthesis of leukotriene A_4 by human leukocytes, *J. Biol. Chem.* **257**:13515.

Mathé, A. A., Strandberg, K., and Yen, S-S., 1977, Prostaglandin release by slow reacting substance from guinea-pig and human lung tissue, *Prostaglandins* **14**:1105.

Molski, L. F. P., Naccache, P. H., Borgeat, P., and Sha'afi, R. I., 1981, Similarities in the mechanisms by which formyl-methionyl-leucyl-phenylalanine, arachidonic acid and leukotriene B_4 increase calcium and sodium influxes in rabbit neutrophils, *Biochem. Biophys. Res. Commun.* **103**:227.

Morris, H. R., Taylor, G. W., Piper, P. J., Samhoun, M. N., and Tippins, J. R., 1980a, Slow-reacting substances (SRS's): The structure identification of SRS's from rat basophil leukaemia (RBL-1) cells, *Prostaglandins* **19**:185.

Morris, H. R., Taylor, G. W., Piper, P. J., and Tippins, J. R., 1980b, Structure of slow-reacting substance from guinea-pig lung, *Nature* **285**:104.

Morris, H. R., Taylor, G. W., Jones, C. M., Piper, P. J., Tippins, J. R., and Samhoun, M. N., 1981, Structure elucidation, biosynthesis and biodegradation of SRS-A from lung, in: *SRS-A and Leukotrienes* (P. J. Piper, ed.), Research Studies Press, John Wiley & Sons Ltd., Chichester, pp. 19–44.

Murphy, R. C., Hammarström, S., and Samuelsson, B., 1979, Leukotriene C: A slow-reacting substance from murine mastocytoma cells, *Proc. Nat. Acad. Sci. U.S.A.* **76**:4275.

Murphy, R. C., Pickett, W. C., Culp, B. R., and Lands, W. E. M., 1981, Tetraene and pentaene leukotrienes—Selective production from murine mastocytoma cells after dietary manipulation, *Prostaglandins* **22**:613.

Narumiya, S., Salmon, J. A., Cottee, F. H., Weatherby, B. C., and Flower, R. J., 1981, Arachidonic acid 15-lipoxygenase from peritoneal polymorphonuclear leukocytes, *J. Biol. Chem.* **256**:9583.

Orange, R. P., Murphy, R. C., Karnovsky, M. L., and Austen, K. F., 1973, The physicochemical characteristics and purification of slow-reacting substance of anaphylaxis, *J. Immunol.* **110**:760.

Orange, R. P., Murphy, R. C., and Austen, K. F., 1974, Inactivation of slow-reacting substance of anaphylaxis (SRS-A) by arylsulfatases, *J. Immunol.* **113**:316.

Örning, L., and Hammarström, S., 1980, Inhibition of leukotriene C and leukotriene D biosynthesis, *J. Biol. Chem.* **255**:8023.

Örning, L., Hammarström, S., and Samuelsson, B., 1980, Leukotriene D: A slow reacting substance from rat basophilic leukemia cells, *Proc. Nat. Acad. Sci. U.S.A.* **77**:2014.

Örning, L., Bernström, K., and Hammarström, S., 1981, Formation of leukotrienes E_3, E_4 and E_5 in rat basophilic leukemia cells, *Eur. J. Biochem.* **120**:41.

Piper, P. J. (ed.), 1981, *SRS-A and Leukotrienes,* Research Studies Press, John Wiley and Sons, Ltd., Chichester, pp. 271–277.

Piper, P. J., and Samhoun, M. N., 1981, The mechanism of action of leukotriene-C_4 and leukotriene-D_4 in guinea-pig isolated perfused lung and parenchymal strips [of the lung] of guinea-pig, rabbit and rat, *Prostaglandins* **21**:793.

Piper, P. J., Samhoun, M. N., Tippins, J. R., Morris, H. R., Jones, C. M., and Taylor, G. W., 1981, SRS-A and SRS: their structure, biosynthesis and actions, *Int. Arch. Allergy Appl. Immunol.* **66**(Suppl. 1):107.

Pistorius, E. K., and Axelrod, B., 1974, Iron, an essential component of lipoxygenase, *J. Biol. Chem.* **249**:3183.

Rådmark, O., Malmsten, C., Samuelsson, B., Goto, G., Marfat, A., and Corey, E. J., 1980, Leukotriene A. Isolation from human polymorphonuclear leukocytes, *J. Biol. Chem.* **255**:11828.

Reed, P. W., and Lardy, H. A., 1972, A 23187: A divalent cation ionophore, *J. Biol. Chem.* **247**:6970.

Rouzer, C. A., Scott, W. A., Cohn, Z. A., Blackburn, P., and Manning, J. M., 1980, Mouse peritoneal macrophages release leukotriene C in response to a phagocytic stimulus, *Proc. Nat. Acad. Sci. U.S.A.* **77**:4928.

Rouzer, C. A., Scott, W. A., Griffith, O. W., Hamill, A. L., and Cohn, Z. A., 1981, Depletion of glutathione selectively inhibits synthesis of leukotriene C by macrophages, *Proc. Nat. Acad. Sci. U.S.A.* **78**:2532.

Samuelsson, B., 1981, Leukotrienes: Mediators of allergic reactions and inflammation, *Int. Arch. Allergy Appl. Immunol.* **66**(Suppl. 1):98.

Samuelsson, B., 1983, Leukotrienes: Mediators of immediate hypersensitivity reactions and inflammation, *Science* **220**:568.

Samuelsson, B., Borgeat, P., Hammarström, S., and Murphy, R. C., 1979, Introduction of a nomenclature: Leukotrienes, *Prostaglandins* **17**:785.

Scott, W. A., Zrike, J. M., Hamill, A. L., Kempe, J., and Cohn, Z. A., 1980, Regulation of arachidonic acid metabolism in macrophages, *J. Exp. Med.* **152**:324.

Sha'afi, R. I., Molski, L. F. P., Borgeat, P., and Naccache, P. H., 1981a, Deactivation of the effects of F-Met-Leu-Phe and leukotriene B_4 on calcium mobilization in rabbit neutrophils, *Biochem. Biophys. Res. Commun.* **103**:766.

Sha'afi, R. I., Naccache, P. H., Molski, L. F. P., Borgeat, P., and Goetzl, E. J., 1981b, Cellular regulatory role of leukotriene B_4: its effects on cation homeostasis in rabbit neutrophils, *J. Cell. Physiol.* **108**:401.

Sheard, P., 1981, Effects of anti-allergic compounds on SRS-A and leukotrienes, in: *SRS-A and Leukotrienes* (P. J. Piper, ed.), Research Studies Press, John Wiley and Sons Ltd., Chichester, pp. 209–218.

Showell, H. J., Naccache, P. H., Borgeat, P., Pickard, S., Vallerand, P., Becker, E. L., and Sha'afi, R. I., 1982, Characterization of the secretory activity of leukotriene B_4 toward rabbit neutrophils, *J. Immunol.* **128**:811.

Smedegård, G., Hedqvist, P., Dahlén, S.-E., Revenäs, B., Hammarström, S., and Samuelsson, B., 1982, Leukotriene C_4 affects pulmonary and cardiovascular dynamics in monkey, *Nature* **295**:327.

Snyderman, R., and Goetzl, E. J., 1981, Molecular and cellular mechanisms of leukocyte chemotaxis, *Science* **213**:830.

Sok, D.-E., Pai, J.-K., Atrache, V., Kang, Y.-C., and Sih, C. J., 1981, Enzymatic inactivation of SRS-Cys-Gly (Leukotriene D), *Biochem. Biophys. Res. Commun.* **101**:222.

Stevens, F. C., Brown, D. M., and Smith, E. L., 1970, Some properties of soybean lipoxygenase, *Arch. Biochem. Biophys.* **136**:413.

Theorell, H., Holman, R. T., and Åkeson, Å., 1947, Crystalline lipoxidase, *Acta Chem. Scand.* **1**:571.

Ueno, A., Tanaka, K., Katori, M., Hayashi, M., and Arai, Y., 1981, Species differences in increased vascular permeability by synthetic leukotriene C_4 and D_4, *Prostaglandins* **21**:637.

Walsh, C., 1979, *Enzymatic Reaction Mechanisms*, W. H. Freeman and Company, San Francisco, p. 520.

Wasserman, S. I., Goetzl, E. J., and Austen, K. F., 1975, Inactivation of slow reacting substance of anaphylaxis by human eosinophil arylsulfatase, *J. Immunol.* **114**:645.

Willoughby, D. A., Giroud, J. P., Dirosa, M., and Velo, G. P., 1973, The control of inflammatory response with special reference to the prostaglandins, in: *Prostaglandins and Cyclic AMP* (R. H. Kahn and W. E. M. Lands, eds.), Academic Press, New York, pp. 187–206.

12
DIGESTION AND ABSORPTION OF LIPIDS

12.1. Introduction and History

A stormy and interesting history has led to our present knowledge of fat digestion and absorption. The elucidation of these processes has evolved slowly over the last 100 years, with a more intense effort at clarification within the last 20 years (Friedman and Nylund, 1980). As in the case of the elucidation of the physiology and biochemistry of other substances, both the development of adequate methods and a large number of observations were needed before an understanding of this complex process was achieved. However, some uncertainties still require resolution.

Approximately 90% of dietary fat consists of triacylglycerols (triglycerides, TG), the rest being made up of free and esterified cholesterol, plant sterols, and phospholipids. Most of the TG of animal or plant origin contain saturated and unsaturated, medium- to long-chain fatty acids (C_{14}–C_{18}). The richest sources of TG containing unsaturated, long-chain fatty acids, e.g., 18:1 and 18:2, are the edible oils of plant origin. Dairy products, such as milk and butter, provide some TG with short-chain fatty acids (C_4–C_8); coconut oil is rich in medium-chain fatty acids (C_8–C_{12}) (see Chapter 5).

One of the problems of fat digestion is that, as opposed to other nutrients, the TG are water-insoluble, whereas in the alimentary tract the accessory factors of fat digestion are in an aqueous medium. However, nature had a superb solution to this seemingly "insoluble" problem by designing pancreatic lipase, which is active only at an oil–water interface.

As early as 1879, Munk proposed that fat was absorbed as an emulsion. He concluded this from the observation that fatty acids (from pork fat) could be well emulsified with less than the stoichiometric amounts of Na_2CO_3 and that the triacylglycerols (from pork fat) could similarly be dispersed into a milky emulsion in a 7% solution of serum albumin. He assumed that such emulsions could also be formed in the intestines. Munk noted that the intestinal lymph vessels of dogs were tightly distended with a milky fluid after a fat meal. He made the further important observation that the lymph collected from thoracic duct cannulae of dogs fed free fatty acids (of pork fat) contained largely triacylglycerols with very little free fatty acid. Munk concluded that the fatty acids are absorbed and converted to "fat" during passage into the thoracic duct. He commented that the origin of the glycerol needed for the synthesis remained obscure ("Woher der Organismus das zur Synthese erforderliche Glycerin nimmt, bleibt vor der Hand noch dunkel." Cf. Munk, 1879, p. 374). Pflüger, the leading physiologist of the period, argued that the existence of pancreatic lipase and its ready action on fat emulsions were proof enough that triacylglycerols are completely hydrolyzed before absorption (Pflüger, 1900).

Over the years a number of short-lived theories of fat absorption were proposed or reexamined. One theory postulated that the TG are transported as an emulsion across the gut wall into the lymphatic vessels and another, the lipolytic theory, proposed a total hydrolysis of the TG into glycerol and fatty acids before absorption (Johnston, 1968). It was postulated that the function of lipase was to free the fatty acids for subsequent emulsification by the bile acids (Verzár and McDougall, 1936).

In the middle 1940s, Frazer proposed the "partition hypothesis" of fat digestion and absorption (Frazer, 1946). It is difficult to understand in retrospect why Frazer's hypothesis engendered as much controversy as it did in the early 1950s; certainly it stimulated much research, particularly on the mode of action of lipase and the role of bile acids. Frazer's experimental observations and hypothesis present a fairly accurate description of the intraluminal phase of fat digestion. Frazer's hypothesis contained the following postulates:

(i) Hydrolysis of triacylglycerols by pancreatic lipase *in vitro* or *in vivo* is incomplete and leads to the formation of free fatty acids and mono- and diacylglycerols, and to only a very small amount of free glycerol. Although no satisfactory explanation could be given for the partial hydrolysis, it was recognized that it was not due to inhibition of the enzyme by the end products of hydrolysis (Frazer and Sammons, 1945). It was shown later that pancreatic lipase cleaves fatty acyl groups from positions 1 and 3 of triacylglycerols at rates many times greater than that for the 2-acyl group.

(ii) Emulsification of fat into negatively charged particles with diameters of 0.5 μm or less must precede absorption. (It was shown that liquid paraffin emulsified to particle size of 0.5 μm was well absorbed from the intestine of rats and that the chylomicrons in blood, representing absorbed fat, were negatively charged.)

(iii) As the pH in the upper intestine is about 6.5, neither bile salts nor soaps alone could adequately emulsify fat. It was shown that taurocholate/taurodeoxycholate activated pancreatic lipase and that the ternary mixture of bile salts, fatty acids, and monoacylglycerols (MG)—the latter two being products of lipolysis—provided an excellent emulsifying system.

(iv) Emulsified glycerides are absorbed as particles into the lymphatics while free fatty acids (or their salts) enter the portal vein on a separate pathway. Actually, this is true only for short- and medium-chain fatty acids up to C_{12} under usual rates of absorption (Carlier and Bezard, 1975) but at low absorption rates long-chain unsaturated fatty acids may be converted to phospholipids, which are transported predominantly via the portal system (Saunders et al., 1982).

Frazer's hypothesis did not provide an insight into the cellular phase of fat absorption, i.e., the resynthesis of triacylglycerols (TG) in the intestinal epithelial cells and their migration into the lacteals, nor did it fully account for the complex role of bile acids.

Introduction of thoracic duct cannulation, the use of isotopically labeled triacylglycerols, the examination of intestinal contents during fat digestion, and investigation of the composition of lymph defined in detail the intraluminal events of fat digestion and the special role of intestinal epithelial cells in fat absorption. The use of doubly labeled triacylglycerols, e.g., TG containing [^{14}C]glycerol and fatty acids with conjugated double bonds absorbing UV light at 235 nm, or [^{14}C]glycerol + [^{3}H]triacylglycerol (Reiser et al., 1952, and others), established unequivocally that (i) in the intestine, most of the TG had been hydrolyzed to MG before absorption; (ii) while essentially all the fatty acids (long-chain) of the TG fed appeared in the thoracic duct lymph, only about two-thirds of the glycerol of the fed TG was found in the lymph; (iii) no more than about 3% of TG was absorbed in unhydrolyzed form; (iv) when labeled long-chain fatty acids and glycerol were fed separately, all the fatty acids appeared in the lymph, mostly in the form of triacylglycerols, but labeled glycerol was excluded from the lymph. These observations established that about one-third of the TG was completely hydrolyzed to free fatty acids and glycerol, and that all the absorbed fatty acids (long-chain) and monoacylglycerols were reconverted to triacylglycerols before appearing in the lymph. Still to be accounted for were: (i) the action of the lipase(s) in the alimentary tract, (ii) the role of the bile acids and their conjugates in fat digestion and absorption, (iii) the physicochemical states of the free fatty acids and monoacylglycerols undergoing absorption, (iv) the mechanisms of recombination of free fatty acids and monoacylglycerols into triacylglycerols, (v) the cellular structures involved in fat absorption and resynthesis into triacylglycerols, and (vi) the formation of chylomicrons appearing in thoracic duct lymph and passing into the blood.

12.2. Present Theories of Digestion and Absorption—Intraluminal Phase

Although a "lingual" lipase, and a lipase secreted by glands adjacent to the pharynx (Hamosh et al., 1975), and a gastric lipase (Schønheyder and Volqvartz, 1946) have been described, their role in fat digestion and absorption is probably minimal, except in infants, as these upper-tract lipases hydrolyze mostly triacylglycerols containing the short- and medium-chain fatty acids (e.g., of milk). Certainly, in the absence of pancreatic lipase for whatever reasons, fat is poorly absorbed and results in steatorrhea.

The entry of food from the stomach into the duodenum stimulates the secretion of enteric hormones (among them cholecystokinin), which in turn cause the gallbladder to contract and to deliver stored bile into the duodenum; the secretion of fresh hepatic bile is also stimulated. These hormones also stimulate secretion of pancreatic juice (Sheehy and Floch, 1964), which contains the triacylglycerol lipase and other hydrolytic enzymes. Pancreatic lipase and the salts of bile acid conjugates (glycocholic, taurocholic, and glyco- and taurodeoxycholic acids) (see Chapter 15) are the most important factors in fat digestion and absorption.

Pancreatic lipase, which can act only at an oil–water interface, catalyzes the following reactions:

$$\text{triacylglycerol} + H_2O \rightarrow \text{1,2-diacylglycerol} + \text{fatty acid}$$
$$\text{1,2-diacylglycerol} + H_2O \rightarrow \text{2-acylglycerol} + \text{fatty acid}$$

Its rate of hydrolysis with a 2-acylglycerol (2-monoglyceride) is very low, which explains the partial hydrolysis of triacylglycerols observed *in vitro* and *in vivo*. This specificity of pancreatic lipase is almost entirely positional and is independent of the nature of fatty acids at positions 1 and 3 of the glyceride except that, as has been reported from the laboratory of Desnuelle (Entressangles *et al.*, 1961), a short-chain fatty acid at either of these positions may be preferentially hydrolyzed. The fact that, *in vivo*, approximately one-third to one-quarter of ingested TG are completely hydrolyzed can be accounted for largely by the action (relatively slow) of isomerases in the gut that convert a 2-acylglycerol into 1-acylglycerol, which is then promptly hydrolyzed by the lipase.

Pancreatic lipase has been purified from several species but the best characterized is the porcine enzyme (cf. Brockman, 1981). All the definitive experiments on the mode of action of pancreatic lipase have been carried out with this enzyme, which exists in two forms, labeled L_A and L_B. Both are glycoproteins with a mean molecular weight of 50,000 (reported ranges 45,000–52,000). Both forms contain 4 mol of mannose and 3 mol of *N*-acetylglucosamine (Garner and Smith, 1972). They differ from one another only in that L_A also contains one *N*-acetylneuraminic acid (sialic acid) per mol, which explains the slightly different electrophoretic mobilities of the two forms.

The first purification of the two isozymes of porcine pancreatic lipase to homogeneity was carried out in Desnuelle's laboratory (Verger *et al.*, 1969), by conventional methods of protein purification. Newer methods of purification exploit the high affinity of pancreatic lipase for a hydrophobic surface, e.g., siliconized glass beads (Brockman, 1981).

Pancreatic lipase is specific for water-insoluble TG containing long-chain fatty acids; it can act only at an oil–water (hydrophobic–water) interface. However, it can also hydrolyze water-soluble TG, such as tripropionyl- and tributyrylglycerol, if it is provided with a hydrophobic surface, such as siliconized glass beads, onto which it can adsorb. However, under such conditions it is rapidly and irreversibly inactivated (Momsen and Brockman, 1976a). Also, under such conditions sodium taurodeoxycholate at low concentrations (up to 0.3 mM) increased the stability of pancreatic lipase fivefold.

At higher concentrations (0.3–0.8 mM) bile salts form a strong complex with the enzyme, 4 mol of bile salts being bound per mol of the lipase with a dissociation constant of $1.4 \times 10^{-15}M^4$ (Momsen and Brockman, 1976b). At such concentrations, bile salts inhibit lipase action since they displace the enzyme from the oil–water interface to the aqueous phase (Borgström, 1975). This untoward effect of high concentrations of bile salts is counteracted by colipase, a small-molecular-weight, heat-stable polypeptide ($M_r = 10,000$) which forms a tight complex with the lipase (dissociation constant $5 \times 10^{-7}M$) and restores its interfacial recognition (Chapus et al., 1975; Momsen and Brockman, 1976b; Patton and Carey, 1979). Colipase, secreted by the pancreas, together with lipase, binds to lipase so strongly that special measures are needed to obtain a lipase free from it (Brockman, 1981). Lipase, tested in the artificial system of hydrophobic glass beads plus tripropionylglycerol in the absence of bile salt and colipase, has a broad pH optimum between 6.5 and 8.5, with pK_a values of 5.5 and 9.3. The combined effects of colipase and sodium taurodeoxycholate narrow the pH optimum of the enzyme to 7.0–7.5, closer to the prevailing pH in the small intestine (about 6.5), with a change of its pK_a value to 6.2 and 8.4 (Momsen and Brockman, 1976a).

The lipase–colipase–bile salt complex may be regarded as the enzyme unit which interacts with the emulsion of fat (Masoro, 1977). The products of hydrolysis are mainly 2-acylglycerols, some 1-acylglycerol (formed by isomerization of 2-acylglycerol), free fatty acids, and free glycerol. As was first demonstrated by Frazer (1946), and confirmed by subsequent investigators, monoacylglycerols and free fatty acids with bile acids form the perfect emulsifying system (at pH 6.9) for even unhydrolyzed TG. This system breaks down fat particles to a size less than 0.6 μm and thereby increases the surface area of TG by a factor of approximately 10,000 (Sickinger, 1975). This emulsification of fat is of great importance as the rate of lipolysis by pancreatic lipase is a function of the surface area offered to the enzyme.

The bile salts have a function beyond stabilizing lipase and promoting emulsification. They also form water-soluble mixed micelles with monoacylglycerols and free fatty acids; triglycerides are entrapped in only low concentrations (3%) in such micelles, but other amphiphilic molecules, and small amounts of apolar substances such as mineral oil, may be absorbed following inclusion in the micelles.

The nature of fatty acids also controls, to some extent, the quantitative aspects of fat absorption. Thus, the longer-chain saturated fatty acids are less readily, or less completely, absorbed than are the shorter-chain or the more highly unsaturated fatty acids (Hofmann and Borgström, 1963). In fact, tristearin is very poorly absorbed and it is interesting to note that a triacylglycerol containing stearate in the 2-position is more completely absorbed than one in which it occupies either the 1- or 3-positions. It is possible, in this case, to speculate that monostearin is more readily absorbed than is free stearic acid.

Unsaturated fatty acids and monoacylglycerols require a lower bile acid concentration for micellar solubilization than do saturated fatty acids and their monoacylglycerols. TG are not incorporated into the micelles in significant amounts, and this property is probably responsible for their low absorption rate.

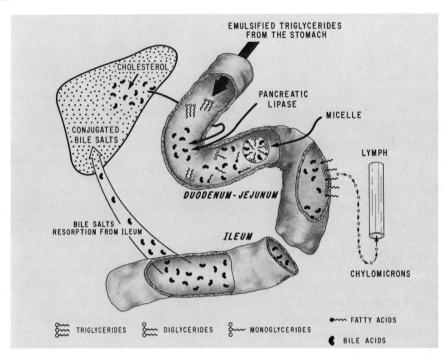

FIGURE 12.1. Diagrammatic representation of intraluminar events during fat absorption. The free fatty acids and acylglycerols separate from the bile salts which are reabsorbed in a different segment of the small intestine. Based on drawing supplied by the courtesy of R. L. Hamilton.

Micelles, water-soluble macromolecular aggregates composed of fatty acids, MG, and bile salts, shuttle lipids between the emulsion and the absorptive surface. These events are summarized diagrammatically in Fig. 12.1.

12.3. Cellular Events of Fat Absorption

Examination of intestinal epithelium by electron microscopy provided new information concerning cellular events during fat absorption, although there is still much uncertainty concerning the entry of the hydrolyzed and emulsified fat into the epithelial cells and in what structures the fatty acids and monoacylglycerols are reincorporated into TG. The earliest electron microscopic pictures showed that the brush border of the epithelial cells seen with the light microscope consisted of numerous microvilli with a diameter of 0.08–0.1 μm, and a length of 0.72–0.75 μm, separated from one another by spaces of not more than 0.02–0.05 μm. These microvilli are covered by a three-layered geometrically asymmetric membrane, 100–115 Å thick, and protrude into the lumen of the intestine from the apical portion of the cells without any direct channels between the microvilli and the cytoplasm (Granger and Baker, 1950; Sjöstrand, 1963). Palay and Karlin

(1959) showed that fat droplets could pass between the microvilli. The diameter of such droplets was less than 0.1 μm, much less than the assumed diameter of emulsified fat in the intestine. Some of Palay and Karlin's electron micrographs show small fat droplets between the microvilli and deep in the well between the microvilli in contact with the plasma membrane. These workers also described the appearance of small membranous vesicles containing fat droplets in the apical portion of the epithelial cells during fat absorption and suggested active pinocytosis as the mechanism of the absorption.

However, Sjöstrand (1963), who also described fat-containing vesicles in the apical region of the epithelial cells during early stages of fat absorption, rejected the idea that these vesicles could be pinocytotic in origin because (i) he could not find such vesicles immediately under the brush border, nor were there any invaginations sprouting in between the microvilli, and (ii) the structure of the membranes surrounding these droplets—although triple-layered—was symmetric and distinctly different from the asymmetric plasma membrane of the microvilli.

Sjöstrand and Borgström (1967) fed either [^{3}H]trioleoylglycerol or [9,10-^{3}H]-oleic acid in vegetable oil, dispersed in skimmed milk by gastric intubation, to rats which had been fasted for 24–48 hr, and then fractionated by step-density gradient centrifugation the homogenates of intestinal epithelial cells 30–60 min after feeding. Examination of the fractions in the electron microscope showed that the most buoyant fraction consisted of the apical vesicles and Golgi membranes. This same fraction contained most of the labeled lipid administered, mainly in the form of triacylglycerol (71%), some free fatty acid (11%), diacylglycerol (7%), and monoacylglycerol (1%). Heavier fractions, consisting of smooth- and rough-surfaced membranes, fragments of brush border, and microvilli, contained only little of the [^{3}H]lipid. However, the distribution of ^{3}H in these heavier fractions was different from that seen in the apical vesicles: 33–49% was in TG, 22–40% in free fatty acids, 7% in DG, and 2–3% in MG. Sjöstrand and Borgström (1967) noted that the epithelial cells lining the macrovilli did not absorb fat synchronously: the cells at the tip of the villi accumulated fat in the apical vesicles sooner than did the cells at the base (the crypt) of the villi. The electron micrographs of Sjöstrand (1963) show that, whereas soon after a fat meal (corn oil) most of the fat is in the apical region of the epithelial cells, after two hours fine fat droplets of the size of chylomicrons are in the intercellular spaces and in the lacteals. He suggested that the chylomicrons may be formed in the intercellular spaces by the emulsification of large fat droplets extruded from the apical region of the epithelial cells. This emulsification may be assisted by protein secreted from the basal portions of the cells.

All the early electron microscopic studies on fat absorption were carried out with animals that had been given huge doses of fat (e.g., 2 ml of vegetable oil to a 250-g rat). Thus, the early pictures may have represented responses to abnormal fat meals. Recently, Marenus and Sjöstrand (1982a,b), in a reinvestigation of the problem, fed to unanesthetized mice 0.8-ml of emulsions of Mazola® oil dispersed in skim milk containing 1.4%, 7%, and 14% of fat. If the mice in these experiments weighed 35 g, the 14% fat meal would correspond to the consumption of about 200 g fat in a single meal by a 70-kg man. Marenus and Sjöstrand observed that, at the earliest times (4–6 min) after the feeding of any of the fat diets, the Golgi apparatus in the apical region of the epithelial cells responded with an enormous swelling and was the only cellular structure that contained fat. They concluded that the Golgi apparatus is the primary site of triacylglycerol accumulation and that the appearance of fat in apical vesicles is a secondary phenomenon (Marenus and Sjöstrand, 1982a). After meals containing 1.4% fat, at all times after the feeding, only the Golgi vesicles contained fat. In animals that received the 7% and 14% fat meals, the columnar epithelial cells underwent drastic structural changes. Marenus and Sjöstrand (1982b) summarized their observations thus: "Forty to sixty minutes after such meals fat had accumulated in the Golgi apparatus, in apical vesicles, in smooth and rough endoplasmic reticulum, in small smooth membrane vesicles, and as large masses free in the cytoplasm."

A summary of the reactions of digestion, absorption, and release of fat in the intestinal epithelial cells is shown schematically in Fig. 12.2.

As the lipolytic products pass from the micelles in the intestinal lumen into

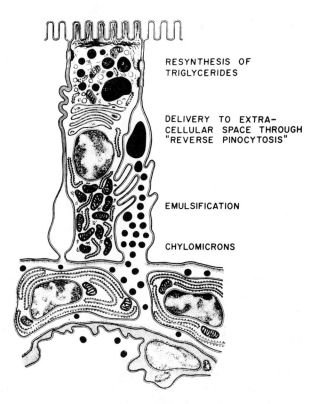

FIGURE 12.2. Schematic reconstruction from electron micrographs of cellular events of fat absorption and appearance of chylomicrons in intercellular spaces leading into lacteals. From Sjöstrand (1963), with permission of the author and Elsevier Biomedical Press BV, Amsterdam.

the interior of the mucosal cell they must cross the unstirred water layer which is immediately adjacent to the lipid–protein membrane of the microvilli. The intestinal uptake of lipolytic products is mediated by a non-energy-dependent passive process (Johnston and Borgström, 1964), but the exact mechanism by which the lipolytic products pass from the micellar phase into the mucosal cell is unknown.

Short-chain fatty acids are transported directly into the portal blood as albumin-bound fatty acid. While the medium-chain fatty acids may be either reesterified into TG within the absorptive cells or pass directly through the cells into the portal circulation, the long-chain fatty acids are converted into TG and become part of chylomicrons (Carlier and Bezard, 1976), although with very small doses they may be absorbed into the portal system. Glycerol, being water-soluble, diffuses through the unstirred water layer and is carried through the absorptive cell membrane. Lipolytic products are absorbed from the proximal small intestine, whereas bile salts leave the micelle and are absorbed from the ileum.

The mid-portion of the intestine is the most favorable area for hydrolysis and solubilization of dietary lipid. Although it is believed that fat absorption is virtually complete in the jejunum, the ileum may also contribute to this process (Wollaeger, 1973). The mucosa of the proximal ileum responds to high-fat feeding by hypertrophy and this change is associated with a more complete uptake of fatty acids (Balint *et al.*, 1980). Thus, during the process of absorption, fatty acids and monoacylglycerols dissociate from the bile salt micelles. The factors leading to dissociation have not yet been defined.

In the intestinal mucosal cell the long-chain fatty acids are activated to their CoA derivatives through the action of fatty acyl:CoA ligase, which is inactive toward the medium-chain fatty acids and is located in the endoplasmic reticulum (Senior and Isselbacher, 1960). Carbohydrate metabolism provides the ATP needed for fatty acid activation.

Three pathways are available for the formation of TG. In the monoacylglycerol pathway, the 2-monoacylglycerol formed during the hydrolysis of triacylglycerol is further acylated. The second pathway, which is inhibited by monoacylglycerol, involves the action of glycerol kinase in the formation of (R)-sn-glycerol 3-phosphate from glycerol derived either from glycolysis or from hydrolysis of monoacylglycerol by monoacylglycerol lipase. The glycerophosphate is then acylated by means of specific acyltransferases to form 1,2-diacylglycerol 3-phosphate (phosphatidic acid). Phosphatidic acid may thereafter be converted into either phospholipids or TG (see Chapters 14, 17). The rate-limiting step in the absence of monoacylglycerol is the formation of glycerol 3-phosphate.

A third pathway involves dihydroxyacetone phosphate, a good precursor of triacylglycerol during fatty acid absorption which also stimulates the incorporation of [^{14}C]glucose into glycerides by fatty acids in intestinal epithelial cells, and leads to the same conclusion, i.e., that the products of glycolysis may furnish the glycerol for glyceride synthesis. However, the discovery that glycerol kinase is present in the intestinal cells indicates that glycerol itself may also be utilized for glyceride synthesis. The fact that it is usually not used for this purpose during fat absorption may be due to the rapid absorption of glycerol at a site different from that followed by the fatty acids and also to the inhibition of phosphatidic acid phosphatase by monoacylglycerol.

On the other hand, the monoacylglycerol pathway involves the sequential transfer of acyl groups from acyl-CoA first to 2-monoacylglycerol and then to the resulting diacylglycerol (Rao and Johnston, 1966).

There is no apparent preference for unsaturated acyl-CoA over the saturated acyl-CoA in the acylation of 2-monoacylglycerol. The rate-limiting step is the formation of acyl-CoA. The rate of acylation of the 2-monoacylglycerol was found to be considerably greater than that for the 1- or 3-monoacylglycerol. There is also an active hydrolase in the intestinal cells, resulting in hydrolysis of those monoacylglycerols that are not rapidly acylated. The net result seems to be the preservation of the 2-acyl group throughout digestion and absorption, with a fairly complete reshuffling of the 1- and 3-acyl groups.

The fatty acid composition of the TG synthesized via the monoacylglycerol pathway is very similar to that of the dietary TG. It has been calculated that 85% of TG are synthesized through this pathway, with the remainder originating via the glycerol 3-phosphate pathway (Kern and Borgström, 1965).

Insulin has no direct effect on intestinal fatty acid esterification but insulin-induced hypoglycemia leads to a decrease in esterification in the small intestine, possibly because of the low availability of glycerol from glycolysis.

A monoacylglycerol lipase has been found in the epithelial cells of the small intestine. Its action results in the regulation of TG synthesis by providing the required fatty acids when the ratio of free fatty acids to monoacylglycerols differs greatly from the normal 2:1. Approximately 18% of the 2-monoacylglycerols absorbed are split intracellularly in this way (Paris and Clément, 1968).

The intestinal epithelium is anatomically and functionally heterogeneous. The cells undergo changes in metabolic activity as they migrate up the side of the villus toward the tip, following their formation in the "crypt" region (Ockner and Isselbacher, 1974). It has been suggested that, in villus tips, most of the TG is synthesized through the monoacylglycerol pathway, while the glycerol 3-phosphate pathway is more active in the crypt region, which, however, is more involved in phospholipid synthesis than in TG synthesis (Johnston, 1976).

12.4. Synthesis of Lipoproteins

The endoplasmic reticulum (ER) of the cell is closely associated with the resynthesis of the lipid particle (Sjöstrand and Borgström, 1967). It has been demonstrated that monoacylglycerol transferases are localized on the smooth ER, while the acyltransferases of the second pathway are preferentially attached to the rough ER (Higgins and Barnett, 1971).

The Golgi complex is also involved in the formation of chylomicrons and the final transport of the lipid from the cell. It is believed that the lipid droplets contained within the ER vesicles are actually nascent chylomicron particles, i.e.,

TG, phospholipids (PL), cholesterol, and protein complexes (see Chapter 13). The final step therefore would be a fusion of the vesicles with the plasma membrane and an extrusion or "exocytosis" of chylomicrons into the intercellular space (Strauss, 1966). However, more current evidence implicates the Golgi complex in the actual packaging of chylomicrons and their release into the intercellular space (Redgrave, 1973).

Chylomicrons are composed of about 70–90% TG, 4–8% PL, 3% cholesterol, 4% cholesteryl esters, and 2% protein (Friedman and Nylund, 1980) (see Chapter 13). The enzymes required for the synthesis of these particles are generally associated with the ER. The ER utilizes the TG, PL, cholesterol, and cholesteryl esters along with a specific protein, apoprotein B, to generate the chylomicron. Its formation is enhanced by Ca^{2+} (Saunders and Sillery, 1979).

In addition, chylomicrons contain a small amount of carbohydrate which is added (as is the case for other lipoproteins) in the Golgi complex (Lo and Marsh, 1970). Therefore, this structural part of the cell plays a role in building chylomicrons as well as in providing a vehicle for their transport out of the cell.

Intracellular lipid transport and discharge of chylomicrons from absorptive cells are handled by microtubules which are about 240 Å in diameter. They may be involved in the transfer of lipid from ER vesicles to the Golgi complex as well as in the final discharge of chylomicrons (Glickman *et al.*, 1976). During fat absorption, intestinal lymph shows high concentrations of both chylomicrons and very-low-density-lipoproteins (VLDL). The small intestine is capable of synthesizing both VLDL and chylomicrons during fasting as well (Shiau, 1981). VLDL have a lower TG and higher phospholipid and protein content than do chylomicrons (see Chapter 13).

An important aspect of intracellular lipid transport is the continuous membrane protein synthesis by the ER, since both Golgi and ER membranes are either degraded or incorporated into plasma membranes. It has been postulated that the lipid transport defect seen in the disease abetalipoproteinemia (Dobbins, 1966) is related to a defect in membranes of the ER or Golgi complex.

Chylomicrons are large particles, approximately 0.5–1.0 μm in diameter. Once outside the cell they diffuse to the central lacteals; the lipid passes into the lymphatic endothelium, both by passive diffusion through endothelial gaps and by an active transport within pinocytic vesicles (Sabesin, 1976). Eventually chylomicrons are transported in the lymphatic system to the superior vena cava via the thoracic duct and reach the sites of utilization or storage through the systemic blood.

The complete digestion and absorption process is quite rapid. Electron microscopic examination has shown that fat absorption resulting in the secretion of chylomicrons into the intercellular space is achieved in approximately 12 min (Jersild, 1966). Approximately 95% of the dietary lipid is absorbed by normal human subjects.

Zinc deficiency negatively affects the formation of chylomicrons (Koo and Turk, 1977). Alcohol decreases fat absorption, possibly by causing retention of the lipids in the stomach (Boquillon, 1976).

12.5. Absorption of Phospholipids

Phospholipids comprise a small but distinct portion of dietary lipid. When they reach the intestinal lumen they are exposed to the action of phospholipases which are present in the pancreatic juice. Phospholipase A_1 hydrolyzes the ester bond in position 1 of glycerol whereas phospholipase A_2 acts at position 2. A lysolecithin is formed in both cases (see Chapter 17). Phospholipase A_2 is present in the pancreatic juice as a zymogen which first must be activated by trypsin. These enzymes require Ca^{++} as a cofactor (Friedman and Nylund, 1980). For enzymatic hydrolysis to occur, it is obligatory for phospholipids to be dispersed into small micelles, and bile acids are essential for this process. In general, dietary lecithin is hydrolyzed by phospholipase A_2 to 1-acylglycerophosphocholine (lysolecithin) which, on entering the mucosal cell, is partly reacylated to lecithin; the remaining lysolecithin is further hydrolyzed to glycerophosphocholine, glycerophosphate, glycerol, and inorganic phosphate. Fatty acids and glycerophosphate are reassembled to TG via the glycerol 3-phosphate pathway (Parthasarathy *et al.*, 1974). Both the fatty acids and the highly polar lysophosphoglycerides are incorporated into the micelles, pass into the mucosal cells, and participate in the synthesis of chylomicrons. At least one-third of the dietary lecithin is incorporated into chylomicron lecithin.

In addition, phospholipids are synthesized intracellularly via the glycerol 3-phosphate pathway. These differ greatly from the dietary diacylglycerols both in composition and in the distribution of the fatty acids (Sickinger, 1975).

Bile acids and dietary phosphatidylcholine play an important role in TG transport out of the intestinal mucosa by providing surfactant compounds for the chylomicron envelope and by supporting mucosal protein biosynthesis. Fat absorption is impaired in the absence of phosphatidylcholine (O'Doherty *et al.*, 1973) or, in the rat, in the absence of free choline in the diet.

On the other hand, in pancreatic insufficiency, when lecithin is not hydrolyzed, mixed micelles are abnormal and the absorption of both fatty acids and bile salts is depressed. Whereas lysolecithin is associated with normal rates of fatty acid absorption, lecithin has been found to reduce the rate by 40% in jejunal and ileal segments. Unhydrolyzed lecithin in pancreatic insufficiency also aggravates steatorrhea (Saunders and Sillery, 1976; Ammon *et al.*, 1979).

12.6. Absorption of Cholesterol

The average American diet supplies approximately 400–700 mg of cholesterol/day. However, intestinal cholesterol is derived not only from the diet, but also from intestinal secretions and bile. Bile usually transports from 750 to 1250 mg cholesterol/day (Bennion and Grundy, 1975); the contribution from the sloughed mucosal cells is insignificant. Cholesterol in bile is unesterified but a portion of dietary cholesterol may be esterified with fatty acids. Cholesteryl esters are hydrolyzed by pancreatic cholesterol esterases which can also catalyze the synthesis of cholesteryl esters. However, this is prevented by the bile salts present in the intestinal lumen, which promote hydrolysis and which, in addition, protect the enzyme from proteolytic digestion (Vahouny et al., 1964).

Free cholesterol must be solubilized prior to absorption. This is achieved by its incorporation into mixed micelles containing conjugated bile acids, fatty acids, monoacyglycerols, and lysolecithin (Hofmann and Borgström, 1964). In the absence of bile acids the absorption of cholesterol is negligible (Watt and Simmonds, 1976).

In contrast to the absorption of TG, cholesterol absorption is limited. Only from 30% to 60% of dietary and biliary cholesterol are absorbed (Grundy and Mok, 1977). Both the quantity of cholesterol that can be solubilized by the micelles and the amount of cholesterol entering the intestine may affect total absorption. Increasing the cholesterol intake results in increased total absorption, although the percentage absorbed is less. After the ingestion of 3,000 mg/day, the absorption is approximately 1,000 mg/day (Grundy, 1979). It is assumed that the absorption of dietary and endogenous cholesterol is identical (Kudchodkar et al., 1973). Although in laboratory animals excess cholesterol is converted to bile acids, which in turn promote further absorption, this does not seem to occur in man (Quintao et al., 1971).

The uptake of cholesterol into micelles is influenced by the quantity of dietary fat (Swell et al., 1955). Increased amounts of cholesterol are solubilized by micelles enlarged by the presence of increased amounts of fat. Increased unsaturation of fat does not seem to decrease absorption despite the earlier movement of the micelle through the unstirred water layer (Grundy and Ahrens, 1970). Passage through the lipid of the mucosal cell membrane occurs by a passive diffusion. In the mucosa, cholesterol is reesterified, largely with unsaturated fatty acids, incorporated into chylomicrons and transferred into intestinal lymph.

Approximately 200–300 mg of plant sterols, which are structurally similar to cholesterol, are usually ingested daily but they are absorbed only in trace amounts and in an unesterified state. In large amounts, they inhibit cholesterol absorption (Salen et al., 1970), possibly because they either displace cholesterol from mixed micelles, or because they compete with cholesterol for uptake within the mucosal cell membrane or perhaps because they inhibit cholesterol esterification. It has been suggested that, in the intestine, plant sterol esters are hydrolyzed and the resulting free sterol decreases the solubility of cholesterol in the oil and micellar phases forming a mixed crystal, with a consequent decrease in cholesterol absorption (Mattson et al., 1977).

Although pectin interferes with the absorption of cholesterol, other complex carbohydrates differ in their effects on cholesterol metabolism. The antibiotics, puromycin and neomycin, affect cholesterol and fatty acid absorption, resulting in the reduction of plasma lipid levels by approximately 25% (Sedaghat et al., 1975; Vahouny et al., 1977).

While the large intestine does not contribute significantly to the absorption of exogenous cholesterol in the rat (Roy et al., 1978), the gallbladder (in guinea pigs) does absorb cholesterol (Neiderhiser et al., 1976).

12.7. Medium-Chain Triacylglycerols

Medium-chain TG (MCT) (containing fatty acids with chain-lengths of 6–12 carbons) occur naturally in some oils such as palm kernel or coconut; these contain approximately 48% of the C_{12} acid as well as additional amounts of fatty acids with shorter chains. Their smaller molecular size and greater water solubility lead to differences in their digestion, absorption, and subsequent transport from the epithelial cells of the small intestine.

MCT are hydrolyzed more rapidly and in larger amounts in the stomach (Cohen *et al.*, 1971). Since their emulsification is facilitated by the lower interfacial tension of the small TG molecule, micellar solubilization is not necessary (Clark and Holt, 1968). Some absorption may take place even in the absence of pancreatic lipase; the intact TG penetrate into the epithelial cells of the small intestine and into the capillaries leading to the portal vein. MCT are also absorbed in the colon (Valdivieso and Schwabe, 1966). Unlike the long-chain fatty acids, the medium-chain fatty acids are absorbed as free fatty acids, and do not require activation with CoA. The monoacylglycerols are hydrolyzed by the enzyme, monoglyceride lipase, present in the epithelial cells (Senior and Isselbacher, 1963). Only a very small portion of the medium-chain fatty acids are resynthesized to TG in the epithelial cells and most of them leave the intestinal epithelial cells via the portal vein (Borgström, 1955).

12.8. Fat Malabsorption

A defect in any of the processes involved in fat absorption leads to a malabsorption syndrome. Steatorrhea, or loss of fat in the feces, may be a result of pancreatic malfunction, which in turn may be caused by destruction of pancreatic tissue as is evidenced by the diseases chronic pancreatitis, carcinoma of the pancreas, or cystic fibrosis. A defective lipolysis resulting from pancreatic lipase insufficiency prevents adequate solubilization in bile acid micelles.

Pancreatic lipase deficiency leads to only a moderate disturbance of absorption of MCT, and bile salt deficiency does not affect the lipolysis, solubilization, or absorption of MCT. As a result, MCT have potential therapeutic applications in patients with lipid malabsorption syndromes, such as pancreatic insufficiency, cystic fibrosis, sprue, and short bowel syndrome (Greenberger *et al.*, 1967).

A decreased micellar solubilization may also be caused by biliary obstruction, ileal resection, and drugs, such as cholestyramine, which bind bile acids. Since the amount of lipid absorbed depends on the availability of an adequate mucosal surface, any abnormality in this area results in fat malabsorption such as that in gluten enteropathy, tropical sprue, Whipple's disease, and amyloidosis.

A congenital disease, abetalipoproteinemia, is characterized by an inability to synthesize the protein moiety associated with chylomicrons. Therefore, TG accumulate in the mucosal cells. Diseases that block the intestinal lymphatics also result in steatorrhea.

Fat malabsorption and steatorrhea also occur in essential fatty acid-deficient rats (Clark et al., 1973). Apparently under these conditions there is a reduction of the lipid-esterifying capacity in the mucosa, with a delay in the removal of newly synthesized TG. Phospholipid availability for chylomicron formation is also reduced under these conditions.

REFERENCES

Ammon, H. V., Thomas, P. J., and Phillipis, S. F., 1979, Effect of lecithin on jejunal absorption of micellar lipids in man and on their monomer activity in vitro, *Lipids* **14**:395.

Balint, J. A., Fried, M. B., and Imai, C., 1980, Ileal uptake of oleic acid: evidence for adaptive response to high fat feeding, *Am. J. Clin. Nutr.* **33**:2276.

Bennion, L. J., and Grundy, S. M., 1975, Effects of obesity and caloric intake on biliary lipid metabolism, *J. Clin. Invest.* **56**:996.

Boquillon, M., 1976, Effect of acute ethanol ingestion on fat absorption, *Lipids* **11**:848.

Borgström, B., 1955, Transport form of ^{14}C-decanoic acid in porta and inferior vena cava blood during absorption in the rat, *Acta Physiol. Scand.* **34**:71.

Borgström, B., 1975, On the interactions between pancreatic lipase and colipase and the substrate, and the importance of bile salts, *J. Lipid Res.* **16**:411.

Brockman, H. L., 1981, Triglyceride lipase from porcine pancreas, *Methods Enzymol.* **71**:619.

Carlier, H., and Bezard, J., 1975, Electron microscope autoradiographic study of intestinal absorption of decanoic and octanoic acids in the rat, *J. Cell. Biol.* **65**:383.

Carlier, H., and Bezard, J., 1976, Electron microscopy of medium chain fatty acid absorption, *Gastroenterology* **70**:460.

Chapus, C., Sari, H., Sémériva, M., and Desnuelle, P., 1975, Role of colipase in the interfacial adsorption of pancreatic lipase at hydrophilic interfaces, *FEBS Lett.* **58**:155.

Clark, S. B., and Holt, P. R., 1968, Rate-limiting steps in steady-state intestinal absorption of trioctanoin-1-^{14}C. Effect of biliary and pancreatic flow diversion, *J. Clin. Invest.* **47**:612.

Clark, S. B., Ekkers, T. E., Singh, A., Balint, J. A., Holt, P. R., and Rodgers, J. B., Jr., 1973, Fat absorption in essential fatty acid deficiency: a model experimental approach to studies of the mechanism of fat malabsorption of unknown etiology, *J. Lipid Res.* **14**:581.

Cohen, M., Morgan, R. G. H., and Hofmann, A. F., 1971, Lipolytic activity of human gastric and duodenal juice against medium and long chain triglycerides, *Gastroenterology* **60**:1.

Dobbins, W. O., 1966, An ultrastructural study of the intestinal mucosa in congenital β-lipoprotein deficiency with particular emphasis upon the intestinal absorptive cell, *Gastroenterology* **50**:195.

Entressangles, B., Paséro, L., Savary, P., Sarda, L., and Desnuelle, P., 1961, Influence de la nature des chaines sur la vitesse de leur hydrolyse par la lipase pancréatique. *Bull. Soc. Chim. Biol.* **43**:581.

Frazer, A. C., 1946, The absorption of triglyceride fat from the intestine, *Physiol. Rev.* **26**:103.

Frazer, A. C., and Sammons, H. G., 1945, The formation of mono- and diglycerides during the hydrolysis of triglyceride by pancreatic lipase, *Biochem. J.* **39**:122.

Friedman, H. I., and Nylund, B., 1980, Intestinal fat digestion, absorption, and transport. A review, *Am. J. Clin. Nutr.* **33**:1108.

Garner, C. W., and Smith, L. C., 1972, Porcine pancreatic lipase. A glycoprotein, *J. Biol. Chem.* **247**:561.

Glickman, R. M., Perrotti, J. L., and Kirsh, K., 1976, Intestinal lipoprotein formation: effect of colchicine, *Gastroenterology* **70**:347.

Granger, B., and Baker, R. F., 1950, Electron microscope investigation of the striated border of intestinal epithelium, *Anat. Rec.* **107**:423.

Greenberger, N. J., Ruppert, R. D., and Tzagournis, M., 1967, Use of medium chain triglycerides in malabsorption, *Ann. Int. Med.* **66**:727.

Grundy, S. M., 1979, Dietary fats and sterols, in: *Nutrition, Lipids, and Coronary Heart Disease: A*

Global View. Nutrition in Health and Disease, Volume 1 (R. I. Levy, B. M. Rifkind, B. H. Dennis, and N. Ernst, eds.), Raven Press, New York, pp. 89–118.

Grundy, S. M., and Ahrens, E. H., Jr., 1970, The effects of unsaturated dietary fats on absorption, excretion, synthesis and distribution of cholesterol in man, *J. Clin. Invest.* **49:**1135.

Grundy, S. M., and Mok, H. Y. I., 1977, Determination of cholesterol absorption in man by intestinal perfusion, *J. Lipid Res.* **18:**263.

Hamosh, M., Klaeveman, H. L., Wolf, R. O., and Scow, R. O., 1975, Pharyngeal lipase and digestion of dietary triglyceride in man, *J. Clin. Invest.* **55:**908.

Higgins, J. A., and Barnett, R. J., 1971, Fine structural localization of acyltransferases, *J. Cell. Biol.* **50:**102.

Hofmann, A. F., and Borgström, B., 1963, Hydrolysis of long-chain monoglycerides in micellar solution by pancreatic lipase, *Biochim. Biophys. Acta* **70:**317.

Hofmann, A. F., and Borgström, B., 1964, The intraluminal phase of fat digestion in man, *J. Clin. Invest.* **43:**247.

Jersild, R. A., Jr., 1966, A time sequence study of fat absorption in the rat jejunum, *Am. J. Anat.* **118:**135.

Johnston, J. M., 1968, Mechanism of fat absorption, in: *Handbook of Physiology,* Section 6: *Alimentary Canal,* Volume III. *Intestinal Absorption* (C. F. Code, ed.), American Physiological Society, Washington, D.C., pp. 1353–1375.

Johnston, J. M., 1976, Triglyceride biosynthesis in the intestinal mucosa, in: *Lipid Absorption: Biochemical and Clinical Aspects* (K. Rommel and H. Goebell, eds.), University Park Press, Baltimore, pp. 85–94.

Johnston, J. M., and Borgström, B., 1964, The intestinal absorption and metabolism of micellar solutions of lipids, *Biochim. Biophys. Acta* **84:**412.

Kern, F., Jr., and Borgström, B., 1965, Quantitative study of the pathways of triglyceride synthesis by hamster intestinal mucosa, *Biochim. Biophys. Acta* **98:**520.

Koo, S. I., and Turk, D. E., 1977, Effect of zinc deficiency on intestinal transport of triglyceride in the rat, *J. Nutr.* **107:**909.

Kudchodkar, B. J., Sodhi, H. S., and Horlick, L., 1973, Absorption of dietary cholesterol in man, *Metabolism* **22:**155.

Lo, C., and Marsh, J. B., 1970, Biosynthesis of plasma lipoproteins, *J. Biol. Chem.* **245:**5001.

Marenus, K. D., and Sjöstrand, F. S., 1982a, Sequence of structural changes in columnar epithelium of small intestine during early stages of fat absorption, *J. Ultrastruct. Res.* **79:**92.

Marenus, K. D., and Sjöstrand, F. S., 1982b, The effects of different concentrations of administered fat on the structure of columnar cells in the small intestine, *J. Ultrastruct. Res.* **79:**110.

Masoro, E. J., 1977, Lipids and lipid metabolism, *Ann. Rev. Physiol.* **39:**301.

Mattson, F. H., Volpenhein, R. A., and Erickson, B. A., 1977, Effect of plant sterol esters on the absorption of dietary cholesterol, *J. Nutr.* **107:**1139.

Momsen, W. E., and Brockman, H. L., 1976a, Effects of colipase and taurodeoxycholate on the catalytic and physical properties of pancreatic lipase B at an oil-water interface, *J. Biol. Chem.* **251:**378.

Momsen, W. E., and Brockman, H. L., 1976b, Inhibition of pancreatic lipase B activity by taurodeoxycholate and its reversal by colipase. Mechanism of action, *J. Biol. Chem.* **251:**384.

Munk, I., 1879, Ueber die Resorption der Fettsäuren, ihre Schicksale und ihre Verwerthung im Organismus, *Archiv für Anatomie u. Physiol. Physiologische Abtheilung,* p. 371.

Neiderhiser, D. H., Harmon, C. K., and Roth, H. P., 1976, Absorption of cholesterol by the gallbladder, *J. Lipid Res.* **17:**117.

Ockner, R. K., and Isselbacher, K. J., 1974, Recent concepts of intestinal fat absorption, *Rev. Physiol. Biochem. Pharmacol.* **71:**107.

O'Doherty, P. J. A., Kakis, G., and Kuksis, A., 1973, Role of luminal lecithin in intestinal fat absorption, *Lipids* **8:**249.

Palay, S. L., and Karlin, L. J., 1959, An electron microscopic study of the intestinal villus. II. The pathway of fat absorption, *J. Biophys. Biochem. Cytol.* **5:**373.

Paris, R., and Clément, G., 1968, Biosynthèse de triglycérides à partir de 2-monopalmitine doublement marquée dans la muqueuse intestinale de rat, *Biochim. Biophys. Acta* **152:**63.

Parthasarathy, S., Subbaiah, P. V., and Ganguly, J., 1974, The mechanism of intestinal absorption of phosphatidyl choline in rats, *Biochem. J.* **140**:503.

Patton, J., and Carey, M. C., 1979, Watching fat digestion, *Science* **204**:145.

Pflüger, E., 1900, Der gegenwärtige Zustand der Lehre von der Verdauung und Resorption der Fette und eine Verurtheilung der hiermit verknüpften physiologischen Vivisectionen am Menschen, *Arch. ges. Physiol.* (Pflügers) **82**:303.

Quintao, E., Grundy, S. M., and Ahrens, E. J., Jr., 1971, Effects of dietary cholesterol on the regulation of total body cholesterol in man, *J. Lipid Res.* **12**:233.

Rao, G. A., and Johnston, J. M., 1966, Purification and properties of triglyceride synthetase from the intestinal mucosa, *Biochim. Biophys. Acta* **125**:465.

Redgrave, T. G., 1973, The role in chylomicron formation of phospholipase activity of intestinal Golgi membranes, *Aust. J. Exp. Biol. Med. Sci.* **51**:427.

Reiser, R., Bryson, M. J., Carr, M. J., and Kuiken, K. A., 1952, The intestinal absorption of triglycerides, *J. Biol. Chem.* **194**:131.

Roy, T., Treadwell, C. R., and Vahouny, G. V., 1978, Comparative intestinal and colonic absorption of [4-^{14}C]cholesterol in the rat, *Lipids* **13**:99.

Sabesin, S. M., 1976, Ultrastructural aspects of the intracellular assembly, transport and exocytosis of chylomicrons by rat intestinal absorptive cells, in: *Lipid Absorption: Biochemical and Clinical Aspects* (K. Rommel and H. Goebell, eds.), University Park Press, Baltimore, pp. 113–145.

Salen, G., Ahrens, E. H., Jr., and Grundy, S. M., 1970, Metabolism of β-sitosterol in man, *J. Clin. Invest.* **49**:952.

Saunders, D. R., and Sillery, J., 1976, Lecithin inhibits fatty acid and bile salt absorption from rat small intestine in vivo, *Lipids* **11**:830.

Saunders, D. R., and Sillery, J., 1979, Effects of calcium on absorption of fatty acid by rat jejunum in vitro, *Lipids* **14**:703.

Saunders, D. R., Sillery, J., and McDonald, G. B., 1982, Effect of ethanol on transport from rat intestine during high and low rates of oleate absorption, *Lipids* **17**:356.

Schønheyder, F., and Volqvartz, K., 1946, The gastric lipase in man, *Acta Physiol. Scand.* **11**:349.

Sedaghat, A., Samuel, P., Crouse, J. R., and Ahrens, E. H., Jr., 1975, Effects of neomycin on absorption, synthesis and/or flux of cholesterol in man, *J. Clin. Invest.* **55**:12.

Senior, J. R., and Isselbacher, K. J., 1960, Activation of long-chain fatty acids by rat-gut mucosa, *Biochim. Biophys. Acta* **44**: 399.

Senior, J. R., and Isselbacher, K. J., 1963, Demonstration of intestinal monoglyceride lipase, *J. Clin. Invest.* **42**:187.

Sheehy, T. W., and Floch, M. H., 1964, *The Small Intestine: Its Function and Diseases,* Hoeber Medical Division, Harper and Row, New York.

Shiau, Y. F., 1981, Mechanisms of intestinal fat absorption, *Am. J. Physiol.* **240**:G1.

Sickinger, K., 1975, Clinical aspects and therapy of fat malassimilation with particular reference to the use of medium-chain triglycerides, in: *The Role of Fats in Human Nutrition* (A. J. Vergroesen, ed.), Academic Press, New York, pp. 115–209.

Sjöstrand, F. S., 1963, The fine structure of the columnar epithelium of the mouse intestine with special reference to fat absorption, in: *Biochemical Problems of Lipids* (A. C. Frazer, ed.), BBA Library, Vol. 1, Elsevier, Amsterdam, pp. 91–115.

Sjöstrand, E. S., and Borgström, B., 1967, The lipid components of the small-surfaced membrane-bound vesicles of the columnar cells of the rat intestinal epithelium during fat absorption, *J. Ultrastruct. Res.* **20**:140.

Strauss, E. W., 1966, Electron microscopic study of intestinal fat absorption in vitro from mixed micelles containing linolenic acid, monoolein, and bile salt, *J. Lipid Res.* **7**:307.

Swell, L., Flick, D. G., Field, H., Jr., and Treadwell, C. R., 1955, Role of fat and fatty acids in absorption of dietary cholesterol, *Am. J. Physiol.* **180**:124.

Vahouny, G. V., Weersing, S., and Treadwell, C. R., 1964, Taurocholate protection of cholesterol esterase against proteolytic inactivation, *Biochem. Biophys. Res. Commun.* **15**:224.

Vahouny, G. V., Ito, M., Blendermann, E. M., Gallo, L. L., and Treadwell, C. R., 1977, Puromycin inhibition of cholesterol absorption in rat, *J. Lipid Res.* **18**:745.

Valdivieso, V. D., and Schwabe, A. D., 1966, Absorption of medium lipids from the rat cecum, *Am. J. Dig. Dis.* **11**:474.

Verger, R., DeHaas, G. H., Sarda, L., and Desnuelle, P., 1969, Purification from porcine pancreas of two molecular species with lipase activity, *Biochim. Biophys. Acta* **188**:272.

Verzár, F., and McDougall, E. J., 1936, *Absorption from the Intestine,* Longmans, Green, London.

Watt, S. M., and Simmonds, W. J., 1976, The specificity of bile salts in the intestinal absorption of micellar cholesterol in the rat, *Clin. Exp. Pharmacol. Physiol.* **3**:305.

Wollaeger, E. E., 1973, Role of the ileum in fat absorption, *Mayo Clin. Proc.* **48**:836.

13
TRANSPORT AND REACTIONS OF LIPIDS IN THE BLOOD

One of the most valuable operations performed on experimental animals for the study of lipid metabolism has been thoracic duct cannulation in the rat (Bollman *et al.*, 1948). Since, as detailed in Chapter 12, almost all the products of normal lipid digestion and absorption appear in the lymph, this technique permits the investigator to follow the transfer of dietary lipid into the vascular system with a minimum of dilution by endogenous lipids. By using this technique, or taking advantage of those rare cases in human disease in which the lymph is externalized (thoracic duct fistulae), it is found that shortly after a fat meal, there is a great increase in lipid in the lymph, reaching a peak in a period of time typical of the species and then decreasing to the post- (or pre-) absorptive levels. This elevated lipid burden of the lymph has been measured in several ways: as total lipid, as triacylglycerol (representing about 90% of the increase), and as light-scattering particles (chylomicrons) that can be counted in a light microscope with dark-field illumination.

13.1. Nature and Function of Chylomicrons

As isolated from the lymph after a fat meal, the chylomicrons are spherical particles 1000–5000 Å in diameter, composed of greater than 98% lipid and less than 2% protein, which appears, however, to be a necessary constituent. The high lipid content is reflected in a low density (<0.95), which permits the ready separation of the chylomicrons from most of the other lipoproteins by ultracentrifugation ($S_f > 400$). The chylomicron lipid consists of about 90% triacylglycerol, the remainder being made up of phospholipids and of both free and esterified (fatty acylated) cholesterol. The phospholipids, which, with the protein, are concentrated in a thin surface layer, are necessary for the integrity of the particle; chylomicron suspensions are flocculated by action of phospholipases. These phospholipids appear to have been synthesized from precursors present in the

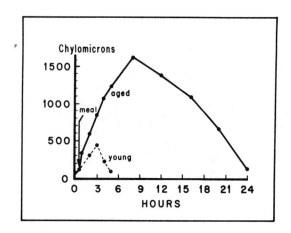

FIGURE 13.1. Changes in chylomicron count ($\times 10^{-6}/\mu l$) or triglyceride concentration of young and old human subjects after a fat meal. From Becker *et al.* (1949).

intestinal epithelial cells, unlike the other lipids, which are derived almost exclusively from dietary sources. The surface area, and hence the size of the particles, is controlled by the ratio of phospholipid to neutral lipid and thus the process that measures out this constituent is of much importance (Robinson and Wing, 1973).

The lipid core has been shown by fatty acid composition and tracer studies to consist of triacylglycerols formed from fatty acids of recently absorbed dietary lipids, a finding consistent with the rise and fall in the concentration of the lymph chylomicrons after a fat meal.

Since the lymph into which the chylomicrons are excreted is carried directly to the blood by the thoracic duct, changes in chylomicron concentration of the lymph are followed by similar changes in that of the blood (Fig. 13.1). The blood chylomicrons, though very similar in appearance and composition to those of the lymph, are not identical, indicating some rapid metabolic or exchange reactions in the blood. In particular, additional protein is added either *in vivo* or by incubation with plasma *in vitro*. Implications of this change are discussed below.

In human blood, the chylomicron triacylglycerols have a half-life of about 20 min, indicating a very rapid mechanism of removal. The nature of the mechanism is revealed by tracer studies in which triacylglycerol fatty acids are found to turn over more rapidly than those of cholesteryl esters or than the glycerol moiety of the triacylglycerols. At least partial hydrolysis of the triacylglycerols, followed by disposal of the fatty acids released, is therefore implicated.

The enzyme responsible for this hydrolysis came to light in experiments in which it was found that lipemic (chylomicron-rich) opalescent serum in dogs or human subjects was cleared by heparin injection (Hahn, 1943). However, heparin had no effect on lipemic serum in a test tube. It was evident, therefore, that *in vivo* the heparin somehow caused release of an agent appropriately termed "clearing factor." When it was found that clearing involved triacylglycerol hydrolysis, the factor was renamed "clearing factor lipase"; and when it was learned that the presence of a lipoprotein was necessary for such triacylglycerol hydrolysis, it was thereafter called "lipoprotein lipase" (LPL).

In most tissues, such as adipose tissue, the ratio of labels of injected, doubly labeled chylomicron triacylglycerol ([^{3}H]glycerol: [^{14}C]fatty acid) is markedly different in triacylglycerol deposited in the tissue, thus indicating complete hydrolysis of the substance before entry into the cells. An apparent contradiction to this rule in liver was resolved by the finding that chylomicrons could pass unchanged into the sinusoids and space of Disse, where they could remain for some time before hydrolysis at the parenchymal cell surface. Thus even in this case hydrolysis precedes true absorption (Shirai et al., 1981). The brain, which appears to contain little lipoprotein lipase, does not absorb triacylglycerol readily.

Chylomicron triacylglycerol hydrolysis does not consistently take place within the cell after absorption nor, generally, in the blood (i.e., before absorption); it is evident, and it has been shown, that the hydrolysis takes place in the capillary endothelium. The function of heparin is to release LPL from this site into the blood, where intravascular hydrolysis of chylomicron triacylglycerol occurs and the lipemic serum is "cleared." When LPL has been exhausted from its capillary location in this manner, more of the enzyme can be extracted from adipose tissue cells, indicating that the enzyme is produced within the cell and then translocated to the capillaries that constitute the major site of its action. Whether LPL hydrolyzes all three acyl groups of triglyceride is uncertain, since a monoglyceride lipase has also been implicated in the hydrolysis of chylomicron triacylglycerol. A simplified picture of this process is seen in Fig. 13.2, which includes some details that will be discussed in later sections.

The glycerol released in the hydrolysis can be utilized in most tissues, such as the liver, for glyceride synthesis. Adipose tissue, however, which does not have an adequate glycerol kinase activity, cannot use the glycerol for this purpose, and the substance therefore, for the most part, is released into the blood. The fatty acids formed in the hydrolysis can either pass into the tissue cells for oxidation or lipid synthesis or, in certain nutritional states to be discussed later, remain in the blood, where they are bound to albumin and transported to other tissues.

Because of its primary importance in facilitating the entry of fatty acids into the cells, lipoprotein lipase would be well suited to act in a regulatory capacity. Several studies have indeed produced evidence that the enzyme functions in this way. Fasting results in increased LPL activity in muscle and decreased activity in adipose tissue, and these changes are reversed with refeeding. Exercise increases

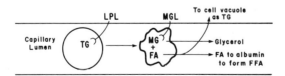

FIGURE 13.2. Schematic representation of hydrolysis of chylomicron triacylglycerol during passage through an adipose tissue capillary. LPL, lipoprotein lipase; MGL, monoacylglycerol lipase; MG, monoacylglycerol; TG, triacylglycerol; FA, fatty acid; FFA, fatty acid–albumin complex, as discussed in text.

muscle LPL, while dietary fat may increase LPL in all locations (Robinson and Wing, 1973). One disease state, familial hyperchylomicronemia (Type I hyperlipoproteinemia) is characterized by decreased LPL activity and greatly increased chylomicronemia, particularly in response to high-fat meals.

13.2. The Nature and Metabolism of Free Fatty Acids

Ultimately, the function of chylomicrons is to provide fatty acids derived from dietary lipids for metabolic purposes. The function can be realized in two ways, both involving lipoprotein lipase. First, as indicated in the case of liver and adipose tissue, the liberated fatty acids may be utilized in the tissue itself for lipid synthesis or oxidation. Second, in the case of adipose tissue, the liberated fatty acids may remain in the blood for transport to other tissues. The original finding that appreciable concentrations of titratable fatty acids are extractable from the blood came as a surprise because concentrations of fatty acids in the range found (about 1 meq/liter) were known to be hemolytic, and it was thus difficult to understand how such concentrations of fatty acids could exist without damage to erythrocytes. This dilemma was resolved when it was found that the fatty acids in the blood are bound to albumin, which functions both to solubilize them for transport and to maintain the actual concentration of free fatty acids* below that producing the hemolytic effect (Gordon and Cherkes, 1956). Although dissociation constants for albumin–fatty acid complexes differ with the fatty acid, typical values show two sites (for 18:1) with $K_d \cong 10^{-8}$, about five with $K_d \cong 10^{-6}$, and 20 or more weaker sites (Goodman, 1958). In general, the second class of sites seems to be most involved in fatty acid transport.

Studies of the changes in blood levels of the FFA (formerly known as UFA and NEFA) have revealed some very interesting properties and relationships. The half-life of FFA in the blood is of the order of 2 min, an indication of the rapidity of uptake of FFA by the tissues in general. A study of uptake by individual tissues reveals that the liver accounts for about 30% of the fatty acids taken up from the blood, converting them to glycerides (mainly secreted as components of lipoproteins) or oxidizing them either completely to H_2O and CO_2, or incompletely to ketone bodies. Uptake by skeletal muscle and heart accounts for most of the remaining utilization, particularly during exercise. Uptake by muscle, and indeed, by all tissues, appears to be in part a function of FFA concentration in the blood or, more exactly, of the ratio of FFA to albumin. The brain also takes up fatty acids from the blood, but since it has little lipoprotein lipase, these substances must be obtained from albumin complexes rather than from triacylglycerols via hydrolysis.

It is therefore found that the FFA concentration of the blood generally drops

---

*Resolution of competing terms for unesterified fatty acids and nonesterified fatty acids was achieved by the term FFA for free fatty acids. It is understood, however, that albumin binding is implied and that actually only minute concentrations of truly free fatty acids are involved.

during passage of the blood through a tissue *in vivo* or through a perfused organ. The exception to this rule is adipose tissue, which is generally responsible for an increase in FFA levels. Additional studies have shown that FFA levels increase during starvation, in uncontrolled diabetes, and after high-fat meals, and that they decrease with carbohydrate refeeding and with administration of insulin. These observations are discussed in detail in Chapter 14.

13.3. Nature and Composition of the Blood Lipids

In most healthy individuals the composition of the blood lipids is remarkably constant in the postabsorptive state. This can be seen in Table 13.1, in which values for a large number of normal human subjects are given, together with the standard errors of determination. Certain ratios that can be derived from these values can be used as measures of normalcy. For example, the ratio of sterol ester to total sterol is normally about 0.75, while that of total sterol to total phospholipid is somewhat less than one. Gross deviations from these values generally indicate some metabolic abnormality. In three such cases, also shown in Table 13.1, examples of serum lipid values of patients with familial lipidoses and other diseases are depicted; these are, of course, extreme cases and values can range from the obviously abnormal to the presumably normal.

As discussed above, a fat-containing meal results, in normal young individuals, in a transient rise in blood lipids, mostly in the form of chylomicron triacylglycerols. Reference to Fig. 13.1 reveals that, as individuals age, the rate of clearance of the lipids decreases, resulting, in some cases, in continuously elevated blood lipid levels (Becker *et al.,* 1949). Fasting also results in an increase of the triacylglycerol fraction, principally in the form of particles somewhat smaller than chylomicrons—the very-low-density lipoproteins (VLDL), shown in various ways to be elaborated largely in the liver, but to some extent in the intestine as well.

13.4. The Blood Lipoproteins

Values given in Table 13.1 for normal serum levels of various classes of lipids actually represent analyses of pooled material derived from the variety of blood lipoproteins that are present. Studies of the complex nature and intricate transformations of these components have proved enormously fruitful in increasing the understanding of many metabolic processes and a wide variety of disease states (Havel *et al.,* 1980).

In Table 13.2 are shown the approximate composition and some of the physical properties of the major blood lipoproteins. The flotation rate and electrophoretic mobility designations reflect the major analytical techniques applied to the lipoproteins. For example, the chylomicrons, with a flotation rate of greater than

TABLE 13.1. Lipid Content of Human Serum[a]

Subjects	Age (years)	Total lipid	Free sterol	Total sterol	Triacylglycerol	Phospholipid	FFA[b] (μeq/liter)	C/P ratio[c]
	Milligrams lipid per 100 ml (means ± s.e.m.)							
			"Normal" values					
Men	16–35	640 ± 15	64 ± 2.4	192 ± 7.3	84 ± 4.4	208 ± 4.4	750 ± 3.4	0.92
Women	16–35	648 ± 21	58 ± 3.5	185 ± 7.1	88	232 ± 8.2	781 ± 45	0.79
			Abnormal values in some diseases					
Hyperglyceridemic	54		203	427	1226	686		
Hypercholesterolemic	39		95	274	70	326		
Biliary cirrhosis	40		1330	136	302	3069		

[a]Data largely from Svanborg and Svennerholm, 1961.
[b]Albumin-bound fatty acids.
[c]C/P, total sterol (C + CE):phospholipid.

TABLE 13.2. Physicochemical Properties of Human Plasma Lipoprotein Classes[a,b]

Ultracentrifugal class	Flotation rate[c]	Density range (g/ml)	Size range (nm)	Electrophoretic mobility	Protein (%)	Percentage of various classes of lipids			
						TG	CE	C	PL
Chylo	$S_f^0 > 400$	0.94	80–500	origin	2	86	3	2	7
VLDL	S_f^0 20–400	0.94–1.006	30–80	pre-β	8	55	12	7	18
IDL	S_f^0 12–20	1.006–1.019	25–30	pre-β and β	15				
LDL	S_f^0 0–12	1.019–1.063	16–25	β	22	6	42	8	22
HDL$_1$	$F_{1.20}$ 9–20	1.063–1.090	10–13	α	30				
HDL$_2$	$F_{1.20}$ 3.5–9	1.090–1.12	8.5–10	α	40	5	20	5	30
HDL$_3$	$F_{1.20}$ 0–3.5	1.12–1.21	7–9	α	55	4	12	4	25

[a] Adapted from Hamilton and Kayden (1975) and Havel et al. (1980).
[b] Abbreviations: Chylomicron (Chylo); very-low-density lipoprotein, VLDL; intermediate-density lipoprotein, IDL; low-density lipoprotein, LDL; high-density lipoprotein 1, HDL$_1$; high-density lipoprotein 2, HDL$_2$; high-density lipoprotein 3, HDL$_3$; triacylglycerol, TG; cholesteryl ester, CE; cholesterol, C; phospholipid (largely phosphatidylcholine), PL.
[c] S_f rate is defined as Svedbergs of flotation, measured at 26°C in a medium of 1.745 M NaCl (density 1.063 g/ml). Flotation rates corrected for the effects associated with concentration dependence are indicated by the symbol S_f^0. F values are flotation rates measured at any other density, signified by a subscript.

S_f 400, do not move from the origin in electrophoretic analysis because of their size. The VLDL, on the other hand, migrate in the electric field just ahead of the β-globulins (thus are designated "pre-β"). The low-density lipoproteins (LDL) migrate with the β-globulins and the high-density lipoproteins (HDL), of which there are three density classes dependent on the lipid:protein ratio, with the α-globulins.

The characteristic densities of the lipoproteins are clearly related to their lipid content (particularly of neutral lipids), with respect to which their flotation rates vary directly. Their electrophoretic mobilities are determined primarily by the apolipoproteins (apoproteins) that are involved. In Table 13.3 are listed some properties of the major apoproteins. It can be seen that in addition to the presumed role of stabilization of the lipoproteins themselves, the apoproteins also have important functions in enzyme activation. Thus APO A-I serves as an activator of lecithin:cholesterol acyltransferase (LCAT), while APO C-II is necessary for activation of LPL. The other apolipoproteins also appear to have specific, though at present less well-known, functions.

The metabolism of the blood lipoproteins is an intricate, somewhat cyclic process that serves functions of transport and disposal of the dietary lipids; of transport of structural lipids and of essential metabolites having similar physical properties; and of contributing to maintenance of the osmotic properties of the blood (see Brewer and Bronzert, 1977).

The triacylglycerol-rich lipoproteins (chylomicrons and VLDL) are secreted principally by the intestine and liver, respectively (see Fig. 13.3). Coming from the intestinal epithelial cells, the chylomicrons consist mainly of diet-derived triacylglycerols stabilized by a surface coating of phospholipid, cholesterol, and protein, mainly APO B and APO A-I. In the lymphatics, and later in the blood, they acquire other apoproteins, in particular those of type C, necessary for their metabolism. The VLDL, secreted by the liver, have a similar lipid composition, with a

TABLE 13.3. Physical Properties of the Major Apoproteins of Human Serum Lipoproteins[a,b]

Apoprotein	Molecular weight	Associated primarily with	Additional actions
A-I	27,000	HDL	Activates LCAT
A-II	17,000	HDL	Prepares VLDL for attack by LPL
B		LDL	Required for secretion of triglyceride-rich proteins
C-I	7000	Chylo	Activates LCAT and LPL
C-II	10,000	VLDL	Activates LPL
C-III	8800	HDL	Inhibits LPL
D	~20,000	HDL	
E	33,000	VLDL, HDL	

[a] Adapted from Brewer and Bronzert (1977).
[b] Abbreviations as in Table 13.2.

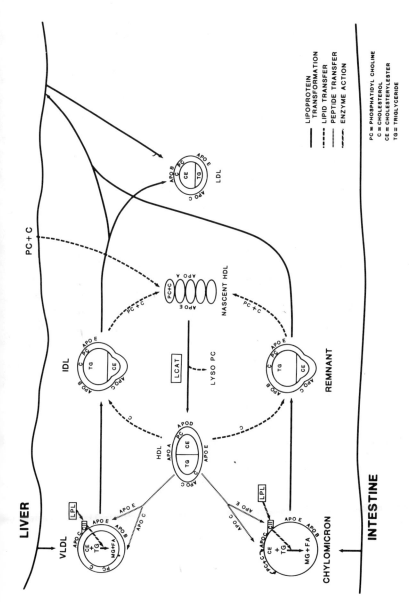

FIGURE 13.3. Schematic diagram illustrating some of the known and proposed pathways of the blood lipoproteins and their component parts. Abbreviations are defined in the figure and are discussed in the text.

protein complement largely of APO B, but quickly add the other apoproteins as they pass into the blood. With this activation, LPL proceeds to catalyze hydrolysis of about 80% of the triacylglycerols of both kinds of particles, leaving less stable aggregates rich in more amphiphilic lipids (phospholipids and cholesterol). These chylomicron remnants and IDL particles are stabilized by the "pinching off" of much of the surface lipids and accompanying apoproteins as stacked lamellar or discoid structures composed largely of phosphatidylcholine and cholesterol with a major protein component of APO A-I. (The bulk of the apolar lipids are left behind as LDL—see below.) Similar discoid bilayers have been shown to be secreted directly by the liver and to make up the major high-density lipoprotein in the blood of patients deficient in lecithin:cholesterol acyltransferase (LCAT), which acts on these particles to catalyze transfer of the acyl group (largely unsaturated) in the 2-position of phosphatidylcholine to the hydroxyl group of the cholesterol, giving a cholesteryl ester (Glomset, 1968) (Eq. 13.1). Fol-

$$\text{phosphatidylcholine} + \text{cholesterol} \xrightarrow{\text{LCAT}} \text{1-acylglycerol-3-phosphocholine} + \text{3-acylcholesterol (CE)} \quad (13.1)$$

lowing this transfer, the lysophosphatidylcholine (Lyso PC), now rather more water-soluble, is transferred from the particle surface to albumin and transported to the liver, where the 2-position is reacylated by an acyltransferase. The cholesteryl ester, on the other hand, a quite hydrophobic lipid, seeks the more compatible interior of the particle, consisting primarily of triacylglycerol. The particle, as a result of these changes, becomes more nearly spherical, acquires additional apoproteins A-I and C, and is known as an HDL. The remnant particle, stripped of most of its triacylglycerol and of excess amphiphilic lipid and protein, becomes an LDL.

The particles resulting from these transformations are, for the most part, either recycled or disassembled by the liver and peripheral tissues. The LDL tend to become bound to specific receptors on the surface of many cells, including the liver parenchymal cells. In normal subjects, they are drawn into the cells by endocytosis and are broken down into their components for reassembly. The high cholesterol content of the LDL, probably after oxygenation of a small fraction of this component, serves to depress the enzyme 3-hydroxy-3-methylglutaryl-CoA reductase (see Chapter 15), thus reducing cholesterol biosynthesis in the cell in a type of feedback inhibition, and activating the enzyme ACAT (acyl-CoA:cholesterol acyltransferase) with consequent production of increased amounts of cholesteryl esters.

The HDL can also transport cholesterol, largely to the liver for metabolism or excretion, and appears to be capable of blocking the LDL receptor sites on the cell surface.

These blood lipoprotein interactions are intimately involved not only in normal human metabolism but also in several familial diseases, and are discussed further in the latter connection in Chapter 19.

REFERENCES

Becker, G. H., Meyer, J., and Necheles, H., 1949, Fat absorption and atherosclerosis, *Science* **110**:529.
Bollman, J. L., Cain, J. C., and Grindlay, J. H., 1948, Techniques for the collection of lymph from the liver, small intestine, or thoracic duct of the rat, *J. Lab. Clin. Med.* **33**:1349.
Brewer, H. B., Jr., and Bronzert, T. J., 1977, Human Plasma Lipoproteins, Fractions. No. 1, Beckman Instruments Co., Palo Alto, California.
Glomset, J. A., 1968, The plasma lecithin:cholesterol acyltransferase reaction, *J. Lipid Res.* **9**:155.
Goodman, DeW. S., 1958, The interaction of human serum albumin with long-chain fatty acid anions, *J. Am. Chem. Soc.* **80**:3892.
Gordon, R. S., Jr., and Cherkes, A., 1956, Unesterified fatty acid in human blood plasma, *J. Clin. Invest.* **35**:206.
Hahn, P. F., 1943, Abolishment of alimentary lipemia following injection of heparin, *Science* **98**:19.
Hamilton, R. L., and Kayden, H. J., 1975, The liver and the formation of normal and abnormal plasma lipoproteins, in: *Biochemistry of Disease,* Volume 5 (F. F. Becker, ed.), Dekker, New York, pp. 531–572.
Havel, R. J., Goldstein, J. L., and Brown, M. S., 1980, Lipoproteins and lipid transport, in: *Metabolic Control and Disease,* 8th ed. (P. K. Bondy and L. E. Rosenberg, eds.), W. B. Saunders, Philadelphia, pp. 373–494.
Robinson, D. S., and Wing, D. R., 1973, Clearing factor lipase and its role in the regulation of triglyceride utilization. Studies on the enzyme in adipose tissue, in: *Advances in Experimental Medicine and Biology,* Vol. 26, *Pharmacological Control of Lipid Metabolism,* Proc. Fourth Int. Conf. on Drugs Affecting Lipid Metabolism (W. L. Holmes, R. Paoletti, and D. Kritchevsky, eds.), Plenum Press, New York, pp. 71–76.
Shirai, K., Barnhart, R. L., and Jackson, R. L., 1981, Hydrolysis of human plasma high density lipoprotein$_2$-phospholipids and triglycerides by hepatic lipase, *Biochem. Biophys. Res. Commun.* **100**:591.
Svanborg, A., and Svennerholm, L. 1961, Plasma total lipid, cholesterol, triglycerides, phospholipids and free fatty acids in a healthy Scandinavian population, *Acta Med. Scand.* **169**:43.

SELECT BIBLIOGRAPHY

Robinson, D. S., 1970, The function of the plasma triglycerides in fatty acid transport, in: *Comprehensive Biochemistry,* Volume 18 (M. Florkin and E. H. Stotz, eds.), Elsevier, Amsterdam, pp. 51–116.
Scanu, A. M., and Landsberger, F. R., eds., 1980, *Lipoprotein Structure,* Vol. 348, *Annals of the New York Academy of Sciences,* The New York Academy of Sciences, New York.
Schettler, G. (ed.), 1967, *Lipids and Lipidoses,* Springer Verlag, New York.

14
TRIACYLGLYCEROL METABOLISM AND THE REACTIONS OF ADIPOSE TISSUE

It has been apparent in the discussions of several of the subjects considered so far that control of lipid metabolism is intimately involved with blood and tissue fatty acid concentration and that control of these concentrations is a matter of considerable importance. Such control resides largely in the interplay of liver and adipose tissue but appears to originate, in most cases, in the latter. Therefore, it is timely to examine some of the reactions of adipose tissue and to consider how these are regulated and how they lead to other regulatory reactions. Since many of these reactions involve glyceride metabolism it is appropriate to examine the biosynthesis and hydrolysis of the triacylglycerols before proceeding to other reactions of the adipose cells.

14.1. Triacylglycerol Biosynthesis

Studies with liver microsomal preparations have shown that early steps in the synthesis of both acylglycerols and phosphoglycerides are the same; these are treated in some detail in Chapter 18 (see Weiss *et al.,* 1960). The first step involves the formation of *sn*-glycerol 3-phosphate, and may proceed by any of several pathways. In tissues such as liver, which have an active glycerol kinase, free glycerol can be phosphorylated (Eq. 14.1). In adipose tissue, which is poorly

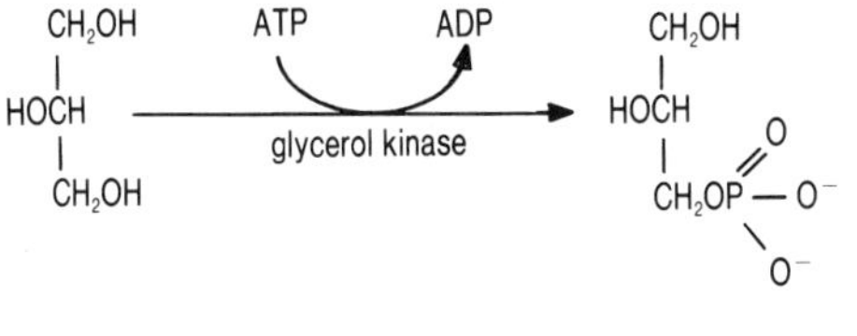

$$\begin{array}{c} CH_2OH \\ | \\ HOCH \\ | \\ CH_2OH \end{array} \xrightarrow[\text{glycerol kinase}]{ATP \quad ADP} \begin{array}{c} CH_2OH \\ | \\ HOCH \\ | \\ CH_2OP(=O)(O^-)O^- \end{array} \quad (14.1)$$

supplied with glycerol kinase, glycerol 3-phosphate is formed via glycolysis, and this tissue is therefore dependent on a supply of glucose for triacylglycerol synthesis.

The glycerylphosphate in liver is acylated in two stages, as detailed under phosphoglyceride biosynthesis. First, a 1-acyltransferase, more or less specific for saturated fatty acids, transfers an acyl group from CoA to the 1-position to give an sn-2-lysophosphatidic acid (Eq. 14.2a) where R_s is usually a saturated acyl

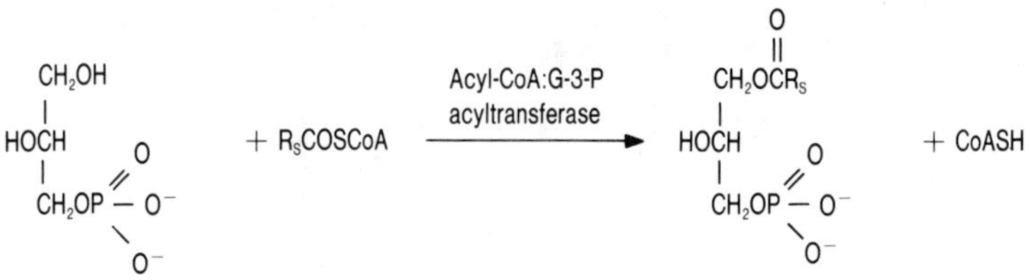

$$\text{glycerol-3-P} + R_s\text{COSCoA} \xrightarrow{\text{Acyl-CoA:G-3-P acyltransferase}} \text{1-acyl-sn-glycerol-3-P} + \text{CoASH} \quad (14.2a)$$

group. The same product may also be formed by acylation of dihydroxyacetone phosphate, followed by reduction of the keto group (Eqs. 14.2b, 2c). As we shall

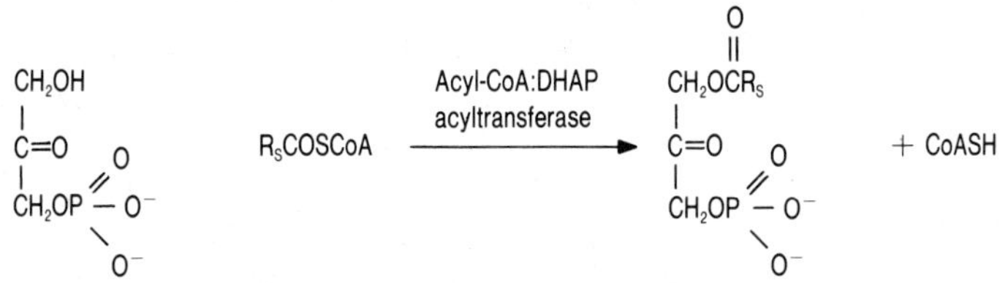

$$\text{DHAP} + R_s\text{COSCoA} \xrightarrow{\text{Acyl-CoA:DHAP acyltransferase}} \text{acyl-DHAP} + \text{CoASH} \quad (14.2b)$$

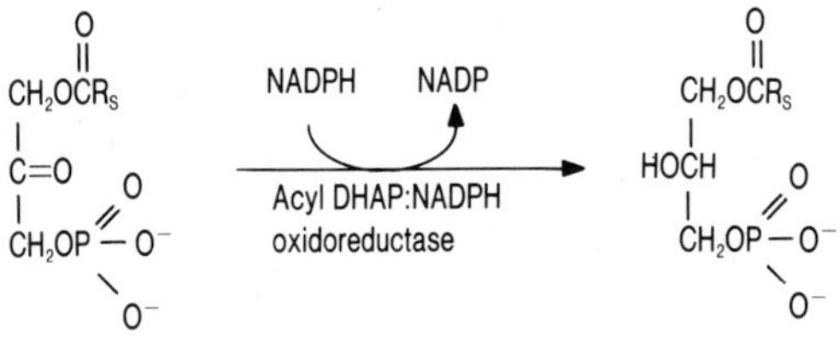

$$\text{acyl-DHAP} \xrightarrow[\text{Acyl DHAP:NADPH oxidoreductase}]{\text{NADPH} \rightarrow \text{NADP}} \text{1-acyl-sn-glycerol-3-P} \quad (14.2c)$$

see later, however, this reaction, particularly in tumor tissue, may lead to the formation of glyceryl ethers.

The second transferase, which is more or less specific for unsaturated fatty acids, transfers an acyl group to the 2-position, yielding a phosphatidic acid, which may, by action of phosphatidic acid phosphatase, be converted to an sn-1,2-diacylglycerol (Eq. 14.2d) where R_u is usually an unsaturated acyl group.

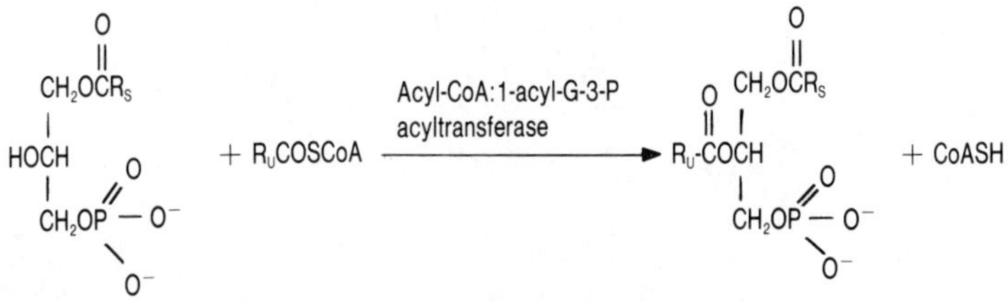

$$\text{1-acyl-sn-glycerol-3-P} + R_u\text{COSCoA} \xrightarrow{\text{Acyl-CoA:1-acyl-G-3-P acyltransferase}} \text{phosphatidic acid} + \text{CoASH} \quad (14.2d)$$

Another mechanism for formation of diacylglycerols is prominent in the intestinal epithelial cells, which are presented with large amounts of monoacylglycerols during fat absorption (see Chapter 12). This involves acylation of the 2-monoacylglycerol formed by the action of pancreatic lipase on dietary triacylglycerol; here the distribution of fatty acyl groups depends importantly on the structure of the dietary fat, since the 2-acyl group is largely retained during digestion and absorption. Although of major importance in the intestine, this pathway is probably of minor importance in other tissues, but it is of interest that it appears to be largely directed (perhaps by compartmentalization) to the production of triacylglycerols, while the phosphatidic acid path is directed toward phosphoglyceride formation. In any event, a third acyl group can now be transferred to the diacylglycerol to complete the synthesis. Most tissues appear to be capable of these reactions, but the major part of the body's triacylglycerol stores (excluding the chylomicrons) is formed in liver and adipose tissue (Dodds *et al.,* 1976).

14.2. Triacylglycerol Hydrolysis

Although lipoprotein lipase is produced in the adipocytes for release to capillary endothelial cells, continued infusion of heparin can exhaust the adipose tissues' ability to produce the enzyme; or inhibitors, such as protamine, can prevent its action. Even under these conditions, however, lipolytic activity of adipose tissue remains high because of the presence of another enzyme, the hormone-sensitive triglyceride lipase. As a result of the activities of this enzyme and of those of the synthetic systems discussed above, the synthesis and hydrolysis of triacylglycerol are continuous processes in adipose tissue. Under normal conditions, however, the systems are well balanced and the triacylglycerol stores in adipose tissue remain relatively constant. Free fatty acids are produced by the hydrolytic reaction, but do not serve to indicate the extensiveness of turnover because they are reutilized in the synthetic process. However, since glycerol is not reutilized in adipose tissue it is released to the blood, where its appearance reflects the rate of hydrolysis independent of that of synthesis (see Fig. 14.1).

Deposition and mobilization of triacylglycerols in adipocytes have been visualized by a combination of electron microscopy and biochemistry (Scow and Chernick, 1970). The adipocyte of white adipose tissue of the well-nourished sub-

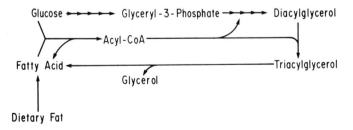

FIGURE 14.1. Schematic representation of deposition and mobilization of triacylglycerols in adipose tissue (see text for discussion).

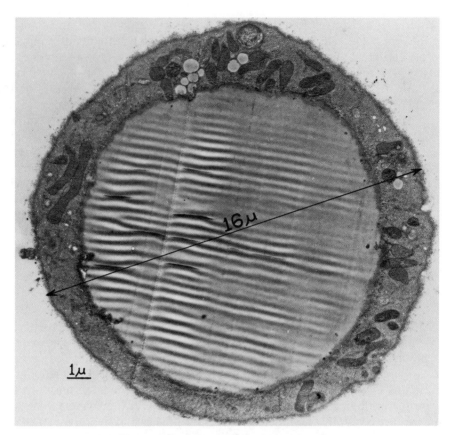

FIGURE 14.2. White adipocyte, magnification, 10,000×. Kindly furnished by Dr. M. Cohen (Schotz et al., 1969).

ject is a large cell, the volume of which is almost entirely occupied by a triacylglycerol-containing central vacuole (see Fig. 14.2). (In starved subjects, however, the adipocyte is difficult to recognize.) During conditions of fat deposition, in which chylomicrons in the adipose tissue capillaries are under hydrolytic attack by lipoprotein lipase (see Chapter 13), small, membrane-enclosed lipid particles are seen, presumably in transit through the cytoplasm for storage as triacylglycerol in the central vacuole. This process appears to be very similar to that observed during fat absorption in the intestinal cells and illustrates the point that the mechanism of fat transport across membranes consists of hydrolysis to fatty acids plus glycerol or monoacylglycerol, transport of these fragments across the membrane, and their reincorporation into glyceride (in this case, triacylglycerol) on the cytoplasmic side (Scow et al., 1972).

During conditions leading to fatty acid mobilization from adipose tissue, passage of lipid droplets across the cytoplasm is not seen, but other signs of fatty acid release such as changes in the mitochondrial membrane have been reported. Under these conditions there is an increase in lipolytic action in the adipocyte, release of fatty acid and glycerol, and decrease in the triacylglycerol content of the cell.

The importance of these reactions requires that they be under the control of several regulatory systems and this is indeed the case (Polheim et al., 1973). Dietary control seems to be largely, but not entirely, concerned with the availability of glucose. Fasting decreases esterification in large part by lowering the concentration of glucose, the source of glycerol 3-phosphate. Thus, fatty acids formed in the hydrolytic reaction are not readily reesterified and are consequently released to the blood as FFA (fatty acid–albumin complex, cf. Chapter 13). Fasting also decreases the availability of dietary fatty acids and the precursors for synthesis and activation of fatty acids in adipose tissue and liver. Refeeding with carbohydrate in normal (non-diabetic) subjects provides a source of carbon for fatty acid synthesis, of ATP for activation and, most important, of glycerylphosphate for esterification of the fatty acids, either of exogenous or endogenous origin, thus promoting storage and preventing release of the major portion. A high-fat meal provides fatty acids which, however, are not readily esterified and stored in adipose tissue without glucose and thus tend to remain in circulation until used in the liver and other tissues for energy.

In addition to the influences of diet, there is evidently some nervous control of adipose tissue metabolism. In studies of this type of control, Clément (1947) used the paired interscapular depots of the rat. The sympathetic nervous connections to one depot were severed, while those of the other were left intact. Under conditions promoting mobilization, such as administration of epinephrine, only the intact depot lost weight. Similarly, with a high-calorie diet, only the intact depot gained weight. Clearly, an intact nervous connection is necessary for both deposition and mobilization. Although the mechanism of this effect is uncertain, one attractive explanation is that stimulation of the sympathetic nervous system of adipose tissue acts largely on the vascular system to enhance blood flow and thus facilitate either deposition or mobilization (Mayerle and Havel, 1969).

Hormonal control of adipose tissue function appears to be very important, and at least three distinct types of such control are recognized. Two hormones inhibit mobilization for quite different reasons. Insulin promotes entry of glucose into the adipocytes and thus ensures maintenance of adequate supplies of glyceryl phosphate required for fatty acid deposition. In the uncontrolled diabetic, hydrolysis therefore exceeds synthesis of triacylglycerol, and fatty acid is released from the depots; normal balance of these processes is restored by insulin injection. Certain prostaglandins (PGE_2) also inhibit fatty acid mobilization. This action appears to be antagonistic to that of epinephrine and to involve, in a sense, a feedback inhibition of mobilization, since certain of the mobilized fatty acids are converted to prostaglandins, which inhibit mobilization. Insulin also appears to have this type of inhibitory action.

Fat mobilization is stimulated by the so-called "fast-acting hormones": epinephrine, norepinephrine, ACTH, TSH (thyroid-stimulating hormone), glucagon, some hypophyseal hormones, and a peptide secreted by cancer cells. The action of these hormones is rapid, of short duration, and involves potentiation of adenylate cyclase, the enzyme that converts ATP to cyclic-AMP (adenosine-3′,5′-

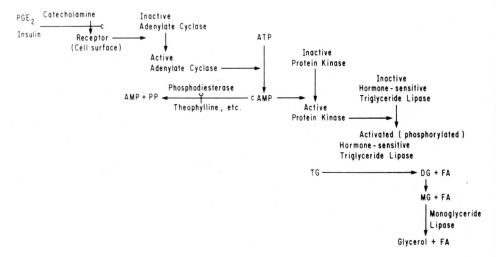

FIGURE 14.3. Hormonal effects on triacylglycerol hydrolysis in adipose tissue (Steinberg, 1972). Abbreviations used: TG = triacylglycerol, DG = diacylglycerol, MG = monoacylglycerol, FA = fatty acid.

monophosphate). Cyclic-AMP activates a protein kinase which phosphorylates and thus activates the hormone-sensitive lipase of adipose tissue, which catalyzes the rate-limiting step of hydrolysis of triacylglycerol to diacylglycerol, then to monoacylglycerol; the action of a monoglyceride lipase then completes the hydrolysis to fatty acids and glycerol. Potentiation of adenylate cyclase by prostaglandins E_1 or E_2 appears to be antagonistic to the effects of the catecholamines. An outline of these reactions is shown in Fig. 14.3.

The "slow-acting" fat-mobilizing hormones, growth hormone and the glucocorticoids, have a lag period of more than an hour, but their effects are of longer duration. Such action is blocked both by inhibitors of DNA-dependent RNA synthesis (e.g., actinomycin D) and by inhibitors of protein synthesis (e.g., puromycin). It is possible that they are involved in the synthesis of adenylate cyclase or another protein involved in cyclic-AMP formation.

14.3. Interrelationship between Adipose Tissue and Liver

Elsewhere than in adipose tissue itself, the main effects of the regulatory mechanisms discussed above are seen in the FFA content of blood. Of course, given the extremely rapid turnover of these substances, very small changes in concentration may result in the transfer of quite large quantities of fatty acid. For this reason, events occurring in adipose tissue have profound consequences in the liver.

With a normal carbohydrate-containing diet, glucose is the main direct energy source of many organs, fatty acid release from adipose tissue is suppressed, and, if caloric intake is not limited, fatty acid and triacylglycerol syntheses are high, with both acetyl-CoA and NADPH in abundant supply (see Chapter 8). With adequate flux of oxaloacetate, fatty acid oxidation in the liver tends to be complete, yielding maximum energy, and ketone body production is not excessive. If caloric intake is high, fat is synthesized and deposited in adipose tissue and, under some circumstances, in the liver. With a high-fat, low-carbohydrate diet or in uncontrolled diabetes, concentrations of FFA from chylomicrons (or in uncontrolled diabetes, from adipose tissue) are high; fatty acids become the main fuel of most organs, since the control of fat oxidation appears to reside partially in the fatty acid/albumin ratio; and fatty acid oxidation in liver is shifted to the low-energy-yield pathway involving increased production of ketone bodies. A similar though more extreme situation exists during fasting. Fat synthesis is inhibited by increased acyl-CoA in the liver, decreased citrate, decreased synthesis of fatty acid synthetase, and low glyceryl-3-phosphate in adipose tissue. Mobilization of fatty acids from adipose tissue is increased and fatty acid oxidation, particularly to ketone bodies, becomes the main energy-producing reaction of the liver. The resulting ketosis and acidosis may be severe, but there is some compensation. Under most circumstances, the liver is subordinated to the needs of certain vital organs, particularly the brain and heart. In this case, when carbohydrate, which is the usual major energy source of the brain, is scarce, the brain can apparently metabolize acetoacetate and hydroxybutyrate as its main fuel. Thus, under these conditions, the liver produces substances for the maintenance of other organs, particularly the brain and heart. A schematic representation of these relationships is seen in Fig. 14.4.

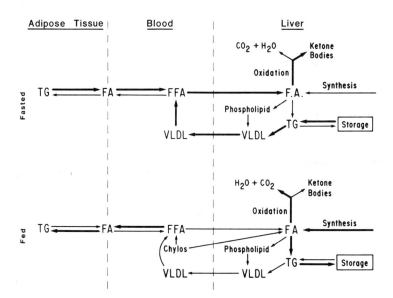

FIGURE 14.4. Schematic representation of some relationships in lipid metabolism of liver and adipose tissue. Abbreviations used: same as in Fig. 14.3, FFA = albumin-bound free fatty acids, VLDL = very low density lipoprotein.

The reactions discussed above apply largely to white adipose tissue, which comprises the large bulk of adipose tissue in most animals. Its main function is to furnish a depot for ready deposit and withdrawal of energy in the form of fatty acids. These reactions also exert a pronounced effect on lipid metabolism in other organs and thus serve an important regulatory function.

A different type of adipose tissue which, though showing some of the same reactions as the white adipose tissue, serves a quite different purpose, is brown fat. The typical cells of brown adipose tissue are quite different from the white adipocytes (Fig. 14.5). Instead of the large central, lipid-storage vacuole, there are many smaller vacuoles, each in very close contact with numerous mitochondria (hence the brown color). The prime function of this tissue appears to be thermogenesis. Following stimulation by the sympathetic nervous system or by norepinephrine, fatty acids are released by the same mechanism as in white adipose tissue. However, instead of being released to the blood for transport to other organs, they are oxidized intracellularly by the associated mitochondria. Thus the released fatty acids have two effects—furnishing substrate for energy production and uncoupling oxidative phosphorylation in the mitochondria, leading to a high rate of respiration. The uncoupling phenomenon has been shown to be a direct result of increased permeability of the inner mitochondrial membranes of the brown adipocytes to protons, and thus to disruption of the "proton pump" (Himms-Hagen, 1972; Lindberg, 1970).

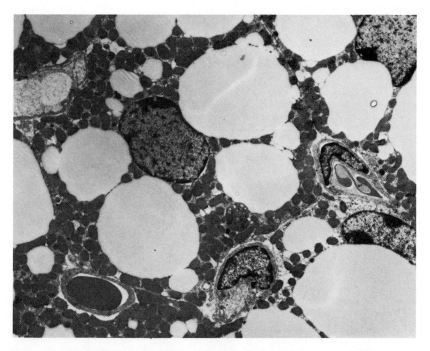

FIGURE 14.5. Brown adipocyte, magnification, 30,000×, kindly furnished by Dr. M. Nechad, Laboratoire de Physiologie Comparée, Université P. et M. Curie, Paris. (Reproduced by Dr. J. Berliner, Laboratory of Biomedical and Environmental Sciences, University of California, Los Angeles.)

This action is of great importance to mammals for two reasons. First, nonshivering thermogenesis, involving increased heat production by brown adipose tissue, is stimulated by adaptation to cold. Adapted animals are thus able to exist more comfortably at low temperatures. Second, the process is also stimulated in most animals by dietary caloric excess, resulting in conversion of the excess calories to heat rather than to fat. Animals in which this response is deficient thus tend to obesity and show low tolerance for cold. This type of thermogenesis is of particular importance in hibernating animals for restoration of normal body temperatures after hibernation and in newly born animals, including human babies, for early maintenance of body temperature.

REFERENCES

Clément, G., 1947, Direct demonstration of the mobilizing action of hypophysial hormones on fat reserves; role of the sympathetic nervous system, *C. R. Soc. Biol.* **141**:317.

Dodds, P. F., Burr, M. I., and Brindley, D. N., 1976, The glycerol phosphate dihydroxyacetone phosphate and monoacylglycerol pathways of glycerolipid synthesis in rat adipose tissue homogenates, *Biochem. J.* **160**:693.

Himms-Hagen, J., 1972, Lipid metabolism during cold-exposure and during cold-acclimation, *Lipids* **7**:310.

Lindberg, O. (ed.), 1970, *Brown Adipose Tissue,* Elsevier, New York.

Mayerle, J. A., and Havel, R. J., 1969, Nutritional effects on blood flow in adipose tissue of unanesthetized rats, *Am. J. Physiol.* **217**:1994.

Polheim, D., David, J. S. K., Schultz, F. M., Wylie, M. G., and Johnston, J. M., 1973, Regulation of triglyceride biosynthesis in adipose tissue and intestinal tissue, *J. Lipid Res.* **14**:415.

Schotz, M. C., Stewart, J. E., Garfinkel, A. S., Whelan, C. F., Baker, N., Cohen, M., Hensley, T. J., and Jacobson, M., 1969, Isolated fat cells: morphology and possible role of released lipoprotein lipase in deposition of lipoprotein fatty acid, in: *Drugs Affecting Lipid Metabolism* (W. L. Holmes, L. A. Carlson, and R. Paoletti, eds.), Plenum Press, New York, pp. 161–183.

Scow, R. O., and Chernick, S. S., 1970, Mobilization, transport and utilization of free fatty acids, in: *Comprehensive Biochemistry* (M. Florkin and E. H. Stotz, eds.), Elsevier, Amsterdam, pp. 19–49.

Scow, R. O., Hamosh, M., Blanchette-Levy, E. J., and Evans, A. J., 1972, Uptake of blood triglycerides by various tissues, *Lipids* **7**:497.

Steinberg, D., 1972, Hormonal control of lipolysis in adipose tissue, in: *Advances in Experimental Medicine and Biology,* Volume 26 (W. L. Holmes, R. Paoletti, and D. Kritchevsky, eds.), Plenum Press, New York, pp. 77–88.

Weiss, S. G., Kennedy, E. P., and Kiyasu, J. Y., 1960, The enzymatic synthesis of triglycerides, *J. Biol. Chem.* **235**:40.

15
BIOSYNTHESIS OF CHOLESTEROL AND RELATED SUBSTANCES

15.1. Discovery of Cholesterol

Cholesterol was discovered around the middle of the eighteenth century. De Fourcroy wrote in 1789 that more than twenty years before him "the late" Poulletier de la Salle obtained from the alcohol-soluble portion of bile stones ". . . une substance feuilletée, lamelleuse, brillante, assez semblable à l'acide boracique." The practice of providing references was apparently not required in the eighteenth century, and De Fourcroy failed to tell readers where Poulletier de la Salle's discovery may have been recorded; we have not been able to trace it.

De Fourcroy confirmed the findings of Poulletier de la Salle and thought that the white crystalline component of bile stones was identical with spermaceti ("blanc de baleine"), a wax found in the head of the sperm whale and some dolphins. Chevreul (1815) reexamined the nature of the crystalline material from bile stones and showed that it could not be identical with spermaceti, or with adipocire (a term also attributed by Chevreul to De Fourcroy and applied to a crystalline waxy material found in bodies buried in wet ground) because it was not saponifiable and because its melting point (Chevreul gave "137°"; present accepted value for cholesterol is 149°C) was much higher than the melting points of spermaceti or adipocire (all in the region of 44–68°C). It was in a publication one year later that Chevreul (1816) recognized that the alcohol-soluble, readily crystallizable component of bile stones was a unique compound and named it "cholesterine" from two Greek words: chole, meaning bile, and stereos, meaning solid. It is of some interest that by the middle of the nineteenth century cholesterin was identified also as a constituent of several normal animal tissues and in atheromatous lesions of arteries. Vogel (1843) shows on plates xi and xxii of his book on pathological histology drawings of crystals of cholesterin obtained from scrapings of aortic lesions, but he called it "Cholestearin." The name cholesterin remained in use up to the early part of the twentieth century, but—after recognition of the substance as a secondary alcohol (Berthelot, 1859; Diels and Abderhalden, 1904)—it was changed, mainly in English-speaking countries, to cholesterol.

Other sterols were discovered much later; among these the first were probably ergosterin (now called ergosterol, the main component of the unsaponifiable lipids in aerobically grown yeast), isolated from rye seeds infected by ergot (Tanret, 1889) and stigmasterin from calabar beans, *Physostigma venenosum* (Windaus and Hauth, 1906). The description of the discovery of many other sterols can be found in the remarkable book, *Steroids,* by Fieser and Fieser (1959).

The structure of cholesterol and bile acids was elucidated during the first part of the twentieth century mainly by the researches of H. Wieland and Windaus extending over a quarter of a century and finally established by crystallography (Bernal, 1932a,b) and revision of Windaus's chemical data by Rosenheim and King (1932) and by Windaus himself (Windaus, 1932; see also Fieser and Fieser, 1959).

Although sterols are widely distributed in the living world, cholesterol is unique in that it is mainly of animal origin, although it is also found in lesser amounts in green plants and fungi. Detailed description of the distribution of the large variety of sterols in all forms of life can be found in Chapter 3 of Myant's book (Myant, 1981, pp. 123–159). The diversity of plant sterols and their biochemistry has been reviewed by Nes (1977).

In vertebrates cholesterol is present in every cell in greater or lesser amounts and in blood plasma partly in the free form, i.e., unesterified, and partly esterified with various fatty acids. Some tissues (skin, liver, adrenals) and body fluids (plasma, bile, and urine) contain also its sulfate ester. In most tissues cholesterol exists mainly (90–95%) in the unesterified form; red cells and normal adult brain probably contain no esterified cholesterol at all, but in liver (35%), skin (55%), plasma (75%), and in the adrenal cortex (75%) an appreciable portion of the total cholesterol is esterified with fatty acids. Cholesterol is an essential constituent of all biomembranes and is an important constituent of the myelin sheaths in the central nervous system and peripheral nerves. In specialized organs it acts as precursor of other products, such as the bile acids in the liver and the steroid hormones in the adrenals, gonads, and placenta. In the adrenals, particularly rich in cholesterol, most of the cholesterol is in the cortex in esterified form as a precursor of steroid hormones.

It has been known for some time that all eukaryotic cells and higher animals, with the exception of insects, synthesize cholesterol or the sterols characteristic for the species. Research in the last decade revealed that sterols are absolutely essential for the growth and maintenance of life of eukaryotic cells of animal origin. Cultured animal cells, in which sterol synthesis is completely inhibited, die within a few days unless they are "rescued" by special measures, as will be discussed in subsequent sections.

Insects are the only forms of higher organisms that cannot make sterols, but sterols are essential in their diets for their normal development, just as several vitamins are needed by vertebrates for health and normal development. Insects need the sterols as essential structural components of their cell membranes and as precursors of an important insect hormone, ecdysone [$(22R)$-2β, 3β, 14α, 22,

25-pentahydroxycholest-en-6-one], which regulates the moulting of the immature forms of the insects.

Cholesterol is required also for growth by some protozoa such as the *Trichomonads,* but *Tetrahymena* can grow without a sterol in a synthetic medium and can synthesize some sterol. It seems that bacteria are the only forms of life that cannot synthesize sterols nor do they need them for growth.

15.2. Definition of Polyisoprenoid Compounds and Their Nomenclature

Although the main topic of this chapter is the biosynthesis of sterols, specifically of cholesterol, and its metabolic transformations, we need to give the discussion a broader basis because sterols are a relatively small group of a very large class of biogenetically related substances, collectively called polyisoprenoid (polyprenyl) or terpenoid compounds, of which many hundreds are known. The biogenetic relation of these substances to one another became known from the study of the biosynthesis of cholesterol: they are all derived from a common precursor, 3-methylbut-3-enyl (isopentenyl) diphosphate.

Polyisoprenoid or polyprenyl substances can be open-chain, partially cyclized, or fully cyclized compounds. They contain a basic structure of a branched-chain C_5-unit, the isoprenoid, or prenyl unit, recognizable in several forms as shown in formulas **1–5**. Methylbutadiene (**4**), isoprene, has been known for a long time (Williams, 1860) as a pyrolytic product of, e.g., rubber, from which the generic name was coined. The term "polyprenyl" substances was proposed specifically for the open-chain polyisoprenoid alcohols and acids (Popják and Cornforth, 1960b). This nomenclature provides easy recognition of the size of the molecules and the number of isoprenoid units they contain. Thus 3,3-dimethylallyl alcohol is called a monoprenol, the C_{15} *trans-trans*-farnesol is *trans-trans*-triprenol, and the bacterial undecaprenyl phosphate is the phosphate ester of a polyprenol containing 11 prenyl (isoprenoid) units and 55 carbon atoms.

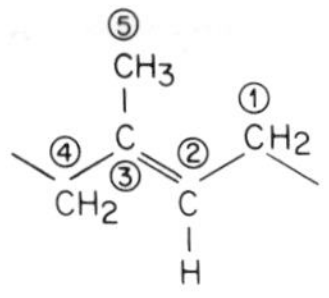

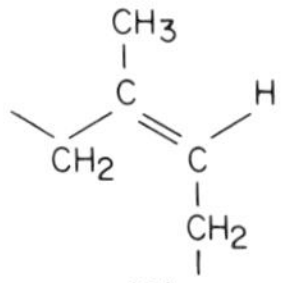

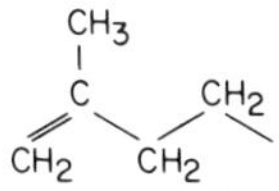

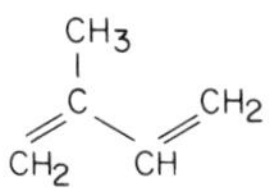

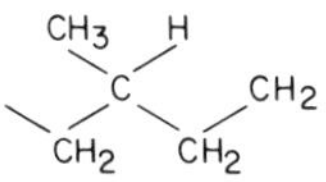

By another terminology the polyisoprenoid substances are also called terpenes or terpenoids, the names being derived from the oil of turpentine, which contains several isoprenoid compounds. By far the largest number of polyprenyl substances are products of plants. Many herbs used in cooking owe their flavor to terpenoid compounds. Fragrant oils, extracted from leaves, flowers, fruits, seeds, roots, and rhizomes of many plants, rich in terpenoids, have been known since ancient times. The Egyptians probably used them in embalming. Many were described in sixteenth century pharmacopoeias as ingredients of medicaments (cf., e.g., Cordus, 1598). In more modern times the fragrant "essential oils" became basic materials for the perfume industry. Bacteria, fungi, insects, and higher animals produce also a variety of these substances.

The first representatives discovered in this class, whose structures were determined towards the end of the nineteenth century, contained 10 carbon atoms and became known as monoterpenes. With the discovery of related substances with 15, 20, 25, 30, and 40 carbon atoms and more, the names were expanded to sesquiterpenes (C_{15}; from the Latin *sesqui*, meaning one and one-half more), diterpenes (C_{20}; tetraprenyls), sesterterpenes (from the Latin, *sestertius*, meaning two and one-half), triterpenes (C_{30}), and tetraterpenes (C_{40}). Thus polyisoprenoid, polyprenyl substances, and terpenes or terpenoids refer to the same broad class of compounds.

It was probably Wallach (1887) who first realized that many compounds of diverse structure could arise by the condensations of isoprene. He described, e.g., the formation of pinene and limonene by the "overheating" of isoprene and even gave some hypothetical structures of some cyclic monoterpenes and sesquiterpenes. Thus Wallach may be credited for the first formulation of the "isoprene rule," which was expanded much later by Ruzicka (1953) and Eschenmoser *et al.* (1955) into the "biogenetic isoprene rule" and which predicted the origins of the cyclic terpenes from open-chain polyprenyl precursors and, more specifically, the origin of the tetra- and pentacyclic triterpenes and sterols, with all their stereochemical features, from squalene.

The simplest representatives of prenols are the two isomeric alcohols 3-methylbut-2-en-1-ol (dimethylallyl alcohol), **6**, and 3-methylbut-3-en-1-ol (isopentenol), **7**. The diphosphate (pyrophosphate) esters of these two prenols are the intermediates in the biosynthesis of all polyprenyl substances.

Geraniol (**8**) and nerol (**9**) are examples of diprenols, or acyclic monoterpene alcohols, and differ from one another in the geometry of the α,β-double bond. They, or their diphosphate esters, are the precursors of many cyclic monoterpenes. The tertiary analogue of diprenols is linalool which occurs in two enantiomeric forms, *R* and *S*. The levorotatory (*R*)-linalool (**10**) is also called licareol, and the dextrorotatory (*S*)-linalool (**11**) is known as coriandrol. The latter has served as starting material for the first chemical synthesis of the natural (*R*)-enantiomer of mevalonolactone (R. H. Cornforth *et al.*, 1962) which—as will be discussed—is the precursor of isopentenyl diphosphate and hence of all polyprenyl or terpenoid compounds.

Trans-trans-farnesol (**12**) is the typical triprenol, or sesquiterpene alcohol; its diphosphate ester is a precursor of longer-chain polyprenols and of squalene. The tertiary alcohol analogue of farnesol, also a plant product, called nerolidol (**13**), exists in two optically active forms of either the R or S absolute configuration, in analogy with linalool. The diphosphate ester of the all-*trans*-tetraprenol (diterpene alcohol), commonly known as geranylgeraniol (**14**, $n = 3$), is the precursor of cyclic diterpenes, such as the giberellins, and of the carotenoids (tetraterpenes) such as phytoene, β-carotene, etc. Two other polyprenols with all-*trans* configuration are also known: solanesol, from green leaves of *Solanacea*, is a nonaprenol (**14**; $n = 8$) and spadicol, from the spadix of *Arum maculatum*, is a decaprenol (**14**; $n = 9$).

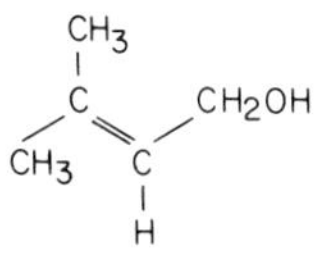

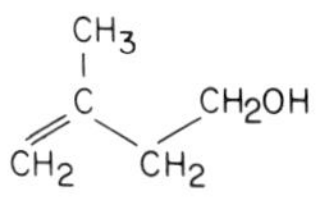

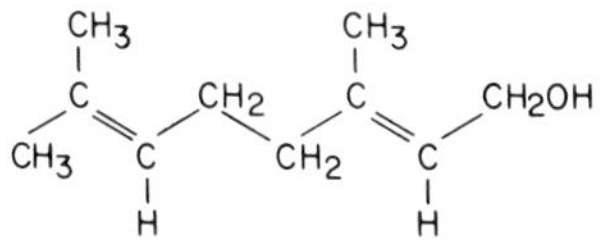

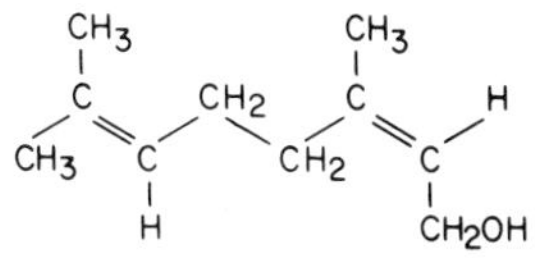

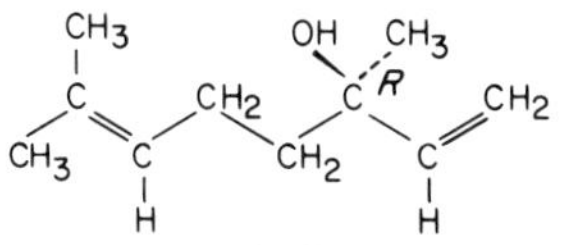

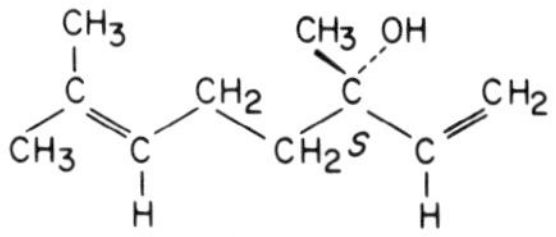

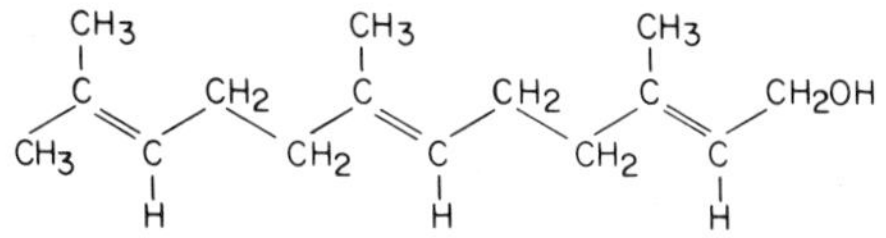

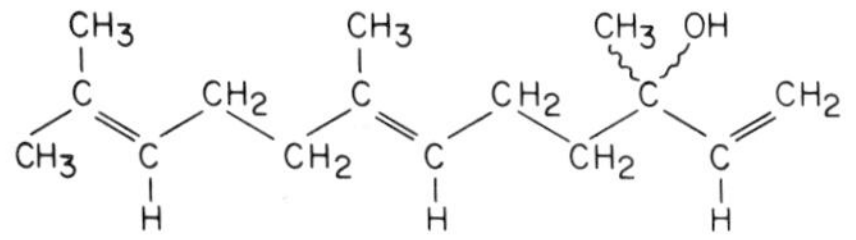

The extreme examples of polyprenols are the dolichols (the name derived from the Greek dolichos, meaning long), discovered in R. A. Morton's laboratory in the early 1960s (Burgos et al., 1963). The first examples, consisting of 20 or 21 prenyl units, were found in all the major organs of man, pig, ox, and rabbit. Subsequently other representatives of polyprenols were found of varying chain-length but with the structural characteristics of the dolichols. The main structural characteristic of these long-chain polyprenols is that at the ω-end they contain two or three prenyl units with E (or *trans*) configuration, followed by varying numbers

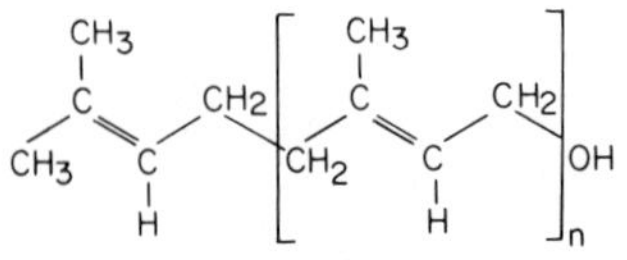

(14)

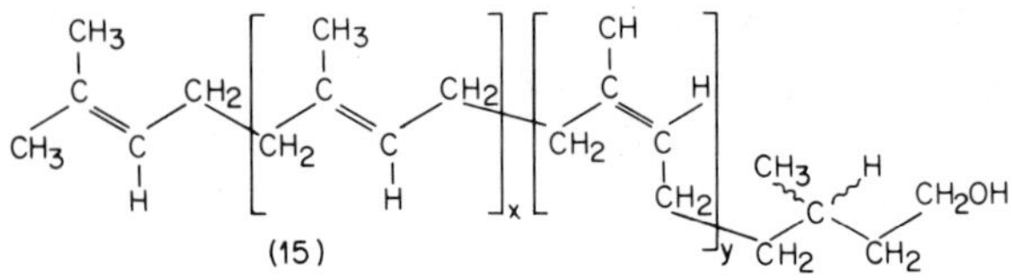

(15)

of prenyl units with the Z (or cis) configuration, and that the α-prenyl unit, bearing the primary hydroxyl group, is almost always saturated. These polyprenols are represented by the general formula 15 in which $x = 2, 3$, and $y = 4–16$. The configuration of the last unsaturated prenyl unit at the ω-end is, of course, indeterminate. C-3 of the saturated α-subunit is an asymmetric center. Burgos et al. (1963) recorded for the human kidney and pig liver dolichol an optical activity of $[\alpha]_{700} = -0.4°$ and $[\alpha]_{360} = -1.8°$; the absolute configuration at C-3, however, has yet to be determined. The phosphate or diphosphate esters of dolichols are vital cofactors in glycosylation reactions, as in the synthesis of the bacterial cell wall peptidoglycans and in glycosylations of proteins in animals. The structures, biosynthesis, and functions of dolichols have been reviewed by Hemming (1974, 1983).

The large cis component of the dolichols is, of course, analogous to the structure of the all-cis rubber of *Hevea brasiliensis* in which the rubber particles consist of long chains of many thousands of prenyl units. Another form of rubber, guttapercha, the inspissated juice of some Malayan trees, e.g., of *Isonandra Gutta* belonging to the order of *Euphorbiaceae,* does not possess the elasticity of hevearubber and is an all-*trans*-polyprenyl substance.

In the examples of polyprenyl substances so far given the prenyl units are clearly recognizable and are joined in a head-to-tail manner, i.e., C-4 of one prenyl unit is linked to C-1 of another (for the numbering of positions in prenyl units see formula 1). However, in many polyprenyl substances tail-to-tail condensations are common, though the reactions by which such linkages are formed are complex, as will be discussed in a later section. The simplest example of such condensation is that found in squalene (16) in which two farnesyl residues are linked tail-to-tail and form a symmetrical molecule. Squalene is commonly referred to as a triterpene, though it might be more correctly called bis-sesquiterpene. Squalene, originally discovered as the main constitutent of shark-liver oil (Heilbron et al., 1926), has a special place in biochemistry, for it is a precursor of animal and plant sterols and of all the tetracyclic and pentacyclic triterpenes.

The carotenoids, products of all photosynthetic organisms, are a very large group among the terpenoids, containing over 400 recognized structures. The simplest representative among these is phytoene (17). Its structure resembles that of squalene except that in its center it contains a conjugated double bond system

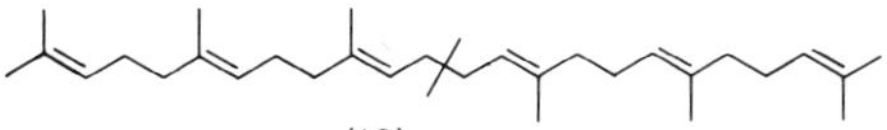

BIOSYNTHESIS OF CHOLESTEROL

(16)

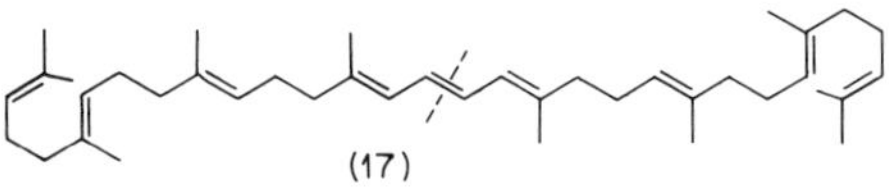

(17)

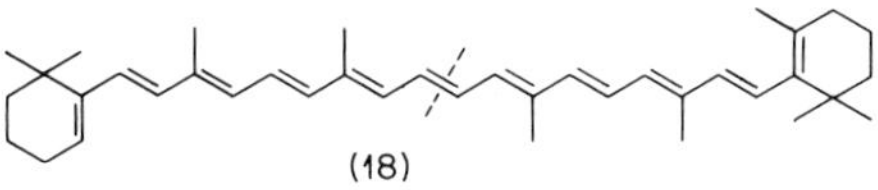

(18)

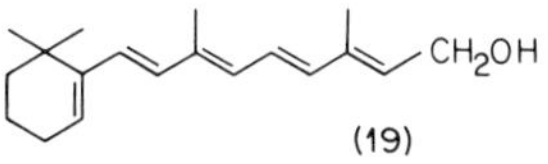

(19)

and that it is formed by the tail-to-tail condensation of two all-*trans*-tetraprenyl (geranylgeranyl) diphosphates. The configuration around the central double bond in 17 can be *trans* or *cis*. There is a whole array of acyclic carotenoids differing from one another by successive introductions of conjugated *trans* double bonds from left to right.

Many carotenoids are cyclized at either end or at both ends of the molecule. In these bicyclic carotenoids the two alicyclic rings are usually linked by an open-chain, fully conjugated system. From among the many known compounds of this kind we chose β-carotene (18) as an example, because of its special biological importance as a precursor of vitamin A (retinol) (19) in all animals needing this vitamin.

Polyprenyls are found not only as individual substances but also as substituents on a variety of natural products. Notable examples among these are chlorophyll, vitamins K_1 and K_2, α-tocopherol (vitamin E), and coenzyme Q (or ubiquinone). All these substances carry a polyprenyl substituent. In chlorophyll the acetic acid side-chain on one of the pyrrole rings is esterified with the partly reduced tetraprenol, phytol (20; hexahydrotetraprenol). The 2-methyl-1,4-naphthoquinone nucleus (21) of vitamins K_1 and K_2 is substituted at position 3 with the phytyl (cf. 20) and the all-*trans* hexaprenyl residues, respectively. α-Tocopherol (22) carries the fully reduced hexahydrotriprenyl residue as substituent at position 2, and coenzyme Q, a component of the respiratory chain in many organisms, has at position 6 an all-*trans* polyprenyl side-chain, varying in

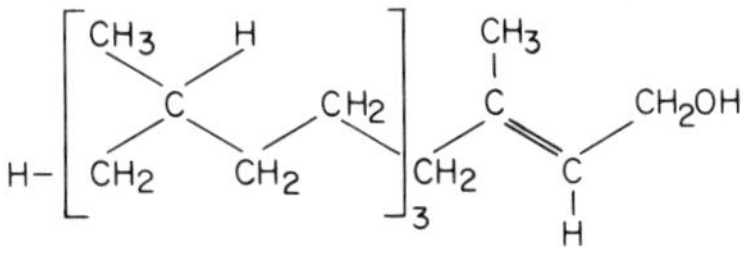

(20)

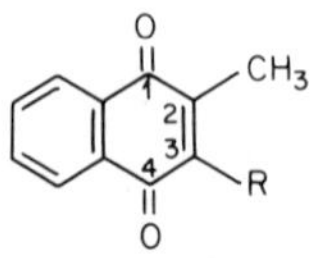

length from 6 to 10 prenyl units condensed in a head-to-tail manner (**23**; $n = 6–10$).

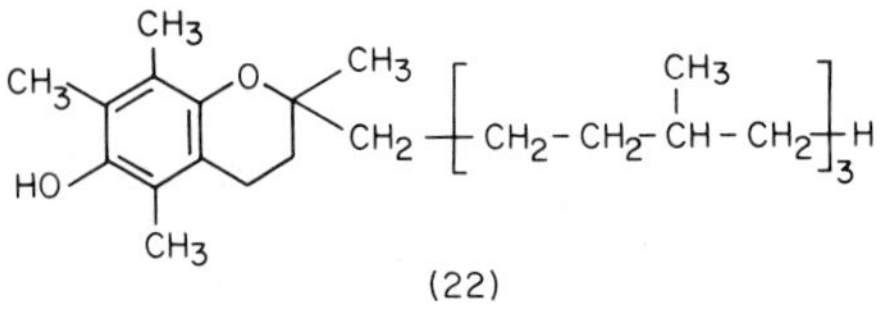

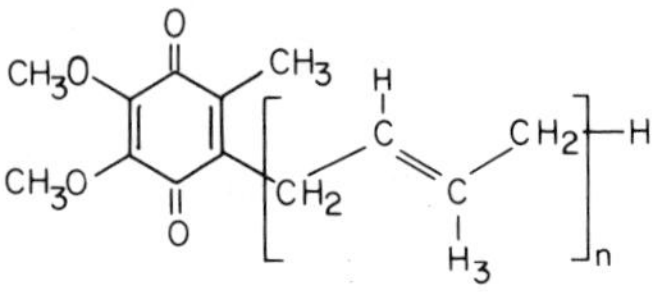

A special class of lipids found in the membranes of *Archaebacteria (Methanogens, Halophiles, Thermoacidophiles)* contains saturated polyprenyl, phytanyl, and diphytanyl residues in ether linkages with glycerol. These lipids are regarded as taxonomic markers for *Archaebacteria* (Woese *et al.*, 1978). In contrast to lipids of eubacteria and higher organisms in which the alkoxy, alkyl, or alkenyl residues are linked to *sn*-1,2 positions of glycerol, in lipids of the *Archaebacteria* the phytanyl residues are at *sn*-2,3 positions of glycerol (Kates, 1972; Kates and Kushwaha, 1978; Bu'Lock *et al.*, 1983). As an example, we show the structure of the *sn*-2,3-diphytanylglycerol ether in formula **24**.

In all the examples of polyprenyls shown so far the prenyl units are all linked head-to-tail, as in all the polyprenols, or tail-to-tail as in the center of squalene and the carotenoids. In the lipids of *Halobacteria* (cf. Bu'Lock *et al.*, 1983) and in the recently discovered *Methanococcus jannaschii* (Comita and Gagosian, 1983), diphytanyl residues were found in which two phytanyl residues are joined head-to-head, as in the bis-(diphytanyl)diglycerol tetraether (**25**) and in the, so far unique, macrocyclic diphytanylglycerol diether (**26**). In the latter two formulas the head-to-head joining of two phytanyl residues is marked with black dots.

In the C_{20} phytanic acid (saturated diterpene acid) and in the corresponding alcohol, phytanol, positions 3,7, and 11 are asymmetric centers. The absolute configuration at all three centers has been deduced to be *R* (cf. **24**) from correlations with the optical rotations of monomethyl branched-chain carboxylic acids of known absolute configuration (Abrahamsson *et al.*, 1964; Kates, 1972).

The few examples of mostly acyclic polyprenyls given here represent but a tiny fraction of the simplest forms of the vast number of natural products in this class. Those seeking more information can find it in the monograph by Newman (1972); in the Specialist Periodical Reports, *Terpenoids and Steroids* and *Biosynthesis*, published by The Chemical Society, London (see Overton, 1971; Geissman, 1972, and succeeding volumes in the series); and in the two volumes, *Biosynthesis of Isoprenoid Compounds* (Porter and Spurgeon, 1981, 1983); or in an earlier monograph by Waller (1969).

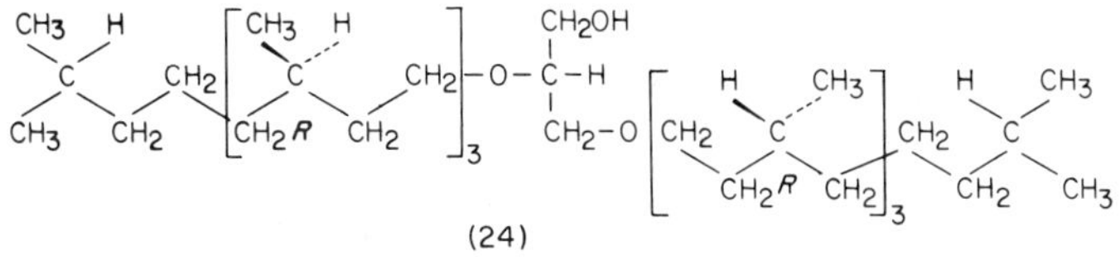

(24)

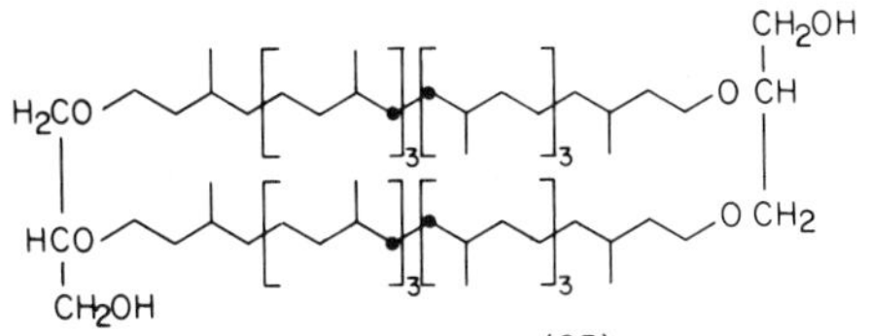

(25)

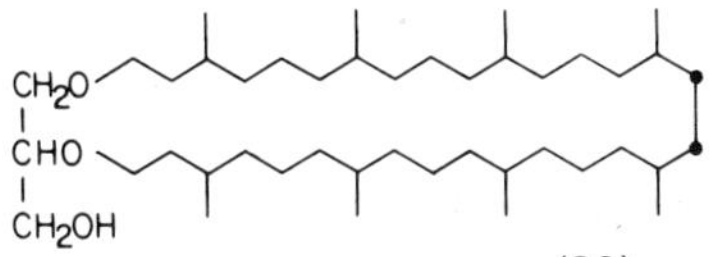

(26)

15.3. Sterols: Nomenclature and Stereochemistry

Among the many cyclized natural products derived from squalene, sterols are the most widely distributed in nature, bacteria being the only forms of life not possessing them. In vertebrates, and especially in Mammalia, the metabolism of the principal animal sterol, cholesterol **(27)**, is intimately linked with that of lipids; it is an integral part of cell membranes, and it is the precursor of steroid hormones and bile acids. Detailed accounts of the chemistry and physical properties of sterols, steroid hormones, and bile acids can be found in the books by Fieser and Fieser (1959), Klyne (1957), Cook (1958a,b), Kritchevsky (1958) and Myant (1981), to name only a few works from among the vast number of accounts on the subject.

We intend to describe here the elements of the nomenclature, structure, and stereochemistry of the sterols commonly encountered, whose biosynthesis will be discussed.

Steroids are compounds having the skeleton of cyclopentano(a)per-

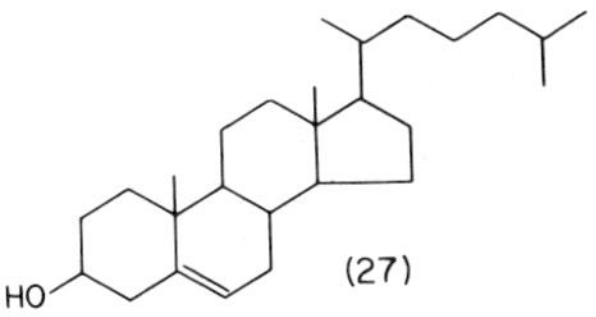

(27)
Cholesterol

hydrophenanthrene nucleus with methyl groups at C-10 and C-13, and an alkyl side-chain or an oxygen function at C-17. Sterols are steroids that carry a hydroxyl group at C-3 and a side-chain at C-17 and retain all or more of the carbon atoms of squalene, but may contain additional carbon atoms on the side-chain, as in many plant sterols and sterols found in yeast and fungi. Steranes represent the fully reduced hydrocarbon skeleton of sterols. The systemic names of the sterols are derived from the terms assigned to a number of such fundamental hydrocarbon units, such as cholestane, ergostane, campestane, poriferastane, stigmastane, lanostane, and cycloartane. The structural meaning of these terms will be explained a little further on. Unsaturation in the structures is indicated by changing the terminal "-ane" to "-ene," "-adiene," or by changing a terminal "-an" to "-en," "-adien," the position of unsaturation being indicated by the number(s), written between hyphens immediately preceding the "-ene" or "-en" ending: for example, cholest-5-ene, cholesta-5,7-diene (not 5-cholestene, or 5,7-cholestadiene). The configuration around a double bond in the side-chain is indicated by the E,Z convention written in parentheses before the name of the compound. The use of the Δ sign (Greek capital delta) is not recommended to designate unsaturation in individual compounds, but may be used when one is referring to a group of sterols of similar unsaturation, e.g., Δ^5-sterols.

To represent the complex structure of sterols, within the two-dimensional limitations of the paper, some arbitrary conventions are required and some compromises must be made particularly with respect to the actual shape of the molecules. Most natural sterols can be related to the C_{27} hydrocarbon cholestane. Its carbon atoms are numbered and its four rings designated by letters as shown in **28**. Thus the two methyl groups, numbered 18 and 19, are attached to carbon atoms 13 and 10, and the branched octyl side-chain to carbon atom 17 of the cyclopentanoperhydrophenanthrene nucleus. In the representation of cholestane by formula **28** it is understood that each straight line segment bears a carbon atom at each end and that all undesignated carbon valencies bear hydrogen atoms. It may be readily seen that cholestane has eight asymmetric centers, at positions 5, 8, 9, 10, 13, 14, 17, and 20, and thus, theoretically, it could exist in $2^8 = 256$ different configurations, but only a few of these exist among natural products. Plant sterols, which have an adventitious substituent, a methyl or ethyl group, not derived from squalene, at C-24 contain an additional asymmetric center. Although all these asymmetric centers could be designated by the R,S convention of Cahn, Ingold, and Prelog (cf. Cahn *et al.,* 1966), two systems have been adopted by IUB–IUPAC Commission on Biochemical Nomenclature (Revised Rules on the Nomenclature of Steroids, to be published). One system applies to substituents on the ring structure and the other to substituents on the side-chain. When the structure of the sterane is projected on the plane of the paper in the orientation shown in **28,** an atom or group attached to a ring is called α if it lies below the plane of the paper, and β if it lies above it pointing towards the observer. Bonds to atoms or groups of α-orientation are shown as broken or dotted lines, and those to atoms or groups of β-orientation are shown with solid, preferably thickened (— or ◄) lines. Thus in formula **28** the methyl groups at C-13 and C-10 and the side-chain at C-17 are all β-oriented and lie on the same side of the tetracyclic skeleton. It may be inferred that the orientation of the hydrogen atom at C-17 in cholestane (28) is α. All sterols carry also a hydroxyl group at C-3 in the β-configuration; this substitution on the ring system creates, of course, another asymmetric center, the hydrogen atom there being in the α-configuration. Formula 28 does not specify the configuration about the other centers of asymmetry which bear single hydrogen atoms (positions 5, 8, 9, 14, and 20). Although this feature of the stereochemistry of steranes or sterols is often not indicated specifically, when necessary it may be shown in either of two ways. In the forthcoming revised nomenclature of steroids it is recommended that these single hydrogen atoms at asymmetric centers on the ring system be written with either a broken line (α-orientation) or with a solid line (β-orientation) as shown in 29a and 30. By an earlier convention, not recommended any more, asymmetric centers bearing a β-oriented hydrogen atom were marked with a heavy dot (29b), the absence of a dot implying α-orientation. The different orientation of the hydrogen atom at C-5 as shown in 29a and 30 determines the relations of rings A and B to one another. In 5α-cholestan-3β-ol (29a) rings A and B are in a *trans* relation and in 5β-cholestan-3β-ol (30), also called coprostanol, they are in a *cis* relation to one another (for details see further on). For this reason, for all steranes and sterols with an asymmetric center at C-5 it is mandatory to specify the 5α- or 5β-configuration, as for the two cholestanols mentioned above.

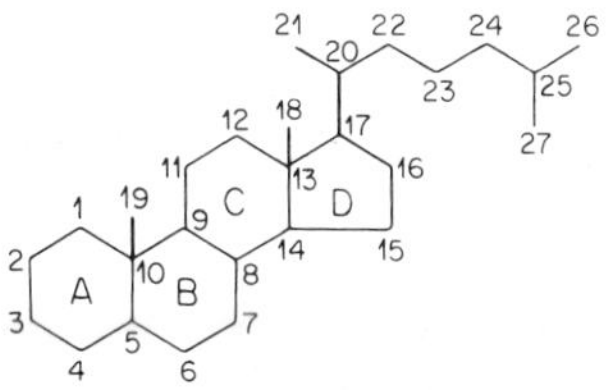

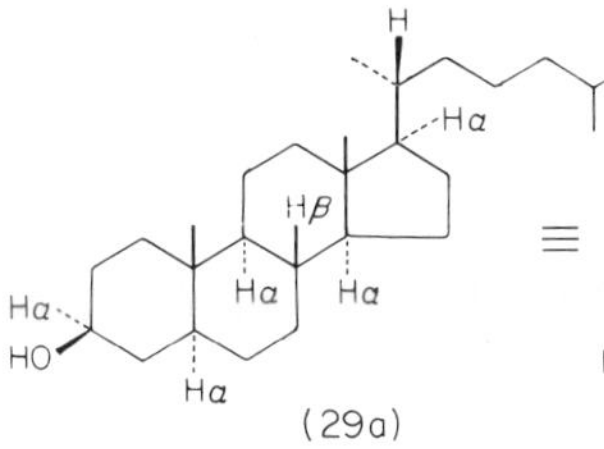

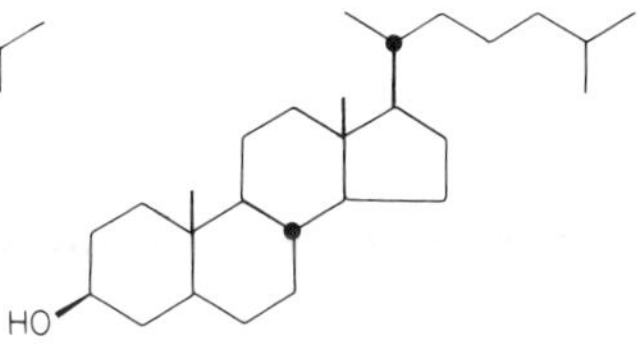

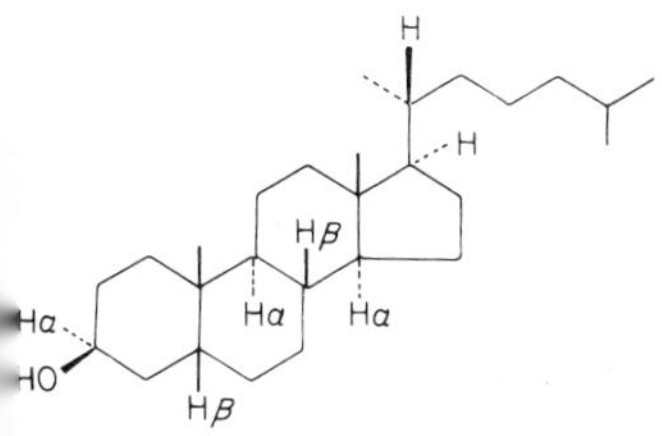

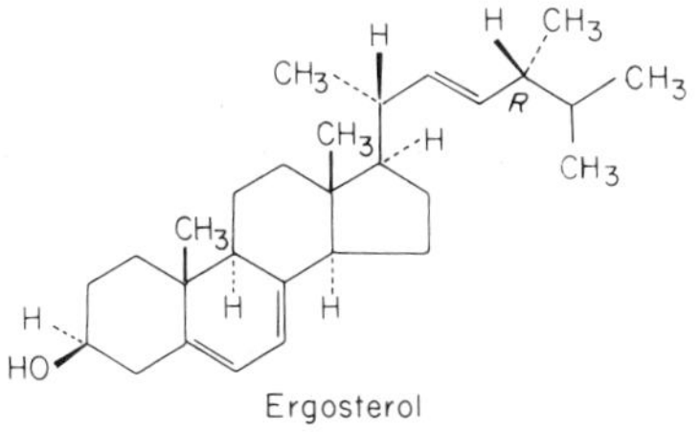

A large number of sterols from plants, yeast, fungi, and sponges bear at C-24 a methyl or ethyl group which creates an asymmetric (chiral) center at that position. The carbon atoms of these adventitious groups, not derived from squalene, are numbered 28 and 29. According to the new recommendations on steroid nomenclature, all chiral centers on the side-chain should be designated by the *R,S* convention. Although this convention is unambiguous, its application occasionally produces seemingly anomalous ascriptions of chirality. This anomaly arises from the fact that the ascription of *R* or *S* configuration depends critically on the ligands attached to the asymmetric center. This anomaly is best illustrated with the structure of ergosterol **(31)** and the partial structures of other sterols (or steranes and sterenes) substituted at C-24 with a methyl or ethyl group. In ergosterol or ergostene **(33)** the absolute configuration at C-24 is designated *R*, but in ergostane **(32),** the fully reduced hydrocarbon core of ergosterol, the absolute configuration at C-24 must be designated *S* even though the spatial orientation of the C-28 methyl group is identical in the two substances. Similar change of ascription of chirality has to be made in the case of stigmastane **(35)** and (22*E*)-stigmast-22-ene **(36).** These anomalies arise because, according to the sequence rule of the *R,S* convention, the doubly bonded ligand in the side-chain of cholestenes has a priority over the ligand at C-25. Conversely, in saturated sterol side-chains with a substituent at C-24, the ligand at C-25 has a priority over the methylene group at C-23. Comparison of the partial structures of campestane **(34)** and (22*E*)-poriferast-22-ene **(37)** further illustrates this particular problem with the *R,S* system.

(32)
Ergostane

(33)
(22 *E*)-Ergost-22-ene

(34)
Campestane

(35)
Stigmastane

(36)
(22 *E*)-Stigmast-22-ene

(37)
(22 *E*)-Poriferast-22-ene

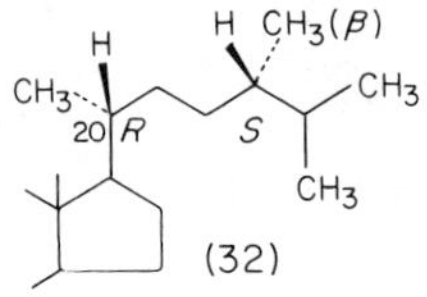

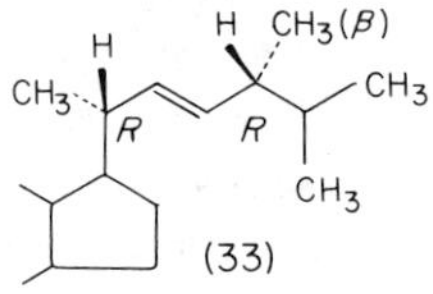

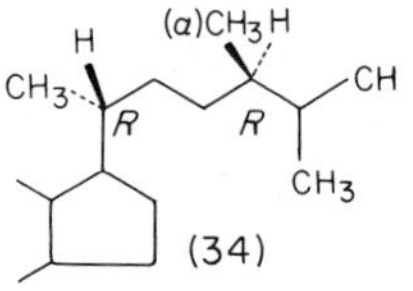

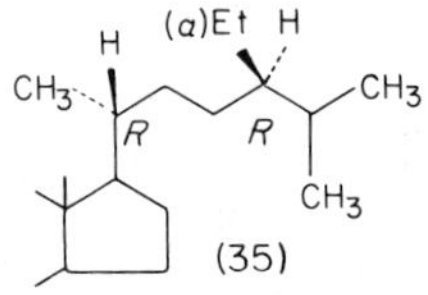

The absolute configuration at C-20 of all natural sterols is designated *R* irrespective of whether the side-chain contains a double bond at C-22, as the ring structure linked to C-20 retains its priority over C-22 even in the face of a 22(23)-double bond.

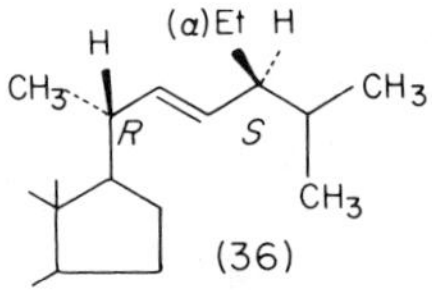

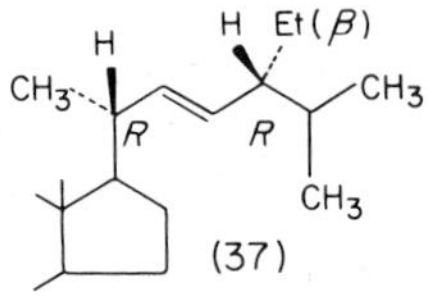

By a former convention substituents on the side-chain were designated α and β, but this α/β notation was unrelated to the similar notation for substituents on the ring structure (cf. Fieser and Fieser, 1959, pp. 337–340; or Myant, 1981, pp. 13–15). As this system is found in the older literature, we present it briefly. According to this notation, the side-chain is rotated around the C-17/C-20 bond so that the long part of the side-chain is extended upwards from C-17 and to the rear as in a Fischer projection. Substituents that are on the left of the main carbon chain are labeled β and those on the right are labeled α. This is illustrated with the partial structures of ergosterol **(38)** and campesterol **(39;** cf. **34)**. This older notation, although now in disfavor, had the advantage—unlike the *R,S* notation—that it was not affected by the presence or absence of the double bond between C-22 and C-23. Its only disadvantage was that it could have been confused with the α/β notation for the orientation of the substituents on the ring structure.

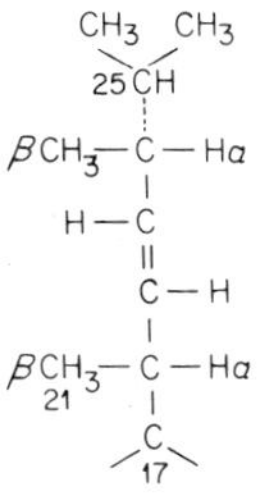

(38) 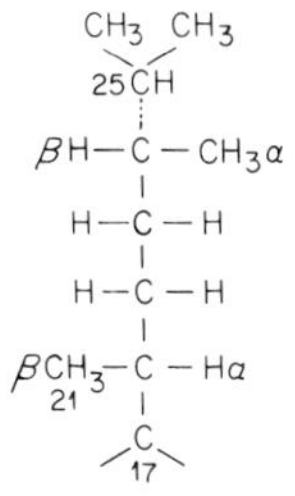 (39)

The C_{27} animal and plant sterols are modified products of two C_{30} primary sterols: the animal sterols are derived from lanosterol **(40)** and the plant sterols from cycloartenol **(41)**. Both have an α-oriented hydrogen atom at C-5, and also three methyl groups more than 5α-cholestanol: one in α-orientation at C-14 and a *gem*-dimethyl group, α and β, at C-4. In cycloartenol the C-19 methyl group, after loss of one hydrogen atom, is fused to C-9 in β-orientation to form a substituted cyclopropane ring with C-9 and C-10. Lanosterol and cycloartenol are, according to the chemical nomenclature, trimethyl sterols, almost implying that the three methyl groups had been adventitiously acquired, although in reality they are direct cyclization products of squalene 2,3-oxide. From a biochemical point of view they are closely related to the tetracyclic triterpenes, such as euphol **(42)** or tirucallol **(43)** from which they differ only in their stereochemistry. Lanosterol and cycloartenol should be classed among the tetracyclic triterpenes. Because the numbers 28 and 29 have been reserved for the carbon atoms of the methyl and

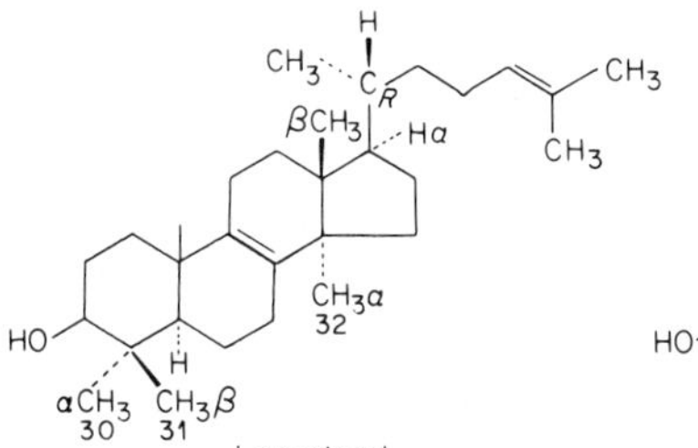

Lanosterol (40)

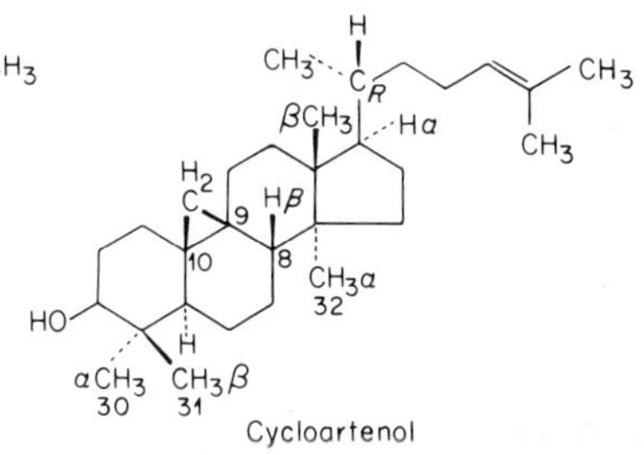

Cycloartenol (41)

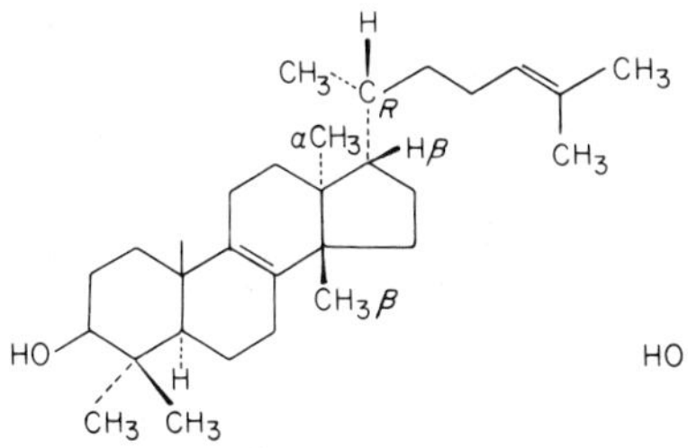

Euphol (42)

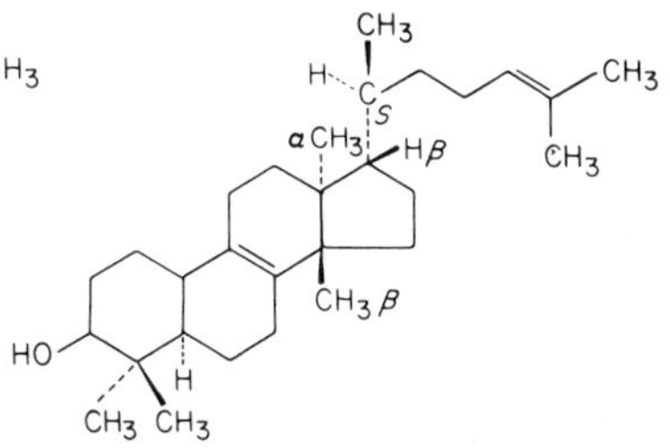

Tirucallol (43)

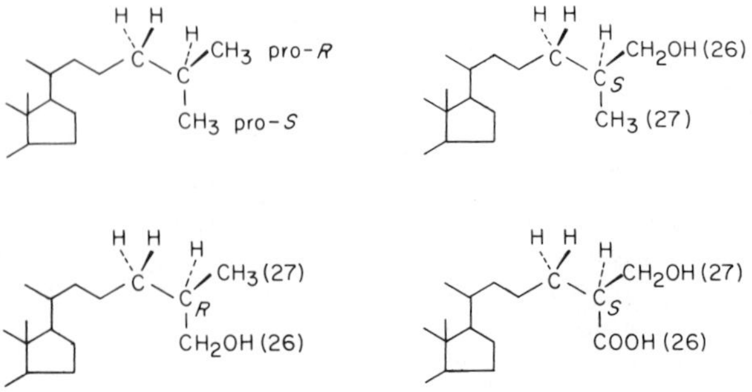

SCHEME 15.1. Effects of substituents on the pro-*R* and pro-*S* terminal methyl groups on the chirality of C-25 and the numbering of those substituted terminal carbon atoms in sterols.

ethyl substituents at C-24 of the plant sterols, the three "additional" methyl groups in lanosterol and cycloartenol are numbered 30 (4α-Me), 31 (4β-Me), and 32 (14α-Me) as shown in formulas **40** and **41**.

In sterols with a saturated side-chain, whether substituted at C-24 or not, C-25 is a prochiral center and hence substitution or oxidation of either of the terminal methyl groups, C-26 or C-27, will confer chirality on C-25. Thus, for example, hydroxylation of the pro-*R* methyl group will create at C-25 the *S*-configuration, and hydroxylation of the pro-*S* methyl group will create at C-25 the *R*-configuration (cf. Popják et al., 1977). By the rules of steroid nomenclature, if either of the terminal methyl groups becomes substituted, that substituted position will be numbered 26 and the other 27. If both become substituted, then the group to be called first in alphabetical order is to be given the lower number (Scheme 15.1).

The reduced tetracyclic nucleus of all steranes with the 5α-configuration, whose partial structures are shown in **32–37**, has the same stereochemistry as 5α-cholestane. The systemic names of the C_{27} sterols are derived from 5α-cholestane as the reference compound, in a manner analogous to that used in naming the various prostaglandins as if they were derived from the imaginary prostanoic acid (cf. Chapter 10). Thus cholesterol **(27)**, by the systemic nomenclature, is cholest-5-en-3β-ol; and desmosterol is cholesta-5,24-dien-3β-ol. The presence of the 5(6)-double bond, of course, abolishes the asymmetric center present at C-5 in 5α-cholestanol; the effect of that double bond on the shape of the molecule will be discussed a little further on.

Although a systemic name for lanosterol could be derived also by the use of cholestane as the reference compound, 5α-lanostane is the recommended fundamental unit for naming this C_{30} sterol, it being understood that the substance has the 4-*gem*-dimethyl and the methyl group at C-14 in α-orientation. Thus the systemic name for lanosterol **(40)** becomes 5α-lanosta-8(9),24-dien-3β-ol instead of the cumbersome 4α,4β,14α-trimethyl-5α-cholesta-8(9),24-dien-3β-ol. The same rules apply to the naming of plant sterols substituted at C-24 with a methyl or ethyl group. Thus the fundamental units for these are ergostane, campestane, stigmastane, and poriferastane, with the understanding that these saturated hydrocarbon structures are substituted at C-24 with a methyl or ethyl group in a specific steric orientation (see formulas **32–37**). Thus the trivial name of ergosterol (**31**) becomes (22*E*)-ergosta-5,7,22-trien-3β-ol, though it could be described also as (22*E*)-(24*R*)-24-methylcholesta-5,7,22-trien-3β-ol. Similarly the systemic name for poriferasterol is (22*E*)-poriferasta-5,22-dien-3β-ol, instead of the unwieldy (22*E*)-(24*R*)-24-ethylcholesta-5,22-dien-3β-ol. For further details concerning rules of sterol and steroid nomenclature the IUPAC–IUB recommendations should be consulted (IUPAC–IUB, 1972).

The planar representation of structures of steroids, steranes, and sterols with the built-in conventions conveys much useful information, except for the actual shape of the molecules, as the rings involved are not planar. Cyclohexane, as seen in the fused rings of cholestane, can exist in two conformations called "chair" and "boat" forms with the normal 109°28′ C–C–C bond angles throughout, instead of the strained 120° that a planar cyclohexane would demand. Models of cyclohexane constructed in the chair (**44**) and boat (**45**) conformations from Dreiding atomic models give readily the nuclear distances of substituents on the ring and account for the greater stability of the chair than the boat conformation. In the chair form the hydrogen atoms fall into two distinct classes: (i) axial (a), which are oriented parallel to the symmetry axis of the ring; and (ii) equatorial (e), which are oriented obliquely away from the ring center at an angle of about 70.5° to the symmetry axis. Thus, each carbon atom bears one axial and one equatorial substituent which alternate in their orientation on adjacent carbon atoms. Measurements show that in the chair conformation the distance between pairs of hydrogen atoms on adjoining carbon atoms, 1a to 2e and 1e to 2e is the same, 2.5 Å; similarly, all the axial hydrogen atoms lying on the same side of the ring are equidistant, all at about 2.5 Å (cf. formula **44**). The two axial hydrogen atoms on adjoining carbon atoms (one β-oriented, the other α-oriented) are 3 Å apart, too far for significant interaction.

In the boat conformation the relations of positions 1 to 2, 3 to 4, 4 to 5, and 6 to 1 are the same as in the chair form, but the 2β and 3β (or 2α and 3α) substituents and the similar pairs of substituents at positions 5 and 6 are only about 2.3 Å apart. Newman projections along the parallel C–C bonds (C-2–C-3 and C-6–C-5) of the chair and boat forms show further that in the former the substituents are staggered (all *gauche*), whereas in the latter they are eclipsed (doubly

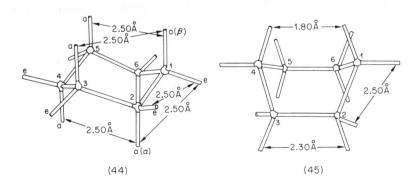

(44) (45)

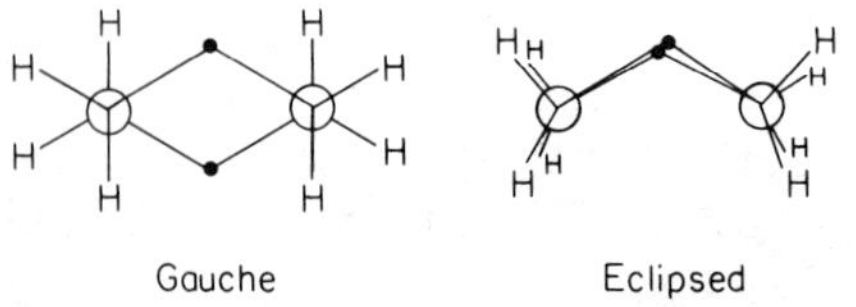

Gauche
(46)

Eclipsed
(47)

cis) and contribute significantly to the instability of the boat form (compare formulas **46** and **47**). This instability is further exaggerated by the proximity of the β-substituents at C-1 and C-4 in the boat form. In the least serious case, when these substituents are hydrogen atoms, the hydrogen nuclei are only about 1.8 Å apart (cf. formula **45**). In the event of bulkier substituents than hydrogen atoms at the bow and stern, e.g., one methyl and the other hydroxyl, the overlapping electron densities create an even more unfavorable condition for the stability of the boat form **(48)**.

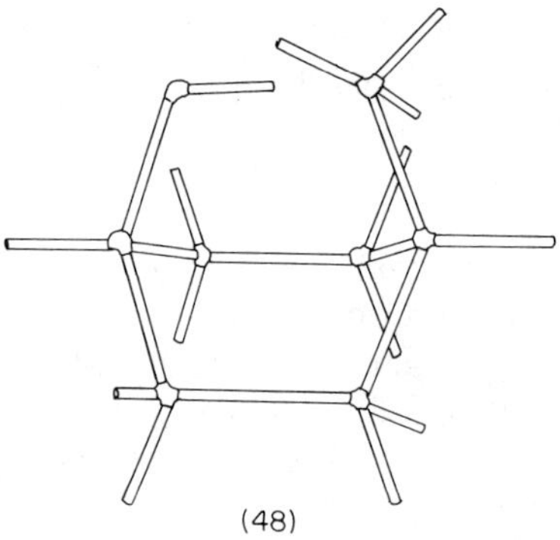

(48)

In the polycyclic sterols the cyclohexane rings are in chair conformations, but modified in rings containing double bonds. The steric relation between two cyclohexane rings sharing one common carbon-to-carbon bond can be either *trans* or *cis,* as seen in 5α-cholestanol **(49)** and 5β-cholestanol **(50)** respectively. In **49** all rings are *trans*-fused, whereas in **50** rings A and B are *cis*-fused. An important difference between the *trans*- and *cis*-junctions is that in the former the substituents at the shared carbon atoms are on opposite sides of the rings—one is α- and the other β-oriented—but both are axial with respect to each ring. At the shared carbon atoms of *cis*-fused rings the substituents are on the same side of the molecule and are axial to one and equatorial to the other ring. For example, the C-19 methyl group in 5β-cholestanol **(50)** is axial to ring B, but equatorial to ring A; the relation of the 5β-hydrogen atom to rings A and B is the reverse.

In contrast to the cyclohexane rings, the homologous five-membered D ring of steroids might be expected, from the standpoint of carbon-to-carbon bond-angle strain alone, to be fairly happy in a planar conformation, as the 108° angles of a regular pentagon depart only little from the normal tetrahedral 109°28′. How-

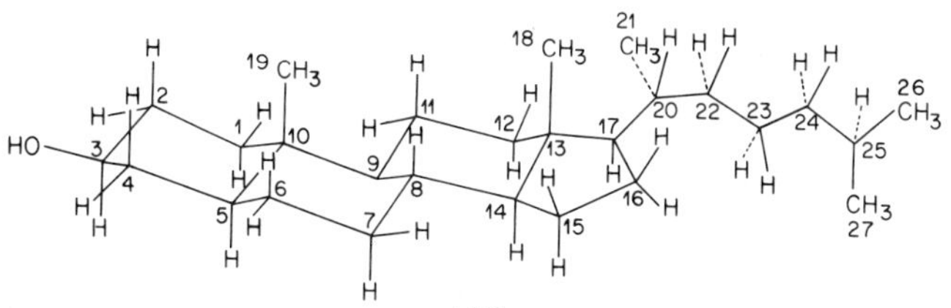

(49)

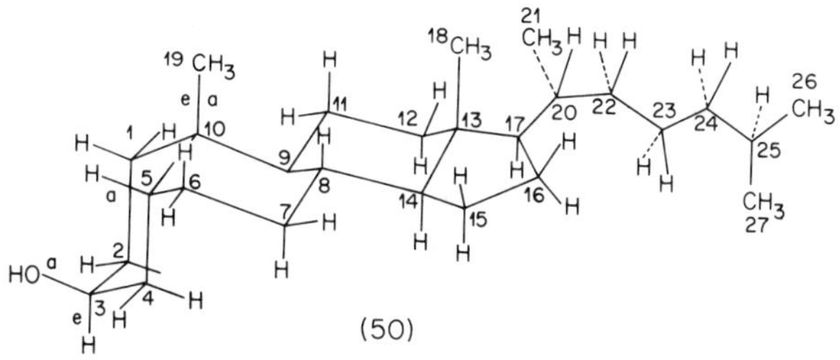

(50)

ever, as planar cyclopentane is an assembly of *cis* conformational arrays, substituents of any pair of linked carbon atoms would obviously be fully eclipsed and their crowding thus severe. Relief is provided by moving one or two of the carbon atoms out of the plane of the others. The substituted cyclopentane ring in many substances assumes either the "envelope" or "half-chair" conformation, depending on steric demands of the substituents. X-ray crystallography revealed that the five-membered D ring of steroids generally assumes the envelope conformation **(51)**, but that the carbon atom out of the plane of the other four varies, depending on the nature of the substituents at C-17. In sterols and bile acids, bearing a β-oriented alkyl substituent and an α-hydrogen atom at C-17, the "out-of-plane" carbon atom in ring D is C-13 carrying the C-18 methyl group. In 17-oxo steroids (e.g., androsterone) C-14 is "out-of-plane" instead.

The presence of the 5(6)-double bond in cholesterol, as in other Δ^5-sterols, changes considerably the shape of ring B. Since the doubly bonded carbon atoms, 5 and 6, and the four other atoms (C-4, C-7, C-10, and C-6) are coplanar, ring B assumes a twisted chair conformation, with C-8 being out of and above the plane and C-9 somewhat below the plane. The rigidity of the plane-trigonal structures, imposed by the 5(6)-double bond, not only changes the shape of ring B, but also causes a slight tilting of ring A so that its front edge, represented by the C-4 to C-5 bond, is a little above the planes of rings C and D.

In lanosterol **(40)**, in which rings B and C share the 8(9)-double bond, ring C

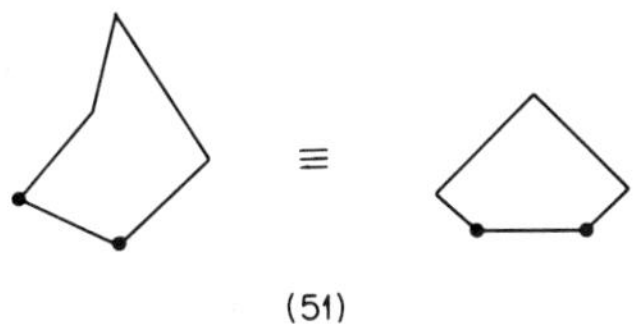

(51)

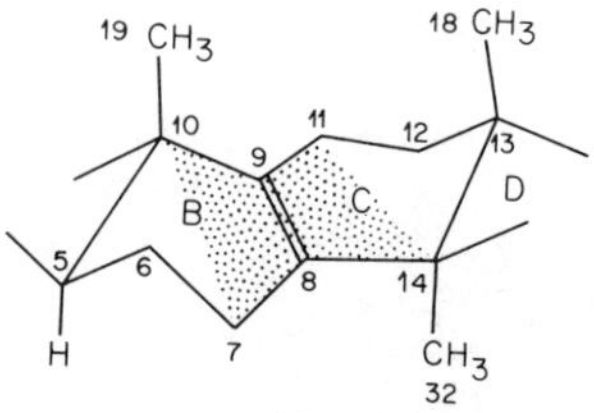

acquires a half-chair conformation and ring B a twisted chair conformation because of the spread of two plane-trigonal structures across two rings; six carbon atoms, numbers 7, 8, 9, 10, 11, and 14, form a common plane (shaded area in formula **52**), with C-13 and C-6 being above and C-5 below the plane.

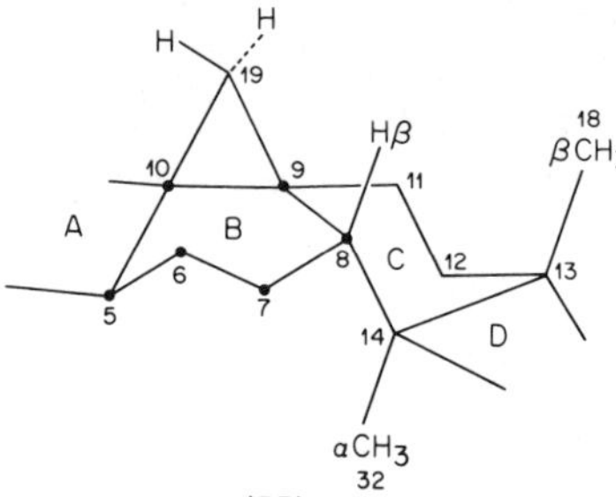

In cycloartenol (**41**) both the cyclopropane ring, over C-9 and C-10, and the hydrogen atom at C-8 are β-oriented; hence rings B and C must be *cis*-oriented to one another (**53**). In addition, ring B assumes a somewhat twisted half-chair conformation on account of the rigidity imposed upon it by the cyclopropane ring.

15.4. Biosynthesis of Cholesterol

It was known by the early 1930s from balance-sheet experiments that animals must synthesize cholesterol, as growing animals, mice, rats, and chicks, kept on sterol-free diet, accumulated as much cholesterol during their growth as their controls which were fed a full diet. The year 1937 is a landmark in biochemistry as that year saw the introduction of deuterium for the study of biosynthetic processes. Sonderhoff and Thomas found in 1937 that when sodium deuterioacetate (CD_3COONa) was added to a medium of yeast growing aerobically, the unsaponifiable lipid fraction of the yeast cells—which consists mostly of ergosterol—contained a much higher concentration of the isotope than any other cell constituent. In the same year, Rittenberg and Schoenheimer reported that when the body water of mice was enriched with D_2O to levels of 2–3% for several weeks, the cholesterol became increasingly labeled with deuterium and that the isotope concentration in the water of combustion of cholesterol ultimately reached a steady value of about one-half that of the body water. They inferred that cholesterol must be synthesized from the condensation of many small molecules.

The value of the use of D_2O was not only to prove the existence of biosynthesis, but the rate of incorporation of deuterium into cholesterol enabled Rittenberg and Schoenheimer also to conclude that the synthesis of total body cholesterol in mice proceeded with a half-life of about three weeks (see Schoenheimer, 1946). For many years afterwards, the rate of uptake of deuterium from the body water into cholesterol, and other body constituents too, was used as a measure of the rate of the synthesis of the substance. Thus, in experiments on rabbits it could be calculated from the rate of appearance of deuterium in cholesterol that in the liver and the intestine of adult rabbits approximately 50 mg and 100 mg, respectively, of cholesterol were synthesized per 100 g fresh tissue per day (Popják and Beeckmans, 1950). Today the use of D_2O has been superseded by the application of ^3H_2O.

15.4.1. Acetate as a Precursor of Cholesterol

Bloch and Rittenberg (1942a,b, 1944) fed various substances labeled with deuterium to mice and rats and found that only deuterioacetate, or substances that were transformed to acetate, gave significant labeling of cholesterol. These observations were confirmed and extended a little later when the carbon isotopes, ^{13}C and ^{14}C, became available, enabling the first phase of extensive investigations into the biosynthesis of sterols.

Some doubts were expressed at first as to whether acetate was the only primary building unit from which sterols were synthesized. These doubts were, however, dispelled by an experiment of Ottke et al. (1951) who made use of a mutant strain of the mold Neurospora crassa which required acetate as an essential growth factor since it could not break down glucose to acetate. When this strain of Neurospora was grown in the presence of $CH_3{}^{13}COOH$ and $^{14}CH_3COOH$ and unlabeled glucose, the concentrations of ^{13}C and ^{14}C in the carbon atoms of ergosterol were very nearly the same as the concentrations of ^{13}C and ^{14}C in the correspondingly labeled acetates. In contrast, the isotope content of the total carbon of the mold was so low as to indicate that two-thirds of this must have originated from glucose.

The use of doubly labeled acetate, $^{13}CH_3{}^{14}COOH$, gave further revealing information. When cholesterol biosynthesized from such doubly labeled substrate was degraded by thermal fission of cholesteryl chloride to a mixture of isooctane + isooctene (from the side-chain) and a hydrocarbon, $C_{19}H_{30}$ (from the ring structure), it became apparent that both the side-chain and the ring structure contained the two carbon atoms of acetate, but in different proportions (Little and Bloch, 1950). The ratio of $^{13}C/^{14}C$ in the whole cholesterol molecule was 5:4, in the side-chain it was 5:3, and in the ring structure the ratio was 10:9. The best interpretation of these results was that out of the 27 carbon atoms of cholesterol, 15 originated from the methyl carbon and 12 from the carboxyl carbon atom of acetate, an interpretation that was subsequently fully confirmed.

When these observations were made there were no biochemical reactions known whereby the complex molecule of cholesterol could be built up from such a simple molecule as acetate. The only choice left to the interested investigators was to prepare specimens of cholesterol biosynthesized from [1-^{14}C]acetate and [2-^{14}C]acetate and then to degrade carbon by carbon the two, presumably differently labeled, specimens of cholesterol in the hope that the pattern of distribution of the two carbon atoms in the cholesterol molecule might reveal a particular biochemical mechanism. The localization of the 15 "methyl" and 12 "carboxyl" carbon atoms of acetate in cholesterol, biosynthesized mostly in liver slices from the ^{14}C-labeled acetates, was carried out in the laboratories of Bloch in the United States and of Cornforth and Popják in England between 1951 and 1957 (cf. Bloch, 1953, 1954, 1965; Popják, 1955; Kritchevsky, 1958; Fieser and Fieser, 1959; Popják and Cornforth, 1960a). It is sad to reflect that today, with the advent of ^{13}C-NMR spectroscopy, similar work could be completed within a few weeks and without any degradation of cholesterol. When a specimen of cholesterol biosynthesized from [1-^{14}C]acetate was degraded chemically it was found that 12 of its carbon atoms were equally labeled and 15 others contained no isotope. Conversely when cholesterol biosynthesized from [2-^{14}C]acetate was degraded, the 15 carbon atoms which did not acquire label from [1-^{14}C]acetate were heavily labeled and the 12 others which acquired label from [1-^{14}C]acetate were also labeled but their isotope content was only 10–20% of that in the 15 heavily labeled positions. Of course, it is well understood that [2-^{14}C]acetate (or, correctly, [2-^{14}C]acetyl-CoA) can readily give rise to [1-^{14}C]acetyl-CoA through the intermediates of the tricarboxylic acid cycle, oxaloacetate and pyruvate. The pattern of distribution of acetate carbon atoms in cholesterol is shown in Fig. 15.1 where "m" denotes an

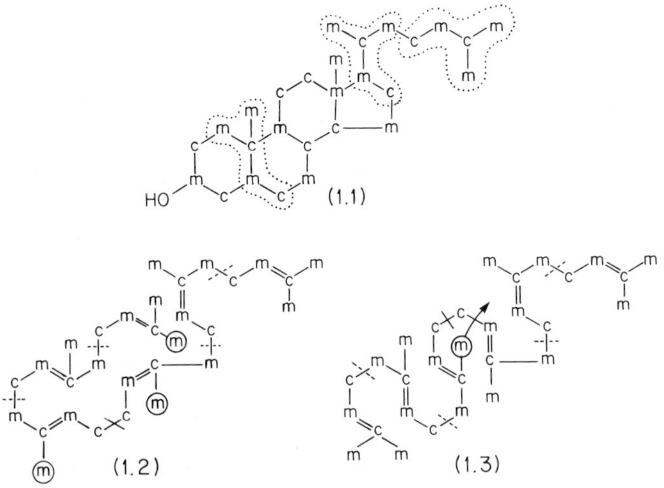

FIGURE 15.1. Distribution of acetate carbons found in cholesterol (**1.1**); a repeating pattern of five carbon atoms, surrounded by dotted lines, is recognizable in three places in the molecule. Formula **1.2** shows the expected distribution if squalene had been folded and cyclized according to Robinson (1934), and formula **1.3** shows the expected pattern if squalene is folded before cyclization according to Woodward and Bloch (1953).

acetate methyl and "c" an acetate carboxyl carbon. There are three places in the molecule where a repeating pattern of $\overset{m}{\underset{m}{>}}c\!-\!m\!-\!c$ can be recognized. Structurally this is one form of the isoprenoid or prenyl unit discussed in Section 15.2. The arrangement of acetate carbons in this pattern was first recognized in Bloch's laboratory (Wüersch et al., 1952) and prompted Bloch (1954) to revive the old hypothesis that the branched-chain hydrocarbon, squalene (cf. formula **16**, Section 15.2), might be a precursor of sterols (Heilbron et al., 1926; Channon, 1926; Robinson, 1934).

15.4.2. Squalene and Lanosterol as Intermediates in Cholesterol Biosynthesis

The revival of the squalene hypothesis by Bloch (1954) was undoubtedly the most fruitful idea concerning the mechanism of formation of cholesterol from acetate. This hypothesis required among other things: (i) that squalene must be synthesized from acetate by all organs and cells that synthesize cholesterol; (ii) that squalene must be convertible into sterol; (iii) that the distribution of acetate carbons in squalene must be such as to be compatible with the distribution found in cholesterol, assuming a chemically and biologically feasible cyclization mechanism for the hydrocarbon; and (iv) that a C_{30} sterol be found that contained all the 30 carbon atoms of squalene. All these requirements were met in quick succession.

Langdon and Bloch (1953a,b) succeeded in isolating [^{14}C]squalene from the liver of rats which were fed squalene and [^{14}C]acetate and showed that the purified [^{14}C]squalene when fed to mice was indeed converted into cholesterol, approximately 15% of the administered dose having been recovered in body cholesterol. Thus squalene appeared to be a much better precursor of cholesterol than any other substance previously tried (Table 15.1). The synthesis of squalene from [^{14}C]acetate was also demonstrated in pig liver perfused with [^{14}C]acetate (Schwenk et al., 1954) and in vitro by liver slices and various structures of the hen's ovary [theca interna and externa, the granulosa cell layer of the Graafian follicle (Popják, 1954)]. When biosynthetic [^{14}C]squalene was chemically degraded it was found that in each of its prenyl units the acetate carbons were distributed in the same pattern, $\overset{m}{\underset{-m}{>}}c\!=\!m\!-\!c\!-$, as was found in the side-chain of cholesterol (Cornforth and Popják, 1954).

If squalene had been folded and cyclized according to Robinson's hypothesis, the pattern of acetate carbons in the sterol might have been as shown in **1.2** of Fig. 15.1. Robinson's hypothesis put the line of symmetry of squalene between C-6 and C-7 of the sterol; hence both of these carbon atoms should have originated from a carboxyl carbon of acetate. Further, according to this scheme, C-13 of cholesterol should also be derived from an acetate carboxyl. Comparison of this hypothetical pattern of acetate carbons to that found in cholesterol (**1.2**)

TABLE 15.1 Conversion of [^{14}C]Squalene, Biosynthesized in Liver Slices from [^{14}C]Acetate, into Digitonin-precipitable Sterols in Mice[a]

Source	[^{14}C]Squalene recovered as percent of dose fed	
	In sterols	In squalene
Liver	4.5 ± 0.4	Trace
Intestine	2.2 ± 0.1	1.3 ± 0.5
Carcass	2.5 ± 0.3	2.1 ± 0.3
Feces	0.7 ± 0.1	36.7 ± 10

[a]Two groups of three mice were each fed in four equally divided doses, 10.2 mg [^{14}C]squalene mixed with an equal amount of olive oil over 24 hr; the animals were killed 19 hr after the last feed (from the notebooks of G. Popják, 1954).

shows definite discrepancies. The first discrepancy was noted by Woodward and Bloch (1953) who were able to deduce that both C-18 and C-13 of cholesterol originated from the methyl carbon atom of acetate. In consequence they proposed another, "harmonica-like," folding of squalene and postulated the migration of a methyl group from either position 8 or 14 to 13 and the subsequent elimination of three methyl groups in the formation of cholesterol (**1.3**). The important difference between Robinson's hypothesis and the hypothesis of Woodward and Bloch is that the latter placed the line of symmetry of squalene between positions 11 and 12 of the sterol structure.

The Woodward–Bloch hypothesis had further important implications. First it suggested that a trimethyl C_{30} sterol might be an intermediate in the biosynthesis of cholesterol. The best candidate for this role was lanosterol (lanosta-8(9),24-dien-3β-ol; see formula **40** in Section 15.3) whose structure was deduced just the year previous to the proposal by Woodward and Bloch (Voser *et al.*, 1952). On the theoretical side, the "harmonica-like" folding of squalene resulted in proximity of double bonds and allowed a new stereochemical interpretation of the cyclization of squalene and placed the ideas of a common biological origin for sterols and tetracyclic triterpenes on a solid basis, as will be discussed later.

15.5. Search for the Origin of Biological Prenyl (Isoprenoid) Units; the Discovery of Mevalonic Acid

The pattern of distribution of acetate carbons in the prenyl units of the side-chain of cholesterol (Wüersch *et al.*, 1952) and in those of squalene (Cornforth and Popják, 1954) stimulated much research into the biosynthesis and metabolism of C_6 and C_5 branched-chain acids and their biosynthesis from [^{14}C]acetate (cf. Coon et al., 1959; Rudney, 1959). 3-Hydroxy-3-methylglutaric acid, discovered in 1952 as a plant product, was the prime suspect for being the source of the isoprenoid units (Bloch, 1954) in polyprenyl biosynthesis. Cornforth and Popják wrote in 1954 in their paper on the biosynthesis of squalene from acetate: "It does not seem improbable that the biosynthesis of isoprenoid units proceeds by the reaction of acetoacetyl-CoA with acetyl-CoA . . .", and postulated the synthesis of some derivative of 3-hydroxy-3-methylglutarate (HMG; **2.1** in Fig. 15.2), in a reaction analogous to the synthesis of citrate from oxaloacetate and acetyl-CoA.

Between 1952 and 1956 HMG became a strong candidate as the source of prenyl units; however, all attempts to prove the involvement of HMG as such or of some other C_5 branched-chain acid in sterol biosynthesis were fruitless (for review, see Popják, 1958).

The discovery of the donor of prenyl units in polyprenyl and sterol biosynthesis was made almost by accident. A group of researchers led by Karl Folkers at the Merck, Sharp and Dohme laboratories were studying the metabolism of *Lactobacillus acidophilus,* a mutant that had an absolute requirement for acetate as a growth factor. They found that the acetate in the culture medium could be replaced by "distiller's solubles," the water-soluble residue of fermentation liquor after the alcohol had been distilled off. The structure of the acetate-replacing factor turned out to be that of the optically active 3,5-dihydroxy-3-methylvaleric acid, first named "divalonic acid," then renamed mevalonic acid, which readily changes in mildly acidic conditions to its δ-lactone, mevalonolactone (Wright et al., 1956; Folkers et al., 1959). The similarity of the structure of mevalonic acid **(2.2; 2.3)** to that of 3-hydroxy-3-methylglutaric acid prompted the Merck group to label it with ^{14}C in position 2 and to test it as a sterol precursor. Tavormina et al. (1956) reported that the [2-^{14}C]mevalonic acid was converted in liver homogenates into digitonin-precipitable sterols with an efficiency of 43%. Since the synthetic material was undoubtedly a racemic mixture, its very high utilization for sterol synthesis meant the quantitative, or nearly quantitative, conversion of one enantiomer into sterols. The Merck group synthesized also [1-^{14}C]mevalonic acid and showed that C-1 of the substance was not used in sterol synthesis, but was eliminated as CO_2 (Tavormina and Gibbs, 1956).

The use of mevalonate was eagerly taken up by many researchers investigating diverse aspects of polyprenyl synthesis, first and foremost that of squalene and cholesterol.

Mevalonate has a chiral center at C-3; its absolute configuration was first

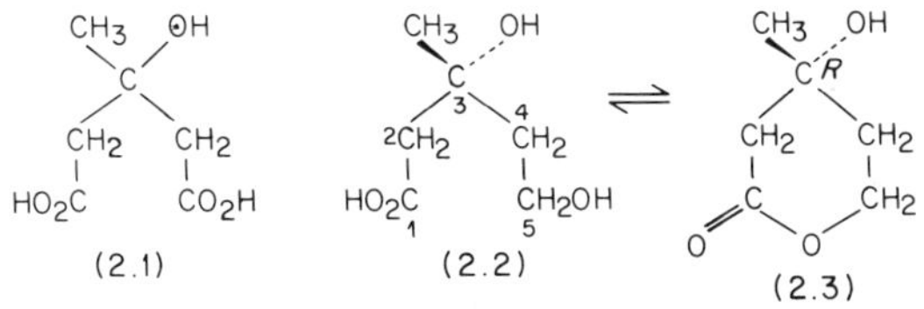

FIGURE 15.2. Structures of 3-hydroxy-3-methylglutaric acid **(2.1)**, mevalonic acid **(2.2)**, and mevalonolactone **(2.3)**.

deduced to be *R* by Eberle and Arigoni (1960) by converting quinic acid to the unnatural *S*-isomer of mevalonolactone, and then more directly by Cornforth *et al.* (1962) by the separate synthesis of the *R*- and *S*-enantiomers from (*S*-) and (*R*)-linalool (coriandrol and licareol), respectively. The chemistry and the diverse syntheses of mevalonolactone labeled in a variety of ways have been reviewed by Cornforth and Cornforth (1970) (see also Cornforth and Popják, 1969).

By the time mevalonate was discovered, an important technical advance was made by the introduction of liver homogenates, made in a smooth-walled Potter-Elvehjem type of homogenizer, which—after removal of mitochondria—converted [^{14}C]acetate aerobically into cholesterol and anaerobically into squalene (Bucher, 1953; Bucher and McGarrahan, 1956). The Bucher-homogenate of liver became a most useful preparation as it also converted [2-^{14}C]mevalonate under aerobic conditions mainly into cholesterol and under anaerobic conditions mainly into squalene. Chemical degradation of squalene biosynthesized from [2-^{14}C]mevalonate showed that the squalene contained only six labeled positions, one in each of the two terminal isopropyl groups and four within the carbon chain of the hydrocarbon and none in the branched methyl groups, as shown in Fig. 15.3 (Cornforth *et al.*, 1957, 1958; Dituri *et al.*, 1957). On the assumption that the cyclization of squalene to lanosterol proceeded according to the Woodward-Bloch hypothesis, it was possible to predict the labeling pattern of lanosterol and cholesterol biosynthesized from [2-^{14}C]mevalonate as shown in **3.2** and **3.3** in Fig. 15.3, and this prediction was corroborated by Isler *et al.* (1957).

The observations on the biosynthesis of squalene from [2-^{14}C]mevalonate provided the first proof that mevalonate through some, at that time unknown, intermediates gave directly the prenyl units used in the synthesis of squalene.

The discovery of mevalonate had the great advantage that the exceptionally long chain of reactions in cholesterol biosynthesis could be broken up into four shorter sequences: (i) acetate to mevalonate; (ii) mevalonate to squalene; (iii) squalene to lanosterol; and (iv) lanosterol to cholesterol. Of these, the first two sequences were taken up with great vigor in several laboratories and studied

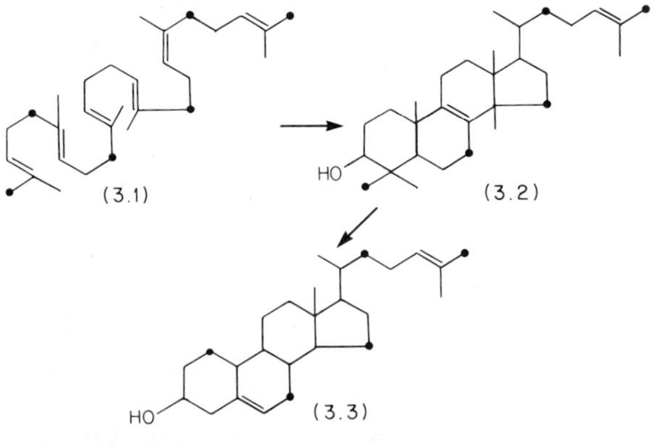

FIGURE 15.3. Distribution of C-2 mevalonate found in squalene (**3.1**) and predicted distribution in lanosterol (**3.2**) and cholesterol (**3.3**) biosynthesized from [2-^{14}C]mevalonate.

mostly in yeast extracts, or yeast autolysates, and in the Bucher-type of liver homogenates. The progress was so rapid that by 1960 the intermediates and broad outlines of the first two sequences became known.

Acetate to Mevalonate

The overall details of mevalonate biosynthesis became known largely from the work of Rudney and Lynen and their associates between 1957 and 1959, though the intimate knowledge of the enzymes involved, their distribution and properties, are of more recent origin.

Mevalonate is formed in two stages: the first stage is the synthesis of 3-hydroxy-3-methylglutaryl coenzyme A (HMG-CoA); the second stage is the reduction of HMG-CoA to mevalonate. These two stages are probably common to all forms of life synthesizing polyprenols.

HMG-CoA is synthesized from acetyl-CoA and acetoacetyl-CoA in two cell compartments, in the cytosol and in mitochondria by the enzyme HMG-CoA synthetase (EC 4.1.3.5). In the liver of birds and of the rat, 20–40% of the total HMG-CoA synthetase is in the cytosol and the remainder is in the mitochondria. The cytosolic and mitochondrial synthetases differ in their catalytic properties and in their isoelectric points. The mitochondrial synthetase has a pI of 7.2 and is inactivated by Mg^{2+}, whereas the cytoplasmic synthetases are activated by Mg^{2+}. The liver of birds (chicken, turkey, pigeon) contains four cytoplasmic HMG-CoA synthetases differing from one another in their isoelectric points which range from pI 5.2 to 6.2; synthetase II with a pI of 5.4 is the most abundant (68%). Rat liver contains only one cytosolic synthetase with a pI of 5.4 with properties similar to those of enzyme II of the chicken liver (Clinkenbeard *et al.*, 1975a,b; Reed *et al.*, 1975).

The mechanism of HMG-CoA synthetases is shown in Fig. 15.4. The first step in the synthesis is the acetylation of a specific cysteinyl residue in the protein with the release of CoA—SH; the acetyl-S-enzyme then reacts with acetoacetyl-CoA and forms an enzyme-S-hydroxymethylglutaryl-S-CoA intermediate which can be trapped when the reaction with the acetyl-S-enzyme is carried out at $-25°C$ in 25% ethanol (Miziorko *et al.*, 1975; Miziorko and Lane, 1977). The enzyme-bound HMG-CoA is rapidly released on warming up the reaction mixture to 0°C. The reaction mechanism revealed by these experiments explains the earlier observations made by Rudney (1959) and by Steward and Rudney (1966) that in the synthesis of HMG-CoA from acetyl-CoA and acetoacetyl-CoA, the coenzyme A associated with the second substrate is the one that appears in HMG-CoA, the CoA of acetyl-CoA being released.

HMG-CoA synthetase shows twofold stereospecificity. (i) The absolute configuration of HMG-CoA at C-3 must be *S* as its reduction yields (3*R*)-mevalonate. Thus it can be deduced that at the catalytic site the acetyl-S-enzyme reacts with the *si*-face at C-3 of acetoacetyl-CoA, i.e., the face of the molecule on which the methyl group, the carbonyl oxygen atom, and the $-CH_2CO-S-CoA$ group of acetoacetyl-CoA are seen in an anticlockwise order (Fig. 15.5; cf. Popják, 1970, pp. 186–187). (ii) It is also stereospecific as regards the methyl group of the acetyl-CoA. Cornforth *et al.* (1974) have shown with the aid of chiral acetic acids that

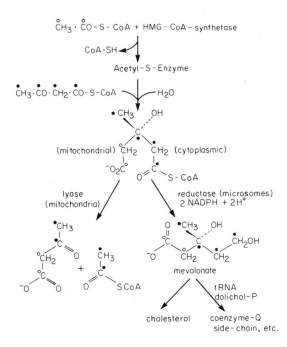

FIGURE 15.4. Synthesis of 3-hydroxy-3-methylglutaryl-CoA from acetyl- and acetoacetyl-CoA by cytoplasmic and mitochondrial HMG-CoA synthetases. The different fates of the cytoplasmic and mitochondrial HMG-CoAs are also shown.

the condensation of acetoacetyl-CoA with acetyl-CoA proceeds with an inversion of configuration of the hydrogen atoms of the acetyl-CoA (Fig. 15.5).

A fundamental difference between the cytosolic and mitochondrial synthetases is that the products synthesized in the two compartments have different fates. HMG-CoA synthesized in the cytosol becomes the substrate for the microsomal HMG-CoA reductase (EC 1.1.1.34) which generates mevalonate, whereas the mitochondrial HMG-CoA becomes the substrate for the mitochondrial HMG-CoA lyase (EC 4.1.3.4) (cf. Coon *et al.*, 1959), which cleaves HMG-CoA into acetyl-CoA and free acetoacetate.

The cleavage of HMG-CoA by the mitochondrial lyase is not the reversal of its synthesis as the carbon atoms of the acetyl-CoA were originally C-1 and C-2 of the acetoacetyl-CoA, and C-1 and C-2 of the free acetoacetate represent the two carbon atoms of the original acetyl-CoA (cf. Fig. 15.4).

Another source of HMG-CoA also arises in the mitochondria from the metabolism of leucine. Leucine gives, through transamination, first α-ketoisocaproic acid, which after decarboxylation and dehydration yields 3,3-dimethylacrylyl-CoA. The latter, after carboxylation with CO_2 and ATP by a biotin-enzyme, is converted to *trans*-3-methylglutaconyl-CoA, which is then hydrated in a reversible reaction to HMG-CoA (Lynen, 1958). This HMG-CoA, like the one synthesized in the mitochondria from acetyl-CoA and acetoacetyl-CoA, falls prey to the mitochondrial HMG-CoA lyase and generates free acetoacetate and acetyl-CoA (Fig. 15.6). This is the reason why leucine is a ketogenic amino acid. Contrary to

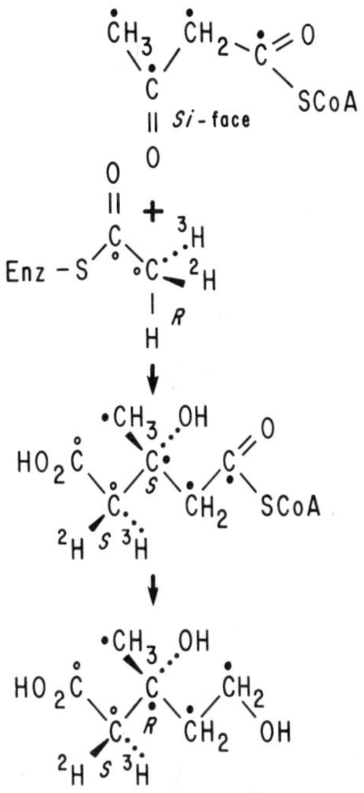

BIOSYNTHESIS OF CHOLESTEROL

FIGURE 15.5. Stereospecificity of the reaction catalyzed by HMG-CoA synthetase.

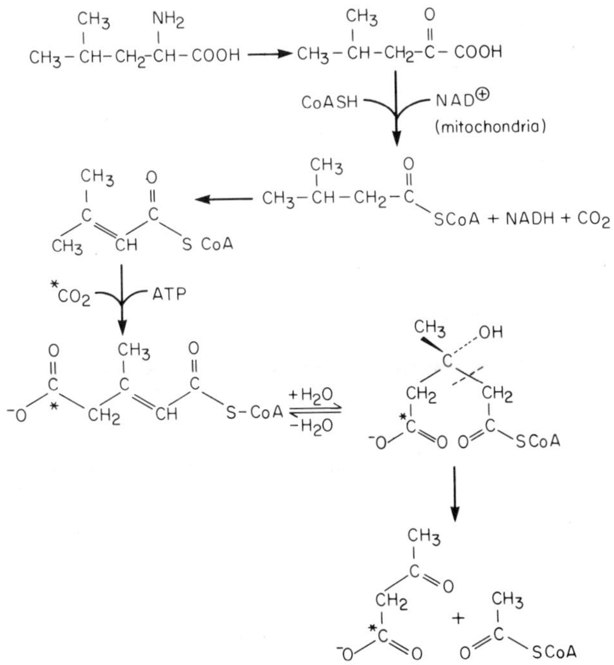

FIGURE 15.6. Mitochondrial origin of HMG-CoA from leucine and the cleavage of HMG-CoA into acetyl-CoA and free acetoacetate (see also Fig. 15.4).

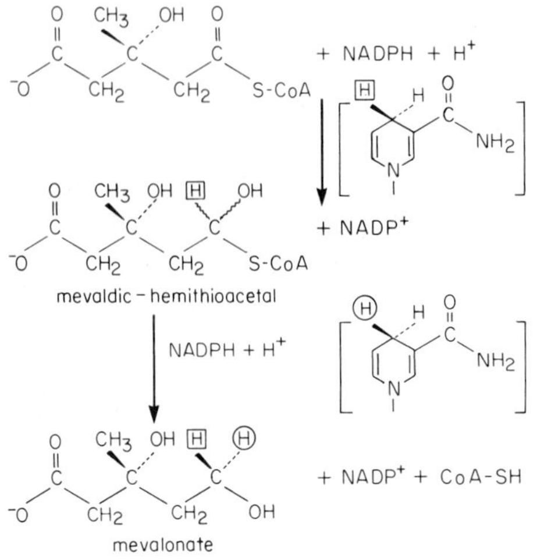

FIGURE 15.7. Two-step reduction of HMG-CoA by HMG-CoA reductase. In both steps the pro-*R* hydrogen atom from C-4 of NADPH is transferred to the substrate.

earlier beliefs, the HMG-CoA derived from the metabolism of leucine is not converted to mevalonate. Carbon atoms of leucine do find their way into cholesterol, but only after degradation of leucine in the mitochondria.

A survey of the organs of chickens and rats revealed that only the liver and kidneys contain mitochondrial HMG-CoA lyase. It is probable that the mitochondrial HMG-CoA lyases of these two organs are the main source of blood acetoacetate, which was thought until recently to serve only as an energy or carbon source, in organs other than the liver or kidney, after reconversion to acetoacetyl-CoA by the succinyl-CoA:3-oxo-acid CoA-transferase (cf. Chapter 7). However, recent investigations have shown that the developing brain, the lactating mammary gland, adipose tissue, and the liver of rats contain substantial amounts of a cytosolic acetoacetyl-CoA synthetase which can activate free acetoacetate probably by the reaction:

acetoacetate + CoA−SH + ATP → acetoacetyl-CoA + AMP + PP_i.

Thus, the free acetoacetate generated in the mitochondria of the liver can become a substrate for the cytosolic HMG-CoA synthetase and provide carbon atoms for sterol synthesis (Bergstrom *et al.,* 1982; 1984).

The second stage of mevalonate formation is the reduction of the cytoplasmic HMG-CoA with NADPH by the enzyme HMG-CoA reductase (EC 1.1.1.34) as shown in Fig. 15.7. This enzyme is widely distributed in many organs and in all organisms that synthesize polyprenyl substances. It was soon recognized after the discovery of mevalonic acid in extracts of yeast and liver (Knappe *et al.,* 1959; Rudney, 1959; Bucher *et al.,* 1959), but it received special attention only during the last 15 years, mainly because it is a highly regulated enzyme with a short half-life.

HMG-CoA reductase is an integral protein of endoplasmic reticulum (microsomes) from which it can be solubilized in a variety of ways (freezing and thawing and by the use of detergents). The reaction catalyzed by this enzyme involves, in two steps, the reduction of the thioester group of HMG-CoA with two moles of NADPH. HMG-CoA reductase is one of the many "A-side" specific oxidoreductases (cf. Popják, 1970), which means that it transfers stereospecifically in both steps of reduction the pro-R hydrogen atom from C-4 of the dihydronicotinamide ring of the coenzyme to the substrate (Dugan and Porter, 1971).

The intermediate in the two-step reduction remained a puzzle for several years; mevaldic acid was the obvious candidate for this role, but it could not be trapped in the reaction catalyzed by HMG-CoA reductase nor could the enzyme reduce mevaldic acid to mevalonate. Popják and Cornforth (1960a) suggested first that mevaldic-CoA hemithioacetal, or an enzyme-bound mevaldic-hemithioacetal, could be the intermediate in the reduction. Rétey et al. (1970) showed subsequently that the mevaldic acid, generated from (R,S)-5,5-dimethoxy-3-hydroxy-3-methylvaleric acid, reacted readily with CoA—SH to form the hemithioacetal and that the mevaldic-S-CoA hemithioacetal was readily reduced, following Michaelis–Menten kinetics, to mevalonate. Further, Blattmann and Rétey (1971) found that the reduction of the mevaldic-S-CoA hemithioacetal, made from (R,S)-mevaldic acid and CoA—SH, with [(4R)-4-^{3}H]NADPH resulted exclusively in the formation of [(3R)-(5S)-5-^{3}H]mevalonate. Considering that the reaction of (R,S)-mevaldic acid with CoA—SH would generate four diastereomers, the stereospecificity of the observed reaction proves that only one of the four isomers acted as substrate with HMG-CoA reductase. If (3S)-mevaldic-S-CoA hemithioacetal is indeed the intermediate in the reduction of HMG-CoA by the reductase, the enzymatically generated hemithioacetal should contain another asymmetric center at C-5 of the hemiacetal, R or S: its determination would be of much interest as it would reveal the mechanism of reduction with NADPH. If the displacement of CoA from the hemiacetal by the incoming hydride ion from NADPH were an S_N2 type of reaction, the absolute configuration at C-5 of the hemiacetal predictably should be R.

Although free mevaldic acid is definitely excluded from the metabolic pathway leading to mevalonate, it proved useful in studies on the stereochemistry of squalene biosynthesis. An enzyme was found in liver extracts that reduced mevaldate with either NADH or NADPH to mevalonate, and was known for many years as mevaldate reductase (Schlesinger and Coon, 1961; Knauss et al., 1962).

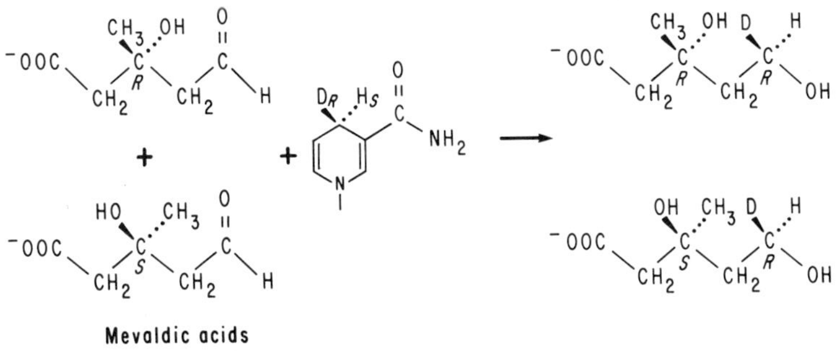

FIGURE 15.8. Reduction of (R,S)-mevaldic acid with [(4R)-4-D$_1$] NADH to [(3R,S)-(5R)-5-D$_1$] mevalonate.

This enzyme is an "A-side" specific dehydrogenase, but reduces (R,S)-mevaldate to (R,S)-mevalonate. However, when [(4R)-4-D]NADH (or NADPH) is used with mevaldate reductase, it generates [(3R,S)-(5R)-5-D$_1$]mevalonate (Donninger and Popják, 1966; Ngan and Popják, 1975). As the (3R)- and (3S)-mevalonates can be separated by the use of mevalonate kinase (see later), the [(3R)-(5R)-5-D$_1$]mevalonate, the first specimen of mevalonate to be labeled stereospecifically with a hydrogen isotope, became a valuable tool in exploring the stereochemistry of the biosynthesis of farnesyl diphosphate and squalene. Mevaldate reductase is now known to be a nonspecific general aldehyde reductase (Fig. 15.8).

15.5.1. Regulation of 3-Hydroxy-3-methylglutaryl Coenzyme A; the Nature of the Enzyme

It has been known since the early 1950s that cholesterol synthesis from [^{14}C]acetate in the liver is profoundly suppressed by cholesterol feeding (Taylor and Gould, 1950; Gould, 1954) and by fasting or restriction of caloric intake (Tomkins and Chaikoff, 1952), but is greatly increased in animals with a bile fistula (cf. Weis and Dietschy, 1969). Clearly, there must be a severely controlled step in cholesterol synthesis. After the discovery of mevalonate and HMG-CoA reductase it was understood that the main controlling step is between HMG-CoA and mevalonate and that the changes in rates of cholesterol synthesis correlate with a fall and rise in activities of HMG-CoA reductase in the liver. This enzyme acquired consequently the epithet of "the rate-limiting enzyme of cholesterol biosynthesis."

Particularly great interest was aroused in this enzyme after 1969 when it was discovered almost simultaneously in three laboratories that in the liver of mice and rats the reductase exhibited a diurnal rhythm of rise and fall with an approximately tenfold amplitude. In rats fed *ad libitum* and kept at night in the dark for 12 hours and then in the light for 12 hours, reductase levels began to rise shortly after the beginning of the dark period and reached a maximum in the middle of the dark period, then gradually fell to the minimum in the middle of the light period (for review, see Rodwell *et al.*, 1976). It is probable that this circadian rhythm has nothing to do with light and dark *per se*, but depends on the eating habits of the rat. The rat is a nocturnal animal and eats in the dark. One of the physiological responses to eating is an increased production and excretion of bile. Since the bile acids are derived from the transformation of cholesterol and since bile also contains cholesterol, the liver becomes somewhat depleted in its cholesterol content after eating. The increase in the levels of HMG-CoA reductase is apparently a compensating mechanism for such depletion. It has been possible to demonstrate also with isolated cells, such as human leukocytes and isolated hepatocytes, that the reductase could be induced in the cells under conditions when the cells lost cholesterol. The level of induction correlated well with the amount of cholesterol lost from the cells into incubation medium (Fogelman *et al.*, 1977).

The mechanism of the daily rise and fall in the levels of HMG-CoA reductase was investigated in many laboratories and generally it was concluded that the rise resulted from increased synthesis of the enzyme, and the fall from cessation of the synthesis. This conclusion was supported by the observation that the administration of cycloheximide (an inhibitor of protein synthesis) to rats completely abolished the rise and the diurnal cycle (Shapiro and Rodwell, 1969). The half-life of the enzyme has been estimated to be as short as 2–4 hr. More recently, it has been shown that the diurnal rhythm of the reductase coincided with a diurnal rhythm of the reductase protein and the reductase-specific mRNA (Clarke et al., 1984).

New synthesis of enzyme protein and its degradation are not the only means by which the activity of the reductase is regulated. It was discovered in 1973 by Beg and colleagues that incubation of microsomes with ATP and Mg^{2+} ions completely inhibited the reductase. The observation strongly suggested that this enzyme, like several others, may be regulated by phosphorylation and dephosphorylation. From subsequent research in several laboratories a new scheme of regulation of HMG-CoA reductase emerged (Hunter and Rodwell, 1980). Figure 15.9 shows that HMG-CoA reductase is inactivated by a phosphorylated active reductase kinase (EC 2.7.1.X), and is activated by a phosphatase. The reductase kinase and the phosphatase are present in the microsomes and also in the cytosol. The reductase kinases are protein phosphorylases and transfer isotope from $[\gamma-^{32}P]$-ATP onto a seryl residue of the reductase. The reductase kinases need both ATP and ADP to phosphorylate and hence inactivate the reductase; it seems probable that the ATP and ADP bind to different sites on the reductase kinase and that ADP acts as its allosteric effector. The cytosolic reductase kinase also phosphorylates serum albumin in the presence of ADP and ATP. Both for inactivation of HMG-CoA reductase and for phosphorylation of bovine serum albumin, the ability of nucleoside diphosphates to replace ADP decreases in the order ADP > CDP > dADP > UDP. GDP cannot replace ADP (Harwood et al., 1984). The existence of the reductase kinase-kinase, inferred previously (Fig. 15.9), is at present uncertain; it is also uncertain whether the active form of reductase kinase is phosphorylated, and this is not mentioned by Harwood et al. (1984). It is possible that the function of the putative reductase kinase-kinase may not be other than to provide ADP as an allosteric activator of the reductase kinase.

The phosphatases, which reactivate the phosphorylated HMG-CoA reduc-

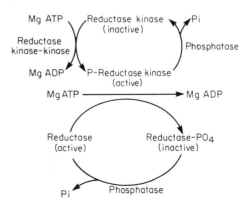

FIGURE 15.9. Regulation of HMG-CoA reductase by phosphorylation–dephosphorylation.

tase, are highly sensitive to NaF and inorganic pyrophosphate (PP_i). When cell extracts or microsomes are prepared with buffers containing NaF or PP_i for assay of HMG-CoA reductase, much lower activities are found than in extracts or microsomes prepared without the inhibitors. Hunter and Rodwell (1980) have surveyed the microsomal HMG-CoA reductase in nine vertebrates and concluded that most of the reductase in the liver of the animals was present in an inactive or latent form *in vivo*. The fraction present in the active form, for a given species, was quite constant, but varied from species to species from 20% to 45%. The total reductase activity is obtained when the microsomes are prepared without NaF and are incubated at 37°C for 20–30 min before assay, as the microsomes contain the activating phosphatase(s). Sitges *et al.* (1984) have described the partial purification of three phosphatases that release $^{32}P_i$ from homogeneous ^{32}P-labeled HMG-CoA reductase. These three phosphatases differed from one another by their molecular weights: phosphatase Ex had $M_r = 90,000$, I_M had $M_r = 75,000$, and II_M had $M_r = 180,000$. All three enzymes also dephosphorylated ^{32}P-labeled phosphorylase *a*, but only phosphatase II_M showed activity with glycogen synthetase. It is not known at present what is the physiological mechanism of the regulation of HMG-CoA reductase activity by phosphorylation–dephosphorylation; Harwood *et al.* (1984) suggest that this may be determined by ATP/ADP ratios in cells.

Inhibitors of HMG-CoA Reductase

It was mentioned earlier that feeding of cholesterol to animals profoundly suppressed synthesis of cholesterol from [^{14}C]acetate, but not from [2-^{14}C]mevalonate (Gould and Popják, 1957), and that the effect of cholesterol feeding was shown to have resulted from the suppression of HMG-CoA reductase. It is very probable that in those early experiments impure preparations of the sterol were used. Cholesterol is highly susceptible to autoxidation and unless great care is taken in its purification by several recrystallizations and purification through the dibromide and storage in an inert atmosphere, its preparations usually contain a variety of oxygenated forms. Kandutsch and his colleagues discovered in 1974 that it was not cholesterol that suppressed HMG-CoA reductase, but its oxygenated forms (for review, see Kandutsch *et al.*, 1978). Among the most active compounds were 25-hydroxy-, 22-hydroxy-, and 7-ketocholesterol (Fig. 15.10); 6-hydroxycholestanol was also a potent suppressor of the reductase. Most of the experiments with the oxygenated sterols were done with various cultured cell lines, mostly fibroblasts. It was found invariably that, in the presence of a few micrograms of the oxygenated sterols per ml in the culture medium, the cells rounded up and died after a few days. However, the cells could be rescued by the addition to the culture medium of cholesterol, or low-density lipoproteins in amounts that would provide 50 μg of cholesterol per ml (for review, see Schroepfer, 1981). Such observations provided irrefutable proof that cholesterol is essential for the life of animal cells.

The oxygenated sterols have no effect *in vitro* on the reductase either in isolated microsomes or on the purified enzyme; they are effective only *in vivo*. Cav-

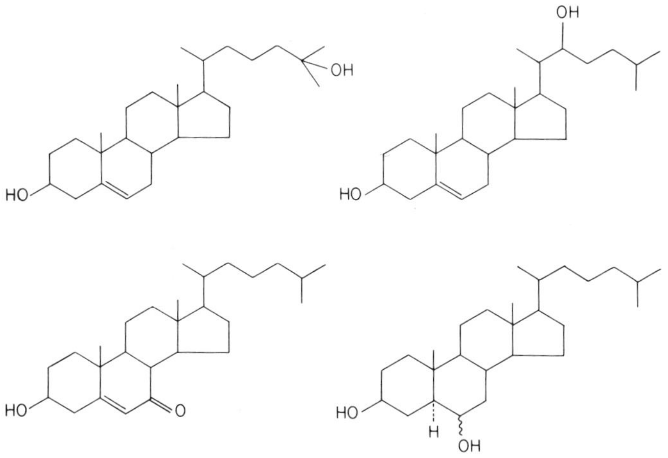

FIGURE 15.10. Oxygenated sterols that suppress HMG-CoA reductase.

enee *et al.* (1981) showed that the oxygenated sterols were effective only in nucleated cells, but had very little effect on the reductase in cytoplasts (cells enucleated with cytochalasin B). Hence it is highly probable that the oxygenated sterols inhibit the transcription of the HMG-CoA reductase gene, although this has not been demonstrated yet.

Another type of inhibitor of HMG-CoA reductase was discovered in 1976 by Endo and his colleagues in Japan. This substance, referred to either as ML-236B or compactin, is an antibiotic and was isolated from *Penicillium citrinum* and *Penicillium brevicompactum*. The structure of this compound (Fig. 15.11) is that of a substituted hexahydronaphthalene carrying the lactone ring of β,δ-dihydroxyheptanoic acid at position 1 and the 2'-methyl-1'-oxobutoxy group at position 8 and a methyl group at position 2. In 1980 a team at the Merck, Sharp and Dohme Research Laboratories reported the isolation of a variant of compactin from *Aspergillus terreus,* and they named it mevinolin (Alberts *et al.,* 1980). This substance differs from compactin only by the presence of an additional methyl group at position 6 of the hexahydronaphthalene nucleus. The lactone ring in compactin and in mevinolin resembles that of mevalonolactone except that mevalonolac-

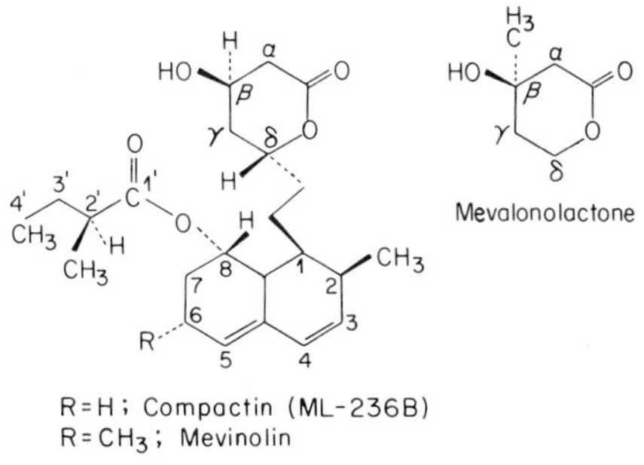

FIGURE 15.11. Structures of compactin and mevinolin. The lactone ring in these two compounds resembles mevalonolactone.

R=H; Compactin (ML-236B)
R=CH$_3$; Mevinolin

tone contains a methyl group in the β-position instead of the hydrogen atom in the two antibiotics. Probably because of this partial structural resemblance of the antibiotic to mevalonolactone, Endo et al. (1976) tested compactin as a possible inhibitor of HMG-CoA reductase. In contrast to the behavior of the oxygenated sterols which inhibits the reductase only *in vivo*, compactin inhibits the enzyme both *in vivo* and *in vitro;* it is an extremely potent competitive inhibitor of the enzyme as regards HMG-CoA, but a noncompetitive inhibitor with NADPH. The value of the inhibition constant, K_i, of compactin is 1 nM, as compared to the K_m value of about 5 μM for HMG-CoA. The inhibitory properties of mevinolin are similar to those of compactin, except that mevinolin is twice as potent as compactin.

The ultimate effects of the oxygenated sterols and of compactin or mevinolin, either in the whole animal or in cultured cells, are quite different. The specific activity of the reductase in microsomes isolated from the liver of animals or from cultured cells treated with oxygenated sterols is very low, i.e., the enzyme is either in an inactive form or is depleted. After administration of compactin or mevinolin the reductase is completely inhibited, but is lying in a latent form. The addition of compactin or mevinolin to the culture medium of normal fibroblasts or hepatoma cells, at concentrations of 2.5–5.0 μM, completely suppresses sterol synthesis from [^{14}C]acetate, but not from [2-^{14}C]mevalonate. When such inhibited cells are washed to remove the inhibitors and are disrupted and assayed for the reductase, it is found that they contain levels of the reductase, depending on the time of exposure of the cells to the inhibitors, five to ten times higher than the normal cells not treated with compactin or mevinolin. The appearance of this latent HMG-CoA reductase, induced by compactin, can be prevented by cycloheximide. It could not be completely suppressed by the addition of LDL or cholesterol to the medium, but could be suppressed by mevalonate or a combination of LDL + mevalonate (Brown et al., 1978; Brown and Goldstein, 1980).

It was demonstrated by Edwards et al. (1977a) that mevalonate was also a potent suppressor of HMG-CoA reductase *in vivo* and *in vitro* in isolated hepatocytes. A single large dose of mevalonate given to rats by stomach tube abolished at least two of the diurnal (midnight) rises in the reductase and reduced the activity of the enzyme in liver microsomes to about 5% of that seen at the nadir of the enzyme at noon in control animals. Mevalonate, like the oxygenated sterols, is effective only *in vivo* in whole animals or in cultured cells; it has no effect whatever on the isolated microsomal enzyme. For the action of mevalonate whole cells are needed as it has no effect on the reductase in cytoplasts (Popják et al., 1983, 1985).

The great interest in the effects of compactin on cultured cells is that, in contrast to the cells treated with, e.g., 25-hydroxycholesterol, whose death can be prevented either by LDL or cholesterol added to the medium, the death of cells exposed to high concentrations of compactin (>2.5 μM) can be prevented only by mevalonate, or by LDL + mevalonate, and not by LDL or cholesterol alone. The death of cells treated with compactin underlined once again the essential role of cholesterol synthesis in the life of eukaryotic cells, or the essential role of mevalonate, or of some substance derived from it, in the life of the cell.

Popják et al. (1985) presented evidence that the mechanism of action of mevalonate is to provide not only a precursor of sterol synthesis, but also a suppressor of the transcription of the HMG-CoA reductase gene. They reconciled the effects of compactin or mevinolin and of mevalonate by the hypothesis that in the presence of compactin or mevinolin the cells are deprived of mevalonate and hence neither sterols nor the putative repressor of the reductase gene can be formed, so that the transcription of reductase gene can continue uninhibited and leads to more synthesis of the specific mRNA and its translation into reductase protein. On the other hand, mevalonate provides substrate for synthesis of sterols and other vital polyprenyl substances and a repressor of the HMG-CoA reductase gene. Edwards et al. (1983) have also shown that mevalonate not only inhibited the synthesis of the reductase, but also enhanced the rate of its degradation and suggested that some product of mevalonate metabolism was responsible for the enhanced degradation.

The use of compactin, and lately, of mevinolin in cell cultures and whole animals brought out new knowledge about the nature of the HMG-CoA reductase protein and its regulation. The reductase purified from rat-liver microsomes without the use of a proteinase inhibitor to a state of homogeneity (Edwards et al., 1980) is a dimer, $M_r = 104,000$, with subunits of $M_r = 52,000$. It has a specific activity of about 20,000 units/mg, a pH optimum of 6.8, and a K_m value for (R,S)-HMG-CoA of 5.6 μM and for NADPH of 51 μM. Antibodies produced against this enzyme fully precipitated this form of the enzyme. However, when isolated hepatocytes were pulse-labeled with [^{35}S]methionine and the cells quickly lysed, the same antiserum precipitated a polypeptide which on examination by sodium dodecyl sulfate (SDS)/polyacrylamide gel electrophoresis gave an M_r of 94,000 (Edwards et al., 1983). Thus it appeared that the subunit of endogenous microsomal HMG-CoA reductase was a much larger molecule than thought previously and that the previously isolated enzyme with subunit mass of 52,000 Da was a degradation product of the native microsome-bound enzyme.

This conclusion was reinforced by study of cells that were selected for their resistance to compactin. From an original CHO-K1 cell culture a mutant cell line, UT-1, was established that could live in the presence of 40 μM compactin and contained 100–1000 times more reductase than the original CHO cells (Chin et al., 1982a). This great increase in the enzyme, amounting to 1–2% of total cellular protein, was accompanied by an enormous proliferation of endoplasmic reticulum, which in cross-sections gave a crystalloid pattern of 60–100 nm diameter rings neatly packed side by side in parallel rows. When the membranes of UT-1 cells were freeze-thawed without a protease inhibitor, to solubilize the reductase, and the enzyme immunoprecipitated with antiserum, two bands with M_r of 50,000 and 55,000 were obtained by SDS/polyacrylamide gel electrophoresis. When the freeze-thawing was carried out in the presence of leupeptine, a protease inhibitor, the subunit of immunoprecipitated enzyme had an M_r of 62,000. Chin et al. (1982a) concluded that the 50,000- and 55,000-Da bands were derived by proteolysis of a single subunit of 62,000 Da. In subsequent work Chin et al. (1982b) found that when they pulse-labeled the UT-1 cells with [^{35}S]methionine and then rapidly lysed the cells with boiling SDS, the polypeptide precipitated by the immune serum had an M_r of 90,000. Also when they translated in a protein-synthesizing system the poly(A)$^+$RNA of the UT-1 cells with [^{35}S]methionine as marker, they obtained by immunoprecipitation a radioactive polypeptide with M_r = 90,000. In the same publication Chin *et al.* (1982b) reported the isolation of a 1.2-kilobase cDNA for 3-hydroxy-3-methylglutaryl-CoA reductase constructed in a plasmid coded pRed-10. The mRNA of UT-1 cells that hybridized to the pRed-10 plasmid directed the synthesis *in vitro* of a 90,000-Da protein that was precipitated by an antireductase antibody. This cDNA hybridized with 4.6- and 4.2-kilobase poly(A)$^+$RNAs of the UT-1 cells. Fortunately it also hybridizes with the 4.6- and 4.2-kilobase poly(A)$^+$RNAs of rat liver (Tanaka *et al.*, 1982; Clarke *et al.*, 1984) and also with the poly(A)$^+$RNA of a cultured rat hepatoma cell line (Popják *et al.*, 1985).

Further examination of the HMG-CoA reductase of UT-1 cells, and that from another cell line, called C100 (derived from baby hamster kidney cells), also resistant to compactin (Brown and Simoni, 1984), led to the conclusion that the enzyme is a transmembrane glycoprotein containing "high mannose" chains, including Man$_6$(GlcNAc)$_2$, Man$_7$(GlcNAc)$_2$, and Man$_8$(GlcNAc)$_2$, with the active site of the enzyme facing the cytoplasm and the carbohydrate-bearing site oriented toward the lumen of the endoplasmic reticulum (Liscum *et al.*, 1983; Brown and Simoni, 1984).

15.5.2. From Mevalonate to Isopentenyl Diphosphate

The intermediary stages of squalene synthesis from isotopically labeled mevalonate have been studied in yeast extracts in the laboratories of Bloch and Lynen, and in liver enzyme preparations in the laboratory of Popják. The progress was so rapid that by 1960 a composite picture emerged and was reviewed in detail (Popják and Cornforth, 1960a; Popják, 1963). Although subtle differences between the yeast and liver enzyme system were noted, the biochemical reactions of squalene synthesis in the yeast and in the liver cell turned out to proceed through the same intermediates. The entire squalene-synthesizing system of the yeast cell is present in a soluble fraction of yeast autolysate, but in the liver both a soluble protein fraction, precipitating between 30% and 60% saturation with ammonium sulfate, and the microsomal particles are needed. The yeast autolysate can synthesize only squalene, but not sterols from (R,S)-[2-^{14}C]mevalonate, whereas the liver system synthesizes squalene anaerobically and mostly sterols in the presence of oxygen. The coenzyme requirements of the yeast and liver systems are ATP, reduced adenine dinucleotide phosphates, and a divalent metal ion, Mg^{2+} or Mn^{2+}. The yeast system functions best with NADPH and Mn^{2+}, whereas in the liver system both NADH and NADPH were needed for maximum activity and Mg^{2+} ions were preferred to Mn^{2+}. The need for ATP in the conversion of mevalonate into squalene suggested that some steps of phosphorylation are involved in the synthesis.

The first two steps of the transformation of mevalonate in yeast and liver systems, and in many others synthesizing polyprenyl substances (cf. Beytia and

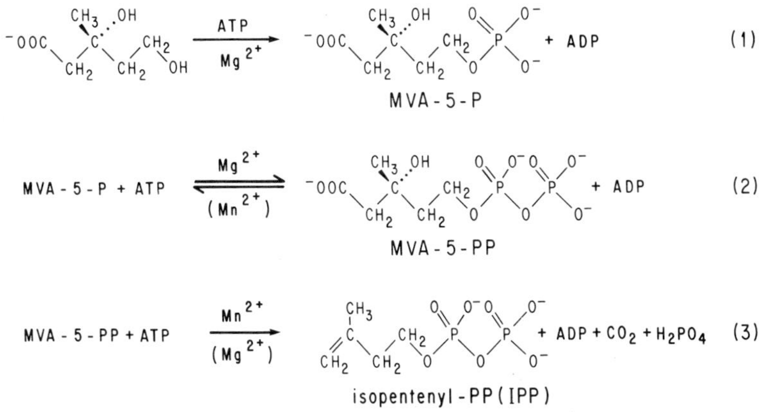

$$\text{}^-OOC\diagdown\underset{CH_2}{\underset{|}{C}}\diagup\underset{CH_2}{\underset{|}{CH_2}}\diagdown OH \quad\xrightarrow{\underset{Mg^{2+}}{ATP}}\quad \text{}^-OOC\diagdown\underset{CH_2}{\underset{|}{C}}\diagup\underset{CH_2}{\underset{|}{CH_2}}\diagdown O-P\diagdown O^- \quad + \text{ ADP} \qquad (1)$$

MVA-5-P

$$\text{MVA-5-P + ATP} \quad\underset{(Mn^{2+})}{\overset{Mg^{2+}}{\rightleftarrows}}\quad \text{MVA-5-PP} + \text{ADP} \qquad (2)$$

$$\text{MVA-5-PP + ATP} \quad\xrightarrow[(Mg^{2+})]{Mn^{2+}}\quad \text{isopentenyl-PP (IPP)} + \text{ADP} + CO_2 + H_2PO_4 \qquad (3)$$

FIGURE 15.12. Transformations of mevalonate leading to isopentenyl diphosphate. Reaction (1) is catalyzed by mevalonate kinase; reaction (2) by phosphomevalonate kinase; and reaction (3) by diphosphomevalonate decarboxylase.

Porter, 1976), are catalyzed by typical kinases, mevalonate kinase (EC 2.7.1.36) and phosphomevalonate kinase (EC 2.7.4.2).

As is shown in Fig. 15.12, the first reaction, catalyzed by mevalonate kinase with ATP as coenzyme, leads to the formation of mevalonate 5-phosphate with release of ADP. Mg^{2+} or Mn^{2+} ions, and to lesser extent Ca^{2+} ions, are needed to activate the enzyme. At low concentrations (<1 mM) the liver enzyme is more active with Mn^{2+} than with Mg^{2+}, but maximum activity is obtained with 4-mM Mg^{2+}, at which concentration Mn^{2+} is inhibitory. Its pH optimum is 7.3. Mevalonate kinase from liver is a sulfhydryl enzyme and needs either cysteine or 2-mercaptoethanol to maintain its activity; it has an M_r = 100,000. It was the first of the enzymes of sterol synthesis to be purified to homogeneity from hog liver (Beytia et al., 1970). Mevalonate kinase cannot act on the lactone, only on the anion, and preparations of mevalonolactone have to be hydrolyzed first with NaOH or KOH before use as substrate with the enzyme. However, with isolated cells, such as hepatocytes, the lactone can be used without previous hydrolysis because the cells apparently contain a lactonase that hydrolyzes the lactone. In fact, the lactone is taken up by the hepatocytes much faster and is utilized about ten times faster for cholesterol synthesis than the anion (Edwards et al., 1977b).

Mevalonate kinase is absolutely specific for the natural 3R-enantiomer (Lynen and Grassl, 1958; Cornforth et al., 1962) and is a most useful reagent for the resolution of synthetic (R,S)-mevalonate. The mevalonate kinase reaction can be driven to completion by inclusion of an ATP-regenerating system in the reaction mixture (e.g., pyruvate kinase + phosphoenol pyruvate) and the (3R)-mevalonate 5-phosphate can be easily separated from the unused (3S)-mevalonate. The (3R)-mevalonate 5-phosphate can be used as such in many enzyme systems or, if desired, the (3R)-mevalonate can be regenerated from the 5-phosphate by hydrolysis with alkaline phosphatase (Cornforth et al., 1966; Popják, 1969). The method can be used on the millimolar or μmolar scale. A micro-method for the determination of mevalonate in pmol amounts has been developed with mevalonate kinase and used for the quantitative determination of mevalonate in ultrafiltrates of blood plasma. Mevalonate is a normal constituent of blood in rat and man and its levels are related to the activity of HMG-CoA reductase in the liver: high HMG-CoA reductase activity is associated with high levels of blood mevalonate and vice versa (Popják et al., 1979).

The second kinase, phosphomevalonate kinase, converts mevalonate 5-phosphate with Mg-ATP to mevalonate 5-diphosphate and ADP (Fig. 15.12, reaction 2). So far only partially purified preparations of this enzyme from yeast, rat- and pig-liver, and the latex of *Hevea brasiliensis* have been reported (cf. Popják, 1969; Beytia and Porter, 1976). The liver enzyme is activated by Mg^{2+} and can barely function with Mn^{2+}. Zn^{2+}, which activates the yeast enzyme, inhibits the liver phosphomevalonate kinase. This enzyme, like mevalonate kinase, is a sulfhydryl protein and is strongly inhibited by *p*-mercuribenzoate, *N*-ethylmaleimide, and to a lesser extent by iodoacetamide. The phosphomevalonate kinase reaction is freely reversible: incubation of mevalonate 5-diphosphate with Mg-ADP and enzyme yields mevalonate 5-phosphate and ATP. The equilibrium of the reaction is in the forward direction; the ratio of mevalonate 5-diphosphate to mevalonate 5-phosphate is about 7:3.

The conversion of mevalonate 5-diphosphate to squalene in either the yeast or liver enzyme system still requires ATP. It is in this third ATP-consuming reaction that the universal donor of prenyl units for all polyprenyl syntheses is generated, which is 3-methylbut-3-enyl 1-diphosphate, usually referred to as isopentenyl pyrophosphate (Fig. 15.12, reaction 3). The formation of this substance was first recognized by Bloch and by Lynen and their associates (Bloch et al., 1959; Lynen et al., 1959). The enzyme catalyzing this reaction has been named ATP:5-diphosphomevalonate carboxy-lyase (dehydrating), and numbered EC 4.1.1.33, but is commonly referred to as mevalonate-5-pyrophosphodecarboxylase, or pyrophosphomevalonate decarboxylase. This is a very labile enzyme even when partially purified. The enzyme is activated by Mg^{2+}, Mn^{2+}, or Co^{2+}, and to a lesser extent by Ca^{2+}; ATP is its sole coenzyme. In the catalysis CO_2 and ADP are formed at identical rates and without a lag period (Bloch et al., 1959; Popják, 1969). Bloch et al. (1959) concluded that " . . . there is no evidence for the existence of a stable phosphorylated intermediate formed separately before decarboxylation." Nevertheless they postulated the transitory formation of a 3-phosphomevalonate 5-diphosphate and a concerted reaction mechanism for the decarboxylation as shown in Fig. 15.13 (a). An alternative reaction mechanism [Fig. 15.13 (b)] has also been proposed by Lindberg et al. (1962) in which the function of ATP is merely that of an acceptor of the leaving OH group with its electron pair, which is transferred to the terminal phosphate of ATP with displacement of ADP. The latter mechanism was favored by Cornforth et al. (1966) who demonstrated also that the decarboxylation–dehydration of mevalonate 5-diphosphate was a *trans*-elimination of the tertiary hydroxyl and the carboxyl groups.

Until recently only partially purified preparations of the pyrophosphomevalonate decarboxylase were available. A few years ago Shama Bhat and Rama-

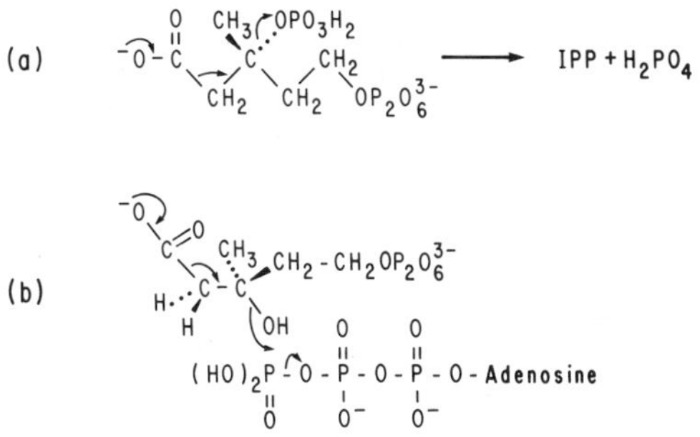

FIGURE 15.13. Possible mechanisms for the decarboxylation of mevalonate 5-diphosphate. In (b) the *trans*-elimination of the carboxyl and tertiary hydroxyl groups are illustrated.

sarma (1980) exploited their observation that *p*-coumaric acid was a competitive inhibitor of the rat-liver decarboxylase and prepared a *p*-coumarate–Sepharose B column. This column bound the decarboxylase firmly, from which, after elution of all other proteins with a pH 7.4 buffer containing ATP, MgCl$_2$, and 2-mercaptoethanol (0.2 mM each), the pyrophosphomevalonate decarboxylase could be eluted as a homogeneous protein with the same buffer containing also 10 mM mevalonate. The molecular weight of this rat-liver enzyme, determined by gel filtration, was 126,000; SDS/polyacrylamide gel electrophoresis gave a single band with M_r = 35,000. Thus the rat-liver enzyme is probably a tetramer. Like all previously reported partially purified pyrophosphodecarboxylases, this rat-liver enzyme proved to be very labile, also losing much of its activity in a few hours.

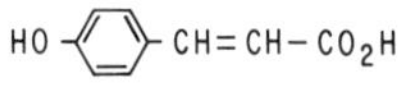

p - Coumaric acid

The mevalonate 5-diphosphate decarboxylase was also purified to homogeneity from chicken liver and, contrary to all previous experience with the similar enzyme from other sources, proved to be quite stable at $-20°C$ in 50% glycerol (Alvear *et al.*, 1982). Unfortunately, no new kinetic or mechanistic information has been gained so far with these pure preparations of the decarboxylase.

15.5.3. From Isopentenyl Diphosphate to Farnesyl Diphosphate

Isopentenyl diphosphate is potentially a bifunctional molecule. By virtue of its terminal vinyl group it is a nucleophile, but after isomerization to 3,3-dimethylallyl diphosphate (Lynen *et al.*, 1959) it assumes electrophilic character (cf. Cornforth, 1959). Thus, longer-chain polyprenyls are formed by the condensation of 3,3-dimethylallyl diphosphate with isopentenyl diphosphate.

The first step after the formation of isopentenyl diphosphate (IPP) is its isomerization to 3,3-dimethylallyl diphosphate (DMAPP) which is catalyzed by isopentenyl diphosphate Δ^3–Δ^2 isomerase (EC 5.3.3.2). The properties and available information on this enzyme have been reviewed by Holloway (1972) and more recently by Poulter and Rilling (1981a). This enzyme, like three previous ones acting on mevalonate, is a soluble cytosolic protein and has been recognized in extracts of yeast, pig liver, pumpkin, orange peel, and pine seedlings. The best preparations available are from pig liver (see also Popják, 1969). The enzyme catalyzes the reversible reaction shown in Fig. 15.14 and is activated by Mn^{2+} in preference to Mg^{2+} at a pH optimum of 6.0–6.3. The enzyme is probably a dimeric protein with an M_r of 83,000. At equilibrium the ratio DMAPP/IPP has been estimated variously to be 5:1 or 10:1. The enzyme is sensitive to iodoacetamide, which at 1 mM causes an 80% and at 2–5 mM a complete inhibition of catalysis.

The reaction promoted by isopentenyl diphosphate isomerase is stereospecific with respect to both the addition of a proton to the terminal vinylic carbon atom and the elimination of a proton from C-2. It was first shown with the aid of mevalonates labeled stereospecifically at C-4 with deuterium (4R and 4S) that the pro-R hydrogen atom at C-2 of IPP was eliminated in the reaction (Cornforth *et al.*, 1966; Popják and Cornforth, 1966). Subsequently Cornforth *et al.* (1972) deduced that the proton added to the terminal doubly bonded carbon atom was added to the *re*-face at C-4 of IPP, i.e., the face on which the diphosphate-bearing group, C-4, and the methyl group are seen in a clockwise order. Hence a concerted reaction mechanism for this isomerase reaction is conceivable without the necessity of invoking the formerly postulated enzyme-bound intermediate (cf. Fig. 15.14).

It was first recognized by Lynen *et al.* (1959) in yeast extracts that condensation between DMAPP (the electrophile) and IPP (the nucleophile) led to the formation of geranyl diphosphate (GPP, also an electrophile), and the latter's condensation with another molecule of IPP yielded the C_{15} farnesyl diphosphate (FPP), which was a precursor of squalene. Corroborating evidence, obtained with a soluble fraction of rat-liver or pig-liver homogenates, was provided by Popják (1959) and by Goodman and Popják (1960). The sequence of reactions leading to the synthesis of FPP was depicted as shown in Fig. 15.15. It was assumed that the reactions involved the elimination of pyrophosphate anion from the allylic substrates, creating a carbonium ion, and alkylation of IPP at C-4 with subsequent elimination of a proton from C-2 of an intermediate electron-deficient species. The reactions depicted in Fig. 15.15 are catalyzed by a single enzyme, prenyltransferase (EC 2.5.1.1), also known as farnesyl diphosphate synthetase.

The elongation of an allylic diphosphate by successive condensations with

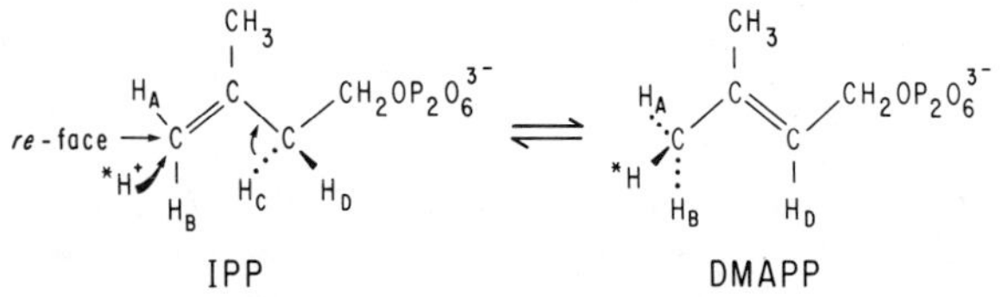

FIGURE 15.14. Mechanism and stereochemistry of the reaction catalyzed by isopentenyl diphosphate isomerase. H_C at C-2 of IPP is the pro-R and H_D is the pro-S hydrogen atom. The proton is added to the *re*-face at C-4, i.e., from above the plane of the paper.

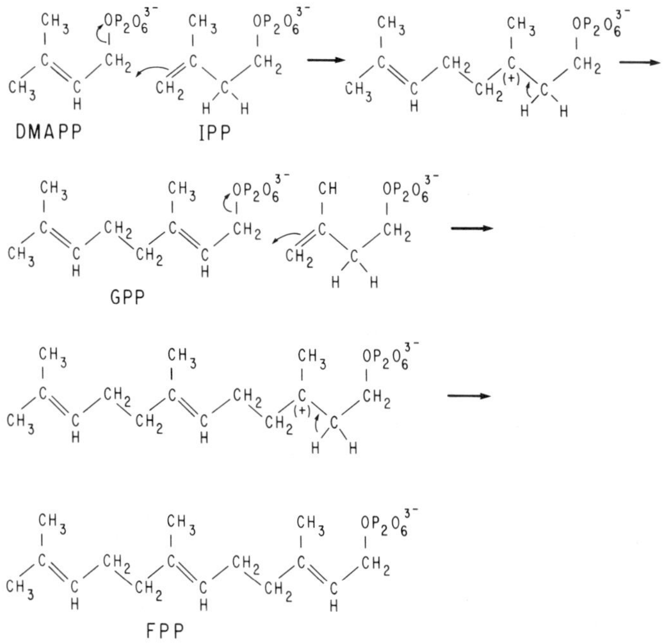

BIOSYNTHESIS OF CHOLESTEROL

FIGURE 15.15. Synthesis of farnesyl diphosphate (FPP) from 3,3-dimethylallyl diphosphate (DMAPP) and isopentenyl diphosphate (IPP) with geranyl diphosphate (GPP) as intermediate. The reactions are catalyzed by prenyltransferase.

IPP is an entirely novel biochemical mechanism of C—C bond formation, since in the biosynthesis of all other types of natural products—fatty acids, sugars, peptides—the C—C bonds are formed by Claisen- or aldol-type of condensation.

Cornforth and Popják and their colleagues examined the several stereochemical questions posed by the reactions catalyzed by prenyltransferase, which were summarized by Popják and Cornforth (1966) and are discussed in detail by Poulter and Rilling (1981a). Specimens of mevalonate labeled stereospecifically with a hydrogen isotope, mostly with deuterium, were used in those studies: $(5R)$-[5-D_1]-, $(4R)$- and $(4S)$-[4-D_1]-, and $(2R)$- and $(2S)$-[2-D_1]mevalonate. With the aid of these speciments of mevalonate, it could be deduced that in the two steps of C—C bond formation in the synthesis of farnesyl pyrophosphate from DMAPP + IPP the configuration at C-1 of the allylic pyrophosphate was inverted, but that the configuration at C-1 of farnesyl diphosphate (FPP) remained unchanged—R, as at C-5 of the starting $(5R)$-[5-D_1] mevalonate (Fig. 15.16). The simplest hypothesis to explain the observations was that the displacement of pyrophosphate ion was synchronous with the attachment of the new C_5 unit, i.e., a concerted reaction of the S_N2 type in which inversion of configuration is the rule.

The mechanism of the mevalonate 5-diphosphate decarboxylase reaction has been described earlier. From that study it was known that $(3R)$-[$(2R)$-2-D_1] mevalonate gave a [4-D_1]isopentenyl diphosphate in which the deuterium atom was

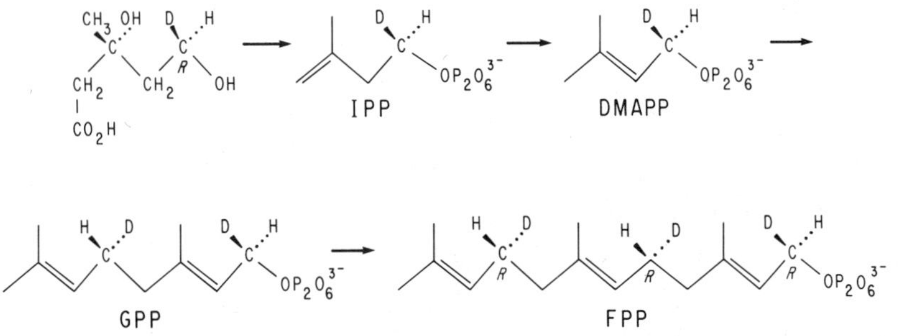

FIGURE 15.16. Stereochemistry of the prenyltransferase reaction studied with the aid of [(5R)-5-D$_1$]-mevalonate: the carbon-to-carbon bond formations proceed with inversion of configuration at C-1 of the allylic substrate. Although the orientation of deuterium atoms at C-5 and C-9 of farnesyl-PP is the opposite of that at C-1, the absolute configuration at C-5 and C-9 have to be assigned the R chirality according to the sequence rule of the R,S-convention.

in a *cis* position to the pyrophosphate-bearing group. When the [(2R)-2-D$_1$] mevalonate was used for the synthesis of farnesyl diphosphate, it was found that it contained two chiral centers at C-4 and C-8 with an absolute configuration of R (Fig. 15.17). The only possible interpretation of this finding was that the alkylation of IPP by the addition of an allylic residue (dimethylallyl, geranyl) to IPP was onto the *si*-face at C-4 of IPP, i.e., the face on which the $-CH_2CH_2OP_2O_6^{3-}$, the vinylic C-4, and the methyl groups are seen in an anticlockwise order, this face being behind the plane of the paper in Fig. 15.18.

The use of [(4R)-4-D$_1$]- and [(4S)-4-D$_1$] mevalonates on the other hand showed that in the formation of both GPP and FPP the pro-R hydrogen atom from C-2 of IPP (originally the pro-S-hydrogen atom at C-4 of mevalonate) was the one eliminated just as was indicated in the isomerization of IPP to DMAPP (cf. Fig. 15.14), i.e., the elimination of hydrogen is from the side of the molecule onto which the allylic residue is added. The elimination of the pro-R hydrogen atom from C-2 of IPP is apparently the rule in the synthesis of polyprenyls with the E-configuration (all-*trans*). In biosynthesis of the Hevea rubber which has the Z-configuration (all-*cis*) it is the pro-S hydrogen atom from C-2 of IPP (the pro-R hydrogen atom at C-4 of mevalonate) which is eliminated. The same rules apply in the biosynthesis of dolichols which contain prenyl units of both the E-

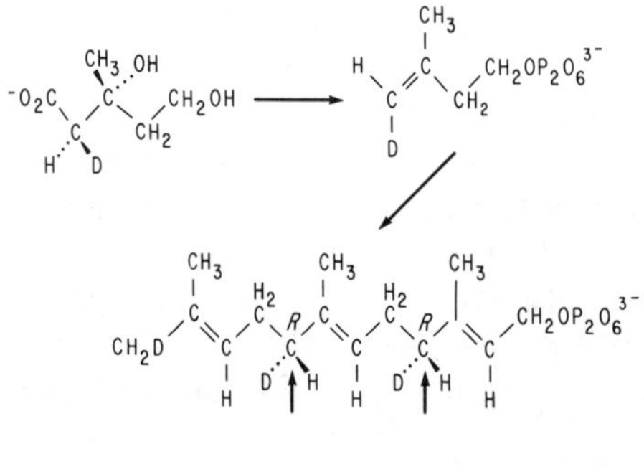

FIGURE 15.17. Synthesis of farnesyl diphosphate from [(2-R)-2-D$_1$]mevalonate. The farnesyl diphosphate contains two chiral centers. This observation meant that the allylic residue was coupled to the *si*-face at C-4 of isopentenyl diphosphate. See also Fig. 15.18.

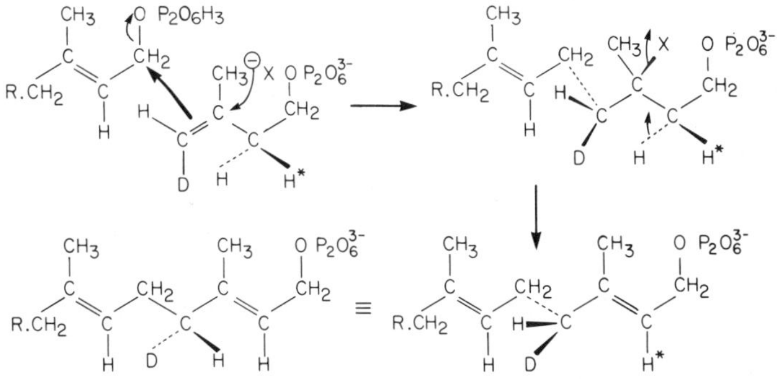

FIGURE 15.18. Interpretation of the mechanism of prenyltransferase reaction according to Cornforth *et al.* (1966). See text.

and Z-configuration: in segments with the *E*-configuration the 2-pro-*R*-and in segments with the *Z*-configuration the 2-pro-*S*-hydrogen atom of IPP is eliminated (cf. Hemming, 1983).

The unambiguous proof that in the synthesis of a polyprenyl (farnesyl diphosphate) with the *E*-configuration the addition of the allylic residue to C-4 of IPP and the elimination of the hydrogen from C-2 of IPP took place on the same side of IPP precluded the formulation of the synthesis as a concerted reaction, since in such processes at a double bond it is usually found that the "withdrawal and donation of the electrons proceed on opposite sides of the double bond" (Popják and Cornforth, 1966). For this reason the existence of an electron-donating X group that would be temporarily attached to C-3 followed by a *trans*-elimination of X and of the pro-*R* hydrogen from C-2 (Fig. 15.18) was postulated. The nature of X could not be defined and was thought to be either a group on the enzyme, such as an enzyme$-S^{-1}$, or the pyrophosphate of IPP itself.

This mechanism came under great scrutiny and attack by Poulter and Rilling (1981a) who conducted experiments with fluorinated analogues of substrates (*E* and *Z* trifluoromethylbut-2-en-1-yl diphosphate, 2-fluorogeranyl diphosphate, 2-fluoroisopentenyl diphosphate) and special experiments designed to trap an X-bound intermediate in the 1'-4 condensation catalyzed by the avian prenyl transferase. They could find no evidence for the existence of such an intermediate and firmly concluded that the X-hypothesis should be abandoned. They envisage that the 1'-4 condensation proceeds by elimination of the pyrophosphate group with creation of the allylic carbonium ion, which is followed by its condensation with C-4 of IPP. Poulter and Rilling (1981a, p.189) argue that "Formation of a covalent bond in concert with loss of the pyrophosphate group is not needed to preserve the chirality of the carbinyl carbon of the allylic substrate." They further believe that a substantial energy-barrier to rotation around the C-1—C-2 bond in the allylic cation that would preserve the chirality at C-1 of the allylic cation may exist. Short of burying oneself inside the enzyme, it is difficult to decide on the precise reaction mechanism. Poulter and Rilling still admit that the X-group mechanism was a "logical approach for rationalizing the 1'-4 condensation reactions . . ." but they think that "the stereochemical experiments did not constitute proof."

The enzyme catalyzing the synthesis of farnesyl diphosphate from 3,3-dimethylallyl diphosphate and isopentenyl disphosphate through the intermediacy of geranyl diphosphate (see Fig. 15.15), farnesyl diphosphate synthetase (prenyltransferase) is the best-studied enzyme among the many participating in sterol biosynthesis and among other prenyltransferases. It has been purified to a state of homogeneity from several sources, from baker's yeast *(Saccharomyces cerevisiae),* from the fungus *Phycomyces blakesleeanus,* and from the liver of pig, man, and chicken. The enzyme from the latter source readily crystallizes. All the purified enzymes seem to be similar dimeric proteins with molecular weights ranging from 74,000 (human) and 85,000 (chicken). The porcine enzyme exists in two interconvertible forms, A and B, which differ in their thiol oxidation–reduction states. Form A contains six titratable SH-groups, whereas form B contains only four per dimer. The B-form is convertible to form A by incubation with 2-mercaptoethanol, and incubation of form A with oxidized glutathione (GSSG) changes it to form B. Since the molecular weights of the two forms are similar, it is probable that intramolecular disulfide-bond formation is responsible for the A → B conversion. The chicken and human enzymes, both dimers, exist in only one form. The porcine and human enzymes are inhibited by phenylglyoxal, an arginine-modifying reagent; the evidence obtained suggests that the modified arginine residues, two per subunit, are located at the binding site for the allylic substrate. The porcine and human enzymes immunologically cross-react, though they are not identical, as on Ouchterlony double-diffusion plates spurs were formed between the precipitin lines of the B-form of porcine enzyme and the human enzyme. The binding site for the allylic substrates of these two enzymes contains also a free SH-group (Barnard *et al.,* 1978; Barnard and Popják, 1980, 1981). The amino acid compositions of the chicken, porcine, and human prenyltransferases are remarkably similar; in comparison with the average composition of 108 families of proteins, these liver enzymes are relatively rich in hydrophobic and aromatic amino acid residues and have a high content of Asx + Glx, but are poor in polar residues and half-cystines.

15.5.4. From Farnesyl Diphosphate to Squalene

Farnesyl diphosphate added to yeast liver microsomes with NADPH or NADH (or both) is converted to squalene in high yield (Lynen *et al.,* 1959; Goodman and Popják, 1960). In this case an interesting change in the cellular distribution of enzymes of squalene synthesis can be noted: whereas all the enzymes converting mevalonate to farnesyl diphosphate are soluble cytosolic proteins, squalene synthetase is attached to endoplasmic reticulum.

The synthesis of squalene from two molecules of farnesyl diphosphate generated many hypotheses to be subsequently abandoned for new ones. One of these hypotheses (Cornforth and Popják, 1959) was seemingly supported by the data of Rilling and Bloch (1959) who reported that squalene biosynthesized by a yeast autolysate from [5-D_2]mevalonate contained 10 atoms of deuterium instead of the theoretically possible 12, the two missing atoms having been lost from the two central carbon atoms of squalene. Rilling and Bloch (1959) reported further that when squalene was biosynthesized from [2-^{14}C]mevalonate in a medium of 99% D_2O, four atoms of deuterium were incorporated into the hydrocarbon; two of these were attached to the terminal carbon atoms at each end of the molecule and the other two to the central carbon atoms of squalene.

On reexamination of the biosynthesis of squalene in a liver enzyme system from [5-D_2]mevalonate and from [D_6-^{14}C]farnesyl diphosphate—obtained biosynthetically from [5-D_2-2-^{14}C]mevalonate—it was found that the squalene contained 11 atoms of deuterium (not 10), the one missing atom of the theoretically possible 12 having been lost from one of the central carbon atoms of squalene. Further, when squalene was biosynthesized by liver microsomes from [D_6-$^{14}C_3$] farnesyl diphosphate in ^3H_2O, no tritium was incorporated into the squalene. However, when ^{3}H-labeled NADPH was used with the doubly labeled farnesyl diphosphate, up to 0.8 µg atom of labeled hydrogen per µmol of squalene was transferred to one of the central carbon atoms of the hydrocarbon from the coenzyme (Popják et al., 1961a,b). These observations were reinforced by the use of chemically synthesized [1-$^3H_2$-2-^{14}C]- and [1-D_2-2-^{14}C]-*trans-trans*-farnesyl diphosphates in incubations with liver microsomes (Popják et al., 1961c; 1962). The squalene biosynthesized from [1-D_2-2-^{14}C] farnesyl diphosphate contained three atoms of deuterium, all three being on the two central carbon atoms of the hydrocarbon. The ^{3}H/^{14}C ratios in three specimens of squalene biosynthesized from [1-$^3H_2$-2-^{14}C] farnesyl diphosphate were 0.75, 0.76, and 0.77 relative to a ratio of 1.0 in the starting substrate, confirming the loss of one hydrogen atom from C-1 of one farnesyl diphosphate. Thus it appeared that the symmetrical squalene molecule was synthesized by some asymmetric process in the sense that the two participating molecules of farnesyl diphosphate were differently treated: one lost a hydrogen atom and the other one did not. Further the lost hydrogen atom was replaced by a hydrogen atom from NADPH. Thus the specimens of squalene biosynthesized from [5-D_2]mevalonate and [1-D_2-^{14}C]farnesyl diphosphate were asymmetrically labeled on one of the two central carbon atoms, thus: $-CHDCD_2-$.

In further experiments it was shown that the loss of the one hydrogen atom from C-1 and its replacement by a hydride ion from NADPH, or NADH, were stereospecific. The conclusions from the subsequent experiments, summarized by Popják and Cornforth (1966), were: (i) the hydrogen eliminated from C-1 of one of the two farnesyl diphosphate molecules participating in squalene biosynthesis was the pro-*S* hydrogen atom; (ii) the lost hydrogen atom was replaced stereospecifically by the pro-*S*-hydrogen atom from C-4 of the dihydronicotinamide ring of NADPH or NADH (the "B-side" of these coenzymes); (iii) the configuration at C-1 of the second farnesyl diphosphate molecule which did not participate in the (seeming) hydrogen exchange of the first one was inverted during synthesis of squalene (Fig. 15.19(A) and (B)).

Even before all the stereochemical features of the head-to-head condensation of two molecules of farnesyl diphosphate to form squalene became known, it was

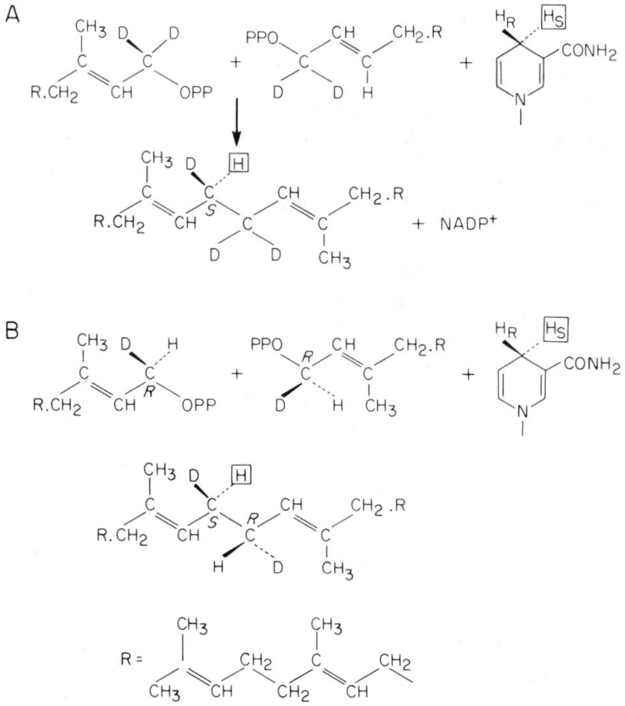

FIGURE 15.19. (A) Overall reaction of the synthesis of squalene from [1-D_2]farnesyl diphosphate; one of the central carbon atoms of squalene is a chiral center with absolute configuration *S*. (B) Stereochemistry of the synthesis of squalene from [(1*R*)-1-D_1]farnesyl diphosphate: the two central carbons are enantiotopically labeled. The succinic acid obtained from the four central carbon atoms was a *meso*-(2*S*, 2'*R*)-compound.

thought that this reaction could not be a one-step process, and specifically that "the removal of the hydrogen cannot involve a molecule of farnesyl pyrophosphate but must occur in a later intermediate derived from it" (Popják *et al.*, 1962, p. 61).

In 1966 Rilling reported the isolation of a C_{30} pyrophosphate-containing substance from incubations of yeast microsomes with [1-$^3H_2$-^{14}C]farnesyl diphosphate in the absence of NADPH. He found that the $^3H/^{14}C$ ratio in the new substance relative to that in the starting substrate indicated the loss of one labeled hydrogen atom out of four. Incubation of this new substance with NADPH and microsomes gave squalene. This substance became known as presqualene pyrophosphate and its dephosphorylated alcohol as presqualene alcohol. The structure of presqualene pyrophosphate was deduced first by Epstein and Rilling (1970) by mass spectrometric and NMR examination of presqualene alcohol obtained by LiAlH$_4$ reduction of the pyrophosphate ester and by the comparison of the NMR spectrum of presqualene alcohol with that of *trans*-chrysanthemol obtained by LiAlH$_4$ reduction of ethyl chrysanthemumate. They showed that presqualene alcohol derived from the pyrophosphate ester which was synthesized from [1-D_2]-

farnesyl diphosphate contained only three atoms of deuterium. Thus the loss of the one hydrogen atom from C-1 of one farnesyl diphosphate molecule, noted previously to occur during squalene synthesis, was shown to occur during the formation of presqualene pyrophosphate. Epstein and Rilling (1970) concluded that presqualene alcohol was a substituted cyclopropylcarbinol. Edmond *et al.* (1971) fully confirmed and extended the findings of Epstein and Rilling (1970) and were able to assign the relative positions of the substituents on the cyclopropane ring, thus reducing the possible number of stereoisomers from eight to two, 1R, 2R, 3R or 1S, 2S, 3S. The structure of presqualene alcohol (Fig. 15.20) is *cis-* (1,2-anti-1,3-anti)-3-(2′,6′,10′-trimethylundeca-2′,6′,10′-trienyl)-2-(2″,6″-dimethylnona-2″,6″-dienyl)-2-methyl-1-hydroxymethylcyclopropane; presqualene pyrophosphate is the pyrophosphate ester of the alcohol. The structure was confirmed by chemical synthesis independently in three laboratories (Altman *et al.,* 1971; Campbell *et al.,* 1971; Coates and Robinson, 1971).

Presqualene pyrophosphate is barely soluble in aqueous alkali, but is soluble in benzene, chloroform, or ethyl acetate. It is optically active with specific rotation in $CHCl_3$ at 588 nm of $+51°$ and at 286 nm of $+596°$, showing a normal optical rotatory dispersion curve (Popják *et al.,* 1969). The absolute configuration of presqualene alcohol shown in Fig. 15.20 is 1R, 2R, 3R as deduced by Popják *et al.* (1973).

To obtain presqualene pyrophosphate with the correct stereochemistry it is necessary to postulate that the two FPP molecules are aligned parallel on the enzyme and that C-1 of one FPP is added to the 2-*si*, 3-*re* face of the other, the donor of the alkyl group losing and the acceptor retaining the pyrophosphate group (Fig. 15.21). In this figure H_A and H'_A are the pro-R hydrogen atoms at C-1 of the two FPP molecules. It was shown that the hydrogen atom at C-3 of the cyclopropane ring was formerly the pro-R hydrogen atom at C-1 of one of the FPP molecules and that the absolute configuration at C-1 of the second FPP mol-

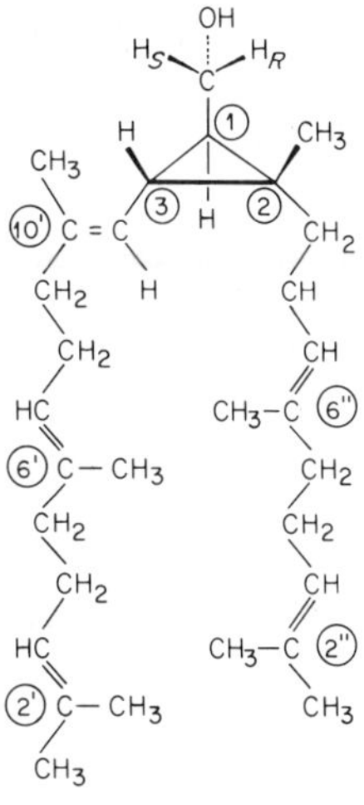

FIGURE 15.20. Structure of presqualene alcohol. The absolute configurations at the three asymmetric centers on the cyclopropane ring are all R. H_S and H_R on the carbinol carbon were originally the pro-S and pro-R hydrogen atoms at C-1 of that farnesyl diphosphate molecule which does not lose a hydrogen atom. The hydrogen atom at position 3 on the cyclopropane ring was originally the pro-R hydrogen atom of the second farnesyl diphosphate molecule.

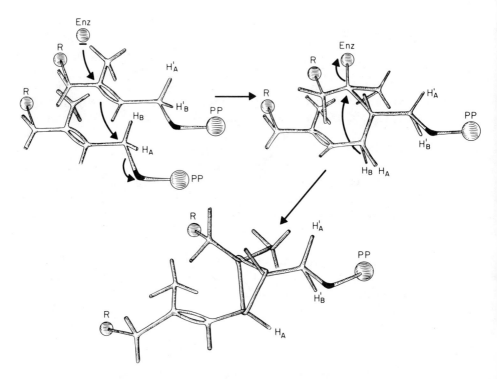

FIGURE 15.21. Hypothetical mechanism of presqualene pyrophosphate synthesis according to Edmond *et al.* (1971). With permission of the authors and the copyright holder, the American Society of Biological Chemists.

ecule was retained on the carbinyl carbon of presqualene pyrophosphate. Moreover, experiments with [1-^{18}O]FPP—biosynthesized from [5-^{18}O]mevalonate (Cornforth and Gray, 1975)—showed that the presqualene alcohol after hydrolysis by microsomal phosphatase(s) contained the same atom % of excess ^{18}O as the substrate [1-^{18}O]FPP (Popják *et al.*, 1975).

Cornforth (1973) conjectured that presqualene pyrophosphate was not a true intermediate in squalene biosynthesis, but merely an artifact created by depriving the enzyme of NADPH. This argument has been refuted in several laboratories by several experiments from which the inescapable conclusion in the words of Corey and Volante (1976) is "that presqualene pyrophosphate is an *essential intermediate* in squalene biosynthesis in liver; that is, there is no pathway from mevalonate to squalene which does not go through this intermediate."

There has been much speculation about the mechanism of the isomerization and reduction of presqualene pyrophosphate to squalene (for review, see Poulter and Rilling, 1981b; Popják and Agnew, 1979). Perhaps the most acceptable model is that of Poulter *et al.* (1974), arrived at from solvolysis experiments with C_{10} analogues of presqualene pyrophosphate (Fig. 15.22). In this model events are started by the elimination of the pyrophosphate group from a conformer of presqualene pyrophosphate in which the C(1)—oxygen bond is *trans* (antiparaplanar)

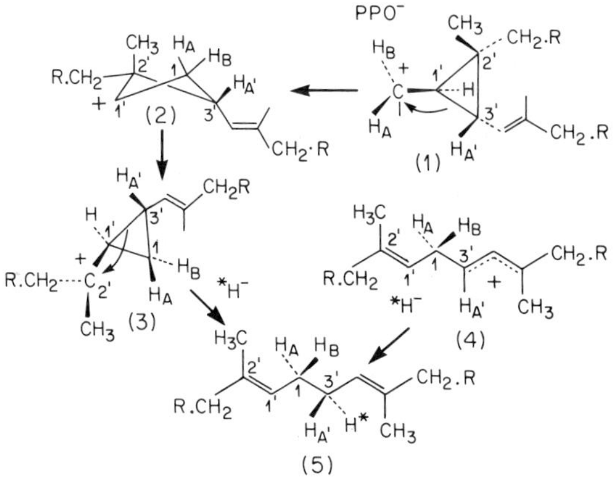

FIGURE 15.22. Possible mechanism of isomerization and reduction of presqualene pyrophosphate to squalene according to Poulter et al. (1974).

to the C(1′)−C(3′) cyclopropane bond with the generation of the cyclopropyl carbinyl cation and the pyrophosphate anion held as an ion pair on the enzyme. The orientation of the hydrogen atoms at C(1) would be maintained because of the large rotational barrier about the C(1)−C(1′) bond. The enzyme now directs the stereospecific migration of the C(1′)−C(3′) bond with the creation of the cyclobutyl cation [(2) in Fig. 15.22]. The stereospecific transfer of hydride ion from NADPH would then proceed with an inversion of configuration at C(3′) in (3) or (4) (Fig. 15.22).

A simpler model of the transformation of presqualene pyrophosphate was presented by Edmond et al. (1971). In this model a two-step isomerization of presqualene pyrophosphate to squalene was assumed (Fig. 15.23). In the first step ring expansion and migration of pyrophosphate leads to the tautomeric *trans*-cyclobutyl pyrophosphate with inversion of configuration around the carbinyl carbon of presqualene pyrophospate. This cylobutyl pyrophosphate is, in fact, a ligand-stabilized form of the cyclobutyl cation (2) shown in Fig. 15.22. In the second step reductive cleavage by NADPH leads to ring-cleavage, generation of the

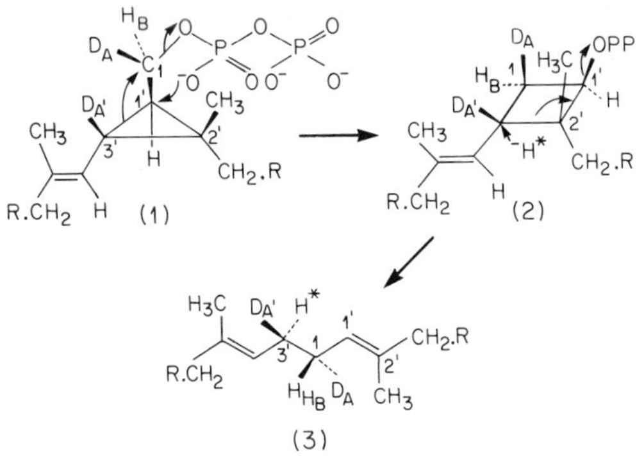

FIGURE 15.23. Hypothetical mechanism for conversion of presqualene pyrophosphate into squalene. After Edmond et al. (1971).

double bond, and elimination of the pyrophosphate anion. This model as well as that of Poulter *et al.* (1974) fully account for the overall stereochemistry of the biosynthesis of squalene from farnesyl diphosphate.

The enzyme catalyzing the synthesis of squalene from farnesyl diphosphate is called squalene synthetase, but presumably because it is still ill defined it has not yet been given an EC number. Squalene synthetase is an intrinsic microsomal protein. Squalene synthesis can be described by two partial reactions shown as (1) and (2) in Scheme 2, while the overall reaction is given by (3) in Scheme 15.2.

(1) $2 \times$ farnesyl-PP $\xrightarrow{Mg^{2+}}$ presqualene-PP + H$^+$ + PP$_i$

(2) presqualene-PP + NADPH $\xrightarrow{Mg^{2+}}$ squalene + PP$_i$ + NADP$^+$

(3) $2 \times$ farnesyl-PP + NADPH $\xrightarrow{Mg^{2+}}$ squalene + 2PP$_i$ + H$^+$ + NADP$^+$

SCHEME 15.2. Reactions catalyzed by squalene synthetase. NADH can substitute for NADPH in reaction (2) or (3), but NADPH gives generally a higher reaction rate.

In contrast to all other reductive biochemical reactions with NADH or NADPH, in which both the hydride ion from the coenzymes and a proton from the medium are taken up by the reductant, in the squalene synthetase reaction only the hydride ion from the coenzyme appears in the product.

The assays for the partial reactions or the overall reaction are simple, but require isotopically labeled substrates, [^{14}C]- and [1-^{3}H$_2$]FPP, together with analytical method for separation of products and ^{3}H$^+$ from substrates. Release of ^{3}H from [1-^{3}H$_2$]FPP into the medium gives a measure of the synthesis of presqualene-PP, and isolation of [^{14}C]squalene or [^{3}H]squalene, after alkaline hydrolysis of the reaction mixture, gives a measure of the overall reaction. Because of the presence of phosphatases in microsomes, which hydrolyze even some of the substrates, and because of NADH and NADPH oxidases in microsomes, neither measurement of PP$_i$ released from the substrates nor consumption of the reduced coenzymes can be used with any of the available preparations for assay of squalene synthetase (cf. Popják and Agnew, 1979; Poulter and Rilling, 1981b).

One of the unsolved problems associated with squalene synthetase is the question whether the two half-reactions of squalene synthesis are catalyzed by two independent proteins which happen to be associated with common membrane particles or by one protein with two catalytic sites, and also whether two distinct binding sites exist for the two FPP molecules. Kinetic experiments carried out with washed yeast microsomes strongly favored the view that a single protein with two catalytic sites was involved, although they did not prove it. Also solubilization of the squalene synthetase from yeast microsomes with sodium deoxycholate and its sedimentation by sucrose density-gradient centrifugation in the presence of deoxycholate, or chromatography on Sephadex G-200 equilibrated with buffer containing deoxycholate, failed to separate the two activities. The solubilized squalene synthetase, in the presence of deoxycholate, is a relatively small molecule with an M_r of 55,000. However on removal of deoxycholate with cholestyramine it aggregates into particles sedimenting at 100,000 × g. The process is entirely reversible on addition of 0.2% deoxycholate. Squalene synthetase loses much of its activity during chromatography on Sephadex G-200 in the presence of deoxycholate. However, it can be readily reactivated by the addition of phosphatidylcholine or phosphatidylethanolamine, but not by phosphatidylserine or phosphatidylinositol. Thus squalene synthetase seems to be a proteolipid (Agnew and Popják, 1978a,b).

Ortiz de Montellano et al. (1976a,b) provided convincing evidence for the existence of two distinct binding sites. They found that 2-methylfarnesyl-PP and 3-desmethylfarnesyl-PP, although inhibitors of squalene synthetase, reacted on the enzyme with farnesyl-PP (but not with themselves) to form 11-methylsqualene and 10-desmethylsqualene, respectively. They showed that these analogues acted only as acceptors of the triprenyl unit from that farnesyl-PP molecule which loses one α-proton. When the analogues were labeled with ^{3}H on C-1, all the ^{3}H was retained in 11-methyl- and 10-desmethylsqualene.

15.5.5. Cyclization of Squalene to Lanosterol

After the early demonstration of conversion of squalene into lanosterol and the latter's degradation to cholesterol (cf. Bloch, 1965), the mechanism of squalene cyclization to lanosterol became of prime interest. Much of the knowledge about this cyclization can be attributed to the extraordinary perception of the late Leopold Ruzicka (1953) who postulated that the structures of all cyclic terpenes, monoterpenes, diterpenes, sterols, tetra- and pentacyclic triterpenes—whose structures cannot even be cleanly dissected into prenyl units—can be derived from the appropriate open-chain olefinic precursors, geraniol, nerol, farnesol, and squalene, by the application of mechanistic rules of organic chemistry. His postulate became known as the "biogenetic isoprene rule" and it was left to future generations to prove it or disprove it by experiment.

Eschenmoser et al. (1955) extended the biogenetic isoprene rule to sterols and tetracyclic triterpenes and assumed that all originated by cyclization of squalene. For an understanding of the cyclization mechanisms it is necessary to recount briefly the assumptions made by Eschenmoser et al. (1955). They supposed that, in all cyclizations of olefins, additions to double bonds proceed in an antiparallel *(trans)* manner, which means that during the configuration-determining phase of the reaction the four centers participating in the addition process (ACCB in Fig. 15.24) remain coplanar in a plane perpendicular to the plane of

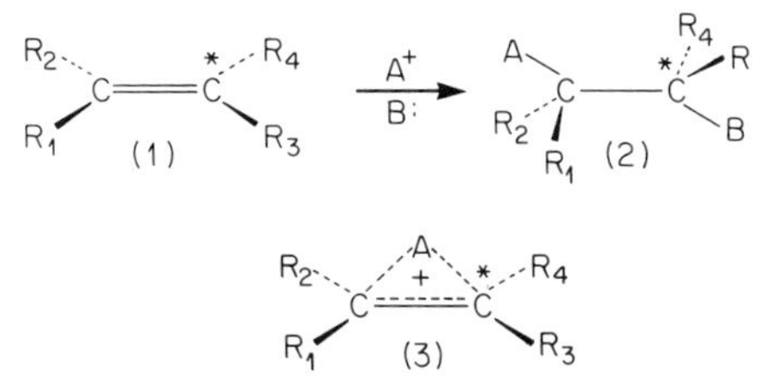

FIGURE 15.24. Antiparallel additions to double bond according to Eschenmoser et al. (1955).

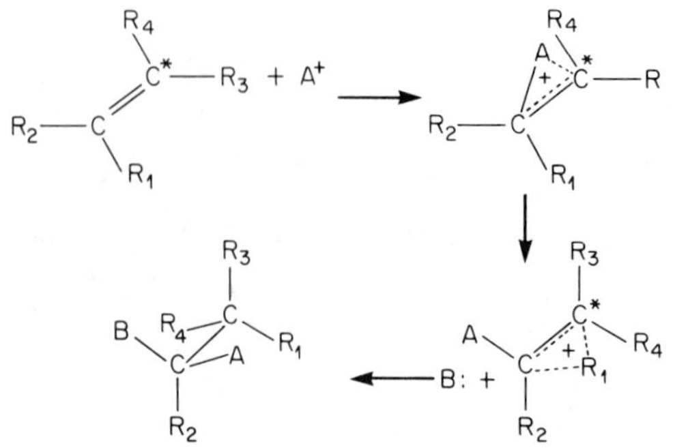

FIGURE 15.25. Rearrangements of substituents around double bond during antiparallel additions to it by 1,2 shifts according to Eschenmoser et al. (1955).

the original double bond. This type of antiparallel addition results in a well-defined steric configuration of the addition product. Such antiparallel additions could proceed through a cationic intermediate (Fig. 15.24 (3)), which has a "stable" configuration. Eschenmoser et al. (1955) used throughout their discussion such nonclassical carbonium ions to illustrate graphically "reaction planes" or the directions of the attack of addition bases. By the participation of nonclassical carbonium ions, rearrangements of the original four substituents around the double bond can occur as illustrated in Fig. 15.25.

Eschenmoser et al. (1955) pointed out further that the product of the cyclization of an olefin would depend on the conformation of the folding of the olefin preceding cyclization. As is shown in Fig. 15.26 the outcome of the cyclization of

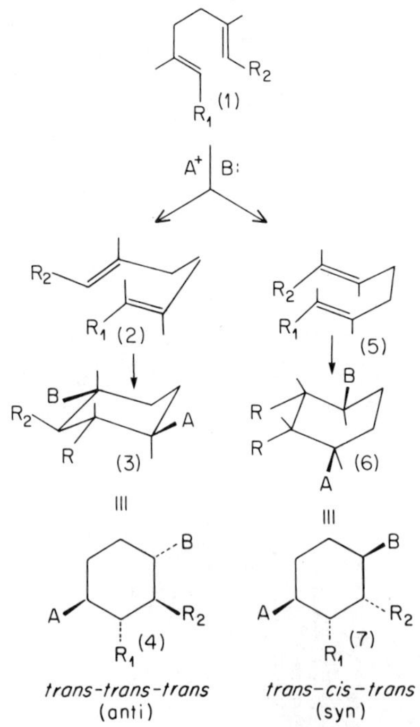

FIGURE 15.26. Cyclization of an olefin from chair (2) or boat (5) conformation according to Eschenmoser et al. (1955).

BIOSYNTHESIS OF CHOLESTEROL 347

the simple olefin (1) would be quite different when the olefin is cyclized from a "chair" or a "boat" conformation. Energetically the cyclization from a chair conformation is the prerequisite.

For the cyclization of squalene to give lanosterol with all of its conformational and stereochemical features a chair-boat-chair-boat conformation of squalene, imposed by the cyclizing enzyme, is required.

Eschenmoser et al. (1955) made their proposals before the discovery of 2,3-oxidosqualene (Corey et al., 1966; van Tamelen et al., 1966) and postulated an HO^+ cation as the initiator of the cyclization. However, that is equivalent to an enzyme-proton initiation of the cyclization of squalene 2,3-oxide.

The cyclization of 2,3-oxidosqualene to lanosterol by the concerted reaction mechanism, a non-stop reaction without the formation of any stable intermediates, as postulated by Eschenmoser et al. (1955) is shown in Fig. 15.27. On initiation of the cyclization by the break of the C(2)—O bond, the neighboring double bond provides the nucleophile to form the C—C bond between C-2 and C-5 and the process is repeated until completion of ring closures leaves an electron deficiency at C-20. This deficiency is then corrected by a series of backward rearrangements—the migration of hydrogen from C-17 to C-20 and from C-13 to C-17α, and by two 1,2-methyl shifts, one from C-14 to C-13β and a second from C-8 to C-14α. The molecule is finally stabilized by the elimination of a proton from C-9 creating the 8(9)-double bond of lanosterol.

The remarkable aspect of this hypothesis is that it stood the test of experiments. The methyl-migration from C-8 to C-14α was demonstrated in Bloch's laboratory (Maudgal et al., 1958), and migration of the methyl group from C-14 to C-13β in the laboratories of Cornforth and Popják (Cornforth et al., 1959). The hydride shifts were also demonstrated, but a little later (Cornforth et al., 1965; Barton et al., 1971).

For the conversion of squalene to lanosterol two enzymes, or enzyme sys-

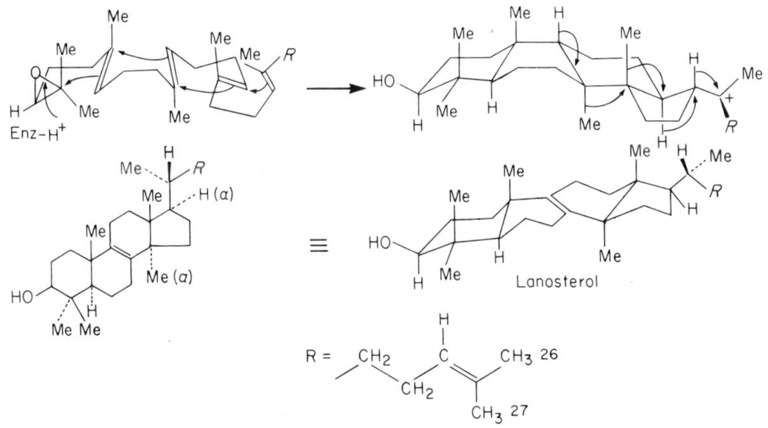

FIGURE 15.27. Cyclization of squalene 2,3-oxide from chair-boat-chair-boat conformation postulated by Eschenmoser et al. (1955).

tems, are needed: (i) squalene epoxidase (squalene monooxygenase (2,3-epoxidizing), EC 1.14.99.7), and (ii) 2,3-oxidosqualene lanosterol- cyclase (EC 5.4.99.7).

The epoxidase can be assayed conveniently with [^{3}H]10,11-dihydrosqualene as substrate, because, although it is epoxidized at either end, the epoxides of 10,11-dihydrosqualene are not cyclized (Corey and Russey, 1966). The epoxidase in the liver is localized in the microsomes and requires NADPH and oxygen for activity and is stimulated by FAD, a small protein (M_r = 44,000) from the soluble supernatant of homogenates (SPF), and a heat-stable factor which is probably a phospholipid (phosphatidylserine, phosphatidylglycerol, or phosphatidylinositol) (Tai and Bloch, 1972). Triton X-100 can replace the two cytoplasmic factors and it also solubilizes the microsomal epoxidase. The solubilized epoxidase was separated into two components, A and B, neither of which alone functions as an epoxidase. A combination of A + B fortified with FAD in the presence of Triton X-100 (0.3%), NADPH, and O_2 provides an effective epoxidase system. Component B is probably identical with the flavoprotein NADPH: cytochrome c reductase. Hemoproteins, such as cytochrome P-450 or cytochrome b_5, are not involved in squalene epoxidase, which is not inhibited by either carbon monoxide, potassium cyanide, or o-phenanthroline (Ono and Bloch, 1975). The liver oxidosqualene lanosterol cyclase also is attached to microsomes from which it was solubilized with deoxycholate and partially purified free of the epoxidase by ammonium sulfate fractionation. It needs no cofactors and converts 2,3-oxidosqualene anaerobically into lanosterol (Dean, 1969; for review, see Schroepfer, 1982).

15.5.6. Conversion of Lanosterol to Cholesterol

The conversion of lanosterol to cholesterol is probably the most complex among all the many reactions so far discussed, with still lingering uncertainties on account of the great technical difficulties encountered in establishing precisely precursor–product relationships among the several possible intermediates. This problem has been reviewed in much detail by Schroepfer *et al.* (1972), Schroepfer (1982), and Myant (1981, pp. 207–213). Schroepfer's reviews contain extensive bibliographies on the subject.

The main features of the transformation of lanosterol into cholesterol involve the removal of three methyl groups, the saturation of the 24(25)-double bond, and the isomerization of the 8(9)-double bond of lanosterol to Δ^5 in cholesterol with changes of some of the nuclear hydrogen atoms.

The magnitude of the problem of deciding the order in which these transformations occur can be best gauged by inspection of Table 1 in Schroepfer's 1982 review in which he lists no less than 57 sterols which have been detected in tissues; among these, 31 have been shown to be convertible into cholesterol. Schroepfer (1982) suggests that the following critieria should be met to establish that a particular substance is an intermediate in the lanosterol → cholesterol conversion: (i) isolation from tissues and unequivocal determination of structure; (ii) demonstration of the enzymatic formation of the substance from a known precursor; (iii) demonstration of convertibility into cholesterol; (iv) demonstration of formation from the postulated immediate precursor; and (v) demonstration of conversion into the next postulated intermediate. It is difficult to meet these criteria because most of the enzymes involved are membrane-bound in microsomes, and none has been characterized; hence one is usually forced to work with crude homogenates. Moreover, common methods of enzymology cannot be applied as the substrates are barely soluble in water and have to be added to the enzyme system either in detergents or organic solvents, which may be deleterious to the enzymes, and the added substrates are likely to be in micelles rather than in solution.

Because of all these problems it is possible only to define the probable pathway or pathways of the lanosterol → cholesterol conversion, particularly as it is likely that more than one pathway exists.

15.5.6.1. Reduction of 24(25)-Double Bond

Reduction of the 24(25)-double bond may occur at any stage of the lanosterol → cholesterol conversion, before or after loss of any of the three "extra" methyl groups. The Δ^{24}-reductase is a microsomal enzyme with NADPH as its coenzyme and it can reduce the Δ^{24}-double bond of lanosterol or of any other Δ^{24}-sterol.

A general rule for sterol reductases, which use NADPH as coenzyme, is that the reduction starts with the addition of a proton from the medium to the least substituted C-atom with the generation of a carbonium ion, which is then reduced by the hydride ion from NADPH:

$$\diagdown C = y \diagup H^+ \longrightarrow \diagdown +C \diagup \text{—y—H} \xrightarrow{\text{NADPH}^*} H^* \text{—} C \text{—y—H}$$

In the reduction of the 24(25)-double bond the proton is added to the 24-pro-S position followed by cis-addition of the hydride ion to C-25. Thus the C-26 methyl group of lanosterol, originally C-2 of mevalonate, becomes the pro-R methyl group attached to the prochiral C-25 (see also Section 15.3, Scheme 15.1).

The sterol Δ^{24}-reductase is selectively inhibited by the drug Triparanol, or MER-29 (for structure see Merck Index, 9th edition, entry no. 9403). Animals treated with Triparanol do not make cholesterol, but accumulate instead desmosterol (cholest-5,24-dien-3β-ol).

15.5.6.2. Removal of the 14α-Methyl Group (C-32)

It was shown in the earliest experiments on the transformation of lanosterol to cholesterol that elimination of the 14α-methyl group as CO_2 was probably the first step of the tranformation (cf. Bloch, 1965). However, it was found later that the 14α-methyl group (C-32 of lanosterol) was eliminated from the 32-aldehyde of lanosterol as formic acid, accompanied by the stereospecific loss of the 15α-hydrogen atom with the creation of the 14(15)-double bond.

The hydroxylation of the 14α-methyl group, its oxidation to the aldehyde, and even the elimination of the aldehyde group requires O_2 and NADPH (Akhtar *et al.*, 1978). Thus, the oxidation of the 32-alcohol to the aldehyde is probably not mediated by an alcohol dehydrogenase. Carbon monoxide inhibits the synthesis of cholesterol from [^{14}C]mevalonate in liver homogenates and causes the accumulation of [^{14}C]lanosterol. This observation suggests the participation of a mixed-function oxidase and cytochrome P-450 in the removal of the 14α-methyl group, supporting the findings of Akhtar *et al.* (1978), who suggested that the conversion of the 32-alcohol into the aldehyde may be mediated by a mixed-function oxidase involving a *gem*-diol intermediate which is dehydrated to the aldehyde (Fig. 15.28). Akhtar *et al.* (1978) proposed two possible ways for the elimination of the 32-aldehyde group: (i) stereospecific hydroxylation at C-15 (Fig. 15.28, (5)) followed by the elimination of formic acid and the 15-OH; and (ii) a reaction analogous to a Baeyer–Williger oxidation at C-14 that would produce the 14-formyllanosterol (Fig. 15.28, (7)), which is then deformylated, accompanied by elimination of the 15α-hydrogen atom.

The reduction of the 14(15)-double bond is catalyzed also by a microsomal Δ^{14}-reductase with NADPH as the coenzyme in accordance with the general rule described a little earlier: the proton is added to C-15 at the β-position and the hydride ion from the coenzyme becomes the 14α-hydrogen.

15.5.6.3. Elimination of the 4,4-gem-Dimethyl Groups

The two methyl groups at C-4 of lanosterol are eliminated in sequence as CO_2, the 4α-methyl group, derived from C-2 of mevalonate, being the first to be

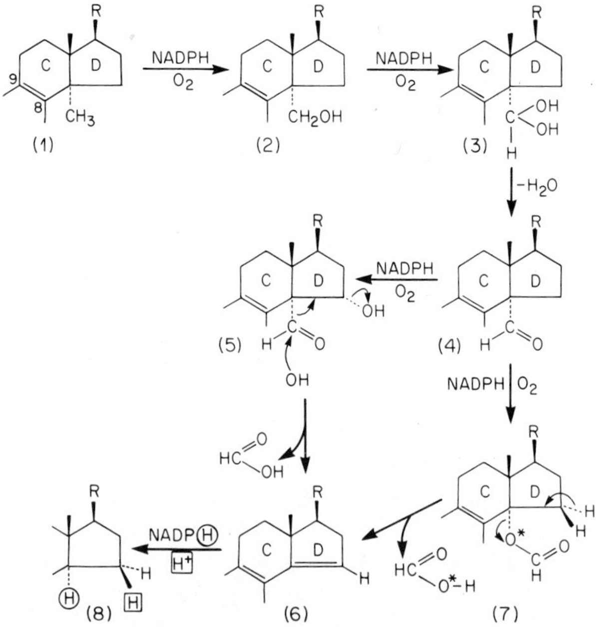

FIGURE 15.28. Possible pathways for the elimination of the 14α-methyl group of lanosterol according to Akhtar *et al.* (1978).

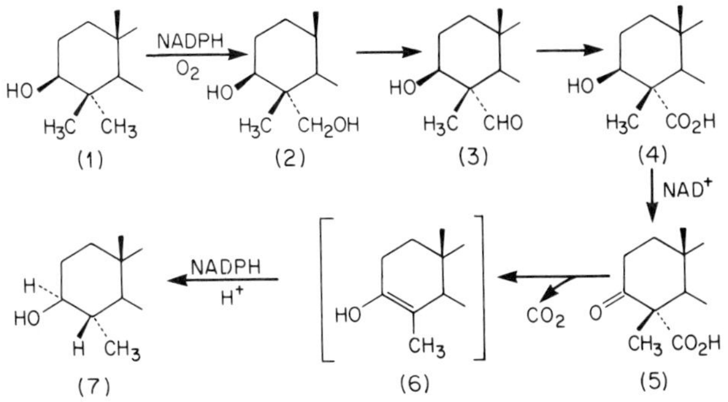

FIGURE 15.29. Probable sequence of reactions in the elimination of the 4α-methyl group as CO_2 during cholesterol biosynthesis.

lost (cf. Gaylor, 1972). The reactions are catalyzed by microsomal enzymes requiring O_2, NADPH, and NAD^+. The loss of each of these methyl groups involves six or seven discrete steps. These steps, illustrated in Fig. 15.29, are initiated by a mixed-function oxidase without the participation of cytochrome P-450. Whether the propagation to the 4α-carboxylate is also a function of this oxidase, or of an alcohol and aldehyde dehydrogenase, is uncertain. The decarboxylation is preceded by the dehydrogenation of the 3β-hydroxy group to the 3-oxo-compound with NAD^+ as coenzyme. The enolate (6) in Fig. 15.29 is conjectural, but could be an intermediate before reduction of the 3-oxo-compound to the 3β-hydroxy-4α-methyl sterol. The notable feature of these steps is that the originally 4β-methyl group is isomerized to 4α and hence, with great economy, the same enzyme system can carry out its elimination.

15.5.6.4. Isomerization of the 8(9)-Double Bond

The rearrangement of the 8(9)-double bond leads to the final stages of the derivation of cholesterol from lanosterol. The three steps involved in the rearrangement of the 8(9)-double bond to the 5 position are catalyzed by microsomal enzymes. The first step is the isomerization of the 8(9)-double bond to 7(8) and proceeds by the uptake of a proton from the medium at the 9α-position and the stereospecific elimination of the 7β-hydrogen. This isomerization is reversible. The product of the reaction is a Δ^7-sterol. The second step is the introduction of the Δ^5-double bond in a Δ^7-sterol and is effected in the presence of O_2 by a dehydrogenase which removes the 5α- and 6α-hydrogen atoms from the Δ^7-sterol, producing in ring B the $\Delta^{5,7}$-diene structure. 7-Dehydrocholesterol (cholesta-5,7-dien-3β-ol) is one of the obligatory intermediates in the biosynthesis of cholesterol. The final step from 7-dehydrocholesterol to cholesterol is the reduction of the 7(8)-double bond with NADPH, a proton being added from the medium to the 8β-position and the hydride ion from NADPH being added to the 7α-position. The rearrangements of the 8(9)-double bond, summarized in Fig. 15.30, may occur at more than one stage of the lanosterol → cholesterol conversion.

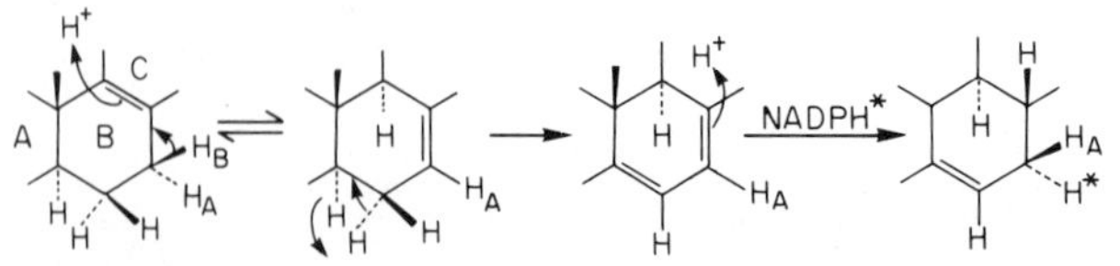

FIGURE 15.30. Rearrangement of the 8(9)-double bond to position 5.

15.5.6.5. Overall Transformation of Lanosterol to Cholesterol

Figure 15.31 shows the most probable intermediary stages of the lanosterol → cholesterol conversion. In Fig. 15.31 only the stable intermediary products that might be found in various tissues under steady-state conditions are shown. There are two streams of intermediates. In one stream the 24(25)-double bond is maintained up to the penultimate step, desmosterol. Under the influence of the drug, Triparanol, this stream dominates and desmosterol is not reduced to cholesterol, because the Δ^{24}-reductase is poisoned. In the second stream the 24(25)-double bond is reduced early or in any other intermediate which might have already been stripped of the 14α-methyl or *gem*-dimethyl groups. In biosynthetic experiments, lanosterol is nearly always accompanied by 24,25-dihydrolanosterol, which can be converted into cholesterol just as readily as is lanosterol.

15.5.6.6. Regulation of Cholesterol Biosynthesis by the LDL-Receptor Pathway

Many species have contributed to the evolution of knowledge on the biosynthesis of cholesterol: the humble yeast (acetic and mevalonic acids plus enzyme systems), the shark (squalene), the sheep (lanosterol), and the thousands of rats whose livers not only provided homogenates to many researchers but also revealed the diurnal cycle of HMG-CoA reductase and thus called attention to an important regulatory step in cholesterol biosynthesis. Apart from the scientists' contributions to unraveling the complex process of sterol biosynthesis, man—by one of the human genetic abnormalities—also made an important contribution to the biochemistry of the regulation of cholesterol metabolism.

Goldstein and Brown (1973) studied the regulation of HMG-CoA reductase in cultured human fibroblasts and found that when the cells were placed in a medium containing lipoprotein-deficient serum, the reductase levels rose to high levels as judged by direct assay of the enzyme or by incorporation of [^{14}C]acetate into cholesterol. The induced high levels of the reductase could be repressed, in a saturable manner, by increasing concentrations of low-density lipoprotein (LDL). On examining fibroblasts from individuals afflicted by the genetic disorder familial hypercholesterolemia, Goldstein and Brown (1973) found that these cells had a high level of HMG-CoA reductase whether they were grown in a lipoprotein-deficient or lipoprotein-containing medium and that LDL had no effect whatever on the reductase in these genetic mutant cells.

From subsequent numerous and elegant studies by Goldstein and Brown and their colleagues an entirely new picture of the regulation of cholesterol metabo-

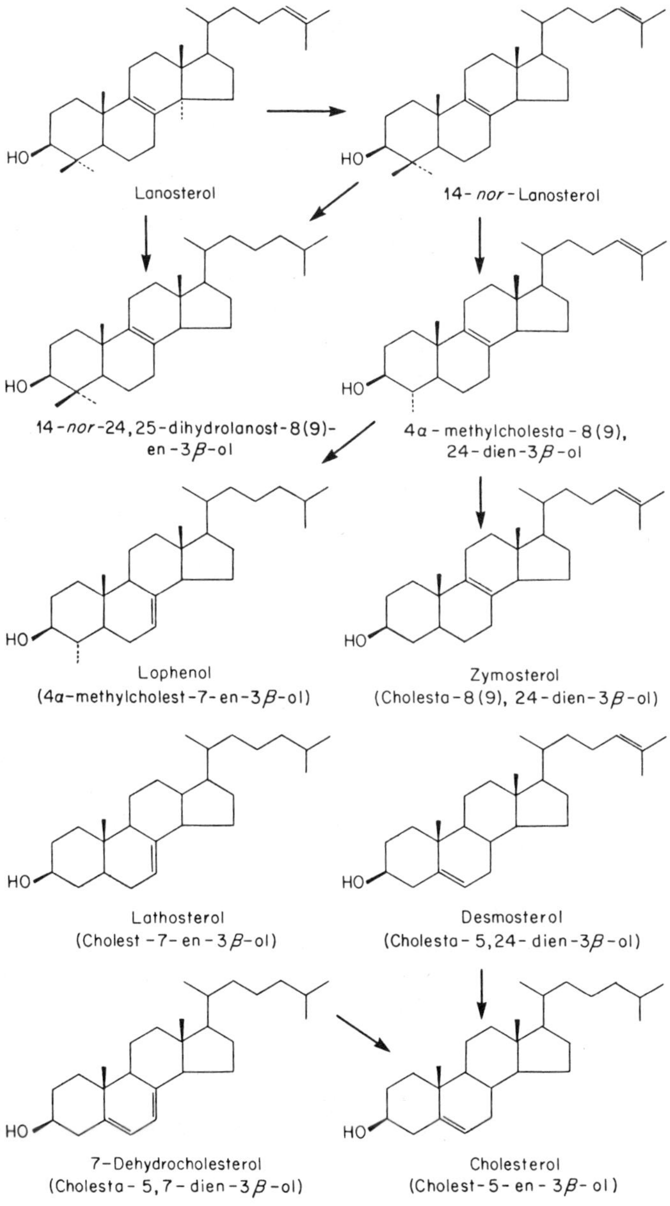

FIGURE 15.31. Probable intermediary stages of the lanosterol → cholesterol conversion.

lism emerged (for reviews, see Brown and Goldstein, 1979; Goldstein and Brown, 1983). These workers discovered that normal fibroblasts possessed on their surfaces, in the so-called "coated pits," specific receptors for LDL, whereas fibroblasts of familial hypercholesterolemics had none or very few. In the normal fibroblasts, the LDL bound to the surface receptors becomes included in endocytotic vesicles formed by the closure of plasma membrane around the coated pits and is delivered to the interior of the cell to the lysosomes. The LDL is then degraded in the lysosomes and the cholesteryl esters of the LDL are also hydrolyzed and trigger three events: (i) they suppress the levels of HMG-CoA reductase, thereby decreasing endogenous sterol synthesis; (ii) they activate a sterol-esterifying enzyme, acyl-CoA: cholesterol acyltransferase (ACAT), which reesterifies the cholesterol delivered to the cells; and (iii) they suppress the resynthesis of LDL receptors. The LDL receptor, as purified from bovine adrenal cortex, is an acidic glycoprotein with an M_r of 164,000 (Schneider *et al.,* 1982). With the aid of fluorescent antibodies to the purified receptor protein, it has been possible to show that the receptors are arranged on the surface of cultured cells in beautiful parallel arrays.

The function of the LDL receptor system is that of regulation—to coordinate intracellular and extracellular levels of cholesterol and maintain a constant level of cellular cholesterol in face of possibly changing blood lipoprotein concentrations.

15.6. The Origin of Bile Acids

Bile is the greenish dark yellow slightly viscous secretion of the liver, delivered through the hepatic and common bile duct into the duodenum through an orifice shared with the pancreatic duct and known in man as the *papilla Vateri.* The hepatic bile, i.e., freshly secreted from the liver, contains about 2 g solid per dl, one-half of which is made up of conjugates of bile acids, 100 mg free cholesterol, and 300 mg phospholipids, and the rest of which consists of bile pigments (bilirubin and biliverdin derived from degradation of hemoproteins, mostly hemoglobin), mucopolysaccharides and some protein (albumin and IgG), and electrolytes. In animals with a gallbladder, as in man, the bladder bile is concentrated four- or fivefold. Man produces up to about one liter of bile per day with the above composition; this would be equivalent to about 10 g of bile acids delivered into the intestine. However, only about 0.8–1.0 g of bile acids are made in the liver per day. The discrepancy is accounted for by the existence of an enterohepatic circulation whereby most of the bile acids and cholesterol delivered to the intestine are reabsorbed. The cholesterol absorbed appears in the chylomicrons; the bile acids are reabsorbed mostly from the ileum (the lower small intestine) into the portal blood.

Synthesis of cholesterol and of bile acids is a very active process in the liver. The human liver manufactures daily 1.0–1.5 g cholesterol, of which 0.8–1.0 g is converted into bile acids. The conversion of cholesterol into bile acids is quantitatively the most important route of elimination of cholesterol from the body. In spite of the enterophepatic circulation and reabsorption of bile acids and sterols from the intestine, in man about 0.8–1.0 g of bile acids and 0.4–0.5 g of neutral sterols are lost in the feces per day. These amounts nicely add up to the daily manufacture.

The neutral sterols lost in the feces are partly cholesterol and partly a bacterial reduction product of cholesterol, 5β-cholestan-3β-ol (coprostanol; cf. formula **50** in Section 15.3). The C_{24}-bile acids and their conjugates are the characteristic components of the bile of birds and mammals. In lower orders (some fishes, amphibians, reptiles, etc.) a variety of C_{27}- and C_{26}-bile alcohols or bile acids are found; their structures reflect an evolutionary process leading to the C_{24} bile acids of the higher orders (cf. Haslewood, 1978).

The structural correlation of bile acids and cholesterol was recognized early in this century; the study of the structure of cholic acid helped greatly in establishing the structure of cholesteral (cf. Fieser and Fieser, 1959, pp. 53–89).

The number of C_{24} bile acids in the bile of vertebrates is relatively small; they are all variously hydroxylated derivatives of the C_{24} cholanic acid (cf. Fieser and Fieser, 1959, pp. 421–443). The important features of cholanic acid are that its A and B rings are *cis*-fused, as in 5β-cholestane and 5β-cholestan-3β-ol (see Section 15.3, formula **15.50**) and that the side-chain bearing the carboxyl group is shorter by three carbon atoms than the side-chain of cholesterol.

There are primary and secondary bile acids. The primary bile acids are those produced in the liver, notably cholic and chenodeoxycholic acid. Secondary bile acids, such as deoxycholic and lithocholic acids, result from the action of intestinal bacteria on the primary bile acids (Fig. 15.32). The secondary bile acids are also absorbed from the intestine in the enterohepatic circulation and are reexcreted in the bile. Human bile is particularly rich in chenodeoxycholic acid (45% of total bile acids), and contains lesser amounts of cholic (31%) and deoxycholic acids (24%). In primates, and other animals, such as the rat, cholic acid is dominant.

A comparison of the structure of cholesterol with that of the bile acids shows that the main changes that have to be made in the structure of cholesterol to convert it into a bile acid, e.g., into cholic acid, are the introduction of new oxygen functions into the nucleus, rearrangement of the 3β-hydroxy group to 3α, reduc-

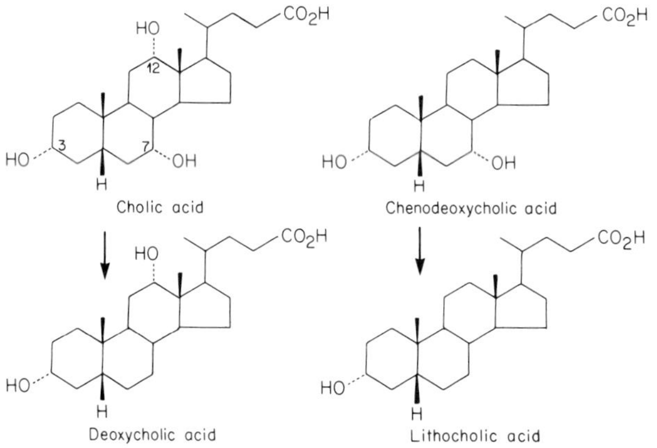

FIGURE 15.32. Two primary bile acids, cholic and chenodeoxycholic, and the secondary deoxycholic and lithocholic acids derived from them.

tion of the nuclear double bond, isomerization of rings A and B from *trans* to *cis* fusion, and shortening of the side-chain by three carbon atoms.

Elucidation of the individual steps giving rise to these changes was achieved mostly by studies *in vivo*, often with bile-fistula animals, and ultimately *in vitro* with cell-free subcellular enzyme preparations with the aid of isotopically labeled precursors. Perhaps the most active group in these studies has been that of Bergström, Danielsson, and others at the Karolinska Institutet in Stockholm. Evidence for the existence of primary and secondary bile acids was first provided by Bergström (1959) from his studies on animals with a bile fistula and on germ-free animals.

From the many studies it is now know that the first step in the transformation of cholesterol into bile acids is the α-hydroxylation at C-7 (cf. Danielsson, 1976, and references cited therein). The cholesterol 7α-hydroxylase is a microsomal enzyme requiring NADPH and O_2; cytochrome P-450 and NADPH:cytochrome P-450 reductase participate in the hydroxylation. This enzyme is rate-limiting in bile acid synthesis in a manner similar to that of HMG-CoA reductase in sterol synthesis and shows exactly the same diurnal rhythm as the reductase (Mitropoulos, 1976), with a maximum in the middle of the dark cycle and a minimum in the middle of the light cycle. Synthesis of bile acids follows the diurnal rhythm of the cholesterol 7α-hydroxylase.

The second step of bile acid synthesis is the oxidation of 7α-hydroxycholesterol to 7α-hydroxycholest-4-en-3-one (Fig. 15.33, (2), (3)), which is either

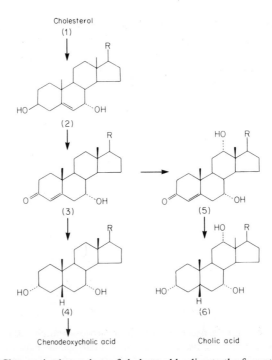

FIGURE 15.33. Changes in the nucleus of cholesterol leading to the formation of bile acids.

reduced to the 5β-cholestane-3α, 7α-diol (Fig. 15.33, (**4**)), the precursor of chenodeoxycholic acid, or is stereospecifically hydroxylated to the 7α,12α-dihydroxycholest-4-en-3-one (Fig. 15.33, (**5**)). The latter is then reduced to 5β-cholestane-3α,7α,12α-triol (Fig. 33, (**6**)), the precursor of cholic acid.

The enzymes that convert either the 7α-hydroxycholest-4-en-3-one or the 7α,12α-dihydroxycholest-4-en-3-one into the 5β-cholestanols are the soluble Δ^4-3-ketosteroid 5-reductase(s) and the 3α-hydroxysteroid dehydrogenase(s) (EC 1.1.1.50).

The remarkable feature of the hydroxylation reactions in bile acid biosynthesis is that at both positions 7 and 12 the α-hydrogen atoms are stereospecifically replaced and that the oxygen of the hydroxyl groups is derived from molecular oxygen. Of course the oxygen of the 3α-hydroxy group is the original oxygen atom of cholesterol.

The oxidation of the side-chain of either the 3α,7α-dihydroxy- or the 3α,7α,12α-trihydroxy-5β-cholestane begins with the hydroxylation of one of the terminal methyl groups by either mitochondria or by microsomes to the 26-hydroxy compounds. These 26-hydroxy compounds are then dehydrogenated by soluble alcohol and aldehyde dehydrogenases, with NAD^+ as coenzyme, to the 5β-cholestan-26-oic acids.

We have pointed out earlier that the reduction of the 24(25)-double bond during steps of cholesterol biosynthesis creates a prochiral center at C-25 and that C-26 of lanosterol (or of desmosterol)—derived from C-2 of mevalonate—becomes the pro-R methyl group, and C-27—derived from C-3' of mevalonate—becomes the pro-S methyl group at C-25 (Popják et al., 1977). According to the IUB–IUPAC recommendations on nomenclature, if either of the terminal methyl groups becomes substituted, that carbon atom becomes C-26. However, as has been pointed out in Section 15.3 of this chapter, substitution of the pro-R methyl group confers on C-25 the S chirality, and substitution of the pro-S methyl group confers on C-25 the R chirality.

According to the evidence presented by Berséus (1965) and Mitropoulos and Myant (1965), the mitochondrial "26-hydroxylase" attacks the pro-S methyl group, and thus the absolute configuration of the two 5β-cholestan-26-oic acids at C-25 is R. However, Gustafsson and Sjösted (1978) have shown that the stereospecificity of the microsomal "26-hydroxylase" is the opposite of that of the mitochondrial hydroxylase; by their evidence the absolute configuration of the 5β-cholestan-26-oic acids at C-25 must be S. This problem has been further discussed by Myant (1981, pp. 237–241).

The mitochondrial and microsomal 26-hydroxylases are typical mixed-function oxidases requiring NADPH and O_2 with the participation of cytochrome P-450 and are inhibited by carbon monoxide. It is uncertain at present which of the hydroxylases is dominant in vivo in bile acid biosynthesis. The mitochondrial 26-hydroxylase has a broad substrate specificity and hydroxylates cholesterol and other C_{27} sterols, whereas the microsomal system of rat liver shows preference for 5β-cholestane-3α,7α-diol and 5β-cholestane-3α,7α,12α-triol, the precursors of chenodeoxycholic and cholic acids. Gustafsson and Sjösted (1978) suggested that the microsomal ω-hydroxylase should be called a 26-hydroxylase and the mitochondrial enzyme should be called a 27-hydroxylase.

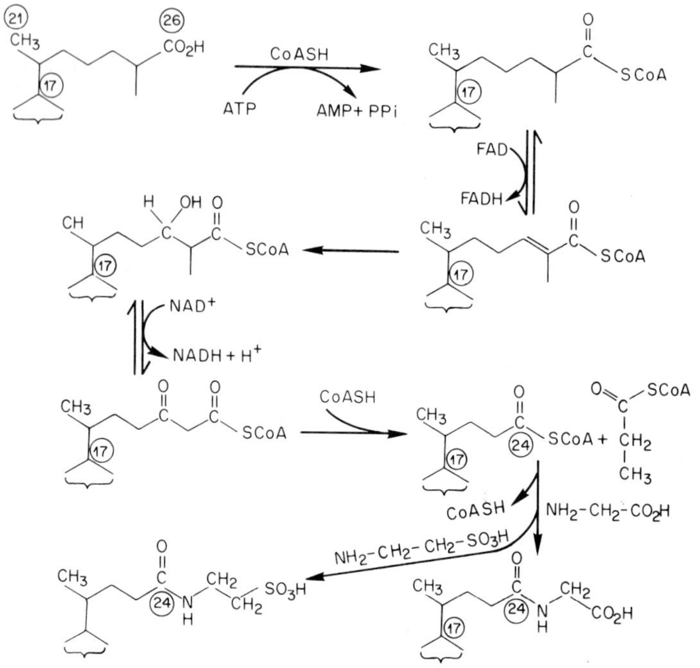

FIGURE 15.34. Shortening of the side-chain in 5β-cholan-26-oic acids by reactions analogous to β-oxidation of fatty acids, and the conjugations of the C_{24} bile acids with glycine or taurine.

It has been assumed that the conversion of the C_{27} 5β-cholestan-26-oic acids to the C_{24} bile acids takes place in mitochondria through the coenzyme A derivatives by reactions analogous to the β-oxidation of the fatty acids (Fig. 15.34), although positive proof of this idea is lacking. The result of the degradation is the loss of the three terminal carbon atoms of the 5β-cholestan-26-oic acid in the form of propionyl-CoA and the formation of the CoA derivatives of cholic and chenodeoxycholic acids. These are conjugated, presumably in the endoplasmic reticulum, with glycine or taurine yielding the glycocholic or taurocholic acid and the glycochenodeoxycholic or taurochenodeoxycholic acid.

The pathways leading to the formation of cholic and chenodeoxycholic acids diverge probably at the stage of 7α-hydroxycholest-4-en-3-one or after the 5β-cholestan-3α,7α-diol has been made. Both substances are rapidly hydroxylated at C-26 by microsomes, and once that position has been hydroxylated, the hydroxylation at C-12 is barred.

Bile acids, or their conjugates, are modified in the intestine by bacteria. The main changes are deconjugations and dehydrations–reductions resulting in conversion of cholic to deoxycholic acid and of chenodeoxycholic to lithocholic acid. These secondary bile acids can be reabsorbed in the enterohepatic circulation and reconjugated in the liver and some, such as taurodeoxycholic acid, can be hydroxylated again at C-7. Deoxycholic acid, one of the major bile acids in the rat, is thus the product of bacterial metabolism of cholic acid, rather than of the rat's

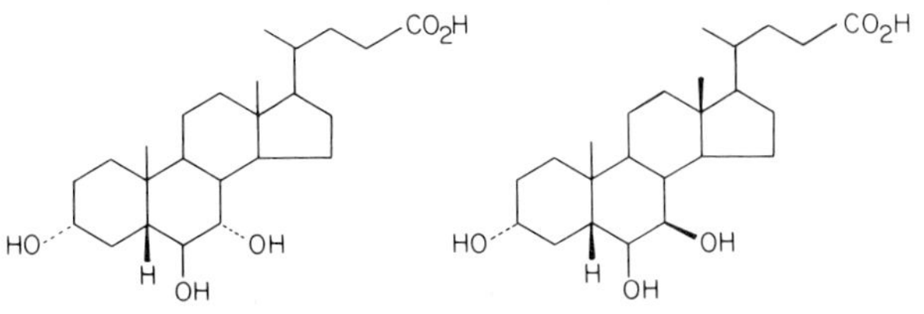

FIGURE 15.35. Muricholic acids derived probably from chenodeoxycholic acid in mice and the Norwegian rat.

own enzyme system. Additional hydroxylations can also occur in the intestine, giving rise to products such as α- and β-muricholic acids in the mouse and the Norwegian rat (Fig. 15.35).

Danielsson (1976) emphasized that the transformations of cholesterol to bile acids have been arrived at indirectly from the study of the metabolic transformations *in vitro* and *in vivo* of very large numbers of sterols and their derivatives. From among the several intermediates postulated, only $3\alpha,7\alpha$-dihydroxy-5β-cholestan-26-oic acid and $3\alpha,7\alpha,12\alpha$-trihydroxy-5β-cholestan-26-oic acid have been isolated from bile, both from human bile. Certainly other metabolic pathways leading from cholesterol to bile acids and bile alcohols must exist, as evidenced by the large variety of "bile salts" found among the vertebrates.

REFERENCES

Abrahamsson, S., Ställberg-Stenhagen, S., and Stenhagen, E., 1964, The higher saturated branched chain fatty acid, in: *Progress in the Chemistry of Fats and Other Lipids*, Vol. 7 (R. T. Holman, ed.), Pergamon Press, Oxford, pp. 1–161.

Agnew, W. S., and Popják, G., 1978a, Squalene synthetase. Stoichiometry and kinetics of presqualene pyrophosphate and squalene synthesis, *J. Biol. Chem.* **253**:4566.

Agnew, W. S., and Popják, G., 1978b, Squalene synthetase. Solubilization from yeast microsomes of a phospholipid-requiring enzyme, *J. Biol. Chem.* **253**:4574.

Akhtar, M., Alexander, K., Boar, R. B., McGhie, J. F., and Barton, D. H. R., 1978, Chemical and enzymic studies on the characterization of intermediates during the removal of the 14α-methyl group in cholesterol biosynthesis, *Biochem. J.* **169**:449.

Alberts, A. W., Chen, J., Kuron, G., Hunt, V., Huff, J., Hoffman, C., Rothrock, J., Lopez, M., Joshua, H., Harris, E., Patchett, A., Monaghan, R., Currie, S., Stapley, E., Albers-Schonberg, G., Hensens, O., Hirshfield, J., Hoogsteen, K., Liesch, J., and Springer, J., 1980, Mevinolin: A highly potent competitive inhibitor of hydroxymethylglutaryl-coenzyme A reductase and a cholesterol-lowering agent, *Proc. Nat. Acad. Sci. U.S.A.* **77**:3957.

Altman, L. J., Kowerski, R. C., and Rilling, H. C., 1971, Synthesis and conversion of presqualene alcohol to squalene, *J. Am. Chem. Soc.* **93**:1782.

Alvear, M., Jabalquinto, A. M., Eyzaguirre, J., and Cardemil, E., 1982, Purification and characterization of avian liver mevalonate-5-pyrophosphate decarboxylase, *Biochemistry* **21**:4646.

Barnard, G. F., and Popják, G., 1980, Characterization of liver prenyl transferase and its inactivation by phenyl glyoxal, *Biochim. Biophys. Acta* **617**:169.

Barnard, G. F., and Popják, G., 1981, Human liver prenyl transferase and its characterization, *Biochim. Biophys. Acta* **661**:87.

Barnard, G. F., Langton, B., and Popják, G., 1978, Pseudo-isoenzyme forms of liver prenyl transferase, *Biochem. Biophys. Res. Commun.* **85**:1097.
Barton, D. H. R., Mellows, G., Widdowson, D. A., and Wright, J. J., 1971, Biosynthesis of terpenes and steroids. Part IV. Specific hydride shifts in the biosynthesis of lanosterol and β-amyrin, *J. Chem. Soc.,* p. 1142.
Beg, Z. H., Allman, D. W., and Gibson, D. M., 1973, Modulation of 3-hydroxy-3-methylglutaryl coenzyme A reductase activity, *Biochem. Biophys. Res. Commun.* **54**:1362.
Bergstrom, J. D., Robbins, K. A., and Edmond, J., 1982, Acetoacetyl-coenzyme A synthetase activity in rat liver cytosol: A regulated enzyme in lipogenesis, *Biochem. Biophys. Res. Commun.* **106**:856.
Bergstrom, J. D., Wong, G. A., Edwards, P. A., and Edmond, J., 1984, The regulation of acetoacetyl-CoA synthetase activity by modulators of cholesterol synthesis *in vivo* and the utilization of acetoacetate for cholesterogenesis, *J. Biol. Chem.* **259**:14548.
Bergström, S., 1959, Bile acids: Formation and metabolism, in : *Ciba Foundation Symposium on the Biosynthesis of Terpenes and Sterols* (G. E. W. Wolstenholme and M. O'Connor, eds.), J. & A. Churchill, London, pp. 185–205.
Bernal, J. D., 1932a, Crystal structures of vitamin D and related compounds, *Nature* **129**:277.
Bernal, J. D., 1932b, Carbon skeleton of the sterols, *Chem. Ind. (London)* **51**:466.
Berséus, O., 1965, On the stereospecificity of 26-hydroxylation of cholesterol. Bile acids and steroids 155. *Acta Chim. Scand.* **19**:325.
Berthelot, M., 1859, Sur plusieurs alcools nouveaux. Combinaisons des acides avec la cholestérine, l'éthal, le camphre de Bornéo et la méconine, *Ann. Chim. Physique,* 2ᵉ Série, **56**:51.
Beytia, E. D., and Porter, J. W., 1976, Biochemistry of polyisoprenoid biosynthesis, *Annu. Rev. Biochem.* **45**:113.
Beytia, E., Dorsey, J. K., Marr, J., Cleland, W. W., and Porter, J. W., 1970, Purification and mechanism of action of hog liver mevalonic kinase, *J. Biol. Chem.* **245**:5450.
Blattmann, P., and Rétey, J., 1971, Zur Wirkungsweise und Stereospezifität der Hydroxymethylglutaryl-CoA-Reduktase, *Hoppe-Seyler's Z. Physiol. Chem.* **352**:369.
Bloch, K., 1953, Über die Herkunft des Kohlenstoff-Atoms 7 in Cholesterin. Ein Beitrag zur Kenntnis der Biosynthese der Steroide, *Helv. Chim. Acta* **36**:1611.
Bloch, K., 1954, Biological synthesis of cholesterol, *Harvey Lectures,* Series 48, Academic Press, New York, pp. 68–88.
Bloch, K., 1965, The biological synthesis of cholesterol, The Nobel Prize lecture, *Science* **150**:19.
Bloch, K., and Rittenberg, D., 1942a, The biological formation of cholesterol from acetic acid, *J. Biol. Chem.* **143**:297.
Bloch, K., and Rittenberg, D., 1942b, On the utilization of acetic acid for cholesterol formation, *J. Biol. Chem.* **145**:625.
Bloch, K., and Rittenberg, D., 1944, Sources of acetic acid in the animal body, *J. Biol. Chem.* **155**:243.
Bloch, K., Chaykin, S., Phillips, A. H., and de Waard, A., 1959, Mevalonic acid pyrophosphate and isopentenylpyrophosphate, *J. Biol. Chem.* **234**:2595.
Brown, D., and Simoni, R. D., 1984, Biogenesis of 3-hydroxy-3-methylglutaryl-coenzyme A reductase, an integral glycoprotein of the endoplasmic reticulum, *Proc. Nat. Acad. Sci. U.S.A.* **81**:1674.
Brown, M. S., and Goldstein, J. L., 1979, Familial hypercholesterolemia: Model for genetic receptor disease, *Harvey Lectures,* Series 73, Academic Press, New York, pp. 163–201.
Brown, M. S., and Goldstein, J. L., 1980, Multivalent feedback regulation of HMG CoA reductase, a control mechanism coordinating isoprenoid synthesis and cell growth, *J. Lipid Res.* **21**:505.
Brown, M. S., Faust, J. R., and Goldstein, J. L., 1978, Induction of 3-hydroxy-3-methylglutaryl coenzyme A reductase activity in human fibroblasts incubated with compactin (ML-236B), a competitive inhibitor of the reductase, *J. Biol. Chem.* **253**:1121.
Bucher, N. L. R., 1953, The formation of radioactive cholesterol and fatty acids from C^{14}-labeled acetate by rat liver homogenates, *J. Am. Chem. Soc.* **75**:498.
Bucher, N. L. R., and McGarrahan, K., 1956, The biosynthesis of cholesterol from acetate-1-^{14}C by cellular fractions of rat liver, *J. Biol. Chem.* **222**:1.
Bucher, N. L. R., Overrath, P., and Lynen, F., 1959, Enzymes controlling cholesterol biosynthesis in livers of fasting rats, *Fed. Proc.* **18**:20.
Bu'Lock, J. D., de Rosa, M., and Gambacorta, A., 1983, Isoprenoid biosynthesis in Archaebacteria,

in: *Biosynthesis of Isoprenoid Compounds,* vol. 2 (J. W. Porter, and S. L. Spurgeon, eds.), Wiley, New York, pp. 159–189.

Burgos, J., Hemming, F. W., Pennock, J. F., and Morton, R. A., 1963, Dolichol: a naturally-occurring C_{100} isoprenoid alcohol, *Biochem. J.* **88**:470.

Cahn, R. S., Ingold, C., and Prelog, V., 1966, The specification of molecular chirality, *Angew. Chem., Int. Ed. Engl.* **5**:385.

Campbell, R. V. M., Crombie, L., and Pattenden, G., 1971, Synthesis of presqualene alcohol, *Chem. Commun.,* p. 218.

Cavenee, W. K., Chen, H. W., and Kandutsch, A. A., 1981, Regulation of cholesterol biosynthesis in enucleated cells, *J. Biol. Chem.* **256**:2675.

Channon, H. J., 1926, The biological significance of the unsaponifiable matter of oils. I. Experiments with the unsaturated hydrocarbon, squalene (spinacene), *Biochem. J.* **20**:400.

Chevreul, M., 1815, Sur plusieurs corps gras, et particulièrment sur leur combinaisons avec les alcalis. Cinquième Mémoire. Des corps qu'on a appelés adipocire, c'est-à-dire, de la substance cristallisée des calculs biliaires humains, du spermacéti et de la substance grasse des cadavres, *Ann. Chim.* **95**:5.

Chevreul, M., 1816, Examen des graisses d'homme, de mouton, de boeuf, de jaguar et d'oie. Sixième Mémoire, *Ann. Chim. Physique,* 2e Série, **2**:339.

Chin, D. J., Luskey, K. L., Anderson, R. G. W., Faust, J. R., Goldstein, J. L., and Brown, M. S., 1982a, Appearance of crystalloid endoplasmic reticulum in compactin-resistant Chinese hamster cells with a 500-fold increase in 3-hydroxy-3-methylglutaryl-coenzyme A reductase, *Proc. Nat. Acad. Sci. U.S.A.* **79**:1185.

Chin, D. J., Luskey, K. L., Faust, J. R., MacDonald, R. J., Brown, M. S., and Goldstein, J. L., 1982b, Molecular cloning of 3-hydroxy-3-methylglutaryl coenzyme A and evidence for regulation of its mRNA, *Proc. Nat. Acad. Sci. U.S.A.* **79**:7704.

Clarke, C., Fogelman, A. M., and Edwards, P. A., 1984, Diurnal rhythm of rat liver mRNAs encoding 3-hydroxy-3-methylglutaryl coenzyme A reductase, *J. Biol. Chem.* **259**:10439.

Clinkenbeard, K. D., Reed, W. D., Mooney, R. A., and Lane, M. D., 1975a, Intracellular localization of the 3-hydroxy-3-methylglutaryl coenzyme A cycle enzymes in liver. Separate cytoplasmic and mitochondrial 3-hydroxy-3-methylglutaryl coenzyme A generating systems for cholesterogenesis and ketogenesis, *J. Biol. Chem.* **250**:3108.

Clinkenbeard, K., Sugiyama, T., Reed, W. D., and Lane, M. D., 1975b, Cytoplasmic 3-hydroxy-3-methylglutaryl coenzyme A synthase from liver. Purification, properties and role in cholesterol synthesis, *J. Biol. Chem.* **250**: 3124.

Coates, R. M., and Robinson, W. H., 1971, Stereoselective total synthesis of ($\pm$)-presqualene alcohol, *J. Am. Chem. Soc.* **93**:1785.

Comita, P. B., and Gagosian, R. B., 1983, Membrane lipid from deep-sea hydrothermal vent methanogen: a new macrocyclic glycerol diether, *Science* **222**:1329.

Cook, R. P. (ed.), 1958a, *Cholesterol: Chemistry, Biochemistry, and Pathology,* Academic Press, New York.

Cook, R. P., 1958b, Distribution of sterols in organisms and in tissues, in: *Cholesterol* (R. P. Cook, ed.), Academic Press, New York, pp. 145–180.

Coon, M. J., Kupiecki, F. P., Dekker, E. E., Schlesinger, M. J., and del Campillo, A., 1959, The enzymic synthesis of branched-chain acids, in: *Ciba Foundation Symposium on the Biosynthesis of Terpenes and Sterols* (G. E. W. Wolstenholme and M. O'Connor, eds.), J. & A. Churchill, London, pp. 62–74.

Cordus, Valerius, 1598, *Dispensatorium Pharmacorum omnium, quae in usu potissimum sunt,* Facsimile reprint, 1969, Konrad Kölbe, Munich.

Corey, E. J., and Russey, W. E., 1966, Metabolic fate of 10,11-dihydrosqualene in sterol-producing rat liver homogenate, *J. Am. Chem. Soc.* **88**:4751.

Corey, E. J., and Volante, R. P., 1976, Application of unreactive analogs of terpenoid pyrophosphates to studies of multistep biosynthesis. Demonstration that "presqualene pyrophosphate" is an essential intermediate on the path to squalene, *J. Am. Chem. Soc.* **98**:1291.

Corey, E. J., Russey, W. E., and Ortiz de Montellano, P. R., 1966, 2,3-Oxidosqualene, an intermediate in the biological synthesis of sterols from squalene, *J. Am. Chem. Soc.* **88**:4750.

Cornforth, J. W., 1959, Biosynthesis of fatty acids and cholesterol considered as chemical processes, *J. Lipid Res.* **1**:3.

Cornforth, J. W., 1973, The logic of working with enzymes, The Robert Robinson Lecture, *Chem. Soc. Rev.* **2**:1.

Cornforth, J. W., and Cornforth, R. H., 1970, Chemistry of mevalonic acid, in: *Natural Substances Formed Biologically from Mevalonic Acid* (T. W. Goodwin, ed.), Academic Press, London, pp. 5–15.

Cornforth, J. W., and Gray, R. T., 1975, Synthesis of [5-^{18}O]mevalonolactone, *Tetrahedron* **31**:1509.

Cornforth, J. W., and Popják, G., 1954, Studies on the biosynthesis of cholesterol. 3. Distribution of ^{14}C in squalene biosynthesized from [*Me*-^{14}C]acetate, *Biochem. J.* **58**:403.

Cornforth, J. W., and Popják, G., 1959, Mechanism of biosynthesis of squalene from sesquiterpenoids, *Tetrahedron Lett.* No. 19:29.

Cornforth, R. H., and Popják, G., 1969, Chemical syntheses of substrates of sterol biosynthesis, *Methods Enzymol.* **15**:359.

Cornforth, J. W., Cornforth, R. H., Popják, G., and Youhotsky Gore, I., 1957, Biosynthesis of squalene and cholesterol from DL-β-hydroxy-β-methyl-δ-[2-^{14}C]valerolactone, *Biochem. J.* **66**:10P.

Cornforth, J. W., Cornforth, R. H., Popják, G., and Youhotsky Gore, I., 1958, Studies on the biosynthesis of cholesterol. 5. Biosynthesis of squalene from DL-3-hydroxy-3-methyl-[2-^{14}C]pentano-5-lactone, *Biochem. J.* **69**:146.

Cornforth, J. W., Cornforth, R. H. Pelter, A., Horning, M. G., and Popják, G., 1959, Studies on the biosynthesis of cholesterol. — 7. Rearrangement of methyl groups during enzymic cyclisation of squalene, *Tetrahedron* **5**:311.

Cornforth, R. H., Cornforth, J. W., and Popják, G., 1962, Preparation of R- and S-mevalonolactones, *Tetrahedron* **18**:1351.

Cornforth, J. W., Cornforth, R. H., Donninger, C., Popják, G., Ryback, G., and Schroepfer, G. J., Jr., 1963, Stereospecific insertion of hydrogen atom into squalene from reduced nicotinamide-adenine dinucleotides, *Biochem. Biophys. Res. Commun.* **11**:129.

Cornforth, J. W., Cornforth, R. H., Donninger, C., Popják, G., Shimizu, Y., Ichii, S., Forchielli, E., and Caspi, E., 1965, The migration and elimination of hydrogen during biosynthesis of cholesterol from squalene, *J. Am. Chem. Soc.* **87**:3224.

Cornforth, J. W., Cornforth, R. H., Popják, G., and Yengoyan, L., 1966, Studies on the biosynthesis of cholesterol. XX. Steric course of decarboxylation of pyrophosphomevalonate and of the carbon to carbon bond formation in the biosynthesis of farnesyl pyrophosphate, *J. Biol. Chem.* **241**:3970.

Cornforth, J. W., Clifford, K., Mallaby, R., and Phillips, G. T., 1972, Stereochemistry of isopentenyl pyrophosphate isomerase, *Proc. Roy. Soc., Ser. B.* **182**:277.

Cornforth, J. W., Phillips, G. T., Messner, B., and Eggerer, H., 1974, Substrate stereochemistry of 3-hydroxy-3-methylglutaryl-coenzyme A synthase, *Eur. J. Biochem.* **42**:591.

Danielsson, H., 1976, Bile acid metabolism and its control, in: *The Hepatobiliary System: Fundamental and Pathological Mechanisms* (W. Taylor, ed.), Plenum Press, New York, pp. 389–404.

Dean, P. D. G., 1969, Enzymatic cyclization of squalene 2,3-oxide, *Methods Enzymol.* **15**:495.

De Fourcroy, 1789, De la substance feuilletée & cristalline contenue dans les calculs biliaires, & de la natur des concrétions cystiques cristallisées, *Ann. Chim.* **3**:242.

Diels, O., and Abderhalden, E., 1904, Zur Kenntniss des Cholesterins, *Chem. Ber.* **37**:3092.

Dituri, F., Gurin, S., and Rabinowitz, J. L., 1957, The biosynthesis of squalene from mevalonic acid, *J. Am. Chem. Soc.* **79**:2650.

Donninger, C., and Popják, G., 1966, Studies on the biosynthesis of cholesterol. VIII. The stereospecificity of mevaldate reductase and the biosynthesis of asymmetrically labelled farnesyl pyrophosphate, *Proc. Roy. Soc. Ser. B.* **163**:465.

Dugan, R. E., and Porter, J. W., 1971, Stereospecificity of the transfer of hydrogen from reduced nicotinamide adenine dinucleotide phosphate in each of the two reductive steps catalyzed by β-hydroxy-β-methylglutaryl coenzyme A reductase, *J. Biol. Chem.* **246**:5361.

Eberle, M., and Arigoni, D., 1960, Absolute Konfiguration des Mevalolactons, *Helv. Chim. Acta* **43**:1508.

Edmond, J., Popják, G., Wong, S-M., and Williams, V. P., 1971, Presqualene alcohol. Further evidence on the structure of a C_{30} precursor of squalene, *J. Biol. Chem.* **246**:6254.

Edwards, P. A., Popják, G., Fogelman, A. M., and Edmond, J., 1977a, Control of 3-hydroxy-3-methylglutaryl coenzyme A reductase by endogenously synthesized sterols *in vitro* and *in vivo, J. Biol. Chem.* **252**:1057.

Edwards, P. A., Edmond, J., Fogelman, A. M., and Popják, G., 1977b, Preferential uptake and utilization of mevalonolactone over mevalonate for sterol biosynthesis in isolated rat hepatocytes, *Biochim. Biophys. Acta* **488**:493.

Edwards, P. A., Lemongello, D., Kane, J., Shechter, I., and Fogelman, A. M., 1980, Properties of purified rat hepatic 3-hydroxy-3-methylglutaryl coenzyme A reductase and regulation of enzyme activity, *J. Biol. Chem.* **255**:3715.

Edwards, P. A., Lan, S-F., Tanaka, R., and Fogelman, A. M., 1983, Mevalonolactone inhibits the rate of synthesis and enhances the rate of degradation of 3-hydroxy-3-methylglutaryl coenzyme A reductase in rat hepatocytes, *J. Biol. Chem.* **258**:7272.

Endo, A., Kurada, M., and Tanzawa, K., 1976, Competitive inhibition of 3-hydroxy-3-methylglutaryl coenzyme A reductase by ML-236A and ML-236B fungal metabolites, having hypocholesterolemic activity, *FEBS Lett.* **72**:323.

Epstein, W. W., and Rilling, H. C., 1970, Studies on the mechanism of squalene biosynthesis. The structure of presqualene pyrophosphate, *J. Biol. Chem.* **245**:4597.

Eschenmoser, A., Ruzicka, L., Jeger, O., and Arigoni, D., 1955, Zur Kenntnis der Triterpene. 190. Mitteilung. Eine stereochemische Interpretation der biogenetischen Isoprenregel bei den Triterpenen, *Helv. Chim. Acta* **38**:1890.

Fieser, L. F., and Fieser, M., 1959, *Steroids,* Reinhold, New York.

Fogelman, A. M., Seager, J., Edwards, P. A., and Popják, G., 1977, Mechanism of induction of 3-hydroxy-3-methylglutaryl coenzyme A reductase in human leukocytes, *J. Biol. Chem.* **252**:644.

Folkers, K., Shunk, C. H., Linn, B. O., Robinson, F. M., Wittreich, P. E., Huff, J. W., Gilfillan, J. L., and Skeggs, H. R., 1959, Discovery and elucidation of mevalonic acid, in: *Ciba Foundation Symposium on the Biosynthesis of Terpenes and Sterols* (G. E. W. Wolstenholme and M. O'Connor, eds.), J. & A. Churchill, London, pp. 20–45.

Gaylor, J. L., 1972, Microsomal enzymes of sterol biosynthesis, *Adv. Lipid Res.* **10**:89.

Geissman, T. A. (Senior Reporter), 1972, *Biosynthesis,* Vol. 1., The Chemical Society, London.

Goldstein, J. L., and Brown, M. S., 1973, Familial hypercholesterolemia: Identification of a defect in the regulation of 3-hydroxy-3-methylglutaryl coenzyme A reductase activity associated with overproduction of cholesterol, *Proc. Nat. Acad. Sci. U.S.A.* **70**:2804.

Goldstein, J. L., and Brown, M. S., 1983, Familial hypercholesterolemia, in: *The Metabolic Basis of Inherited Disease,* 5th ed. (J. B. Stanbury, J. B. Wyngaarden, D. S. Fredrickson, J. L. Goldstein, and M. S. Brown, eds.), McGraw-Hill, New York, pp. 672–712.

Goodman, DeW. S., and Popják, G., 1960, Studies on the biosynthesis of cholesterol: XII. Synthesis of allyl pyrophosphates from mevalonate and their conversion into squalene with liver enzymes, *J. Lipid Res.* **1**:286.

Gould, R. G., 1954, Sterol metabolism and its control, in: *Symposium on Atherosclerosis,* Publication 338, National Academy of Sciences–National Research Council, Washington, D. C., pp. 153–168.

Gould, R. G., and Popják, G., 1957, Synthesis of cholesterol *in vivo* and *in vitro* from DL-β-hydroxy-β-methyl-δ-[2-^{14}C]-valerolactone, *Biochem. J.* **66**:51P.

Gustafsson, J., and Sjösted, S., 1978, On the stereospecificity of microsomal "26"-hydroxylation in bile acid biosynthesis, *J. Biol. Chem.* **253**:199.

Harwood, H. J., Jr., Brandt, K. G., and Rodwell, V. W., 1984, Allosteric activation of rat liver cytosolic 3-hydroxy-3-methylglutaryl coenzyme A reductase kinase by nucleoside diphosphates, *J. Biol. Chem.* **259**:2810.

Haslewood, G. A. D., 1978, *The biological importance of bile salts,* North-Holland, Amsterdam.

Heilbron, I. M., Kamm, E. D., and Owens, W. M., 1926, The unsaponifiable matter from the oils of elasmobranch fish. Part I. A contribution to the study of the constitution of squalene (spinacene), *J. Chem. Soc.* p. 1630.

Hemming, F. W., 1974, Lipids in glycan biosynthesis, in: *Biochemistry of Lipids,* MTP International Review of Science, Biochemistry Series One, Vol. 4 (T. W. Goodwin, ed.), Butterworths, London, pp. 39–97.

Hemming, F. W., 1983, Biosynthesis of dolichols and related compounds, in: *Biosynthesis of Isopre-*

noid Compounds, Vol. 2 (J. W. Porter and S. L. Spurgeon, eds.), John Wiley & Sons, New York, pp. 305–354.

Holloway, P. W., 1972, Isopentenylpyrophosphate isomerase, in: *The Enzymes,* Vol. 6, 3rd ed. (P. D. Boyer, ed.), Academic Press, New York, pp. 565–572.

Hunter, C. F., and Rodwell, V. W., 1980, Regulation of vertebrate liver HMG-CoA reductase via reversible modulation of its catalytic activity, *J. Lipid Res.* **21**:399.

Isler, O., Rüegg, R., Würsch, J., Gey, K. F., and Pletscher, A., 1957, Zur Biosynthese des Cholesterins aus β,δ-Dihydroxy-β-methyl-valeriansäure, *Helv. Chim. Acta* **40**:2369.

IUPAC–IUB, 1972, Definitive rules for nomenclature of steroids (1971), *Pure Appl. Chem.* **31**:283. See also in: Collected Tentative Rules & Recommendations of the Commission on Biochemical Nomenclature, IUPAC–IUB and Related Documents, 2nd ed., 1975, American Society of Biological Chemists, Bethesda, Maryland, pp. 80–96.

Kandutsch, A. A., Chen, H. W., and Heiniger, H-J., 1978, Biological activity of some oxygenated sterols, *Science* **201**:498.

Kates, M., 1972, Ether-linked lipids in extremely halophilic bacteria, in: *Ether Lipids. Chemistry and Biology* (F. Snyder, ed.), Academic Press, New York, pp. 351–398.

Kates, M., and Kushwaha, S. C., 1978, Biochemistry of extremely halophilic bacteria, in: *Energetics and Structure of Halophilic Microorganisms* (S. R. Caplan, and M. Ginzburg, eds.), Elsevier/North Holland, Amsterdam, pp. 461–480.

Klyne, W., 1957, *The Chemistry of the Steroids,* Methuen, London.

Knappe, J., Ringelmann, E., and Lynen, F., 1959, Über die β-Hydroxy-β-methyl-glutaryl-Reduktase der Hefe. Zur Biosynthese der Terpene IX, *Biochem. Z.* **332**:195.

Knauss, H. J., Brodie, J. D., and Porter, J. W., 1962, Studies on mevaldic acid reductase of rat liver, *J. Lipid Res.* **3**:197.

Kritchevsky, D., 1958, *Cholesterol,* John Wiley, New York.

Langdon, R. G., and Bloch, K., 1953a, The biosynthesis of squalene, *J. Biol. Chem.* **200**:129.

Langdon, R. G., and Bloch, K., 1953b, The utilization of squalene in the biosynthesis of cholesterol, *J. Biol. Chem.* **200**:135.

Lindberg, M., Yuan, C., deWaard, A., and Bloch, K., 1962, On the mechanism of formation of isopentenylpyrophosphate, *Biochemistry* **1**:182.

Liscum, L., Cummings, R. D., Anderson, R. G. W., DeMartino, G. N., Goldstein, J. L., and Brown, M. S., 1983, 3-Hydroxy-3-methylglutaryl-CoA reductase: A transmembrane glycoprotein of the endoplasmic reticulum with N-linked "high mannose" oligosaccharides, *Proc. Nat. Acad. Sci. U.S.A.* **80**:7165.

Little, H. N., and Bloch, K., 1950, Studies on the utilization of acetic acid for the biological synthesis of cholesterol, *J. Biol. Chem.* **183**:33.

Lynen, F., 1958, Verzweigte Carbonsäuren als Baustoffe der Polyisoprenoide, *Proc. Internat. Symposium on Enzyme Chemistry,* Tokyo and Kyoto, 1957, Maruzen, Tokyo.

Lynen, F., and Grassl, M., 1958, Zur Biosynthese der Terpene, II. Darstellung von (—)-Mevalonsäure durch bakterielle Racematsspaltung, *Hoppe-Seyler's Z. Physiol. Chem.* **313**:291.

Lynen, F., Agranoff, B. W., Eggerer, H., Henning, U., and Möslein, E. M., 1959, γ,γ-Dimethyl-allylpyrophosphat und Geranyl-pyrophosphat, biologische Vorstufen des Squalens. Zur Biosynthese der Terpene, VI, *Angew. Chem.* **71**:657.

Maudgal, R. K., Tchen, T. T., and Bloch, K., 1958, 1,2-Methyl shifts in the cyclization of squalene to lanosterol, *J. Am. Chem. Soc.* **80**:2589.

Mitropoulos, K. A., 1976, Diurnal variation in bile acid biosynthesis, in: *The Hepatobiliary System: Fundamental and Pathological Mechanisms* (W. Taylor, ed.), Plenum Press, New York, pp. 409–427.

Mitropoulos, K. A., and Myant, N. B., 1965, Evidence that oxidation of the side chain of cholesterol by liver mitochondria is stereospecific, and that the immediate product of the cleavage is propionate, *Biochem. J.* **97**:26C.

Miziorko, H. M., and Lane, M. D., 1977, 3-Hydroxy-3-methylglutaryl-CoA synthase. Participation of acetyl-S-enzyme and enzyme-S- hydroxymethylglutaryl-SCoA, *J. Biol. Chem.* **252**:1414.

Miziorko, H. M., Clinkenbeard, K. D., Reed, W. D., and Lane, M. D., 1975, 3-Hydroxy-3-methylglutaryl coenzyme A synthase. Evidence for an acetyl-S-enzyme intermediate and identification of a cysteinyl sulphydryl as the site of acetylation, *J. Biol. Chem.* **250**:5768.

Myant, N. B., 1981, *The Biology of Cholesterol and Related Steroids*, William Heinemann Medical Books, London.
Nes, W. R., 1977, The biochemistry of plant sterols, in: *Advances in Lipid Research*, Vol. 15 (R. Paoletti and D. Kritchevsky, eds.), Academic Press, New York, pp. 233–324.
Newman, A. A. (ed.), 1972, *Chemistry of Terpenes and Terpenoids*, Academic Press, New York.
Ngan, H-L., and Popják, G., 1975, Stereochemistry of the reaction catalyzed by mevaldate reductase, *Bioorg. Chem.* **4**:166.
Ono, T., and Bloch, K., 1975, Solubilization and partial characterization of rat liver squalene epoxidase, *J. Biol. Chem.* **250**:1571.
Ortiz de Montellano, P. R., Castillo, R., Vinson, W., and Wei, J. S., 1976a, Squalene synthetase. Differentiation between the two substrate binding sites by a substrate analogue, *J. Am. Chem. Soc.* **98**:2018.
Ortiz de Montellano, P. R., Castillo, R., Vinson, W., and Wei, J. S., 1976b, Squalene biosynthesis. Role of the 3-methyl group in farnesyl pyrophosphate, *J. Am. Chem. Soc.* **98**:3020.
Ottke, R. C., Tatum, E. L., Zabin, I., and Bloch, K., 1951, Isotopic acetate and isovalerate in the synthesis of ergosterol by Neurospora, *J. Biol. Chem.* **189**:429.
Overton, K. H. (Senior Reporter), 1971, *Terpenoids and Steroids*, Vol. 1, The Chemical Society, London.
Popják, G., 1954, Biosynthesis of squalene and cholesterol *in vitro* from acetate-1-^{14}C, *Arch. Biochem. Biophys.* **48**:102.
Popják, G., 1955, *Chemistry, Biochemistry and Isotopic Tracer Technique*, Royal Institute of Chemistry, Lectures, Monographs and Reports No. 2, London.
Popják, G., 1958, Biosynthesis of cholesterol and related substances, *Annu. Rev. Biochem.* **27**:533.
Popják, G., 1959, The biosynthesis of allylic alcohols from [2-^{14}C]mevalonate in liver enzyme preparations and their relation to synthesis of squalene, *Tetrahedron Lett.* No. 19:19.
Popják, G., 1963, Polyisoprenoid synthesis with special reference to the origin of squalene, in: *Biosynthesis of Lipids*, Proceedings of the Fifth International Congress of Biochemistry, Moscow, 1961, Vol. VII; International Union of Biochemistry Symposium Series, Vol. 27 (G. Popják, ed.), Pergamon Press, Oxford, pp. 207–235.
Popják, G., 1969, Enzymes of sterol biosynthesis in liver and intermediates of sterol biosynthesis, *Methods Enzymol.* **15**:393.
Popják, G., 1970, Stereospecificity of enzymic reactions, in: *The Enzymes, Kinetics and Mechanism*, Vol. 2 (P. D. Boyer, ed.), Academic Press, New York, pp. 115–215.
Popják, G., and Agnew, W. S., 1979, Squalene synthetase, *Mol. Cell. Biochem.* **27**:97.
Popják, G., and Beeckmans, M-L., 1950, Extrahepatic lipid synthesis, *Biochem. J.* **47**:233.
Popják, G., and Cornforth, J. W., 1960a, The biosynthesis of cholesterol, *Adv. Enzymol.* **22**:281.
Popják, G., and Cornforth, R. H., 1960b, Gas–liquid chromatography of allylic alcohols and related branched-chain acids, *J. Chromatogr.* **4**:214.
Popják, G., and Cornforth, J. W., 1966, Substrate stereochemistry in squalene biosynthesis. The first Ciba Medal lecture, *Biochem. J.* **101**:553.
Popják, G., Goodman, DeW. S., Cornforth, J. W., Cornforth, R. H., and Ryhage, R., 1961a, Mechanism of squalene biosynthesis from mevalonate and farnesyl pyrophosphate, *Biochem. Biophys. Res. Commun.* **4**:138.
Popják, G., Goodman, DeW. S., Cornforth, J. W., Cornforth, R. H., and Ryhage, R., 1961b, Studies on biosynthesis of cholesterol. XV. Mechanism of squalene biosynthesis from farnesyl pyrophosphate and from mevalonate, *J. Biol. Chem.* **236**:1934.
Popják, G., Cornforth, J. W., Cornforth, R. H., and Goodman, DeW. S., 1961c, Synthesis of 1-T$_2$-C^{14}- and of 1-D$_2$-2-C^{14}-*trans-trans*-farnesyl pyrophosphate and their utilization in squalene synthesis, *Biochem. Biophys. Res. Commun.* **4**:204.
Popják, G., Cornforth, J. W., Cornforth, R. H., Ryhage, R., and Goodman, DeW. S., 1962, Studies on the biosynthesis of cholesterol. XVI. Chemical synthesis of 1-H$_2^3$-2-C^{14}- and 1-D$_2$-2-C^{14}-*trans-trans*-farnesyl pyrophosphate and their utilization in squalene biosynthesis, *J. Biol. Chem.* **237**:56.
Popják, G., Edmond, J., Clifford, K., and Williams, V., 1969, Biosynthesis and structure of a new intermediate between farnesyl pyrophosphate and squalene, *J. Biol. Chem.* **244**:1897.
Popják, G., Edmond, J., and Wong, S-M., 1973, Absolute configuration of presqualene alcohol, *J. Am. Chem. Soc.* **95**:2713.

Popják, G., Ngan, H-L., and Agnew, W., 1975, Stereochemistry of the biosynthesis of presqualene alcohol, *J. Bioorg. Chem.* **251**:279.
Popják, G., Edmond, J., Anet, F. A. L., and Easton, N. A., Jr., 1977, Carbon-13 NMR studies on cholesterol biosynthesized from [^{13}C] mevalonates, *J. Am. Chem. Soc.* **99**:931.
Popják, G., Boehm, G., Parker, T. S., and Edmond, J., 1979, Determination of mevalonate in blood plasma of man and rat. Mevalonate "tolerance" test in man, *J. Lipid Res.* **20**:716.
Popják, G., Hadley, C., and Meenan, A., 1983, Regulation of HMG-CoA reductase and cholesterol synthesis in H4-II-E-C3 cultured hepatoma cells and their cytoplasts, *J. Lipid Res.* **24**:1411.
Popják, G., Clarke, C., Hadley, C., and Meenan, A., 1985, The role of mevalonate in regulation of cholesterol synthesis and 3-hydroxy-3-methylglutaryl coenzyme A reductase in cultured cells and their cytoplasts, *J. Lipid. Res.* **26**:831.
Porter, J. W., and Spurgeon, S. L. (eds.), 1981/83, *Biosynthesis of Isoprenoid Compounds*, Vols. 1 and 2, John Wiley and Sons, New York.
Poulter, C. D., and Rilling, H. C., 1981a, Prenyl transferases and isomerases, in: *Biosynthesis of Isoprenoid Compounds*, Vol. 1 (J. W. Porter and S. L. Spurgeon, eds.), John Wiley and Sons, New York, pp. 161–224.
Poulter, C. D., and Rilling, H. C., 1981b, Conversion of farnesyl pyrophosphate to squalene, in: *Biosynthesis of Isoprenoid Compounds*, Vol. 1 (J. W. Porter and S. L. Spurgeon, eds.), John Wiley and Sons, New York, pp. 413–441.
Poulter, C. D., Muscio, O. J., and Goodfellow, R. J., 1974, Biosynthesis of head-to-head terpenes. Carbonium ion rearrangements which lead to head-to-head terpenes, *Biochemistry* **13**:1530.
Reed, W. D., Clinkenbeard, K. D., and Lane, M. D., 1975, Molecular and catalytic properties of mitochondrial (ketogenic) 3-hydroxy-3-methylglutaryl coenzyme A synthase, *J. Biol. Chem.* **250**:3117.
Rétey, J., von Stetten, E., Coy, U., and Lynen, F., 1970, A probable intermediate in the enzymic reduction of 3-hydroxy-3-methylglutaryl coenzyme A, *Eur. J. Biochem.* **15**:72.
Rilling, H. C., 1966, A new intermediate in the biosynthesis of squalene, *J. Biol. Chem.* **241**:3233.
Rilling, H. C., and Bloch, K., 1959, On the mechanism of squalene biogenesis from mevalonic acid, *J. Biol. Chem.* **234**:1424.
Rittenberg, D., and Schoenheimer, R., 1937, Deuterium as an indicator in the study of intermediary metabolism. XI. Further studies on the biological uptake of deuterium into organic substances, with special reference to fat and cholesterol formation, *J. Biol. Chem.* **121**:235.
Robinson, R., 1934, Structure of cholesterol, *J. Soc. Chem. Ind.* **53**:1062.
Rodwell, V. W., Nordstrom, J. L., and Mitschelen, J. J., 1976, Regulation of HMG-CoA reductase, in: *Advances in Lipid Research* (R. Paoletti and D. Kritchevsky, eds.), Academic Press, New York, pp. 1–74.
Rosenheim, O., and King, H., 1932, The ring system of sterols and bile acids, *Nature* **130**:315.
Rudney, H., 1959, The biosynthesis of β-hydroxy-β-methyl-glutaryl coenzyme A and its conversion to mevalonic acid, in: *Ciba Foundation Symposium on the Biosynthesis of Terpenes and Sterols* (G. E. W. Wolstenholme and M. O'Connor, eds.), J. & A. Churchill, London, pp. 75–94.
Ruzicka, L., 1953, The isoprene rule and the biogenesis of terpenic compounds, *Experientia* **9**:357.
Schlesinger, M. J., and Coon, M. J., 1961, Reduction of mevaldic acid to mevalonic acid by a partially purified enzyme from liver, *J. Biol. Chem.* **236**:2421.
Schneider, W. J., Beisiegel, U., Goldstein, J. L., and Brown, M. S., 1982, Purification of the low density lipoprotein receptor, an acidic glycoprotein of 164,000 molecular weight, *J. Biol. Chem.* **257**:2664.
Schoenheimer, R., 1946, *The Dynamic State of Body Constituents*, Harvard University Press, Cambridge, Mass.
Schroepfer, G. J., Jr., 1981, Sterol biosynthesis, *Annu. Rev. Biochem.* **50**:585.
Schroepfer, G. J., Jr., 1982, Sterol biosynthesis, *Annu. Rev. Biochem.* **51**:555.
Schroepfer, G. J., Jr., Lutsky, B. N., Martin, J. A., Hontoon, S., Fourkans, B., Lee, W.-H., and Vermilion, J., 1972, Recent investigations on the nature of sterol intermediates in the biosynthesis of cholesterol, *Proc. Roy. Soc., Ser. B* **180**:125.
Schwenk, E., Todd, D., and Fish, C. A., 1954, Studies on the biosynthesis of cholesterol. VI. Companions of cholesterol-C^{14} in liver perfusions, including squalene-C^{14}, as possible precursors in its biosynthesis, *Arch. Biochem. Biophys.* **49**:187.
Shama Bhat, C., and Ramasarma, T., 1980, Purification & properties of mevalonate pyrophosphate decarboxylase of rat liver, *Indian J. Biochem. Biophys.* **17**:249.

Shapiro, D. J., and Rodwell, V. W., 1969, Diurnal variation and cholesterol regulation of hepatic HMG-CoA reductase activity, *Biochem. Biophys. Res. Commun.* **37**:867.

Sitges, M., Gil, G., and Hegardt, F. G., 1984, Partial purification from rat liver microsomes of three native phosphatases with activity towards HMG-CoA reductase, *J. Lipid Res.* **25**:497.

Sonderhoff, R., and Thomas, H., 1937, Die enzymatische Dehydrierung der Trideutero-essigsäure, *Liebigs Ann. Chem.* **530**:195.

Stewart, P. R., and Rudney, H., 1966, The biosynthesis of β-hydroxy-β-methylglutaryl coenzyme A in yeast. IV. The origin of the thioester bond of β-hydroxy-β-methylglutaryl coenzyme A, *J. Biol. Chem.* **241**:1222.

Tai, H-H., and Bloch, K., 1972, Squalene epoxidase of rat liver, *J. Biol. Chem.* **247**:3767.

Tanaka, R. D., Edwards, P. A., Lan, S-F., Knöppel, E. M., and Fogelman, A. M., 1982, The effect of cholestyramine and mevinolin on the diurnal cycle of rat hepatic 3-hydroxy-3-methylglutaryl coenzyme A reductase, *J. Lipid. Res.* **23**:1026.

Tanret, C., 1889, Sur un nouveau principe immédiat de l'ergot de seigle, l'ergostérine, *C. R. Acad. Sci.* **108**:98.

Tavormina, P. A., and Gibbs, M. H., 1956, The metabolism of β, δ-dihydroxy-β-methylvaleric acid by liver homogenates, *J. Am. Chem. Soc.* **78**:6210.

Tavormina, P. A., Gibbs, M. H., and Huff, J. W., 1956, The utilization of β-hydroxy-β-methyl-δ-valerolactone in cholesterol biosynthesis, *J. Am. Chem. Soc.* **78**:4498.

Taylor, C. B., and Gould, R. G., 1950, Effect of dietary cholesterol on rate of cholesterol synthesis in the intact animal measured by means of radioactive carbon, *Circulation* **2**:467.

Tomkins, G. M., and Chaikoff, I. L., 1952, Cholesterol synthesis by liver. I. Influence of fasting and of diet, *J. Biol. Chem.* **196**:569.

van Tamelen, E. E., Willet, J. D., Clayton, R. B., and Lord, K. E., 1966, Enzymic conversion of squalene 2,3-oxide to lanosterol and cholesterol, *J. Am. Chem. Soc.* **88**:4752.

Vogel, J., 1843, *Erläuterungstafeln zur pathologischen Histologie mit vorzüglicher Rücksicht auf sein Handbuch der pathologischen Anatomie*, Leopold Voss, Leipzig.

Voser, W., Mijović, M. V., Heusser, H., Jeger, O., and Ruzicka, L., 1952, Über Steroide und Sexualhormone. 186. Mitteilung. Über die Konstitution des Lanostadienols (Lanosterins) und seine Zugehörigkeit zu den Steroiden, *Helv. Chim. Acta* **35**:2414.

Wallach, O., 1887, Zur Kenntniss der Terpene und der ätherischen Oele, *Liebigs Ann. Chem.* **239**:1.

Waller, G. R., 1969, Metabolism of plant terpenoids, in: *Progress in the Chemistry of Fats and Other Lipids*, Vol. 10, Part 2 (R. T. Holman, ed.), Pergamon Press, Oxford, pp. 151–238.

Weis, H. J., and Dietschy, J. M., 1969, Failure of bile acid to control hepatic cholesterogenesis: Evidence for endogenous cholesterol feedback, *J. Clin. Invest.* **48**:2398.

Williams, C. G., 1860, "On isoprene and caoutchine", *Proc. Roy. Soc.* **10**:516.

Windaus, A., 1932, Über die Konstitution des Cholesterins und der Gallensäuren, *Z. Physiol. Chem.* **213**:147.

Windaus, A., and Hauth, A., 1906, Über Stigmasterin, ein neues Phytosterin aus Calabar-Bohnen, *Chem. Ber.* **39**:4378.

Woese, C. R., Magrum, L. J., and Fox, G. E., 1978, Archaebacteria, *J. Mol. Evol.* **11**:245.

Woodward, R. B., and Bloch, K., 1953, The cyclization of squalene in cholesterol synthesis, *J. Am. Chem. Soc.* **75**:2023.

Wright, L. D., Cresson, E. L., Skeggs, H. R., MacRae, G. D. E., Hoffman, C. H., Wolf, D. E., and Folkers, K., 1956, Isolation of a new acetate-replacing factor, *J. Am. Chem. Soc.* **78**:5273.

Wüersch, J., Huang, R. L., and Bloch, K., 1952, The origin of the isooctyl side chain of cholesterol, *J. Biol. Chem.* **195**:439.

16
THE AMPHIPHILIC LIPIDS: STRUCTURE, PROPERTIES, AND CONFORMATION

Any lipid (by definition, any substance that is either exclusively or predominantly apolar, i.e., hydrophobic) may, if it also contains at least one functional group that is either formally charged or strongly H-bonding (see Chapter 2), be properly termed amphiphilic. This chapter is, however, concerned primarily with lipids of this type that are usually found in biomembranes.

16.1. Nomenclature and Definitions

The many names that have been applied to this group of substances reflect the fact that none is fully satisfactory. The venerable term "complex lipids" includes the more restricted subgroups of phospholipids or phosphatides, glycolipids, and sphingolipids, but is understood to exclude a variety of other lipids of similar or even greater structural complexity. Moreover, the subgroups are not mutually exclusive: the phosphatidylinositols, for example, are both phospho- and glycolipids; while some sphingolipids are phospholipids and others glycolipids. Hartley (1936) originally coined the adjective "amphipathic" to describe substances having both highly polar (hence hydrophilic) and weakly polar (hydrophobic) parts. The attractiveness of Hartley's term is shown by its having been quickly accepted and widely used for over a decade. However, in 1948, Winsor chose to use the virtually synonymous "amphiphilic," which has since become increasingly popular, and is used here. (In his elegant treatise on certain properties of such substances, Tanford (1980) ascribes his preference for Winsor's adjective simply to its being "more euphonious.")

Even though the term amphiphilic lipid is not explicit in this respect, it is commonly understood that the existence of a restrictive balance of hydrophobicity and hydrophilicity is implied: the single hydroxyl group of the sterols or of the di-fatty-acylglycerols (diglycerides), for example, is not hydrophilic enough to offset the hydrophobicity of the remainders of these substances; on the other hand, the polysaccharidic moieties of the gangliosides are so overwhelmingly hydrophilic that their classification as lipids could be questioned on the basis of solubility characteristics. The "unbalanced" amphiphilic properties of substances of this kind certainly influence their favored location and orientation in the tissues in which they occur—but they are not considered to be truly amphiphilic lipids, and will not be discussed here.

Fatty acid carboxylate (soap) anions and lysophosphatides, which are clearly amphiphilic but powerful disruptors of biomembranes, are also excluded from detailed discussion, which is focused instead on those amphiphilic lipids that are major and essential components of biomembranes. The remarkable stability of biomembranes, which might appear to be fragile aggregates of separate molecules, in fact is directly attributable to the properties of specific amphiphilic lipids, intercalated in eukaryotic cells (as well as in membranes of some, but not all, bacteria) with sterols. Although living systems occasionally take advantage of the surfactant properties of these substances in other ways (e.g., in chylomicrons, in lipoproteins, and in alveolar fluid), their major role is in membrane structure. Indeed, their quantity in most tissues is directly related to biomembrane content. The term "membrane lipids" is therefore both directly meaningful and conveniently exclusive, and will be considered here to be synonymous with "amphiphilic lipids."

The striking differences in biomembrane compatibility of different types of amphiphiles (e.g., soaps and membrane lipids) invite rationalization of these phenomena in terms of structural features. In the observed structures of the common membrane lipids, it will be noticed that, although details of the architecture of the hydrophilic (or polar) and hydrophobic parts of these molecules vary widely, they all contain (i) a single, relatively compact, highly polar hydrophilic moiety (often called the "head" of the amphiphile); and (ii) two long, flexible, hydrophobic (apolar) "tails," i.e., they are bicaudal. (In rare instances, biomembrane lipids have more than two such tails.) In contrast, the biomembrane-lysing amphiphiles differ in having but a single such tail. The basic structural difference between the two classes of amphiphiles is that the polar head of the membrane-disrupting amphiphiles is at one end of its single tail, while that of the membrane-forming amphiphiles is near the center of a single, very long hydrophobic moiety, if the two tails are regarded as being continuous. Differences in the behavior of these two types of amphiphiles in, or in contact with, aqueous media that may be attributed to these structural distinctions will be discussed below.

16.2. Isolation from Tissue

Since the amphiphilic lipids are true lipids in the classical sense of their exhibiting distinctive, ready solubility in solvents of low polarity, they are easily extracted from tissue along with lipids of other types (see Chapter 2), which are then removed by taking advantage of properties unique to the amphiphiles because of the presence in them of considerably more highly polar moieties.

Chromatography of the total lipid mixture on columns or thin layers of silicic acid is a particularly powerful technique for separation of classes of lipids, since distinctions are dependent primarily on the relative strengths of interactions between the H-bond-donating or -accepting hydroxyl groups foresting the surface of this adsorbent and complementary potentialities of functional groups of the lipids. Since the affinity of alkyl and alkenyl groups for SiO_2 is virtually nil, all members of a given class of lipid, differing only in chain-length, branching, and/or saturation of fatty acyl (or related) groups, tend to be eluted together. If some but not all members of a class of lipids bear hydroxylated fatty acyl groups, these will, of course, be more strongly adsorbed; thus the α-hydroxy fatty acyl phrenosins are easily separated from the kerasin type of cerebrosides.

The dry chloroform solution of total lipids obtained by the commonly used Folch–Bligh–Dyer procedure (see Chapter 2) is suitable for such chromatographic fractionation. If such a solution is passed through a column of silicic acid, the original plus additional chloroform readily elutes virtually all of the so-called neutral (i.e., nonamphiphilic) lipids (including the monohydroxylic sterols and dihydroxylic mono-fatty-acylglycerols (monoglycerides)), but not the amphiphilic lipids. Passing diethyl ether through the column is sometimes done at this point to effect more complete removal of neutral lipids from the still strongly retained amphiphilic lipids. The total amphiphilic lipids may now be cleanly eluted from the column with methanol, which hydrogen-bonds strongly with the adsorbent and thus displaces the lipids. However, use of a succession of mixtures of chloroform and methanol in which the proportion of methanol is either serially or continuously increased results in extensive fractionation of the amphiphilic lipids with respect to class. In some cases it proves advantageous, following removal of the neutral lipids with chloroform and/or ether, to pass acetone through the column. This solvent tends to dislodge the glycolipids, but not the phospholipids—a distinction possibly related to the characteristically low solubility of phospholipids in acetone. The resulting glyco- and phospholipid concentrates may then be submitted separately to fractionation with chloroform–methanol mixtures. Variations in adsorbent and/or elutant regimens permit isolation of any class of amphiphilic lipid from others with which it may be admixed in the original tissue extract (see Stein and Slawson, 1966). Alteration of the adsorbability of classes of such lipids differing in basic or acidic properties may also be effected in thin-layer chromatography by addition of aqueous ammonia or of acetic acid to the developing solvent mixtures (see Malins, 1966).

16.3. Structures of Amphiphilic Lipids

16.3.1. Common Features

Practically all of the many membrane amphiphilic lipids that have been isolated and identified to date belong to one or another of two great classes differing in their being derivatives either of glycerol (e.g., the lecithins, **1.1**) or of sphingo-

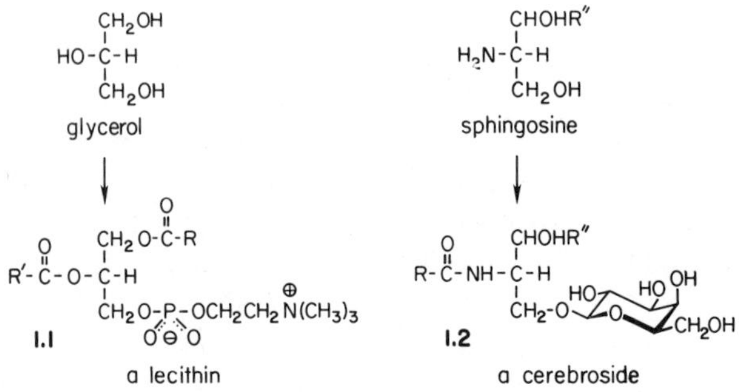

FIGURE 16.1. Similar structural features of a lecithin (**1.1**, derived from glycerol) and of a (galacto)cerebroside (**1.2**, derived from sphingosine: $R'' = Z\text{-}CH=CH(CH_2)_{12}CH_3$).

$$R-\overset{O}{\overset{\|}{C}}- \text{ and } R'-\overset{O}{\overset{\|}{C}}- = \text{fatty acyl moieties.}$$

sine (e.g., the cerebrosides, **1.2**) (see Fig. 16.1*). The sphingosine in the latter class of substances may be replaced by another of a group of structurally closely related bases, known collectively as sphingoids, differing only in details of the structure of group R''.

Although both glycerol and the sphingoids are vicinally trifunctional, the secondary hydroxyl group of the sphingoids is rarely if ever substituted in membrane lipids derived from them. Kochetkov et al. (1963) have reported an exception to this generalization, but neither Klenk and Doss (1966) nor Kishimoto et al. (1968) were able to confirm this finding.

Structural features common to nearly all known glycero- and sphingolipids are summarized in Fig. 16.2. Variations distinguishing one class of amphiphilic lipid from another involve differences in either the hydrophobic (A) or the hydrophilic moieties (B). The close similarity of the two classes with respect to arrangement of hydrophobic and hydrophilic moieties is readily apparent.

With regard to absolute configuration of chiral atoms (C-2 of the glycero-amphiphilic lipids, **2.1**, C-2 and C-3 of the sphingolipids, **2.2**), the structural formulas shown in Fig. 16.2 are Fischer projections. The stereo-specific numbering (*sn*-) of the glycerol-skeleton carbon atoms (shown) is based on Fischer conventions, the central (secondary) hydroxyl group of the prochiral glycerol molecule being placed to the left, and the carbon atoms then numbered from top to bottom. The *sn*-convention for distinguishing the two primary α-carbon atoms of asymmetrically substituted glycerol derivatives, sanctioned by the nomenclature committees of IUPAC–IUB, is now replacing the previously used L-α-designation,

---

*Structural formulas are identified by either of two series of boldface-type numbers: formulas not incorporated into formal figures are numbered consecutively throughout the text of the chapter; those in figures are doubly numbered, with that of the figure preceding and separated by a decimal point from that of the formula within it. Formula **8**, for example, is somewhere between formulas **7** and **9**; and **5.2** appears in Fig. 16.5.

FIGURE 16.2. General structural formulas of amphiphilic glycerolipids (**2.1**) and sphingolipids (**2.2**). Variable moieties A and B represent, respectively, hydrophobic and hydrophilic groups.

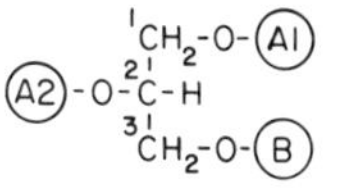

$$\text{glycerolipids} \quad \mathbf{2.1}: \quad \overset{1}{C}H_2\text{-O-}\textcircled{A1},\; \textcircled{A2}\text{-O-}\overset{2}{C}H\text{-},\; \overset{3}{C}H_2\text{-O-}\textcircled{B}$$

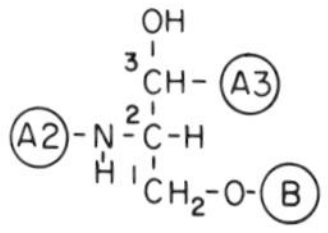

$$\text{sphingolipids} \quad \mathbf{2.2}: \quad \overset{3}{C}H(\text{OH})\text{-}\textcircled{A3},\; \textcircled{A2}\text{-N}\overset{2}{C}H\text{-}(H),\; \overset{1}{C}H_2\text{-O-}\textcircled{B}$$

which was based on configurational analogy with the levorotatory L-glyceraldehyde. Glycero-amphiphilic lipids of *Halobacterium cutirubrum* represent the only exception so far known to this configurational generalization; in these substances the hydrophilic group is in the *sn*-1 position. Because of the Cahn–Ingold–Prelog conventions used in ranking substituents of chiral centers, the absolute configuration of C-2 of almost all glycerophospholipids can be designated as *R*. In the glyceroglycolipids, however, the rank of either group B or A1 may be higher and the configuration at C-2 thus may be either *R* or *S*, despite their being configurationally analogous.

In order to emphasize certain structural similarities between the two classes, the general formula (**2.2**) of the sphingolipids departs from strict adherence to the Fischer convention insofar as only the first three of a long chain of carbon atoms that includes those in the A3 group have been placed in a vertical row, and insofar as the more oxidized end of that chain is at the bottom rather than at the top. Fischer conventions are, however, used to represent the absolute configurations of the chiral centers at C-2 and C-3, which are, in all known cases, *S* and *R*, respectively. Although the configurations of the central C-2s of both the glycerolipids and sphingolipids are analogous and thus appear to reinforce the structural similarities of the two classes, this could well be a coincidence of biosynthesis (see Chapters 17 and 18) rather than a structural feature dictated by optimal function in biomembranes; those of the extremely halophilic bacteria, in which the major (glycero-) amphiphilic lipids are of opposite configuration, appear, for example, to function quite well.

Before discussion of variability of the hydrophobic (A) and hydrophilic (B) moieties, differences in the skeletal cores of the glycerol- and sphingoid-based substances should be pointed out. Several features tend to make the hydrophilic part of the sphingolipids more extended, independently of the nature of group B: presence of the more hydrophilic −NH− instead of the −O− linkage to C-2, and presence of the free hydroxyl group on C-3. This general distinction is amplified in certain sphingolipids. The fatty acyl groups (A2) of many cerebrosides, for example, are α-hydroxylated. Moreover, the phytosphingolipids of plants have an additional hydroxyl group on C-4, the first carbon atom of group A3. The hydrophobic group A3 of the sphingolipids is actually an extension of the core carbon chain, which remains intact during exhaustive hydrolysis. Such treatment of sphingolipids therefore yields, among other products, one mole-equivalent of sphingosine or other sphingoid. Sphingoids, exceptional among lipid hydrolysis products in being hydrophobic primary amines, are also called "long-chain bases" or LCBs.

16.3.2. Variability in Structure of Hydrophobic Groups and in Mode of Attachment

The hydrophobic groups of both glycero- and sphingo-amphiphilic lipids vary in detailed structure in two respects: in their carbon skeletons, reflecting derivation from fatty acids (except, of course, in organisms such as the halophilic bacteria that do not synthesize fatty acids), and in the mode of union of these parts to the rest of the lipid molecule.

Since they are derived from fatty acids, the A groups are, in general, unbranched; contain an even number of carbon atoms (often 16 or 18); and may be mono- or poly-(1,4)-olefinic, the double bond(s) being almost invariably of Z *(cis)* configuration, and usually no closer than 9' in position with respect to the persistent or modified carbonyl carbon atom of the parent fatty acyl group. These generalizations also apply to the A3 group of the sphingolipids, since biosynthesis of the sphingoid base involves condensation of an activated fatty acid molecule with L-serine, the carboxyl group of which is lost in the process, C-1 of the fatty acid becoming C-3 of the sphingoid (see Chapter 18).

In general, however, the biosynthetic processes involved do not draw indiscriminately on a single mixed pool to supply whatever fatty acyl groups may be incorporated into the finished amphiphilic lipids. In the glycero-amphiphilic lipids, for example, more unsaturated fatty acids are involved in genesis of the A2 than of the A1 moiety. Furthermore, in construction of the sphingolipids, little unsaturated but atypically large proportions of fatty acyl groups having an odd number of carbon atoms are incorporated into the intermediate sphingoids, and others of unusually great length form group A2 (involvement of α-hydroxy fatty acyl groups in A2 of cerebrosides is also noteworthy).

With respect to the functionality of the union of hydrophobic groups (A) to the glycerol or sphingoid cores of these lipids, both A1 and A2 in the glycerolipids are usually, and A2 in the sphingolipids apparently always, unmodified fatty acyl groups (sphingolipids are therefore fatty acid amides). Glycerophospholipids in which the A1 is Z-1'-alkenyl instead of fatty acyl (i.e., $-CH=CH-R$ instead of $-COCH_2R$) are widespread in both plants and animals and indeed represent a major type of amphiphilic lipid in some tissues (e.g., brain, heart, and erythrocytes). Because of the ease with which fatty aldehydes ("plasmals," **1**) are released from these "vinyl ethers" by hydrolysis *in vitro* under acidic conditions, such lipids were originally called "plasmalogens" before their structures were fully elucidated. For many years plasmalogens were erroneously believed to be *sn*-1,2-(fatty aldehyde)-acetals of glycerol. Glycero-amphiphilic lipids are also known in which the A1 or both the A1 and A2 groups (e.g., the di-(dihydrophytyl)-glycerophospholipids and -glyceroglycolipids of the extremely halophilic bacteria) are

$$\text{>C-O-CH=CH-R} + H_2O \xrightarrow{(H^\oplus)} \text{>C-OH} + H-\overset{O}{\underset{\|}{C}}-CH_2R$$

1

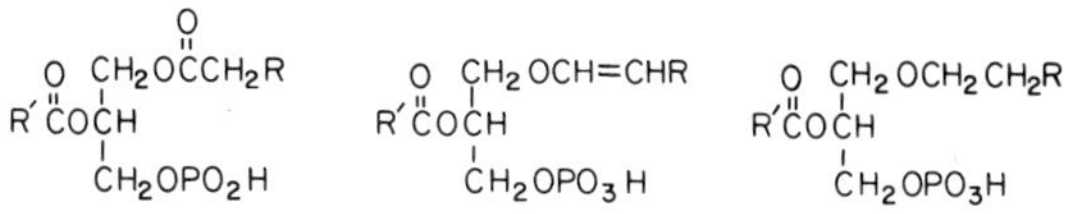

THE AMPHIPHILIC LIPIDS.   375

$$
\begin{array}{ccc}
\begin{array}{l} \quad\;\; O\;\; CH_2O\overset{O}{\overset{\|}{C}}CH_2R \\ R'\overset{\|}{C}O\overset{|}{C}H \\ \quad\;\; CH_2OPO_2H \end{array} &
\begin{array}{l} O\;\; CH_2OCH=CHR \\ R'\overset{\|}{C}O\overset{|}{C}H \\ \quad\;\; CH_2OPO_3H \end{array} &
\begin{array}{l} O\;\; CH_2OCH_2CH_2R \\ R'\overset{\|}{C}O\overset{|}{C}H \\ \quad\;\; CH_2OPO_3H \end{array} \\
3.1 & 3.2 & 3.3
\end{array}
$$

FIGURE 16.3. Structures of phosphatidic (**3.1**), plasmenic (**3.2**), and plasmanic (**3.3**) acids.

alkyl (RCH_2-) rather than acyl ($RCO-$). In contrast to the 1'-alkenoxy ("vinyl ether") groups of the plasmalogens, such alkoxy groups are much less easily hydrolyzed than the corresponding acyloxy moieties. By extension of the long-standing practice of abbreviated naming of several commonly occurring glycerophospholipids as esters of phosphatidic acid (*sn*-1,2-diacylglycerol-3-phosphoric acid, **3.1**), IUPAC–IUB has sanctioned naming of the *sn*-1-(1'-alkenyl) and -1-alkyl analogues as esters of plasmenic (**3.2**) and of plasmanic (**3.3**) acids, respectively (see Fig. 16.3). The important ethanolamine-containing plasmalogenic glycerophospholipids of brain, for example, may be represented by a general structural formula (**2**) and referred to as plasmenylethanolamines (such substances have, in the past, been called phosphatidalethanolamines).

$$
\begin{array}{l}
\quad\;\; O\;\; CH_2OCH=CHR \\
R'\overset{\|}{C}O\overset{|}{C}H \\
\quad\;\; CH_2OPOCH_2CH_2\overset{\oplus}{N}H_3 \\
\mathbf{2} \qquad\; O\overset{\ominus}{\;}O
\end{array}
$$

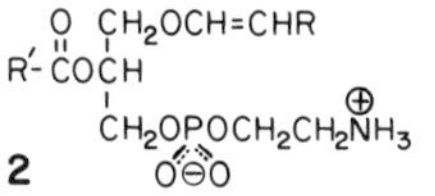

16.3.3. Structure of the Sphingoids

Variability in structure of hydrophobic groups A3 (see Fig. 16.2) of the sphingolipids is related to that of the fatty acyl group involved in the condensation with L-serine and to details of that anabolic process, and is retained by the sphingoids remaining after *in vitro* hydrolytic removal of the A2 and B moieties. These substances (and hence the sphingolipids derived from them) fall into three classes, structural features and IUPAC–IUB-favored names of which are shown in Fig. 16.4. (Carbon atoms C-1 and C-2 of the sphingoid are C-β and C-α, respectively, of L-serine, and C-3, C-4, C-5.... are C-1, C-2, C-3.... of the fatty acyl precursor.) The names cited are used generally, i.e., for any sphingoid having the struc-

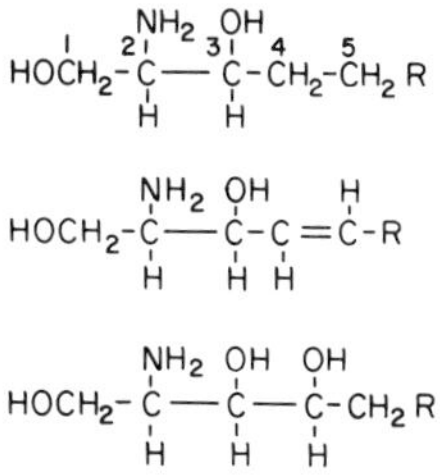

FIGURE 16.4. Structures and names of the major sphingoids. R = long-chain alkyl or alkenyl, normal or branched.

4.1 $HOCH_2-\overset{\overset{NH_2}{|}}{\underset{H}{C}}-\overset{\overset{OH}{|}}{\underset{H}{C}}-CH_2-CH_2\,R$ sphinganine (dihydrosphingosine)

4.2 $HOCH_2-\overset{\overset{NH_2}{|}}{\underset{H}{C}}-\overset{\overset{OH}{|}}{\underset{H}{C}}-C=\overset{H}{\underset{}{C}}-R$ sphingenine (sphingosine)

4.3 $HOCH_2-\overset{\overset{NH_2}{|}}{\underset{H}{C}}-\overset{\overset{OH}{|}}{\underset{H}{C}}-\overset{\overset{OH}{|}}{\underset{H}{C}}-CH_2\,R$ phytosphinganine

tural features shown, regardless of the chain-length of R. Specific chain-length is indicated when necessary or desirable by an appropriate prefix: thus octadecasphinganine is the 18-carbon homologue (**4.1**, R = n-$C_{13}H_{27}$). Before the existence of homologues was recognized, the name "sphingosine" was given to the principal (and first isolated) sphingoid of animal tissues, octadecasphingenine (see **4.2**), and "dihydrosphingosine" to the corresponding sphinganine.

Absolute configurations of the chiral centers of these substances are represented in Fig. 16.4 as true Fischer projections turned 90° counterclockwise; substituents above and below the chiral carbon atoms are therefore closer to the observer than those to the right or left. In systematic nomenclature, octadecasphinganine (dihydrosphingosine) is 2S-amino-1,3R-dihydroxyoctadecane; octadecasphingenine (sphingosine) is 2S-amino-1,3R-dihydroxy-4E-octadecene; and octadecaphytosphinganine (see **4.3**) is 2S-amino-1,3S,4R-trihydroxyoctadecane. In the past, the term D-erythro has been used to define the configuration of the sphinganines because they are configurationally analogous to D-erythrose.

The principal sphingoids of animal sphingolipids are 4E-sphingenines of chain-length 16,17,18, and 20 (the "odd" C_{17} is not a trace homologue) accompanied by the analogous sphinganines and by other analogues containing additional olefinic centers probably stemming from the involvement of unsaturated fatty acyl moieties in their biosynthesis and, if so, of Z *(cis)* configuration. Correspondence of the *trans*-unsaturation of the sphingenines in position 4 to the *trans*-α,β-unsaturated intermediates of β-oxidation of fatty acids is coincidental, since the dehydrogenation occurs after rather than before incorporation of the fatty acyl group into the sphingoid molecule (see Chapter 18). The fact that the absolute configuration of C-4 of the phytosphingoids, bearing the additional hydroxyl group, is R, i.e., analogous to that of C-α in many α-hydroxy fatty acids, is also coincidental, it having been shown (see Stoffel *et al.*, 1968) that these substances are derived (in yeast) from sphinganine. C_{17} through C_{20} phytosphinganines occur in plants, together with a C_{18} 8E-phytosphingenine; this 8E unsaturation corresponds in position and configuration to that in a 6E-16:1 fatty acyl precursor, not known to occur commonly in plants. Sphinganines, which lack the 4-hydroxy group, appear to be synthesized in small quantity by plants. Small amounts of phytosphingolipids have been isolated from certain animal tissues (e.g., bovine kidney), but it is possible that these substances are derived intact from ingested plant tissue.

16.3.4. Variability in Structure of Hydrophilic Groups

In both the glycerol- and sphingoid-based classes of amphiphilic lipids, hydrophilicity is provided by incorporation of formally charged (ionic) moieties or of polyhydroxylated moieties, or of a combination of the two. Structural variations of this sort are exemplified by lecithin, a glycerophospholipid; by diglyceride glycosides, which are glyceroglycolipids; by sphingomeylin, a sphingophospholipid; and by the cerebrosides, which are sphingoglycolipids.

Anderson *et al.* (1975) have shown that diatoms contain sphingolipids of

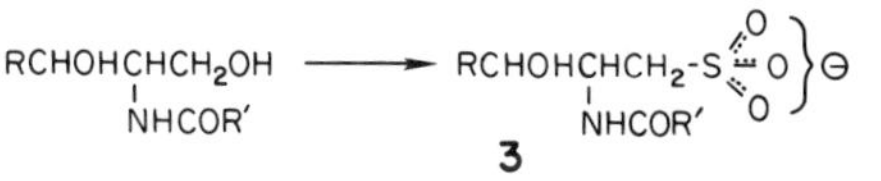

$$\text{RCHOHCHCH}_2\text{OH} \longrightarrow \text{RCHOHCHCH}_2\text{-SO}_3^{\ominus}$$
$$\quad\quad\;\;| \quad\quad\quad\quad\quad\quad\quad\quad\; |$$
$$\;\text{NHCOR}' \quad\quad\quad\quad\quad\quad \text{NHCOR}'$$

3

type **3**, in which hydrophilicity is furnished by replacement of the ceramide 1-hydroxy with a sulfonic acid moiety (virtually completely dissociated and hence bearing a formal negative charge at physiological pH). This transformation is nominally analogous to that involved in conversion of a D-glucosyl to the D-quinovosyl-6-sulfonic acid moiety of the plant sulfonolipids (see below). It has been suggested, however, that these substances may be formed by substitution of cysteine for serine in sphingoid biosynthesis, followed by oxidation of the thiol group. Direct incorporation of cysteic acid would yield the same intermediate aminosulfonic acid. Although inclusion of these unusual amphiphilic lipids with the true sphingolipids is easily justifiable, their hydrolysis yields a zwitterion rather than a true sphingoid base.

16.3.4.1. Glycerophospholipids

With the exception of examples found in haliform bacteria, which differ in configuration of the chiral central carbon of the glycerol core, these substances share features shown in structural formula **4**. Except for the parent phosphatidic (**3.1**) (and the analogous plasmenic (**3.2**) and plasmanic (**3.3**)) acids, in which R = H (see below), these substances are diesters of phosphoric acid. Being strongly acidic ($pK_a \sim 1.3$), they are virtually completely dissociated at any physiological pH and therefore bear an essentially immutable formal negative charge. It is important to recognize that in the dissociated, anionic species, the four substituents of the phosphorus atom are tetrahedrally oriented (rather than coplanar), and that the two "monovalent" oxygen substituents are, in fact, equivalent, with the single formal charge equally shared by them by resonance (see resonance hybrid **5**). Alternatively, the group may be regarded as involving a phosphorus atom with an unexpanded valence shell, in which case both oxygens bear full negative charges, and the phosphorus a formal positive charge (see **6**).

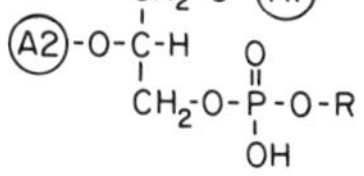

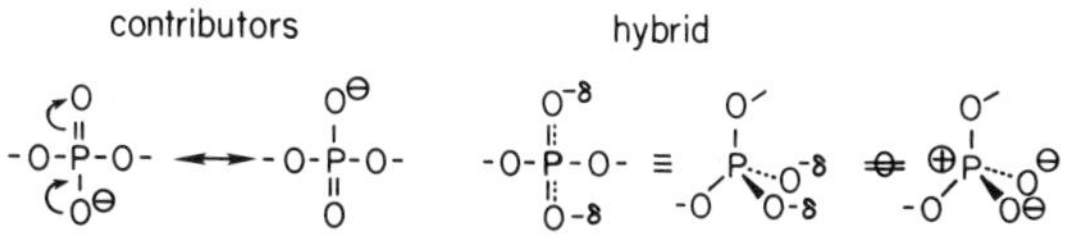

Phosphatidic acids (**4**, R = H; A1 and A2 = fatty acyl), often abbreviated PA (1,2-diacyl-*sn*-glycero-3-dihydrogen phosphoric acid), are monoesters of phosphoric acid and are thus expected to have pK_as of about 1.5 (i.e., to be slightly weaker acids than the diesters) and 6.5. At pH 7 the undissociated (uncharged) species would therefore be negligible, and the anionic species G—OPO$_3^=$ and G—OPO$_3$H$^-$ (where G is an *sn*-glycero–3 moiety) would be present in a ratio of about 3:1. (Resonance of the predominant dianionic species distributes the two formal negative charges equally over three oxygen atoms.) Phosphatidic acids, and the analogous plasmenic and plasmanic acids, are intermediates in the biosynthesis of other glycerophospholipids (see Chapter 17) and therefore occur in tissues, although their steady-state concentrations are normally quite small. They are, however, obtained in substantial quantity from plant material if care is not taken to prevent action of enzymes (phospholipases D) specifically catalyzing conversion of intact lipids to such hydrolysis products.

Many of the known kinds of amphiphilic lipids are esters of phosphatidic acids (i.e., diesters of phosphoric acid) in which the hydrophilic moiety has been elaborated by attachment of a group bearing additional formal charge(s), hydroxyl group(s), or both. Structural formulas and names (as esters of phosphatidic acid) of the best known of these are shown in Fig. 16.5 (as the ionic species dominant at pH 7, and with A1 and A2 = fatty acyl). It is customary to discuss the first three of these (**5.1, 5.2**, and **5.3**), which are closely related biosynthetically (see Chapter 17), in order of discovery.

The phosphatidylcholines (**5.1**; 1,2-diacyl-*sn*-glycero-3-phosphocholines)—often abbreviated to PC—are commonly called lecithins, a name derived from

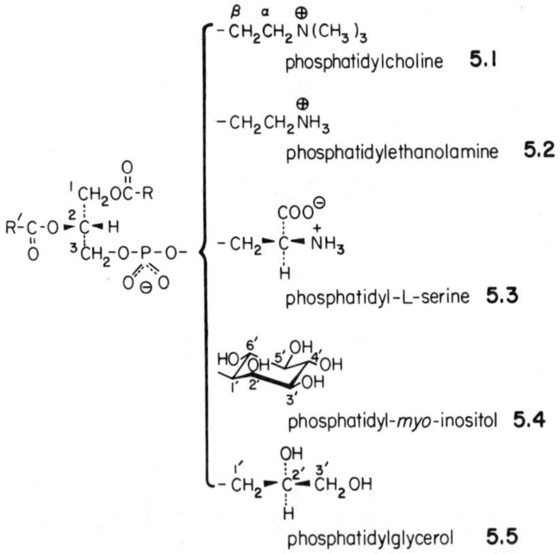

FIGURE 16.5. Structures and names of the major phosphatidyl amphiphilic lipids. Abbreviations PC (for **5.1**), PE (**5.2**), PS (**5.3**), and PI (**5.4**) are commonly used; PG for **5.5**, however, has recently been appropriated to signify the prostaglandins.

the Greek word, "lekithos," for egg yolk, which remains the source of choice for bulk preparation of such material. These substances occur abundantly and nearly ubiquitously in nature, rivaled in this respect only by the phosphatidylglycerols, **5.5** (see below). Since the quaternary ammonium group is neither acidic nor basic, and the phosphate moiety is extremely weakly basic (reflecting the high acidity of diesters of phosphoric acid), the lecithins are true zwitterions, and remain so throughout any conceivable excursions of physiological pH. About half of the fatty-acyl-group content of mammalian PC is saturated (principally 16:0, concentrated in the sn-1 position), most of the rest being oleyl (9Z-18:1); polyene fatty acyl groups tend to be relatively rare in this class of amphiphilic lipid. About 40% of the phospholipids of ox heart are lecithins, of which almost half are of the plasmalogen type. The fully saturated dipalmitoyl (di-16:0) lecithin (a so-called hydrolecithin) is, however, a major surfactant of pulmonary alveolar fluid.

Anderson et $al.$ (1976) have shown that certain diatoms that are quite devoid of PC elaborate a sulfur analogue of these substances, "sulfolecithin," in which $-\overset{\oplus}{S}(CH_3)_2$ is substituted for $-\overset{\oplus}{N}(CH_3)_3$; sulfolecithin comprised 68% of total phospholipids of the species examined by these workers.

A homologue of PC, phosphatidyl-β-methylcholine (**7**), appears to be formed either by decarboxylation of phosphatidylcarnitine (**8**) (Mehendale et $al.$, 1966) or from phosphatidylthreonine (see below) by steps analogous to those involved in the PS → PE → PC conversion. By either pathway the configuration of the additional chiral center would be R, as shown in **7** and **8**.

$$\text{phosphatidyl} - O - \underset{\underset{CH_3}{|}}{\overset{\overset{H}{|}}{C}} - CH_2 \overset{\oplus}{N}(CH_3)_3 \xrightarrow[-CO_2]{?} - O - \underset{\underset{CH_2 COO^{\ominus}}{|}}{\overset{\overset{H}{|}}{C}} - CH_2 \overset{\oplus}{N}(CH_3)_3$$

                                      7   **8**

Phosphatidylcarnitine (**8**) is reported by Mehlman and Wolf (1963) to be an important glycerophospholipid in embryonic tissues of higher animals (chick, rat); these substances are also found in certain larvae raised on carnitine instead of the otherwise required choline (Bieber et $al.$, 1963; Bridges et $al.$, 1965). Phosphatidylcarnitines are zwitterionic anions at physiological pH values and, like other esters of carnitine, presumably have quite high negative free energies of hydrolysis (of the carnityl–phosphate bond).

The phosphatidylethanolamines (PE; **5.2**; 1,2-diacyl-sn-glycero-3-phosphoethanolamines) are still frequently called kephalins (now usually spelled "cephalins"), a name originally assigned to them by Thudicum (1884, pp. 52–64), who showed that PE is present in a fraction of material isolated from brain tissue, separated from lecithin by virtue of its markedly lower solubility in ethanol. Although such material was later shown (see Folch, 1942) to contain more of both phosphatidylinositols (**5.4**) and phosphatidyl-L-serines (**5.3**) than of phosphatidylethanolamines, Thudicum's synonym for PE is still retained. About half of the fatty acyl groups of mammalian PE, like those of PC, are saturated and located almost entirely in the sn-1 position, but they are predominantly stearoyl (18:0) rather than palmitoyl (16:0); moreover, the unsaturated fatty acyl groups (primarily in position *sn*-2) are more richly polyenoic in PEs (10–20%) than in PCs (traces). Plasmenylethanolamines (ethanolamine plasmalogens (2)) occur in substantial quantity in mammalian brain, heart, and erythrocytes.

Since the phosphatidylcholines are formed in part by stepwise triple methylation of phosphatidylethanolamines (see Chapter 17), the *N*-monomethyl and *N,N*-dimethyl intermediates are known. Exhaustive methylation appears, however, to be the normal course of events, and such intermediates are therefore not ordinarily expected to attain a high steady-state concentration.

Aneja *et al.* (1969) have shown that substantial quantities of PE found in many seeds are *N*-fatty-acylated (see **9**) and thus contain a fatty-acid-amide linkage, uncommon except in the sphingolipids. Structurally analogous *N*-fatty-acyl α-amino acids or "lipoamino acids" occur widely in animals and bacteria, but their function is not yet known: see Macfarlane (1964). *N*-Acylation nullifies the basic properties of the erstwhile amino group, and these substances are therefore acidic, i.e., anionic (as shown) at physiological pH values. They are also unusual in being "tricaudal" amphiphiles (i.e., possess three long hydrophobic moieties).

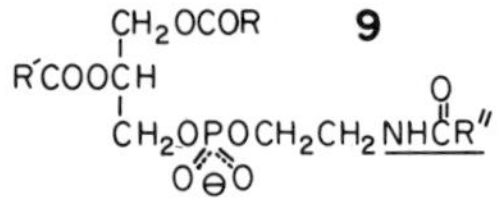

$$\begin{array}{l} \quad\quad\quad CH_2OCOR \quad \mathbf{9} \\ R'COOCH \\ \quad\quad\quad CH_2OPOCH_2CH_2NHCR'' \end{array}$$

Rumen protozoa of sheep have been shown (Coleman *et al.*, 1971) to modify PEs by *N*-α-carboxyethylation, yielding amphiphilic lipids of structure **10**. The close functional analogy of these substances is readily apparent by comparison of the structure of **10** with that of the phosphatidyl-L-serines (**5.3**), a type of phospholipid notably absent in these organisms. Isotopic labeling experiments showed that these substances are formed from PE by addition of three-carbon units derived efficiently from starch, lactate, or pyruvate, but did not discern whether L-alanine might be involved (with nitrogen-atom exchange) as an immediate precursor; the configuration of the added chiral center (as shown in **10**) is, however, in agreement with this possibility.

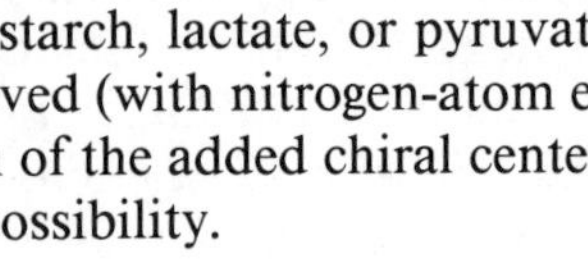

Tetrahymena and a growing number of other microorganisms have been found (see, e.g., Thompson, 1967) to contain, in addition to the ubiquitous PEs, substantial amounts of remarkable analogues differing solely (except for alkyl instead of fatty acyl at position *sn*-1) in the absence of a single atom—the oxygen atom otherwise linking phosphorus to the ethanolamine moiety—and thus having a direct P—C covalent bond (see arrow in structure **11**) 1-Acyl- and 1-(1'-alkenyl) analogues of these 2-acyl-*sn*-glycero–3-phosphono–2'-ethylamines

$$\begin{array}{l}\text{CH}_2\text{OCH}_2\text{R}\\\text{R'COOCH}\\\text{CH}_2\text{OP}-\text{CH}_2\text{CH}_2\overset{\oplus}{\text{NH}}_3\\\quad\;\;\overset{\nearrow\,\searrow}{O\;\overset{\ominus}{O}\;O}\end{array}$ II

are also known. Analogously constituted sphingolipids are found in coelenterates and mollusks. The nonhydrolyzability of the P—C linkage greatly facilitates the identification of such lipids, which yield β-aminoethylphosphonic acid ($\overset{\ominus}{\text{HO}}_3\text{PCH}_2\text{CH}_2\overset{\oplus}{\text{NH}}_3$, "ciliatine") in addition to other products on vigorous acid-catalyzed hydrolysis. These substances are also immune to hydrolytic attack by D-type phospholipases (see Chapter 17), and it has been suggested that organisms distinctive in having "naked" cytoplasmic membranes exposed to the environment have evolved biosyntheses of these ostensibly less vulnerable membrane lipids. A further and possibly more important implication should not be overlooked: this structural variation is apparently not crucial to the physicochemical function of such lipids in biomembranes.

The phosphatidyl-L-serines (PS; 1,2-diacyl-*sn*-glycero–3-phospho-L-serines) tend to accompany PE (e.g., in brain tissue). PE and PS are, in mammalian tissues generally, similarly acylated and possess similar solubility characteristics. Since it is L-($\equiv S$)-serine that is incorporated into this phospholipid, the configuration of the chiral center of the serine moiety of PS is *S,* as indicated in **5.3.** The existence of the homologous phosphatidylthreonine in a few tissues has been reported, but the fact that such substances are rarely encountered suggests that incorporation of serine is highly selective. Presence of the proximate, formally negatively-charged phosphate moiety in PS would be expected, by inductive and/or field effects, to diminish the acidities of the α-carboxyl and α-ammonium groups to some extent (cf. pK_as 2.2 and 9.15, respectively, for serine itself), but the hydrophilic head group of PS, in contrast to those of the zwitterionic PC and PE, is expected to bear a net negative charge in the usual physiological pH range.

The phosphatidylinositols (PI; **5.4;** 1,2-diacyl-*sn*-glycero–3–phospho–1'-L-*myo*-inositols)* occur in both plants and animals (prominently in brain, heart, and liver tissue of mammals). The distribution of fatty acyl groups in mammalian PI, including about 50% 18:0 and 30% polyenes (notably excluding linoleoyl), resembles that in PE rather than PC (see above). *myo*-Inositol, having hydroxyl groups on carbon atoms 1,2,3, and 5 on the same side of the cyclohexane ring, is one of the *meso* (achiral) stereoisomers (note plane of symmetry through C-2 and C-5); "phosphatidylation" of the C-1 hydroxyl group, however, destroys this symmetry, and the inositol group therefore contributes to the optical activity of the PIs. In contrast to the other phospholipids considered up to this point, PIs bear a single negative formal charge solely in the phospho moiety; these substances are therefore anionic at physiological pH and are included among the so-called acidic

---

*The configurational significance of the qualifier "L" in the name of these phospholipids recommended in 1976 by the IUPAC–IUB Commission on Biochemical Nomenclature (1978) is unclear. Like glycerol, *myo*-inositol is prochiral, and a convention analogous to that ("*sn*") adopted for use in unequivocal naming of chiral derivatives of glycerol is needed here as well. See Klyashchitskii *et al.* (1969). Formula **5.4** is configurationally explicit.

phospholipids. The hydrophilicity of PI is therefore a function of both formal charge and polyhydroxylation. The hydrophilicity of PIs is further enhanced by phosphorylations at the 4'-, or 4'- and 5'-positions. These phosphorylations increase hydrophilicity to such an extent that the resulting products are unusual in being efficiently extracted by solvents of low polarity from aqueous solutions only by adjustment of pH to suppress dissociation of the phosphate moieties.

In *Mycobacteria* and other microorganisms, what appears to be an analogous modification accomplished by incorporation of additional polyhydroxy moieties (rather than formally charged groups) is seen in the occurrence of α-D-mannopyranosides of PI, containing from one to at least six mannose units joined at the 2'- and/or 6'-positions of the inositol moiety and/or at 2''- and/or 6''-positions of mannose moieties more closely attached to the inositol group. The observation that certain of these PI mannosides yield, on hydrolysis, fatty acid:phosphate ratios larger than 2:1 shows that additional fatty-acylation occurs, involving hydroxyl groups of inositol or mannose units. Since none of the additional groups attached to the PI in these substances bears formal charges, they remain "acidic" (i.e., anionic), but are unusual among amphiphilic lipids with respect to numbers and arrangement of hydrophilic and hydrophobic moieties.

The phosphatidylglycerols (**5.5**; 1,2-diacyl-*sn*-glycero-3-phospho-*sn*-1'-glycerols) occur in small amounts in animals (mainly in mitochondria), but are the major (20–30%) phospholipids of higher plants, concentrated in chloroplasts; quantities occurring in bacteria are widely variable, but such substances appear to be the sole phospholipids in certain Gram-positive species. Because of their prominence in plants, phosphatidylglycerols probably rival the phosphatidylcholines as the most abundant type of membrane lipid in nature. Since *sn*-glycerol 3-phosphate serves as the source of the second glycerol moiety in biosynthesis of phosphatidylglycerols (see Chapter 17), the two glyceryl groups are esterified via different primary hydroxyl groups, and mild hydrolytic removal of the fatty acyl moieties therefore yields the *meso* (*R,S*)-*sn*-1,3'-diglycerophosphate. Renkonen (see Somerharju *et al.*, 1977) has recently demonstrated the presence, in lysosomes of cultured baby hamster kidney cells, of di-, tri-, and tetra-fatty-acylated derivatives of the stereoisomeric (*S,S*)-*sn*-1,1'-diglycerophosphoric acid; most of the phospholipids of the halophilic bacteria, on the other hand, are derivatives of the enantiomeric (*R,R*)-*sn*-3,3'-diglycerophosphoric acid.

In certain microorganisms, the hydrophilicity and functionality of the phosphatidylglycerols is modified by glucosaminidation of either of the free *sn*-2' or the *sn*-3' hydroxyl groups, yielding 2'- or 3'-*O*-(2''-deoxy-2''-amino-β-D-glucopyranosyl)phosphatidylglycerols (see **6.1**, the 3' isomer); or, presumably via involvement of a variety of α-aminoacyl-tRNAs, by conversion to *sn*-3'-α-aminoacylphosphatidylglycerols, of which the L-lysyl derivative (**6.2**) is shown in Fig. 16.6.

The presence in amphiphilic lipids of an aminosaccharide moiety in which the amino group is free (and hence protonated and bearing a positive formal charge under physiological conditions—see **6.1**) is unusual; cf. the *N*-acylated units of the cytolipins and gangliosides, mentioned below. α-Aminoacylphosphatidylglycerols (e.g., **6.2**) derived from basic amino acids (ornithine, arginine, and histidine, in addition to lysine) have been reported. These substances are unusual among membrane lipids in being predominantly cationic at physiological pH.

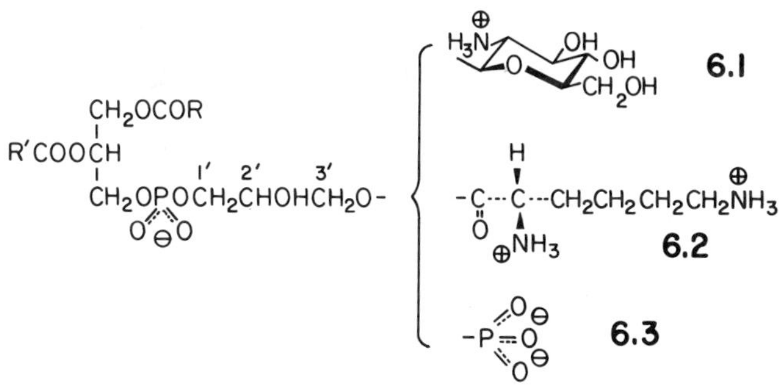

THE AMPHIPHILIC LIPIDS

FIGURE 16.6. Structures of derivatives of phosphatidylglycerols elaborated by certain microorganisms.

The sn-3′-phosphoric acid esters of the phosphatidylglycerols (**6.3**; PGP) are intermediates in the biosynthesis of phosphatidylglycerol itself, and of the cardiolipins (see below), although they do not, in higher animals, attain a high steady-state concentration. Small amounts of such substances have been shown, for example, to be present in brain tissue. The closely analogous, stereoisomeric 2,3-di-(dihydrophytyl)-sn-glycero–1–phospho-sn-3′-glycero-1′-phosphates,* on the other hand, constitute 70–80% of the phospholipids of *Halobacterium cutirubrum;* less halophilic bacteria, that are able to synthesize fatty acids, contain the corresponding 2,3-di-(fatty acyl) analogues (diastereomers of **6.3**). These substances are "acidic," i.e., anionic at physiological pH.

The cardiolipins (CL; **12**; sn-1,3-diphosphatidylglycerols) derive their common name from the fact that they were originally isolated from heart tissue, in which they appear to be concentrated (about 10% of total phospholipids; cf. about 5% of total phospholipids of the whole body); widely variant amounts of such substances have since been shown to occur in other organisms, including plants and microorganisms. Cardiolipins are highly concentrated in mitochondria, particularly in the inner membranes of these organelles, and their concentration in various tissues reflects this: high in heart, little in the brain, none in the erythrocytes. Cardiolipins contain unusually large proportions of linoleoyl (18:2) moieties (fatty acids released on mild hydrolysis: almost all 18:2 and 18:1, in a ratio

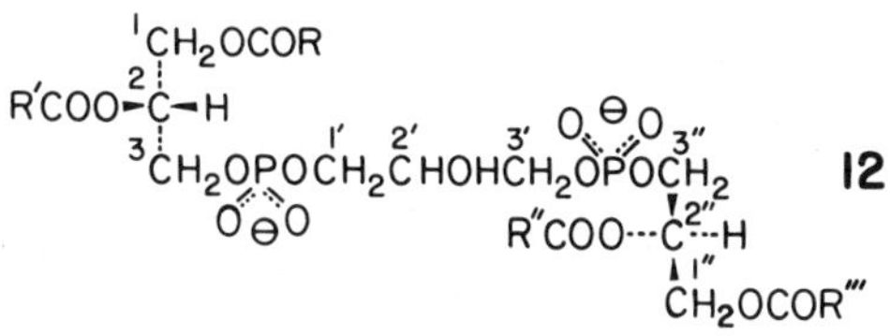

---

*Dihydrophytyl ≡ 3R,7R,11R,15-tetramethylhexadecyl; this polyprenoid alkyl group is, by some, called "phytanyl", but objectionably so, since this term could well be taken, alternatively, to signify the acyl group corresponding to phytanic acid.

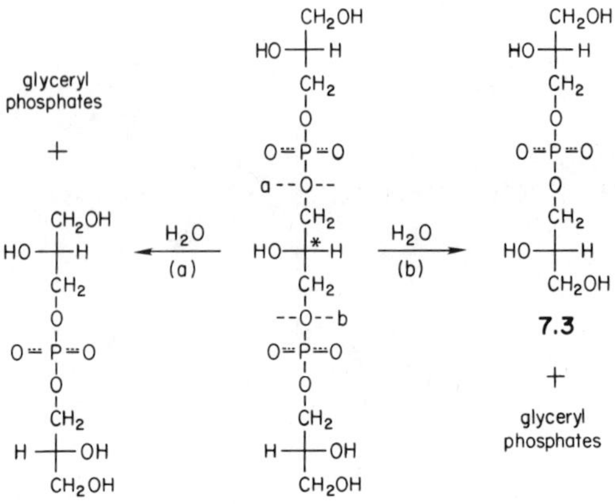

FIGURE 16.7. Bis-glycerophosphates formed by partial alkaline hydrolysis of deacylated cardiolipin [sn-1,3-bis-(sn-glycero-3'-phospho)glycerol; **7.1**]: **7.2** by scission at bond a; **7.3** by scission at bond b. **7.1** may be drawn with the central hydroxyl group to the left (as shown) or to the right; note that this is the only change that would result if the formula were rotated 180° in the plane of the drawing.

of approximately 5:1), and it has been suggested that the growing higher animal is stressed with respect to elaboration of mitochondria if dietary intake of this essential fatty acid is inadequate. Inspection of structure **12** reveals that this anionic amphiphilic lipid is extraordinary in having a comparatively extended hydrophilic group and four long-chain hydrophobic moieties; the possible significance of the abundance of such lipids in the inner membranes of mitochondria, which are distinguished by small radii of curvature, is discussed on p. 394. With the exception of the polysaccharidic sphingolipids, cardiolipins are unusual among lipids in possessing immunological activity.

Even though the central carbon atom (C-2' in **12**) of the cardiolipins bears two identical substituents (ignoring asymmetry of acylation), it is nonetheless chiral by the more rigorous criterion of lacking a plane of symmetry. Deacylation by hydrolysis under mild conditions yields an optically active, enantiomeric sn-1,3-bis-(sn-glycero–3'–phospho)glycerol (**7.1**), shown in Fischer projection in Fig. 16.7. Hydrolysis under conditions strenuous enough to cause breakage of phosphate ester bonds yields, among other products, a mixture of two isomeric bis-glycerophosphates: a *meso* form **7.3** (sn-1,3'-; R,S); and the diastereomeric **7.2** (sn-3,3'-; R,R), enantiomeric with that obtained similarly from Renkonen's phospholipids (Somerharju et al., 1977).

16.3.4.2. Glyceroglycolipids

The major amphiphilic lipids of the photosynthetic membranes of plant chloroplasts are glycosides (rather than phosphate esters) of sn-1,2-diacylglycerols (see Fig. 16.8): the sn-3-β-D-galactopyranosides, **8.1**; the corresponding D-galactopyranosyl(α1″→6′)-β-D-galactopyranosides, **8.2**; and the sn-3-α-D-6′-deoxyglu-

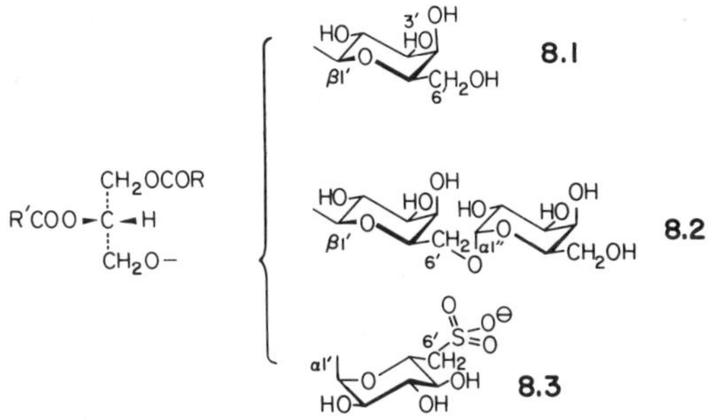

THE AMPHIPHILIC LIPIDS 385

FIGURE 16.8. Structures of the major glycoglycerolipids of chloroplast membranes.

copyranoside-6'-sulfonates (6-deoxyglucose ≡ quinovose), **8.3** (in order of decreasing abundance). In substances of type **8.1** and **8.2**, hydrophilicity is provided by polyhydroxylic moieties, no formally charged group(s) being present. Hydrophilicity is unusually highly enhanced in the sulfonolipids (**8.3**), which are reported to be extractable from apolar solvents by water and thus to have lost the solubility characteristics that are otherwise typical of lipids (see Chapter 2); since the sulfonic acid group (note the direct C−S bond) is strongly acidic, these substances are anionic under physiological conditions.

Higher oligogalactosides (cf. **8.2**), containing up to at least four galactose units, occur in plants as well as in microorganisms, which may also contain analogous glucosides and mannosides.

Amphiphilic lipids of types **8.1** and **8.2** have been detected in the mammalian brain, although the possibility remains that these substances are exogenous. The presence of 3'-sulfate esters of **8.1** (of both diacyl and alkyl, acyl types) in rat (but not in fish or human) brain suggests involvement of the same enzyme system responsible for formation of brain galacto-cerebroside 3'-sulfates. 3-Sulfation of a galactose moiety of an amphiphilic lipid is a major eventuality in at least one microorganism: 25% of the membrane lipids of *Halobacterium cutirubrum* are 3'''-sulfate esters of sn-2,3-di-(dihydrophytyl)-1-(α1'-[D-galactosyl(β'''→6'')-D-mannosyl(α1''→2')]-D-glucosyl)glycerol. The structural formula of this substance (**13**) emphasizes that this microorganism uses glycerol "upside down," the hydrophilic trisaccharide group being at the sn-1 instead of the sn-3 position, and that hydrophobicity is provided by polyprenyl instead of fatty acyl groups.

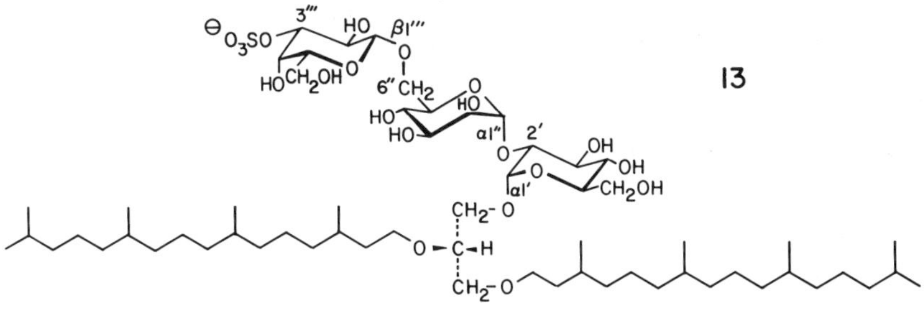

An unusual feature of plant diacylglycerogalactosides (**8.1**) is that they contain predominantly α-linolenoyl (18:3) residues at both *sn*-1 and *sn*-2 positions. Moreover, some plant glycerogalactosides bear a third fatty acyl group at the 6′ position of the galactose moiety and are therefore tricaudal amphiphiles. A small fraction of the mammalian brain galactocerebrosides is reportedly similarly modified.

The prefix X-lyso- applied to names of any of the glycerophospholipids or glyceroglycolipids described above signifies a derivative in which one of the hydroxyl groups of the glycerol core is free, i.e., lacks a fatty acyl (or alkyl or 1′-alkenyl) group in the position (usually *sn*-1 or *sn*-2). Such substances possess the striking property of causing lysis of erythrocytes (hence the prefix), suggesting strongly that their ingress seriously compromises biomembrane integrity. Biomembranes are usually found to contain lyso-amphiphiles in quantity so small that it is reasonable to assume such components to be artifacts of isolation procedures. Remarkably, however, Blaschko *et al.* (1967), found 17% lysolecithin in the phospholipids of the limiting membrane of bovine chromaffin granules, which apparently serve as vesicles of catecholamine storage in the adrenal medulla; this was later shown to be a peculiarity as well of the corresponding subcellular organelles of man.

These lyso-lipids remain amphiphilic but, having lost one of the original two long-chain hydrophobic moieties, are now more hydrophilic and monocaudal (cf. soap anions, which are also highly hemolytic). Such substances, although not expected to be found in quantity in tissue, are intermediates in both the biosynthesis and catabolism of the parent glycerolipids (see Chapter 17). Because of the closely similar reactivities of the two ester linkages, *in vitro* hydrolytic monodeacylation of the diacyl lipids is virtually indiscriminate, although differential hydrolysis is inducible by highly position-specific lipases. On the other hand, plasmenyl esters (e.g., **2**) are readily converted to the corresponding *sn*-1-lysophosphatidyl esters by mild *in vitro* hydrolysis at low pH, taking advantage of the particularly facile hydrolysis of the vinyl ether linkage under such conditions.

16.3.4.3. Sphingophospholipids and Sphingophosphonolipids

These substances, as well as the sphingoglycolipids (to be discussed in the next section), are derivatives of ceramides (fatty acyl amides of sphingoids) (**14**), differing in the nature of substituents of the C-1 hydroxyl group: sphingophospholipids (including the sphingomyelins) are thus ceramide 1-phosphates; sphingophosphonolipids are 1-deoxyceramide 1-phosphonates; and sphingoglycolipids (including the cerebrosides and cerebroside sulfates) are 1-ceramidyl glycosides.

$$\text{RCHOHCHCH}_2\text{OH} \equiv \underset{\text{H}}{\overset{\text{O}}{R'\text{-}\overset{\|}{C}\text{-N}}}\text{-}\underset{\text{CH}_2\text{OH}}{\overset{\overset{\text{OH}}{\overset{|3}{H\text{-}\overset{|}{C}\text{-R}}}}{\underset{|2}{\overset{|}{C}\text{-H}}}} \qquad \mathbf{14}$$
NHCOR′

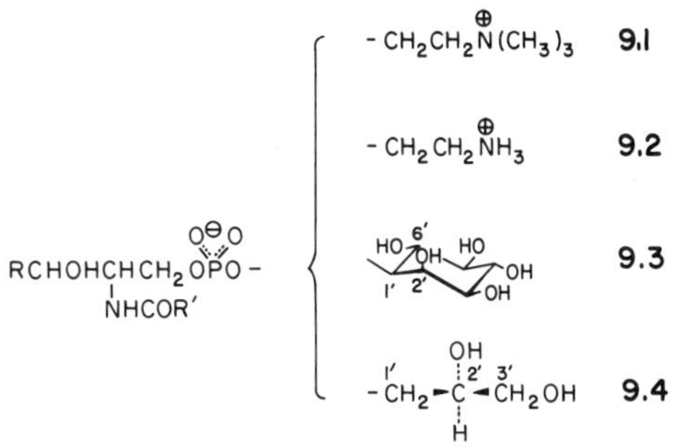

THE AMPHIPHILIC LIPIDS 387

$$\text{RCHOHCHCH}_2\text{O}\overset{\overset{O}{\|}}{\underset{O^{\ominus}}{P}}\text{O}- \begin{cases} -\text{CH}_2\text{CH}_2\overset{\oplus}{\text{N}}(\text{CH}_3)_3 & \textbf{9.1} \\ -\text{CH}_2\text{CH}_2\overset{\oplus}{\text{NH}}_3 & \textbf{9.2} \\ \text{inositol} & \textbf{9.3} \\ -\text{CH}_2-\underset{\text{H}}{\overset{\text{OH}}{\text{C}}}-\text{CH}_2\text{OH} & \textbf{9.4} \end{cases}$$

FIGURE 16.9. Structures of the major sphingophospholipids, including the sphingomyelins (**9.1**).

Within each of these classes there is some variability in the R-group of the sphingoid (see Section 16.3.3) and in the R-group of the fatty acyl moiety.

Since ceramides are intermediates in both the biosynthesis and degradation of the sphingolipids (see Chapter 18), they may be found in small amounts as such in tissues involved in these metabolic processes. Similarly, tissues in which cerebrosides are elaborated by glycosylation of sphingoids prior to fatty-acylation may contain some of the intermediate 1-sphingoidyl glycosides, to which the generic name psychosines has been applied.

Inspection of a summary of the structures of the known sphingophospholipids (see Fig. 16.9) reveals the striking similarity of their hydrophilic moieties to those of the glycerophospholipid analogues (cf. Fig. 16.5).

The sphingomyelins (**9.1**) are, like the lecithins, zwitterionic, and occur in particularly large proportions in the white matter (myelin) of the brain, where the principal sphingoid involved is 4E-octadecylsphingenine, and the fatty acyl moieties are of unusually great chain-length (about two-thirds C_{22}–C_{26}, mainly 24:0 ≡ lignoceroyl and 15Z-24:1 ≡ nervonoyl). Interestingly, the major fatty acyl group in the sphingomyelin of gray matter of the brain is stearoyl (18:0). Sphingomyelins are also important phospholipids in the red-blood-cell membrane, and in kidney tissue, and constitute about 4–10% of the phospholipids of other tissues of the body, where the usual fatty acyl composition is about one-third 16:0 and one-third 24:1. Of great interest is the fact that, while the proportions of phosphatidylcholine (lecithin) and of sphingomyelins present in erythrocyte membranes of different animals are widely variable (but quite species-specific), the sum of them is remarkably constant—suggesting that either type of lipid serves equally well with respect to some generally important property of these biomembranes. It should also be noted that the sphingomyelins of biomembranes tend to be strongly concentrated in the outer layer of these structures.

The ethanolamine analogues (**9.2**) of sphingomyelins are known to occur in many lower forms of life: bacteria, protozoa, mollusks, and shellfish. In such sphingomyelin analogues of certain snails and of the blowfly, the principal sphingoid is sphingenine (**4.2**); in those of the fly, the major fatty acyl groups are 20:0 and 22:0. In such analogues of at least certain bacteria, both the sphingoids and fatty acyl moieties include methyl-branched homologues. Phospholipids of the coelenterate sea anemone (as well as those of certain mollusks and protozoa) have been shown to include some of the corresponding "deoxy" analogues, 1-sphingoido-phosphonoethylamines, **15** (see Horiguchi and Kandatsu, 1959; Rouser et al., 1963).

$$\text{RCHOHCHCH}_2\text{O-P-CH}_2\text{CH}_2\overset{\oplus}{\text{NH}}_3 \quad \textbf{15}$$
$$\underset{\text{NHCOR'}}{}$$

Smith and Lester (1974) have shown that about one-third of the phospholipids of the yeast *Saccharomyces cerevisiae* contain inositol, and about 40% of these substances are ceramide derivatives, of which the major type is *N*-2'-hydroxy-C_{26}-fatty acyl phytosphinganine-1-phospho-1"-*myo*-inositol (**9.3**). D-Mannosides of these substances are also present in small amounts (cf. the mannosides of phosphatidylinositol occurring in other microorganisms, p. 382). Seeds of many plants contain complex polysaccharidic derivatives of **9.3**.

1-Phytosphingophospho-*sn*-1'-glycerol (**9.4**) and the corresponding *sn*-3'-phosphate have been shown to be present in certain bacteria. These substances are accompanied by others bearing a second fatty acyl moiety, involving one of the usually unsubstituted hydroxyl groups of the phytosphingoid (a very rare structural feature), of an α-hydroxy-*N*-fatty acyl moiety, or of the glycerol group; in any case, these substances represent unusual tricaudal amphiphilic lipids. Kemp et al. (1972) have isolated from sheep rumen a strain of bacteria, in which about 30% of the phospholipids is the 3'-amino analogue of **9.4** (3'−OH → 3'−$\overset{\oplus}{\text{NH}}_3$); configuration at C-2', however, was not established.

16.3.4.4. Sphingoglycolipids (Glycosphingolipids)

In terms of abundance in the higher animal, the principal amphiphilic lipids of this type are the ceramide(1←β1')-D-galactopyranosides (**16**), which, together with the corresponding 3'-sulfate esters, comprise about 25% of the lipid of the white matter (axon myelin sheaths) of the brain, rivaled in this quantitative respect only by cholesterol (about 28%). A feature peculiar to these cerebrosides is the prevalence of long-chain (C_{22}–C_{26}) and/or α-D-(2*R*)-hydroxy-fatty acyl groups. Specific names have been assigned, mostly by Thudicum (1884, pp. 115–120), to those bearing certain fatty acyl moieties: cerasin (originally spelled kerasin) (−COR' = lignoceroyl, 24:0); phrenosin or cerebron (cerebronoyl, 2*R*-HO-24:0); nervone (nervonoyl, 15*Z*-24:1); and oxynervone (oxynervonoyl, 2*R*-HO-15*Z*-24:1). These are neutral amphiphiles, the hydrophilic properties of which are attributable not to formal charge but to five or six hydroxyl groups: four in the glycoside moiety, another in the sphingoid base, and a sixth if an α-hydroxy-fatty acyl group is involved. The corresponding β-D-glucoside (4'-epimer of **16**) is widely distributed in tissues other than the brain and accumulates massively, par-

$$\underset{\text{NHCOR'}}{\text{RCHOHCH}-\text{CH}_2-\text{O}}\overset{\text{H} \quad 3' \text{ OH}}{\underset{\beta 1'}{\diagdown}} \overset{\text{HO}}{\underset{\text{O}}{\diagup}} \overset{4'}{\underset{\text{CH}_2\text{OH}}{\diagdown}} \quad \textbf{16}$$

ticularly in the spleen, liver, and bone marrow of victims of Gaucher's disease, which involves a genetically controlled insufficiency or lack of a vital β-glucosidase (see Chapter 18). Glucosidic (phyto)-cerebrosides also occur in higher plants (but not in algae).

About 25% of the brain cerebroside (ceramidyl galactosides) exists normally in the form of the corresponding 3'-sulfate esters (**16**; 3'-OH → 3'-OSO_3^-); these cerebroside sulfates or sulfatides thus constitute about 6% of the total lipid of this neural tissue. The hydrophilicity of these derivatives is enhanced considerably by attachment of the formally charged, anionic group. A smaller fraction of the brain cerebrosides also bear an additional fatty acyl group, esterifying the terminal, primary hydroxyl group of the sugar moiety; these substances (**16**; 6'-OH → 6'-OCOR″) possess three long-chain hydrophobic groups and are unusual among biomembrane amphiphilic lipids in this respect.

With the exception of these last-named, quantitatively minor derivatives, the cerebrosides are bicaudal, and the single sugar moiety appears to have sufficient hydrophilicity to permit their serving as structural units of biomembranes. This indeed seems to be their principal function in myelin. Those that occur in the outer face of the cytoplasmic membranes, however, have a different, very important function: service as a site for glycosidic attachment (to the glucose moiety of such cerebrosides) of one or more of an assortment of hexose, N-acylated hexosamine, and sialic acid units, forming ceramide oligosaccharides with immunological properties. These substances are firmly anchored in the cell membrane by the hydrophobic effect of the apolar groups of their ceramide moieties, but their oligosaccharidic moieties extend into, and are thus readily accessible to, components of fluids bathing the cell. They are intimately involved (in some cases in conjunction with similarly constituted mucopolysaccharides) in such phenomena as cell recognition and specific binding of bacterial toxins and protein hormones. Only a small number of additional saccharidic units are required to convey water solubility on these substances and thus to deprive them of one of the characteristics generally taken to be typical of lipids—even though Bloor (1943), on the basis of the presence of the fatty acyl group, would have placed them in this class. Some of the types of monosaccharidic units involved in these elaborations and a selection of examples of the resulting products are mentioned below.

Glycosidic attachment of additional sugar and/or N-acylated-amino sugar units to cerebrosides gives rise to ceramide oligosaccharides that remain neutral (i.e., uncharged), e.g., galactosyl($\beta 1\rightarrow 4$)glucosyl($\beta 1\rightarrow 1$)ceramides (lactosylceramides or cytolipins H, ex heart) and N-acetylgalactosaminyl ($\beta 1\rightarrow 3$)galactosyl($\alpha 1\rightarrow 4$)galactosyl($\beta 1\rightarrow 4$)glucosyl($\beta 1\rightarrow 1$)ceramides (cytolipins K, ex kidney, identical with globoside, the most abundant substance of this type attached to the outer surface of the human erythrocyte membrane). The terminal galactose unit of cytolipin H tends to be converted to some extent to the corresponding 3-sulfate ester (cf. the analogous modification of the brain galactosyl cerebroside), giving a formally charged (anionic) derivative.

The complex gangliosides, structures of a substantial number of which have been elucidated, are derived from those of the type just described by glycosidic incorporation of one or more units of N-acetylneuraminic acid (NANA ≡ sialic

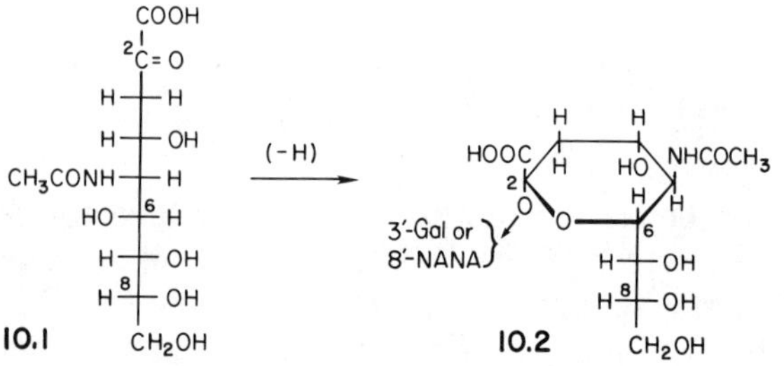

FIGURE 16.10. Structures of N-acetylneuraminic (sialic) acid (**10.1**, Fischer projection of acyclic form) and of the corresponding α2-sialopyranosidyl (NANA) moiety (**10.2**, Haworth projection rotated 180° about vertical axis of the ring) occurring in gangliosides by (α2→3)galactosyl (3'-Gal) or (α2→8) NANA (8'-NANA) linkage.

acid), of which the Fischer projection of the acyclic form (**10.1**) is shown in Fig. 16.10, along with a rotated Haworth projection (**10.2**) of the α2-sialopyranosidyl (NANA) moiety* involved in the gangliosides. (N-Glycolylneuraminic acid analogues accompany gangliosides in a wide variety of animals—not, however, in man.) The α2-sialosidyl groups are in almost all cases attached either to the 3-position of a galactosyl moiety or to the 8-position of another already incorporated sialosidyl (NANA) moiety of the ceramide oligosaccharide.

Although acylation of the amino group of sialic acid effectively eliminates its basic properties, the carboxyl group remains free and virtually fully dissociated at pH 7 (cf. K_a of pyruvic acid, 3×10^{-3}); the gangliosides thus bear one or more negative charges at physiological pH, in contrast to the neutral ceramide oligosaccharides.

Although found in much smaller amounts in many other tissues, gangliosides contribute about 6% of the lipid of the gray matter of the brain, localized in synapses. Brain gangliosides differ from the cerebrosides (and cerebroside sulfates) of white matter not only in being derived from 1-glucosyl- rather than 1-galactosylceramides, but also in the fatty acyl groups involved (80–90% stearoyl, 18:0) as well as in the content of substantial quantities of C_{20} sphingoids.

16.4. Physical Properties of Biomembrane Amphiphilic Lipids

A common structural feature of biomembrane amphiphiles is that they contain at least two long hydrophobic "tails" covalently attached to a compact

---

*Although the α-β distinction of anomers now has a strictly configurational basis [rather than the relative optical activity basis originally proposed by Hudson (1909)], a possible ambiguity in nomenclature persists with respect to choice of reference chiral carbon atom. Configuration of the anomeric carbon atom (C-2) of the NANA moiety is explicit in formula **10.2**, but may be designated either β (L) (using C-6, at the site of ring closure, as reference) or α(D) with reference to the highest-numbered chiral carbon (C-8); the latter is favored by the IUPAC–IUB Commission on Biochemical Nomenclature (1971).

strongly hydrophilic "head," which may be regarded as being situated near the center of these long flexible molecules. Biomembranes also contain proteins and at least some sterol or structurally analogous alcohol (Nes, 1974). Although the proportions of the amphiphilic lipids vary widely in biomembranes, no biomembrane is free of them. They are unique, in contrast to the proteins and sterols, in their spontaneous tendency in the presence of water to form stable lamellar aggregates with structure and physical properties closely resembling those of biomembranes.

16.4.1. The Neat Crystalline State of Membrane Amphiphiles

Membrane amphiphiles crystallize readily from supersaturated solutions in solvents of intermediate polarity. Since crystallization generally involves arrangement of molecules in such a way that intermolecular attractive forces are maximized, the crystal structures of membrane amphiphilic lipids are expected to be closely analogous to those of the fatty acids (see Chapter 4), in which the strongly interacting, highly polar hydrophilic head groups are paired in planes separated by strata of opposed pairs of hydrophobic tails, parallel to each other and usually oblique (rather than normal) to the head-group planes (see p. 50). An easily demonstrable consequence of such structures, having alternate planes of very strong and very weak intermolecular interactions, is the flaky habit of crystals of such substances.

The complex melting behavior of these lipids can be related to the fact that their crystals consist essentially of alternating layers of highly polar (high melting) and of quite apolar (low melting) matter. On the basis of the classical criterion of observation of meniscus formation, the melting points of common membrane amphiphilic lipids are relatively high (see Table 16.1) and quite independent of the specific fatty acyl or other hydrophobic moieties they contain.

TABLE 16.1 Melting Points of Selected Common Classes of Biomembrane Amphiphilic Lipids

Class	m.p. (°C) (Meniscus)
Phosphatidylcholines (**5.1**)	230
Phosphatidylethanolamines (**5.2**)	196
Sphingomyelins (**9.1**)	210
Cerebrosides (**16**)	182[a]
Diacylglycerol glycosides[b]	
β-D-Galactoside (**8.1**)	149–150
β-D-Gal(6-α1)-D-Gal (**8.2**)	225–230

[a]The lignoceroyl (24:0) β-D-galactoside, the corresponding dihydrosphingosine analogue, and the behenoyl (22:0) β-D-glucoside all melt at or near this temperature: the cerebronoyl (2-hydroxy-24:0) galactoside melts at a higher temperature (195°C); and the behenoyl lactoside ($\equiv$ cytolipin H) at a much higher temperature (235°C) (Shapiro and Flowers, 1961).

[b]The melting point of the monogalactoside is that of the dipalmitoyl (16:0) member of the class; the melting point of the Gal-Gal analogue is that of the distearoyl (18:0) member, the Gal-Gal-Gal analogue of which melts at about the same temperature (Gent and Gigg, 1975a,b,c).

When crystals of such substances are warmed during the conventional determination of their melting points, a phase transition (a "softening point") may be observed, which, in contrast to the true melting points, is widely variable but is related to the nature of the hydrophobic moieties involved. The softening points of several synthetic phosphatidylcholines (cf. melting points of the corresponding fatty acids) are, for example: dimyristoyl 24°C (cf. 14:0, m.p. 58.5°C); dipalmitoyl 42°C (cf. 16:0, m.p. 64°C); distearoyl 90°C (cf. 18:0, m.p. 70°C); and dioleoyl $-$22°C (cf. 18:1, m.p. 13°C). The true melting points of all of these substances (about 230°C) are essentially indistinguishable. The nature of the strongly interacting hydrophilic moieties affects not only the melting points but also the softening points of these substances: α-stearoyl-β-oleoylphosphatidylcholine softens at about 40°C, while the corresponding phosphatidylethanolamine softens at about 80°C. The temperatures at which these phase transitions occur can be precisely determined by differential scanning calorimetry, which also reveals that the solid $\rightarrow$ liquid crystal transition at the softening point is much more endothermic than the liquid-crystal $\rightarrow$ liquid transition at the true melting point.

These thermal characteristics of membrane amphiphiles are interpreted in terms of initial loss of order (at the softening point) in the apolar moiety layers of the crystal, followed by loss of order in the polar-group layers at the true melting point.

The stability of bimolecular layers of such substances is not critically dependent on solidity of the internal apolar region, which may therefore possess a wide range of fluidities without compromising the stability and integrity of the bilayer. Many observations have indicated that the fatty-acyl-group composition of amphiphiles in biomembranes is closely controlled, presumably to maintain optimal (not necessarily maximal) internal fluidity of these structures at ambient temperatures of the organism.

16.4.2. Interactions with Water

Formation of Micellar Aggregates

According to Small's (1968) classification of lipids on the basis of their modes of interaction with water, the membrane amphiphilic lipids and soaps are "insoluble swelling amphiphiles," which, although having very limited true solubility in water, spontaneously disperse in water as aggregates of high order. This behavior is manifest, however, only above a certain temperature, individually characteristic of each member of the class, that may be identified as the temperature at which the hydrophobic-group strata of these substances become disordered or liquefied, i.e., at the temperature of transition from the crystalline to the liquid-crystalline state. The "critical micellar temperatures" correspond to the softening points of the membrane amphiphiles and are analogous to the Krafft points of soaps.

The swelling of soaps in contact with water is a commonly observed phenomenon that involves initial penetration of water into the hydrophilic-group interfaces, yielding separated lamellar aggregates that may, depending on the quantity of water available, change successively through rods to spherical micellar aggregates (see Chapter 4). An analogous—and much more spectacular—phenomenon can be observed with amphiphiles under a microscope. When a drop of water is brought into contact with neat membrane amphiphiles at temperatures above their softening points, a spontaneous growth of cylindrical structures into the water droplet is seen. These branching "myelin figures" consist of tightly concentrically nested cylinders of lamellae closely similar in structure to the myelin sheaths of neurons—hence the name. Vigorous agitation of such substances with water leads to formation of milky suspensions of "liposomes", which consist of onion-like assemblages of concentric, roughly spherical, closed lamellae. Liposomes are closely analogous in structure to "myelin figures" and indeed may be considered as having been formed by fragmentation of the cylindrical bodies, followed by spontaneous closures driven by the hydrophobic effect, i.e., minimization of exposure of hydrophobic moieties to water along edges of the broken lamellae. Liposome suspensions become optically clear on ultrasonication, during which the layered assemblages are converted primarily into spherical or near-spherical vesicles, single-layered, closed lamellae of approximately 250-Å diameter containing a few thousand molecules per vesicle. These bodies, enclosing volumes of the original aqueous medium, are readily transferred to different aqueous media and thus used to investigate permeability properties of their singly lamellar walls.

That different membrane amphiphiles are converted in water under these conditions into closed, single-walled vesicles of about the same size indicates a strong resistance of the planar lamellae of these substances to assume less than a limiting radius of curvature. This distinguishing characteristic is related to the fact that soaps form spherical and rod-like micelles, while the membrane amphiphiles do not, and suggests another difference between lamellae of the two types of swelling amphiphiles that is less obvious. The structures of the three types of aggregates that soaps form in aqueous solutions of various concentrations (see Chapter 4) reflect the ease with which these anions assemble in compact arrays having a radius of curvature determined by—and equal to—the maximum length of the hydrophobic moiety. The edges of soap lamellae are therefore rounded off, with carboxylate groups efficiently shielding the hydrophobic interior of the bimolecular sheets here as well as on both faces from contact with water. For essentially the same reason that membrane amphiphiles do not form spherical micelles or rods in water, these substances are unable to finish off edges of lamellae in the same way and thus spontaneously form closed (edgeless) lamellar structures of types exemplified by liposomes, vesicles, and the advancing ends of myelin figures. It would be expected, furthermore, that vesicles and closely analogous biomembranes would, in the event of mechanical rupture, be self-healing.

These qualitative differences in the interactions of soaps and of membrane lipids with water must be related to the fact that soaps are monocaudal, the hydrophilic head group being at one end of a single hydrophobic tail, while the lipids essential to biomembrane structure are usually bicaudal, the head now being situated near the middle of a doubly long hydrophobic moiety. These considerations lead compellingly to the conclusion that splayed-tail conformations of the bicaudal membrane amphiphiles are substantially favored over those involving continuously parallel tails. (With little more justification than that they are easier to draw, configurations of the parallel-tail type are shown in most depictions of cross sections of biomembranes.) The presence of particularly high proportions of the tetracaudal cardiolipins in the inner membranes of mitochondria, extensive invagination of which creates regions of small radius of curvature, may be similarly related to the preferred conformation of these unusually structured amphiphiles in lamellar aggregation.

16.4.3. Formation of Monomolecular Films

In addition to their dispersibility in water as lamellar aggregates, the swelling amphiphiles also form stable monomolecular films at interfaces between an aqueous phase and apolar liquid or gaseous media, the highly hydrophilic head groups of the molecules of such substances being very difficult to remove from the aqueous phase, and the hydrophobic tails being strongly resistant to immersion in it. If sufficient area at a gas (air)–aqueous interface is available, the tails would be expected to lie on the water surface (rather than extending into the gas phase, as is so often depicted), retained by omnipresent London forces; if the area to which a certain number of molecules is confined is reduced, however (as, for example, by use of a Langmuir trough—see Chapter 4), little resistance is occasioned by substitution of London interactions between parts of the apolar hydrophobic tails for those of comparable magnitude between tails and water—until further confinement, once the tails are in van der Waals contact, would necessitate removal of head groups from the aqueous surface, an eventuality that would be stoutly resisted. Monomolecular films of membrane amphiphiles under lateral compression are reasonably assumed to possess structural features and related physical properties closely similar to those of apposed pairs of them present in biomembranes.

It is tempting to entertain the assumption that such monolayers are good models for lamellar bilayers because of the technical simplicity of quantitative studies of how certain characteristics of oriented planar aggregates of such substances are affected by composition of the monolayer (in which cholesterol, for example, may be incorporated) and by the pH or other ion content of the aqueous phase on which they rest (see Shah and Schulman, 1965, 1967; Oldani *et al.,* 1975; Cadenhead, 1977).

It can be shown in this way, for example, that compressed films of dipalmitoylphosphatidylcholine at 25°C have an area of 40–44 $Å^2$ per molecule (i.e., twice that observed for films of palmitic acid); of egg lecithin (primarily *sn*-1-saturated-2-unsaturated-fatty-acylphosphatidylcholine), 62 $Å^2$ per molecule (somewhat greater than the sum of the molecular areas of palmitic and oleic acids); and of yeast lecithin (about 88% di-unsaturated fatty acyl), 72 $Å^2$. Indication of the role of cholesterol in biomembranes is provided by the observation that monomolecular films of membrane amphiphiles containing *cis*-unsaturated fatty acyl groups condense more compactly if admixed with cholesterol (i.e., the area per molecule is less than that calculated on the basis of the sums of areas of the individual components). Membrane lipids that do not contain *cis*-unsaturated fatty acyl groups show this effect to a much smaller extent. It has been shown in other ways that the fluidity of biomembranes is actually enhanced by the improved packing occasioned by the presence of cholesterol, clearly suggesting that the characteristic and often high content of cholesterol in biomembranes reflects the operation of control of this important physical property.

Similarly, monomolecular films of phosphatidylethanolamines, of phosphatidyl-L-serines, and of phosphatidic acids are substantially compacted on aqueous phases of pH 4 (vs. pH 7), while those of phosphatidylcholines are not—a difference clearly related to the basicities of the hydrophilic head groups of these classes of membrane amphiphiles. The compactness of laterally compressed films of the phosphocholine amphiphiles (lecithins and sphingomyelins) is not affected by metal cation content of the aqueous phase; but those of the tetracaudal cardiolipins (**12**) (110 $Å^2$ per molecule; cf. areas of palmitic and oleic acid molecules under these conditions) are contracted 10–13% when calcium ions are added to the aqueous phase supporting these diphosphatidylglycerols.

In connection with the conclusion that the bicaudal membrane amphiphilic lipids tend to be splayed and therefore to resist both compression in monomolecular films and assumption of small radii of curvature in lamellar aggregates, experiments by van Deenen and collaborators (van Deenen et al., 1962; van Deenen, 1965) are of considerable interest. Under conditions that compress films of distearoyl (18:0) or of dipalmitoyl (16:0) lecithins to areas of 40 $Å^2$ per molecule (i.e., about twice that of the corresponding saturated fatty acids), those of dimyristoyl (14:0) lecithin have an area of 50 $Å^2$, and those of didecanoyl (10:0) lecithin, 75 $Å^2$ per molecule. These results may be taken to indicate that a conformational resistance of about 10 kcal/mol must be overcome in order to bring the fatty acyl groups of a lecithin molecule into end-to-end contact.

16.4.4. Conformation of Membrane Amphiphiles

Conformation in the Crystalline State

The marked tendency of amphiphilic lipids to crystallize in bilayers between which intermolecular forces of attraction are very weak makes it difficult to obtain crystals of such substances that are precisely enough ordered in all directions to permit determination of crystal structure by X-ray diffraction analysis. Despite the technical difficulties, the crystal structures of racemic 1,2-dilauroylphosphatidylethanolamine, **5.2** (R = R′ = n-$C_{11}H_{23}$—, plus enantiomer), and of a cerebroside, 1-β-D-galactosyl-N-(2R-hydroxyoctadecanoyl)-sphinganine, **9.3** (R = n-$C_{15}H_{31}$—; R′ = 2R-hydroxy-n-$C_{17}H_{35}$—) have been determined (Hitchcock et al., 1974; Pascher and Sundell, 1977).

The molecular conformation and packing in the crystalline phosphatidylethanolamine are shown in Fig. 16.11, and in crystals of the cerebroside in Fig. 16.12.

Crystals of both substances have the general characteristics of those of

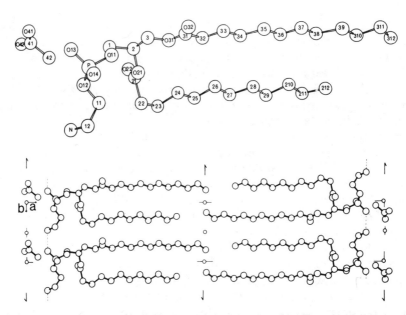

FIGURE 16.11. Conformation of individual molecules (above) and packing of molecules in ab plane (below) in crystals of racemic dilauroylphosphatidylethanolamine (plus acetic acid, 1:1). The isolated four-atom groups seen in both drawings are the molecules of acetic acid. The marked N and P atoms (top drawing) provide landmarks for location of major atoms of the phospholipid molecule. The α-lauroyl (sn-1) group is attached to carbon atom 3, and the phosphoethanolamine group to carbon atom 1 of glycerol (top drawing). From Hitchcock et al. (1974), with permission of the authors.

amphiphilic lipids. The molecules are laid down in stacked (parallel) bilayers, creating planes of apposed, strongly interacting polar groups alternating with others of very weakly interacting hydrophobic group termini. The hydrophobic moieties are in all-anti conformation—except for a short section of one of the two, which must be bent to bring the remainder of it alongside the other—and have parallel axes, almost perpendicular ($\beta = 92°$) to the bilayer planes in the case of the phosphatidylethanolamine, but sharply inclined ($\beta = 49°$) with respect to the corresponding reference plane in the cerebroside crystals. Because parallel, all-anti n-alkyl groups can be packed in a variety of ways little different in stability (see Abrahamsson et al., 1978), it is hardly surprising that crystals of the two substances differ in this respect (for example, the two alkyl groups in the same molecule of the phosphatidylethanolamine have carbon atoms in planes nearly parallel to each other, while in the cerebroside they are approximately normal to each other).

Consequences of the necessity of bending one alkyl group in order to bring it close and parallel to the other are of interest with respect to packing at the apolar-group interface. Bending the β-fatty acyl group of the phosphatidylethanolamine shortens it by about four methylene groups and, since both fatty acyl groups are of equal size in this case, the α-group extends farther (e.g., from the polar interface) than the β; voids in the apolar interface are eliminated by apposition of longer chains in one layer with shorter chains in the other (see bottom

THE AMPHIPHILIC LIPIDS 397

of Fig. 16.11). In the cerebroside crystal, on the other hand, most of the bending occurs in the C_{18} sphinganine (rather than in the C_{18} fatty acyl) group, which is thus foreshortened by about the length of three methylene groups; the sharp inclination of the axes of the chains, however, results in both terminating approximately the same distance from the polar interface, and little interdigitation of methyl termini is involved (see bottom of Fig. 16.12).

Conformation and nonbonding interactions of the polar moieties of these substances in the crystalline state are noteworthy. The phosphoethanolamine

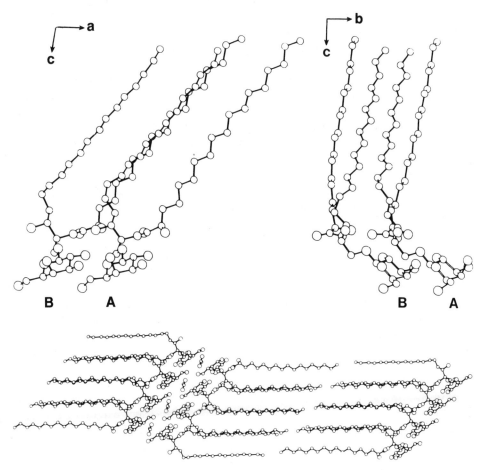

FIGURE 16.12. Conformation of pairs of molecules (above: *ac* plane view at left; *bc* at right) and packing of molecules in ac plane (below) in crystals of 1-β-D-galactosyl-*N*-(2*R*-hydroxyoctadecanoyl)-sphinganine (plus ethanol, 2:1). Molecules are situated alternatively in the lattice in two slightly different orientations, A and B. In the drawing at top left, the nested galactosyl moieties are discernible at the bottom; the oxygen atom of the free hydroxyl group of the sphinganine moiety is seen extending leftward from the lowest carbon atom of the tail at left; the α-hydroxyoctadecanoyl tail is at right; and the amide group is seen on edge, joining the two tails. Location of the isolated ethanol molecules is shown in the bottom drawing. From Pascher and Sundell (1977), with permission of the publishers (Elsevier Scientific Publishers Ireland Ltd.).

group (see top of Fig. 16.11) is approximately at right angles to the virtually linear (all-anti) glycerol-α-fatty acyl group, and parallel to the initial section of the β-fatty acyl group; it lies in the polar face of the bilayer, and is extensively gauche, shortening it and bringing the ammonium group within good hydrogen-bonding distance (2.79 Å) of a strongly negatively biased oxygen atom of the phosphate moiety of a neighboring molecule (see bottom of Fig. 16.11).

Although the cerebroside bears no formally charged groups, its very large hydrophilic region bears ten potentially strong hydrogen-bond-donating and/or-accepting moieties that are used extensively in maintaining this part of the molecule in a tight conformation and orienting it to the rest of the molecule, and the whole molecule to others in the lattice. The amide group hydrogen atom does double duty, being shared with oxygen atoms of the galactosidic linkage and of the 2-hydroxyl group of the fatty acyl group. Although the tetrahydroxy-galactopyranose ring would otherwise be free to rotate about the ether linkage joining it to the rest of the molecule, the ring is locked by additional hydrogen bonds to the same part of an adjacent molecule. The planes of the amide linkage and of the galactose ring are approximately parallel to each other as well as to the bilayer plane (see Fig. 16.12). Despite the compact conformation of the hydrophilic part of these molecules, an acute (49°) inclination of the anticoplanar portions of the alkyl chains is necessary for efficient packing, and the individual molecules thus assume what is described by Pascher and Sundell (1977) as "the shape of a shovel," with an angle of about 130° between handle (the pair of straight, parallel alkyl chains) and blade (the galactopyranose ring).

Conformation in Biomembranes

With knowledge of the structure of crystals of a glycerophospholipid and of a cerebroside, reasonable guesses can be made about the arrangement of molecules of other membrane amphiphiles in this state. However, the crucial question remains: how much of this conformational detail is retained when such molecules are constituents of functional biomembranes, which are in a liquid-crystalline state and have the polar surfaces of the bilayers in contact with aqueous media instead of the polar surface of an adjacent bilayer? Changes in physical properties that occur at the "softening point" of such substances clearly demonstrate the loss of order in the arrangement of their hydrophobic moieties. Furthermore, change in conformation and orientation of hydrophilic moieties effected in the crystalline state by hydrogen bonding (or by structurally equivalent electrostatic interactions between formally charged groups likely to be involved in phosphatidylcholines and sphingomyelins) are highly probable because of the potent hydrogen-bonding potentialities and high dielectric constant of water.

Theoretical conformational analysis applied to membrane amphiphiles of representative structure (Sundaralingam, 1972; McAlister *et al.,* 1973; Vanderkooi, 1973; see also Kreissler and Bothorel, 1978) predicted most features of the crystal structure of a phosphatidylethanolamine, reported shortly thereafter (Hitchcock *et al.,* 1974). These studies were based, however, on the assumption of maximum contact of all-anti-conformation, normal alkyl hydrophobic groups

THE AMPHIPHILIC LIPIDS 399

and serve to show that this condition places considerable constraints on reasonable conformations of other parts of these molecules, including their hydrophilic moieties. This condition is substantially compromised in the liquid-crystalline state more pertinent to biomembranes.

By application of nuclear magnetic resonance techniques (see Lee, 1976; Seeling, 1977), evidence is obtained that some of the structural features of crystalline dilauroylphosphatidylethanolamine persist in the liquid-crystalline state of closely analogous substances. A short section of the β-fatty acyl group of dipalmitoylphosphatidylcholine, for example, is still parallel to the bilayer surface before it turns to become parallel to the α-fatty acyl group, and both then run essentially normal to the bilayer plane, unable to attain the increased conformational freedom characteristic of the liquid state until they have extended about half-way to the center of the bilayer. Attempts to establish the orientation of the polar moiety of such phospholipids in the liquid-crystalline state have given conflicting results. It now seems clear, however, that in neat samples of such substances, even after appreciable diffusion of water vapor into the polar interfaces, the phosphoethanolamine and phosphocholine groups remain essentially parallel to the bilayer surface.

Structures of crystalline membrane amphiphiles indicate that close packing of hydrophobic moieties and strong intermolecular intralayer interactions of the hydrophilic moieties have mutual stabilizing effects. This interrelationship apparently persists in the liquid-crystalline state, where the hydrophilic group interactions remain at least qualitatively unmodified and thus limit the disordering of the now molten apolar groups. If the liquid-crystalline bilayers are dispersed in water, however, the intermolecular linkage by hydrophilic groups almost certainly vanishes, and the integrity of the bilayers (now very closely analogous to biomembranes) is maintained instead by the hydrophobic effect. Moreover, the apolar moieties should now have considerably increased conformational freedom, and conformations other than those permitting the most prompt parallel juxtaposition of the hydrophobic moieties warrant consideration.

Of the three staggered conformations of the sn-C2-C1 array of the diacylglycerophospholipids (shown in Newman projection, with C2 in front), the most crowded (**17**) is actually involved in crystalline racemic dilauroylphosphatidylethanolamine (see Fig. 16.11). The other two are less crowded (note better place-

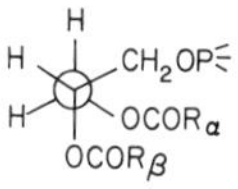

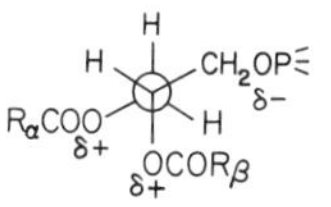

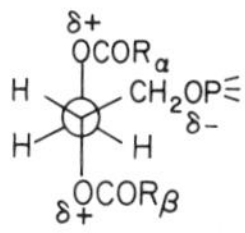

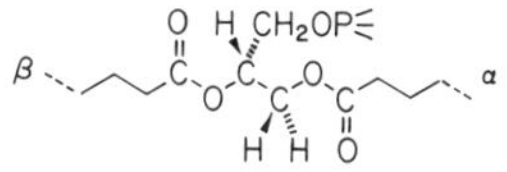

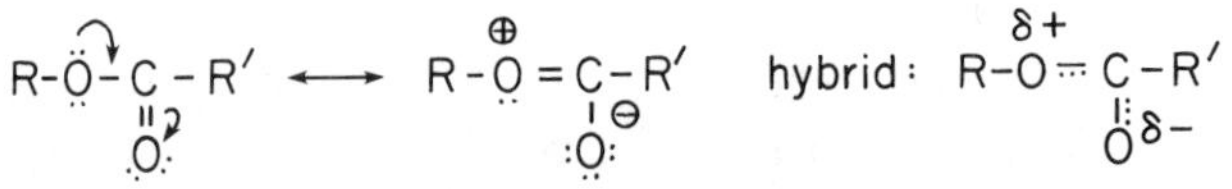

$$R-\overset{\frown}{\underset{\underset{\ddot{O}:}{\|}}{\ddot{O}}}-C-R' \longleftrightarrow R-\overset{\oplus}{\ddot{O}}=\underset{:\overset{\ominus}{O}:}{C}-R' \quad \text{hybrid:} \quad R-\overset{\delta+}{O} \mathrel{\hbox{$=\hskip -3pt =$}} \underset{\underset{\delta-}{O}}{\overset{\|}{C}}-R'$$

FIGURE 16.13. Resonance stabilization in an ester.

ment of the small hydrogen substituents), and on the basis of steric considerations alone, would be about equally preferable. However, the fact that the ester group is substantially resonance-stabilized (Fig. 16.13), placing a positive charge (shown in **18** and **19**) on the oxygen atoms, indicates a clear electrostatic superiority of **19** over **18**. Similar resonance also places a partial positive charge on the vinyl ether oxygen atom, but the oxygen of saturated ether groups is negatively biased because of the greater electronegativity of oxygen than of carbon. Plasmalogens and di-(saturated)-ether analogues would therefore be most stable in a conformation corresponding to **19**; saturated ether fatty acyl analogues, however, should, because of the opposite sign of the charge on the oxygen atoms of interest, be more stable in conformation **18,** and thus exhibit anomalous properties of predictable nature. A side view (**20**) of **19** shows more clearly that this conformation places the two fatty acyl groups in anticoplanar colinearity—in accord with the conclusion that differences in the behavior of monocaudal and bicaudal amphiphiles in aqueous media are explicable on the basis of the latter being more stable in splayed than in parallel configuration.

Similarly, sphingolipid components of biomembranes in the liquid-crystalline state, with bilayer surfaces exposed to aqueous media, should be free of most of the conformational constraints responsible for their configuration in the crystalline state (see Fig. 16.12) and thus tend to have their strong hydrogen-bonding groups interacting with water rather than intra- or intermolecularly. In the likely conformation shown in Fig. 16.14, for example, the anticoplanar conformation of the extended hydrophilic region of a sphingomyelin or of a cerebroside, stretching from the free 3R-hydroxyl group of the sphingosine to the 2'R-hydroxyl group of an α-hydroxy fatty acyl moiety (if present), would have four of its five hydrophilic substituents, including the phosphocholine or glycosidic group, pointed toward or into the aqueous phase, and the two long hydrophobic tails pointed away and very widely spread. It is of interest that the usually present 4E-double bond serves to sustain the general opposite course of the tails two carbon–carbon bonds farther than would be the case for the corresponding dihydrosphingosine (sphinganine) and analogues. Although quite possibly coincidental, the R configuration of both the 3- and 2'-hydroxyl groups of an α-hydroxy–fatty-acylsphin-

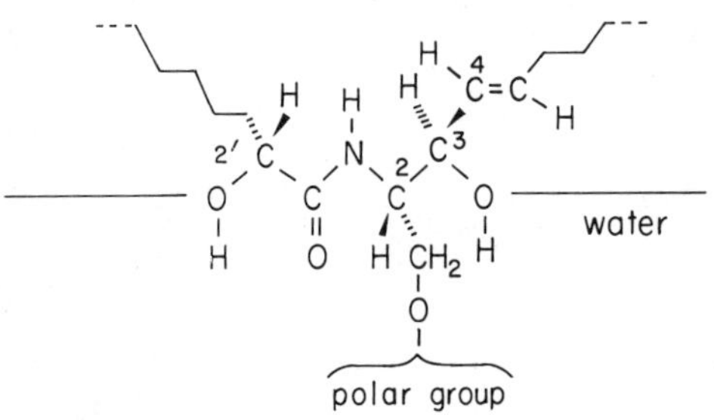

FIGURE 16.14. Conformation of a sphingomyelin (polar group: phosphocholine) or of a cerebroside (polar group: glucose or galactose) at a water-air interface.

golipid serves to accentuate the splaying. Whether or not sphingolipids tend to assume such extended conformations in biomembranes at temperatures above the crystal→liquid-crystal transition point should be indicated by pressure–area characteristics of monomolecular films at the aqueous medium–air interface (see Shah and Schulman, 1967; Oldani et al., 1975). At lower temperatures, such films exhibit the discontinuous, substantial contraction characteristic of the surface pressure having crowded the molecules in the film so as to initiate condensation of the hydrophobic moieties in crystalline order. As discussed above, the change in conformation of these parts of such molecules is likely to affect that of the hydrophilic group as well. This is a consideration of interest with respect to structure and characteristics of the myelin sheath bilayers, about a third of the lipids of which are cerebrosides and cerebroside sulfates with softening points, when neat, above ambient temperature.

REFERENCES

Abrahamsson, S., Dahlén, B., Löfgren, H., and Pascher, I., 1978, Lateral packing of hydrocarbon chains, in: *New Concepts in Lipid Research* (R. T. Holman, ed.), Pergamon Press, Oxford, pp. 125–143.

Anderson, R., Livermore, B. P., Volcani, B. E., and Kates, M., 1975, A novel sulfonolipid in diatoms, *Biochim. Biophys. Acta* **409**:259.

Anderson, R., Kates, M., and Volcani, B. E., 1976, Sulphonium analogue of lecithin in diatoms, *Nature* **263**:51.

Aneja, R., Chadha, J. S., and Knaggs, J. A., 1969, N-Acylphosphatidylethanolamines: occurrence in nature, structure and stereochemistry, *Biochem. Biophys. Res. Commun.* **36**:401.

Bieber, L. L., Cheldelin, V. H., and Newburgh, R. W., 1963, Studies on a β-methylcholine-containing phospholipid derived from carnitine, *J. Biol. Chem.* **238**:1262.

Blaschko, H., Firemark, H., Smith, A. D., and Winkler, H., 1967, Lipids of the adrenal medulla. Lysolecithin, a characteristic constituent of chromaffin granules, *Biochem. J.* **104**:545.

Bloor, W. R., 1943, in: *Biochemistry of the Fatty Acids and Their Compounds, the Lipids,* Reinhold, New York, pp. 37–46.

Bridges, R. G., Ricketts, J., and Cox, J. T., 1965, The replacement of lipid–bound choline by other bases in the phospholipids of the housefly, *Musca domestica, J. Insect Physiol.* **11**:225.

Cadenhead, D. A., 1977, Monomolecular films and membrane structure, in: *Structure of Biological Membranes* (S. Abrahamsson and I. Pascher, eds.), Plenum Press, New York, pp. 63–83.

Coleman, G. S., Kemp, P., and Dawson, R. M. C., 1971, The catabolism of phosphatidylethanolamine by the rumen protozoon *Entodinium caudatum* and its conversion into the N-(1-carboxyethyl) derivative, *Biochem. J.* **123**:97.

Folch, J., 1942, Brain cephalin, a mixture of phosphatides. Separation from it of phosphatidyl serine, phosphatidyl ethanolamine, and a fraction containing an inositol phosphatide, *J. Biol. Chem.* **146**:35.

Gent, P. A., and Gigg, R., 1975a, Synthesis of 1,2-di-O-hexadecanoyl-3-O-(β-D-galactopyranosyl)-L-glycerol (a 'galactosyl diglyceride') and 1,2-di-O-octadecanoyl-3-O-(6-O-octadecanoyl-β-D-galactopyranosyl)-L-glycerol, *J. Chem. Soc. Perkin Trans. 1,* p. 364.

Gent, P. A., and Gigg, R., 1975b, Synthesis of 3-O-[6-O-(α-D-galactopyranosyl)-β-D-galactopyranosyl]-1,2-di-O-stearoyl-L-glycerol, a 'digalactosyl diglyceride', *J. Chem. Soc. Perkin Trans. 1,* p. 1521.

Gent, P. A., and Gigg, R., 1975c, Synthesis of 3-O-{6-O-[6-O-(α-D-galactopyranosyl)-α-D-galactopyranosyl]-β-D-galactopyranosyl}-1,2-di-O-stearoyl-L-glycerol, a 'trigalactosyl diglyceride', *J. Chem. Soc. Perkin Trans. 1,* **1975**: p. 1779.

Hartley, G. S., 1936, *Aqueous Solutions of Paraffin-Chain Salts,* Hermann, Paris.

Hitchcock, P. B., Mason, R., Thomas, K. M., and Shipley, G. G., 1974, Structural chemistry of 1,2 dilauroyl-DL-phosphatidylethanolamine: molecular conformation and intermolecular packing of phospholipids, *Proc. Nat. Acad. Sci. U.S.A.* **71**:3036.

Horiguchi, M., and Kandatsu, M., 1959, Isolation of 2-aminoethane phosphonic acid from rumen protozoa, *Nature* **184**:901.

Hudson, C. S., 1909, The significance of certain numerical relations in the sugar group, *J. Am. Chem. Soc.* **31**:66.

IUPAC–IUB Commission on Biochemical Nomenclature, 1971, Tentative rules for carbohydrate nomenclature. Part I, *Biochemistry* **10**:3983.

IUPAC–IUB Commission on Biochemical Nomenclature, 1978, Nomenclature of phosphorus-containing compounds of biochemical importance, *Biochem. J.* **171**:14.

Kemp, P., Dawson, R. M. C., and Klein, R. A., 1972, A new bacterial sphingophospholipid containing 3-aminopropane-1,2-diol, *Biochem. J.* **130**:221.

Kishimoto, Y., Wajda, M., and Radin, N. S., 1968, 6-Acylgalactosyl ceramides of pig brain: structure and fatty acid composition, *J. Lipid Res.* **9**:27.

Klenk, E., and Doss, M., 1966, Über das Vorkommen von Estercerebrosiden im Gehirn, *Hoppe-Seyler's Z. Physiol. Chem.* **346**:296.

Klyashchitskii, B. A., Shvets, V. I., and Preobrazhenskii, N. A., 1969, On the nomenclature of asymmetrically substituted myoinositol derivatives with particular reference to phosphatidylinositols, *Chem. Phys. Lipids* **3**:393.

Kochetkov, N. K., Zhukova, I. G., and Glukhoded, I. S., 1963, Sphingoplasmalogens. A new type of sphingolipids, *Biochim. Biophys. Acta* **70**:716.

Kreissler, M., and Bothorel, P., 1978, Theoretical conformational analysis of phospholipids: influence of the parallelism of the β and γ alkyl chains on the glycerol moiety conformations. Anchoring of the aliphatic chains at the hydrophobic–hydrophilic interface, *Chem. Phys. Lipids* **22**:261.

Lee, A. G., 1976, Functional properties of biological membranes: a physical-chemical approach, in: *Progress in Biophysics & Molecular Biology,* Vol. 29 (J. A. V. Butler and D. Noble, eds.), Pergamon Press, Oxford, pp. 3–56.

Macfarlane, M. G., 1964, Phosphatidylglycerols and lipoamino acids, in: *Advances in Lipid Research,* Vol. 2 (R. Paoletti and D. Kritchevsky, eds.), Academic Press, New York, pp. 91–125.

Malins, D. C., 1966, Recent developments in the thin-layer chromatography of lipids, in: *Progress in the Chemistry of Fats and Other Lipids,* Vol. 8, Part 3 (R. T. Holman, ed.), Pergamon Press, Oxford, pp. 301–358.

McAlister, J., Yathindra, N., and Sundaralingam, M., 1973, Potential energy calculations on phospholipids. Preferred conformations with intramolecular stacking and mutually tilted hydrocarbon chain planes, *Biochemistry* **12**:1189.

Mehendale, H. M., Dauterman, W. C., and Hodgson, E., 1966, Phosphatidyl carnitine: a possible intermediate in the biosynthesis of phosphatidyl β-methylcholine in *Phormia regina* (Meigen), *Nature* **211**:759.

Mehlman, M. A., and Wolf, G., 1963, Phosphatidylcarnitine, *Arch. Biochem. Biophys.* **102**:346.

Nes, W. R., 1974, Role of sterols in membranes, *Lipids* **9**:596.

Oldani, D., Hauser, H., Nichols, B. W., and Phillips, M. C., 1975, Monolayer characteristics of some glycolipids at the air-water interface, *Biochim. Biophys. Acta* **382**:1.

Pascher, I., and Sundell, S., 1977, Molecular arrangements in sphingolipids. The crystal structure of cerebroside, *Chem. Phys. Lipids* **20**:175.

Rouser, G., Kritchevsky, G., Heller, D., and Lieber, E., 1963, Lipid composition of beef brain, beef liver, and the sea anemone: two approaches to quantitative fractionation of complex lipid mixtures, *J. Am. Oil Chem. Soc.* **40**:425.

Seelig, J., 1977, Deuterium magnetic resonance: theory and application to lipid membranes, *Quart. Rev. Biophys.* **10**:353.

Shah, D. O., and Schulman, J. H., 1965, Binding of metal ions to monolayers of lecithins, plasmalogen, cardiolipin, and dicetyl phosphate, *J. Lipid Res.* **6**:341.

Shah, D. O., and Schulman, J. H., 1967, Interaction of calcium ions with lecithin and sphingomyelin monolayers, *Lipids* **2**:21.

Shapiro, D., and Flowers, H. M., 1961, Synthetic studies on sphingolipids. VI. The total syntheses of cerasine and phrenosine, *J. Am. Chem. Soc.* **83**:3327.
Small, D. M., 1968, A classification of biologic lipids based upon their interaction in aqueous systems. *J. Am. Oil Chem. Soc.* **45**:108.
Smith, S. W., and Lester, R. L., 1974, Inositol phosphorylceramide, a novel substance and the chief member of a major group of yeast sphingolipids containing a single inositol phosphate, *J. Biol. Chem.* **249**:3395.
Somerharju, P., Brotherus, J., Kahma, K., and Renkonen, O., 1977, Stereoconfiguration of bisphosphatidic and semilysobisphosphatidic acids from cultured hamster fibroblasts (BHK cells), *Biochim. Biophys. Acta* **487**:154.
Stein, R. A., and Slawson, V., 1966, Column chromatography of lipids, in: *Progress in the Chemistry of Fats and Other Lipids,* Vol. 8, Part 3 (R. T. Holman, ed.), Pergamon Press, Oxford, pp. 373–420.
Stoffel, W., Sticht, G., and LeKim, D., 1968, Synthesis and degradation of sphingosine bases in *Hansenula ciferrii, Hoppe-Seyler's Z. Physiol. Chem.* **349**:1149.
Sundaralingam, M., 1972, Molecular structures and conformations of the phospholipids and sphingomyelins, *Ann. N.Y. Acad. Sci.* **195**:324.
Tanford, C., 1980, *The Hydrophobic Effect: Formation of Micelles and Biological Membranes,* 2nd ed., John Wiley, New York, p. 14.
Thompson, G. A., Jr., 1967, Studies of membrane formation in *Tetrahymena pyriformis.* I. Rates of phospholipid biosynthesis, *Biochemistry* **6**:2015.
Thudicum, J. L. W., 1884, *A Treatise on the Chemical Constitution of the Brain* (A facsimile edition of the original published by Bailliere, Tindall and Cox, London, with a new historical introduction: "Reflections upon a Classic," by D. L. Drabkin), Archon Books, Hamden, Connecticut, 1962.
van Deenen, L. L. M., 1965, Phospholipids and biomembranes, in: *Progress in the Chemistry of Fats and Other Lipids,* Vol. 8, Part 1 (R. T. Holman, ed.), Pergamon Press, Oxford, pp. 1–127.
van Deenen, L. L. M., Houtsmuller, U. M. T., de Haas, G. H., and Mulder, E., 1962, Monomolecular layers of synthetic phosphatides, *J. Pharm. Pharmacol.* **14**:429.
Vanderkooi, G., 1973, Conformational analysis of phosphatides: mapping and minimization of the intramolecular energy, *Chem. Phys. Lipids* **11**:148.
Winsor, P. A., 1948, Hydrotropy, solubilisation and related emulsification processes. Part I, *Trans. Faraday Soc.* **44**:376.

17
PHOSPHOGLYCERIDE METABOLISM

17.1. Hydrolysis of Phosphoglycerides

After elucidation of most of the details of digestion and absorption of triacylgylcerols, interest turned to analogous processes involving the structurally similar phospholipids. Although it was assumed that they are completely hydrolyzed in the intestine, followed by absorption of resulting hydrolysis products, experiments of Artom and Swanson (1948) cast some doubts on such simple premises. These authors reported that after [^{32}P]phosphatidylcholine (PC) is fed, the radioactivity of the blood phospholipids is consistently higher than when labeled phosphoric acid plus the other components of PC are administered, which clearly implies that complete hydrolysis is not a prerequisite to absorption. These findings were confirmed and elaborated by Scow *et al.* (1967), in a series of experiments in which variously labeled PC was fed to thoracic duct-cannulated rats. PC recovered from the lymph had ratios of labels in the 1-acyl, phosphate, and choline moieties that were largely unchanged, whereas the label associated with the 2-acyl group had been reduced to about 5% of its original level. It is thus evident that the 2-acyl group is largely hydrolyzed in the intestinal lumen and that the residual lysophosphatidylcholine is absorbed and then reacylated in the intestinal cells. Digestion and absorption of phosphoglycerides (assuming PC to be representative) therefore appears to require, as in the case of the triacylglycerols, removal of all but one acyl group, yielding a fragment now efficiently passed into the intestinal cells, where reconstruction (reacylation) takes place.

Considerable knowledge had been obtained earlier on the mode of action of several phospholipases, and this had been used to great advantage (and had led to some confusion) in determination of phospholipid structure. According to current terminology, phospholipases A catalyze the hydrolytic removal of one of the acyl groups from a phosphoglyceride, leaving a lysophosphoglyceride (so called because of the lytic action of these substances on erythrocytes and other cells). Phospholipase A_1 removes the acyl group from carbon *sn*-1 of the phosphoglyceride, and phospholipase A_2 that from carbon atom 2. For many years it was thought that "lecithinase A" (phospholipase A_2 from viper venom) removed the acyl group from the α (*sn*-1) carbon atom (C-1) of lecithin; and, since in general the acyl groups found in the two positions are different, certain features of the structure of lecithin remained in error during this period. That phospholipase A_1 action releases mixtures of fatty acids that are, on average, more saturated than those liberated by phospholipase A_2 reflects positional rather than acyl group saturation specificity. Phospholipase A_2, for example, acts on synthetically saturated "hydro-lecithin" as well as on the natural lecithin itself. Thus, after correction of the lysolecithin structures, the information on which is based the knowledge that the sn-1 and sn-2 positions are usually occupied by saturated and unsaturated acyl groups respectively (see below) was obtained. These specificities are shown diagrammatically in Fig. 17.1.

Phospholipase A_1 is widely distributed in nature, though it is usually difficult to separate it from other phospholipases. Major mammalian sources are the pancreas and the brain, in the latter of which it appears to be the major phospholipase A. The subcellular distributions of A_1 and A_2 are usually different in that A_2 is the major phospholipase of mitochondria, while A_1 is dominant in the endoplasmic reticulum.

Phospholipase A_2 has been known for a long time as a constituent of viper venom and has been used in structure determination of phosphoglycerides because of its positional specificity. It is most active in the hydrolysis of PC and phosphatidylethanolamine (PE); less so in that of phosphatidylserine (PS), phos-

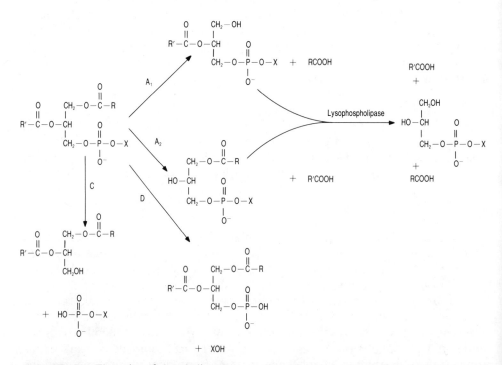

FIGURE 17.1. The action of phospholipases A_1, A_2, C, and D and lysophospholipase in the hydrolysis of the phosphoglycerides. X is choline, ethanolamine, serine, etc.

phatidylinositol (PI), and phosphatidylglycerol (PG); still less so in that of the plasmalogens, and inactive toward sphingolipids. The specificity of the brain particulate phospholipase A_2 is somewhat different in that its rate is maximal with phosphatidylinositol (PI) followed in order by its rates with PC, phosphatidic acid (PA), PE and PS. It is inhibited by its fatty acid products, particularly those unsaturated fatty acids usually resulting from its action (Gray and Strickland, 1982). In most mammalian tissues this enzyme occurs in admixture with other phospholipases.

The lysophospholipases catalyze the hydrolysis of the remaining acyl group of lysophosphoglycerides, yielding a fatty acid and sn-glycerol 3-phosphate. They are widely distributed, but it is not known whether they show positional or other specificity.

Phospholipase B is a nonspecific term that has been applied variously to phospholipase A_1, lysophospholipase and 1,2-deacylase. At present, the last designation appears to be accepted by most scientists in the field and probably describes a single enzymatic activity, though there is still some controversy about this point.

Phospholipase C, which catalyzes the hydrolysis of the bond between diacylglycerol and phosphocholine in phosphatidylcholine (see Fig. 17.1), acts analogously on sphingomyelin, but is somewhat less active toward PE or PI. It is most readily isolated from bacterial toxins (e.g., *Clostridium welchii*) but appears to have a considerably wider distribution. For example, it has been postulated to act on arachidonoylphosphatidylinositols, followed by action of a diglyceride lipase, which provides arachidonic acid for prostaglandin biosynthesis.

Phospholipase D is the major phospholipase in many plant tissues and acts specifically on intact phosphoglycerides to give phosphatidic acid and an alcohol (see Fig. 17.1). It is also found in bacterial and mammalian sources, where it is involved in alteration (by replacement) of the polar head groups of membrane phospholipids.

The phospholipases, like many lipid-active enzymes, exist in both "soluble" and membrane-bound forms. They appear to show maximal activity *in vivo* within or at the surface of membranes and, *in vitro*, at interfaces, such as those present in aqueous solutions of micellar amphiphiles and even in a biphasic milieu of ether and aqueous buffer. They generally show little tendency to act on monomeric (molecularly dispersed) lipids and are, in many cases, active only with aggregates of such substances.

Within the membrane, they are generally most active at the gel–liquid crystal transition state because a substrate–enzyme organization step, required as part of the activation process, is most rapid when the substrate is in the gel state. The subsequent hydrolytic action, however, is most rapid in the liquid–crystal state, as is true with most membrane-bound enzymes.

A related characteristic of phospholipase A_2 may be its activation by non-substrate lipids as well as by its substrates. For example, the enzyme is activated, probably by an allosterically induced conformation change, by both PC and sphingomyelin (a non-substrate). Its activation by calcium ions seems to involve a different mechanism and to include activation by calcium ionophores. In extreme examples such activation can lead to disruption of the cell membrane (Shier and Du Bourdieu, 1982; Chan *et al.,* 1982).

17.2. Biosynthesis of Phosphoglycerides

The complexity and diversity of structures embraced by this class of lipids (see Chapter 16) would lead reasonably to the expectation that their biosynthesis is correspondingly complex. However, as will be seen, this is fortunately not true and the existence of several unifying principles and of common pathways simplifies the whole picture considerably.

17.2.1. The Total Synthesis Pathway

Two classical papers greatly stimulated and directed subsequent fruitful research in this area. In 1953 Kornberg and Pricer, in connection with experiments on fatty acid activation as shown in Eq. 17.1a (for detailed discussion, see Chapter 8),

$$RCOOH + ATP \xrightarrow[Mg^{2+}]{thiokinase} [RCAMP] + PP_i \qquad (17.1a)$$

$$[RCAMP] + CoASH \longrightarrow RCSCoA + AMP$$

observed that when L-α-glycerol (*sn*-3) phosphate is added to microsomal preparations in which such fatty acid activation is taking place, phosphatidic acid is formed, presumably as shown in Eq. 17.1b:

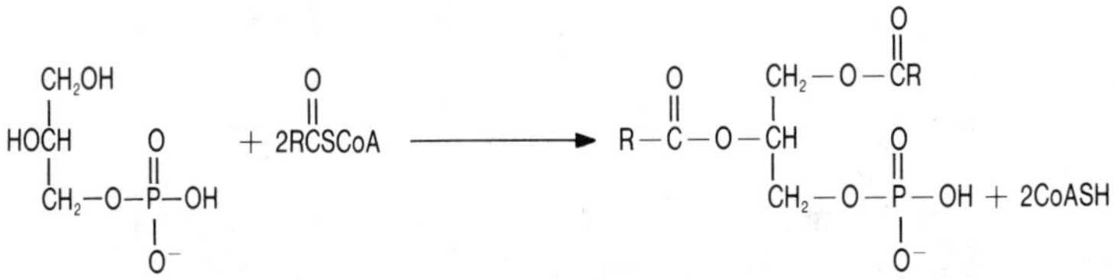

$$\begin{array}{c} CH_2OH \\ | \\ HOCH \\ | \\ CH_2-O-P-OH \\ | \\ O^- \end{array} + 2RCSCoA \longrightarrow \begin{array}{c} O \quad CH_2-O-CR \\ \| \quad | \\ R-C-O-CH \\ | \quad O \\ CH_2-O-P-OH \\ | \\ O^- \end{array} + 2CoASH \qquad (17.1b)$$

Kornberg and Pricer also suggested (but did not demonstrate) that addition of phosphocholine to this system might lead to the formation of PC (Eq. 17.1c):

$$\text{phosphatidic acid + phosphocholine} \rightleftharpoons \text{phosphatidylcholine} + P_i \qquad (17.1c)$$

Kennedy and Weiss (1956) reasoned that if this were true, reversal of the reaction, involving phosphorolysis of PC, should yield phosphatidic acid. However, this reaction did not occur, and Kennedy and his coworkers therefore embarked on

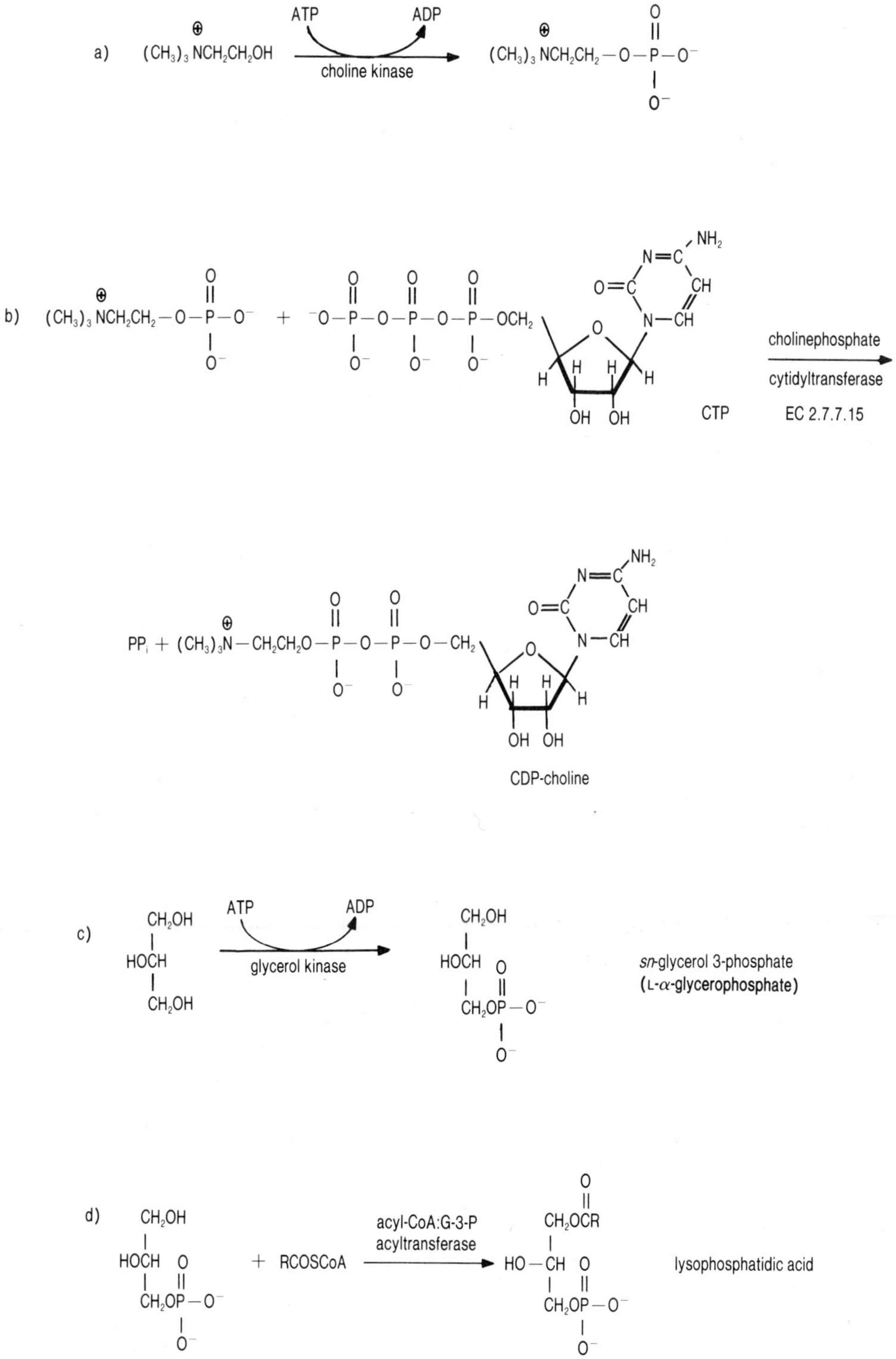

PHOSPHOGLYCERIDE METABOLISM

FIGURE 17.2. The biosynthesis of phosphatidylcholine.

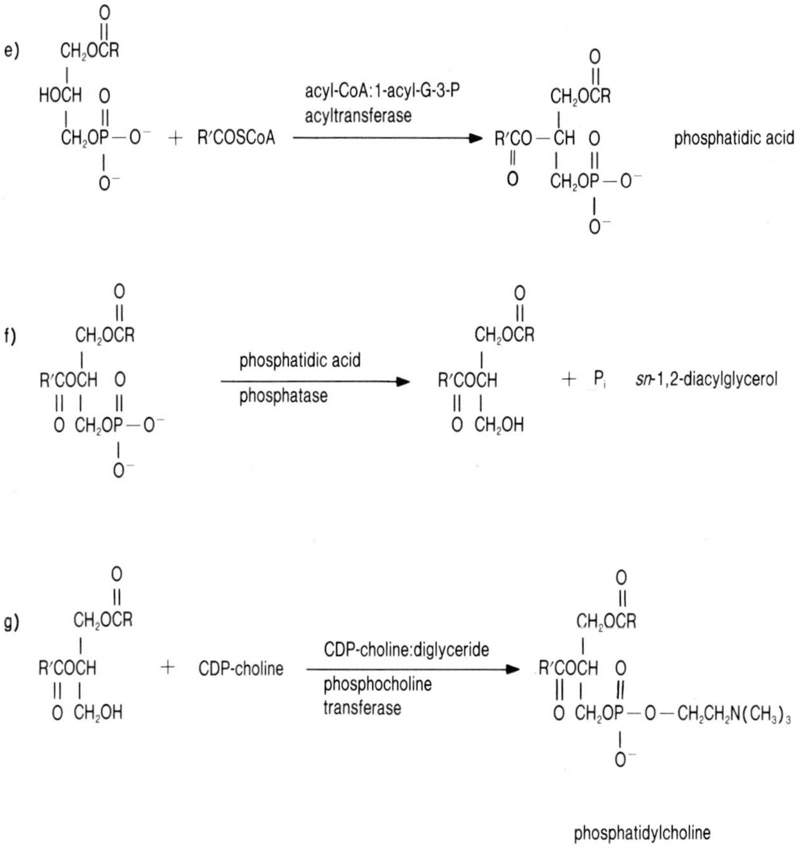

e)
$$\text{CH}_2\text{OCR} \atop \|\ \text{O}$$
$$|$$
$$\text{HOCH} \quad \text{O}$$
$$|\quad \|$$
$$\text{CH}_2\text{OP}-\text{O}^-\ +\ \text{R'COSCoA}$$
$$|$$
$$\text{O}^-$$

→ (acyl-CoA:1-acyl-G-3-P acyltransferase) →

$$\text{CH}_2\text{OCR} \atop \|\ \text{O}$$
$$|$$
$$\text{R'CO}-\text{CH}\quad \text{O}$$
$$\|\quad |\quad \|$$
$$\text{O}\quad \text{CH}_2\text{OP}-\text{O}^-$$
$$|$$
$$\text{O}^-$$

phosphatidic acid

f)
$$\text{CH}_2\text{OCR} \atop \|\ \text{O}$$
$$|$$
$$\text{R'COCH}\quad \text{O}$$
$$\|\quad |\quad \|$$
$$\text{O}\quad \text{CH}_2\text{OP}-\text{O}^-$$
$$|$$
$$\text{O}^-$$

→ (phosphatidic acid phosphatase) →

$$\text{CH}_2\text{OCR} \atop \|\ \text{O}$$
$$|$$
$$\text{R'COCH}$$
$$\|\quad |$$
$$\text{O}\quad \text{CH}_2\text{OH}$$

$+\ P_i\qquad$ sn-1,2-diacylglycerol

g)
$$\text{CH}_2\text{OCR} \atop \|\ \text{O}$$
$$|$$
$$\text{R'COCH}$$
$$\|\quad |$$
$$\text{O}\quad \text{CH}_2\text{OH}$$

$+\ \text{CDP-choline}$ → (CDP-choline:diglyceride phosphocholine transferase) →

$$\text{CH}_2\text{OCR} \atop \|\ \text{O}$$
$$|$$
$$\text{R'COCH}\quad \text{O}$$
$$\|\quad |\quad \|$$
$$\text{O}\quad \text{CH}_2\text{OP}-\text{O}-\text{CH}_2\text{CH}_2\text{N}(\text{CH}_3)_3$$
$$|$$
$$\text{O}^-$$

phosphatidylcholine

(Continuation of FIGURE 17.2)

investigations designed to reveal the actual mechanism of PC biosynthesis. Using a particulate (microsomal) fraction from rat liver homogenate, diglyceride, phosphocholine and commercial ATP, Kennedy and Weiss were able to demonstrate the formation of PC. However, when they repeated these experiments, using carefully purified ATP, PC was not obtained. Careful fractionation of the effective ATP revealed that CTP, present as an impurity, imparted the activity to the mixture. The overall scheme of synthesis by this pathway (the Kennedy pathway) can, therefore, be written as shown in Fig. 17.2.

PE is formed by a series of identical reactions except for involvement of CDP-ethanolamine, and the addition of phosphocholine to ceramide to form sphingomyelin may proceed analogously although, as will be seen below, alternative pathways may be more important. As discussed in Chapter 14 and outlined in reactions c–e of Fig. 17.2 and in Fig. 17.3, diacylglycerol is formed by action of phosphatidic acid phosphatase. This may well be the major pathway for the formation of this phosphoglyceride precursor in the liver and in the intestinal cells, even though the principal pathway of triacylglycerol formation in the intes-

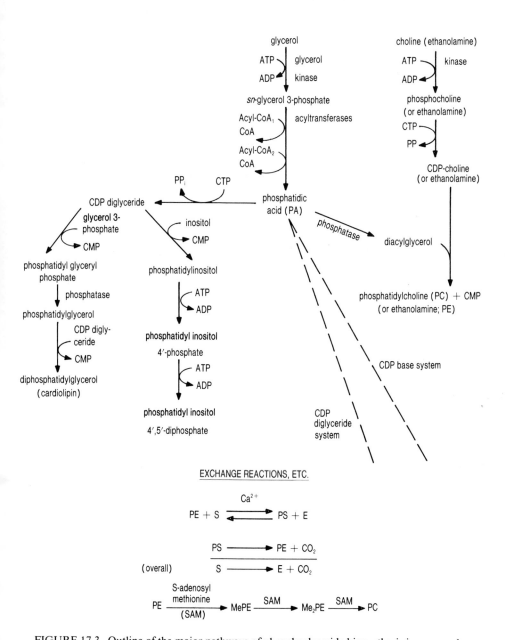

FIGURE 17.3. Outline of the major pathways of phosphoglyceride biosynthesis in mammals.

tine during fat absorption must obviously be by double acylation of 2-monoacylglycerol, diacylglycerol being an intermediate in the process.

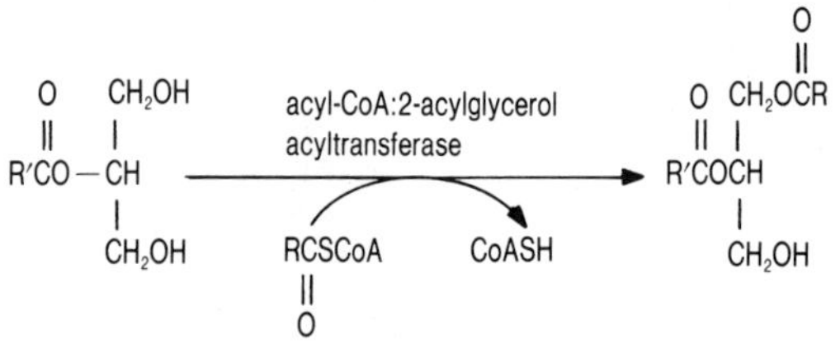

$$\underset{\substack{\text{R'CO—CH} \\ | \\ \text{CH}_2\text{OH}}}{\overset{\text{O} \quad \text{CH}_2\text{OH}}{\underset{|}{\|}}} \xrightarrow[\substack{\text{RCSCoA} \quad \text{CoASH} \\ \|\\ \text{O}}]{\text{acyl-CoA:2-acylglycerol acyltransferase}} \underset{\substack{\text{R'COCH} \\ | \\ \text{CH}_2\text{OH}}}{\overset{\text{O} \quad \text{CH}_2\text{OCR} \\ \| \quad |}{}} \overset{\text{O}}{\underset{}{\|}} \qquad (17.2)$$

There is also some indication that phosphorylation of monoacylglycerol to lysophosphatidic acid may occur, such reactions being enhanced in the presence of monoacylglycerol not only because it is a precursor but also because such substances appear to inhibit some step in the alternative pathway (possibly the action of phosphatidic acid phosphatase).

Adipose tissue, as has been discussed (see Chapter 14), lacks adequate glycerol kinase activity and is therefore dependent on glycolysis for supplies of glycerophosphate.

An alternate route to phosphatidate, starting from dihydroxyacetone phosphate, the other product of action of aldolase on fructose 1,6-diphosphate, is of particular interest because it provides an easily understood explanation for differences in fatty acyl groups at the sn-1 and sn-2 positions, which, by this pathway, are not simultaneously available:

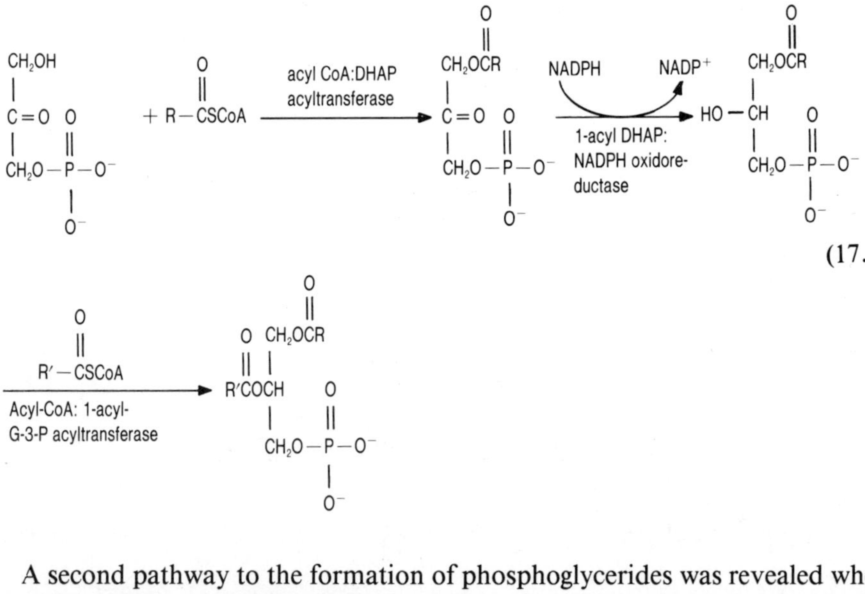

$$(17.3)$$

A second pathway to the formation of phosphoglycerides was revealed when Kennedy and his coworkers (Raetz and Kennedy, 1973) noted that, in *Escherichia coli,* CTP is required for the formation of PS but that serine phosphate is not an intermediate. This finding led to the proposal that the diacylglycerol rather than the serine is activated as the CDP derivative:

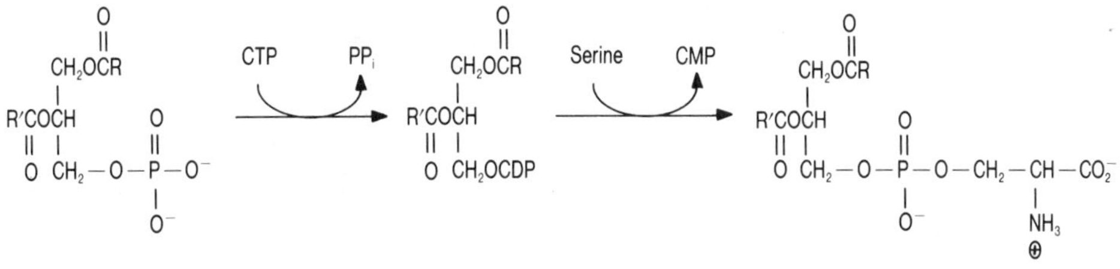

$$\begin{array}{c} \text{CH}_2\text{OCR} \\ | \\ \text{R'COCH} \\ | \\ \text{CH}_2-\text{O}-\text{P}-\text{O}^- \\ | \\ \text{O}^- \end{array} \xrightarrow[]{\text{CTP} \quad \text{PP}_i} \begin{array}{c} \text{CH}_2\text{OCR} \\ | \\ \text{R'COCH} \\ | \\ \text{CH}_2\text{OCDP} \end{array} \xrightarrow[]{\text{Serine} \quad \text{CMP}} \begin{array}{c} \text{CH}_2\text{OCR} \\ | \\ \text{R'COCH} \\ | \\ \text{CH}_2-\text{O}-\text{P}-\text{O}-\text{CH}_2-\text{CH}-\text{CO}_2^- \\ | \qquad\qquad\qquad | \\ \text{O}^- \qquad\qquad \overset{\oplus}{\text{NH}_3} \end{array}$$

(17.4)

Analogous mechanisms are now known to be involved in the formation of the inositol phosphoglycerides and of the phosphatidylglycerols in mammals (see Fig. 17.3) but, surprisingly, not in that of PS in higher organisms (see below). It is curious that use should be made of the CDP derivatives to facilitate formation of either of the two phosphate ester bonds in these structurally similar compounds. A possibly meaningful correlation is seen in the involvement of CDP derivatives of cationic precursors (choline, ethanolamine), but of CDP diglyceride as the active reactant with uncharged or anionic substances (e.g., inositol and glycerol phosphate).

17.2.2. Exchange and Secondary Pathways

Although the CDP-activation pathways account for a large proportion of initial phospholipid biosynthesis, other pathways are involved in formation of certain members of this group of lipids. In fact, it is probable that most of the phospholipids actually found in the various membrane lipid bilayers are products of *in situ* alteration of previously deposited phospholipids (Porcellati and di Jeso, 1971).

The actual mechanism of formation of PS in higher animals appears to be a Ca^{2+}-stimulated exchange reaction primarily involving phosphatidylethanolamine:

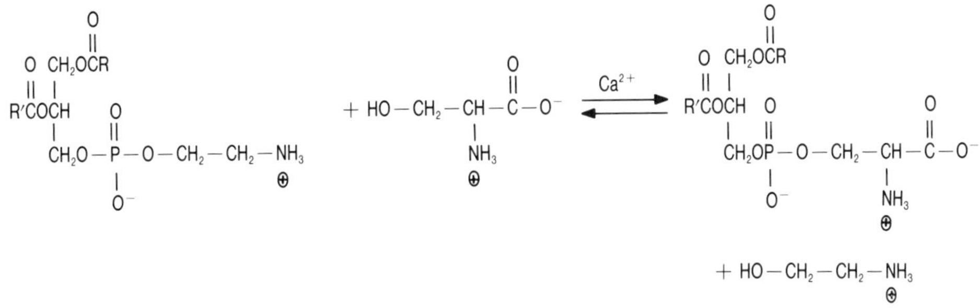

$$\begin{array}{c} \text{CH}_2\text{OCR} \\ | \\ \text{R'COCH} \\ | \\ \text{CH}_2-\text{O}-\text{P}-\text{O}-\text{CH}_2-\text{CH}_2-\overset{\oplus}{\text{NH}_3} \\ | \\ \text{O}^- \end{array} + \text{HO}-\text{CH}_2-\text{CH}-\overset{\text{O}}{\text{C}}-\text{O}^- \underset{}{\overset{Ca^{2+}}{\rightleftharpoons}} \begin{array}{c} \text{CH}_2\text{OCR} \\ | \\ \text{R'COCH} \\ | \\ \text{CH}_2\text{OP}-\text{O}-\text{CH}_2-\text{CH}-\text{C}-\text{O}^- \\ | \qquad\qquad\qquad | \\ \text{O}^- \qquad\qquad \overset{\oplus}{\text{NH}_3} \end{array}$$

$$+ \text{HO}-\text{CH}_2-\text{CH}_2-\overset{\oplus}{\text{NH}_3}$$

(17.5)

Although this appears to be the major exchange, it is not the only one, and there is good evidence that "base" exchange is a common way for regulation of hydrophilic "head" groups of the phosphoglycerides. In the competitive reactions for exchange of different "bases," affinity seems to be in the order serine > ethanolamine > choline. (Relative affinities for inositol or glycerol are not known.) A second reaction, dependent on such exchange, was discovered when it was found that, although labeled ethanolamine in the presence of a liver particulate preparation and Ca^{2+} initially yielded labeled PE as the major product, labeled serine under the same conditions gave about ⅓ labeled PS and ⅔ labeled PE. Since none of the known intermediates in PE biosynthesis via the CDP-ethanolamine pathway were found, it was concluded that PS had been decarboxylated to phosphatidylethanolamine:

$$PS \to PE + CO_2 \qquad (17.6)$$

Thus a cycle of reactions exists, involving the phosphatidyl derivatives, for the decarboxylation of serine to ethanolamine, and this indeed appears to be the principal mechanism for ethanolamine synthesis in higher organisms.

Dependent in part on this reaction is an alternative pathway for the formation of PC. Studies with *Neurospora* have revealed mutants that were unable to form PC but accumulated phosphatidyl-*N*-mono- (PMME) and *N,N*-dimethylethanolamine (PDME) (Hall and Nyc, 1959). That PC could be formed from PE by methylation was confirmed in the rat by use of methyl-labeled S-adenosylmethionine (SAM) (Rehbinder and Greenberg, 1965). At present two distinct enzymes appear to be involved: a soluble enzyme that carries out the first methylation and a microsomal enzyme that promotes the other two:

$$PE \xrightarrow[\text{sol. enzyme}]{\text{SAM}} PMME \xrightarrow[\text{particulate enzyme}]{\text{SAM}} PDME \xrightarrow[\text{particulate enzyme}]{\text{SAM}} PC$$

$$\begin{array}{c}
\text{O} \\
\parallel \\
\text{O} \quad \text{CH}_2\text{OCR} \\
\parallel \quad | \\
\text{R'COCH} \quad \text{O} \\
| \quad \parallel \\
\text{CH}_2\text{OP}-\text{O}-\text{CH}_2-\text{CH}_2-\overset{\oplus}{\text{NH}_3} \\
| \\
\text{O}^-
\end{array} \xrightarrow[\text{sol. enzyme}]{\text{SAM}} \begin{array}{c}
\text{O} \\
\parallel \\
\text{O} \quad \text{CH}_2\text{OCR} \\
\parallel \quad | \\
\text{R'COCH} \quad \text{O} \\
| \quad \parallel \\
\text{CH}_2\text{OP}-\text{O}-\text{CH}_2-\text{CH}_2-\overset{\oplus}{\text{NH}_2\text{CH}_3} \\
| \\
\text{O}^-
\end{array} \qquad (17.7)$$

$$\xrightarrow[\text{particulate enzyme}]{\text{SAM}} \begin{array}{c}
\text{O} \\
\parallel \\
\text{O} \quad \text{CH}_2\text{OCR} \\
\parallel \quad | \\
\text{R'COCH} \quad \text{O} \\
| \quad \parallel \\
\text{CH}_2\text{OP}-\text{O}-\text{CH}_2-\text{CH}_2-\overset{\oplus}{\text{NH}(\text{CH}_3)_2} \\
| \\
\text{O}^-
\end{array} \xrightarrow[\text{particulate enzyme}]{\text{SAM}} PC$$

This appears, indeed, to be the major route by which choline itself is made by higher organisms (ultimately from serine via phosphatidyl intermediates), a process that is of major importance, however, only in the liver. In all other tissues, PC is elaborated to a significant extent via the CDP-choline pathway which requires a source of choline either synthesized (primarily in the liver) or from the diet. The intermediary metabolism of choline is expected to be atypical in at least some respects in the few animals—notably the rat—that require choline in the diet.

As shown in Fig. 17.3, the inositol phosphoglycerides and phosphatidylglycerols are formed via CDP diglyceride. In the first case, the transferase is specific for *myo*-inositol and has the greatest rate for a CDP diglyceride with saturated R and unsaturated R' (Bishop and Strickland, 1970):

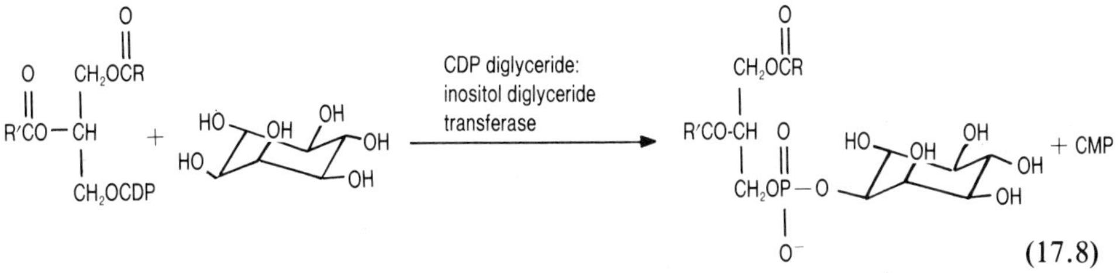

$$(17.8)$$

Further steps in the formation of the inositol phosphoglycerides involve successive phosphorylations to the 4'-monophosphate and the 4',5'-diphosphate in the presence of ATP and Mg^{2+}. The first enzyme is particulate (in synaptosomes, etc.) and the second is present in the cytosol, possibly because its substrate is soluble:

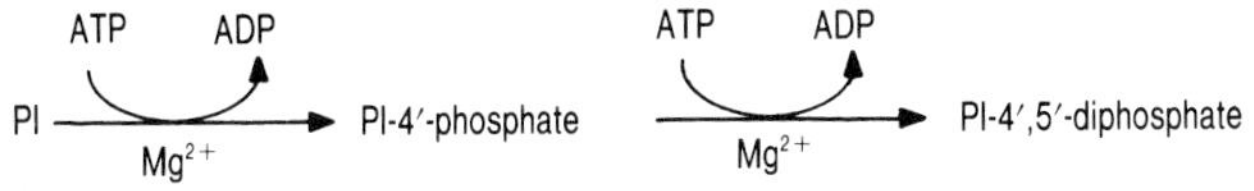

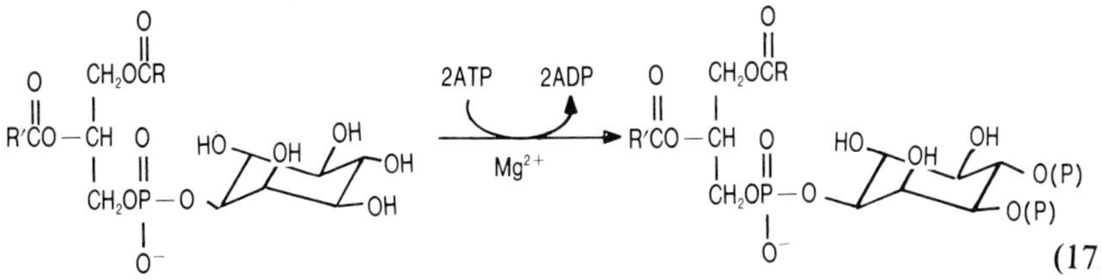

$$(17.9)$$

In the case of the phosphatidylglycerols, CDP diglyceride first reacts with glycerophosphate in a reaction catalyzed by a mitochondrial enzyme (a rarity, since in most other cases fully elaborated phospholipids are transferred to mitochondria from the endoplasmic reticulum) shown in Eq. 17.10 on p. 416. A phosphatase, also mitochondrial, then hydrolyzes the phosphatidylglycerylphosphate, and the resulting phosphatidylglycerol may react with an additional molecule of

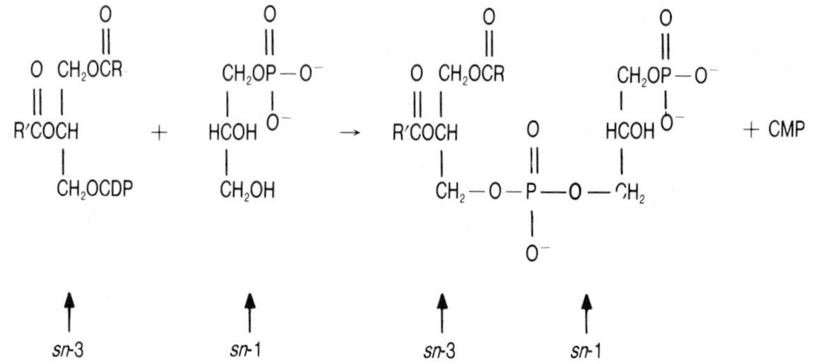

$$\begin{array}{c}\text{O}\\\|\\\text{O CH}_2\text{OCR}\\\|\ \ |\\\text{R'COCH}\\|\\\text{CH}_2\text{OCDP}\end{array}\ +\ \begin{array}{c}\text{O}\\\|\\\text{CH}_2\text{OP}-\text{O}^-\\|\ \ \ \ \ |\\\text{HCOH O}^-\\|\\\text{CH}_2\text{OH}\end{array}\ \longrightarrow\ \begin{array}{c}\text{O}\\\|\\\text{O CH}_2\text{OCR}\\\|\ \ |\\\text{R'COCH}\ \ \ \ \ \ \text{O}\\|\ \ \ \ \ \ \ \ \ \ \|\\\text{CH}_2-\text{O}-\text{P}-\text{O}-\text{CH}_2\\|\\\text{O}^-\end{array}\ \begin{array}{c}\text{O}\\\|\\\text{CH}_2\text{OP}-\text{O}^-\\|\ \ \ \ \ |\\\text{HCOH O}^-\\|\end{array}\ +\ \text{CMP}$$

↑ sn-3 ↑ sn-1 ↑ sn-3 ↑ sn-1

(17.10)

CDP diglyceride to form diphosphatidylglycerol, or "cardiolipin" (the antigenic lipid in the Wasserman test for syphilis), which occurs particularly richly in the inner membranes of mitochondria:

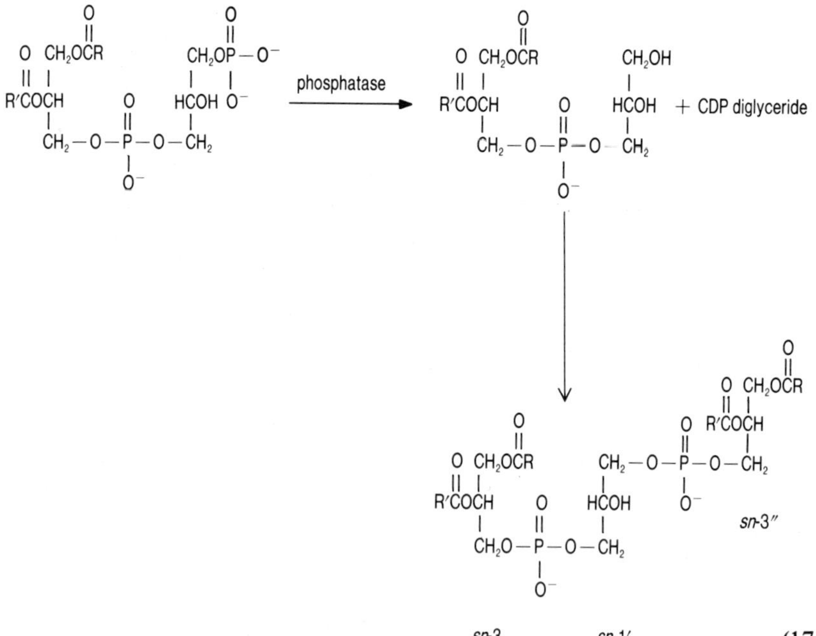

(17.11)

While the total synthesis of the phosphoglycerides is obviously of great importance in the formation of cellular membranes, it is equally important that it be possible to alter the physical and chemical characteristics of the membrane lipids in response to environmental changes more quickly than can be managed by processes involving total synthesis. For example, small poikilothermic organisms could not survive abrupt temperature changes without the ability to alter the aggregate "melting" characteristics of the hydrocarbon moieties of the membrane lipids. To accommodate such stress, a variety of exchange reactions permit rapid modification of membrane lipids *in situ*.

One such reaction, already considered, is the Ca^{2+}-activated "base" exchange, important in the formation of PS. That alteration of the polar head group of membrane phospholipids would modify the transport properties of these structures has been amply demonstrated. Nyc and his coworkers, for example, have shown that a *Neurospora* mutant unable to methylate phosphatidylmonomethylethanolamine to PC is "colonial" rather than filamentous and is unable to transport nicotinic acid at high pH (this becomes obvious in a double mutant requiring nicotinic acid). If intact PC is supplied to these organisms, it is incorporated into their membranes and they can then utilize medium nicotinate (Lie and Nyc, 1962).

The universal response of organisms to temperature decrease with an increase in unsaturation or decrease in chain-length of the membrane lipid fatty acids has been discussed elsewhere (Chapter 9). Such changes require facile exchange of acyl groups of membrane phospholipids with those of the medium. Such exchange reactions have been well documented. The acyl transferases, which have been studied by Lands and his coworkers and others, are widely distributed in most tissues and subcellular fractions and involve the transfer of an acyl group from CoA to either *sn*-1 or *sn*-2 positions of the corresponding lysophosphoglycerides:

$$\begin{array}{c}
\text{CH}_2\text{OCR} \\
| \\
\text{HO-CH} \\
| \\
\text{CH}_2\text{OP-OX} \\
| \\
\text{O}^-
\end{array} + \text{R'CSCoA} \xrightarrow{\text{acyl-CoA:1-acyl glycerol 3-phosphate acyltransferase}} \begin{array}{c}
\text{CH}_2\text{OCR} \\
| \\
\text{R'COCH} \\
| \\
\text{CH}_2\text{OP-OX} \\
| \\
\text{O}^-
\end{array} + \text{CoASH}$$

(17.12)

Such transacylation, including the prerequisite action of a phospholipase A, is quite specific both for the nature of the acyl group and for the position to which it is transferred. Thus, in the case of a 2-acylglycerol 3-phosphate, the rate of transfer of a saturated acyl group is more rapid than that of an unsaturated acyl group. For the 1-acylglycerol 3-phosphate, on the other hand, the rate of transfer of a *cis*-9-unsaturated acyl group is more rapid. In a series of experiments designed to ascertain the structural features to which the acyltransferases actually respond, Lands and his coworkers investigated the rates of transfer of various fatty acids from CoA to either the 1- or 2-positions of the monoacylglycerophosphates (Lands and Merkl, 1963). The results of those studies indicate that the transferase effecting the 1-acylation is sensitive to conformation, transferring a saturated (or *trans*-unsaturated) fatty acid to this position. The enzyme transferring acyl groups to the 2-position, however, appeared to recognize π bonds (*cis* or *trans* ethylenic or acetylenic) located at the 5-, 9-, or 12-positions. Acids with unsaturation at other positions were treated like saturated acids.

The end result of these preferences is the positioning of saturated fatty acids at the 1-position and of certain *cis*-unsaturated fatty acids at the 2-position. These specificities can be overridden, however, by concentration; the formation of the lung surfactant dipalmitoyl PC, for example, is attributable to presence of a high concentration of palmitate present at the site and time of synthesis.

This positional specificity, which seems to be of considerable importance in determining certain details of the structure of the phosphoglycerides, may have origins in addition to the conditions imposed by the phospholipase–acyltransferase sequence. For example, since triacylglycerol absorption from the intestine involves deacylation only as far as the 2-monoacylglycerols, the acyl group at the 2-position, which is generally unsaturated, may be incorporated into the tissue lipids largely unchanged. In the synthesis of phosphoglycerides, a possibility for acyl-group selectivity also exists during the formation of phosphatidic acid. Until recently, it had not been possible to separate the two acyltransferases active in the synthesis of phosphatidic acid from glycerol 3-phosphate. However, in both *E. coli* and rat liver, the separation of these microsomal enzymes has now been accomplished. The first enzyme specifically transfers a saturated acyl group from CoA to the *sn*-1-hydroxyl group of glycerol 3-phosphate. Following this reaction, the second transferase converts the 1-acylglycerol 3-phosphate to the diacyl derivative by transferring a predominantly unsaturated fatty acyl group to the 2-position. Thus the acyl specificity of the phosphoglycerides is a net consequence of (i) features of their direct synthesis; (ii) the structure of a dietary fat and its absorption; and (iii) transacylation of the complete phosphoglyceride.

The actual lipid composition of various membranes thus depends on a number of reactions, each contributing to structural features of lipids that are appropriate to the tissue, cell type, and subcellular membranes involved.

First, the synthesis of the phosphoglyceride involves fatty acylation specificity at the formation of diacylglycerol from monoacylglycerol and at the formation of phosphatidic acid from glycerol 3-phosphate. Second, once formed or deposited in the membrane bilayers, whole molecules can be exchanged by transfer via the phospholipid exchange proteins (PLEP), which appear to have some specificity for particular lipids in particular locations and thus contribute to regulation of the types of phosphoglycerides found in each tissue. Third, the polar head group may be exchanged and thus alter markedly the hydrophilic characteristics of at least portions of the lipid bilayer comprised of these substances. Finally, the fatty acyl groups may be removed and replaced via the phospholipase–acyltransferase route, thus regulating the fluidity and other properties of the hydrophobic portion of the lipid bilayer.

It has been found in the case of the last process that membrane fatty acid alteration (elongation, desaturation) by membrane-bound enzymes may take place entirely within the membrane without release of intermediate fatty acids to equilibrate with an external fatty acid pool. Desaturation of fatty acyl groups may occur, for example, either by release from the phospholipid, desaturation, and reincorporation—all within the membrane—or by desaturation of an acyl group of the intact phospholipid.

In these ways, structural features of the membrane lipids and thus functional characteristics of the membranes themselves are regulated within reasonably narrow limits.

Phospholipase-acyltransferases also function importantly in repair of adventitious oxidative damage. When the unsaturated fatty acyl group (in position 2) is oxidized to an epoxy or hydroxy derivative, phospholipases A_2 and C are activated and the oxidized fatty acyl group thus removed either as such (phospholipase A_2) or as part of a diacylglycerol (phospholipase C)

17.3. Ethers and Alkenyl Ethers

It has been previously pointed out (Chapter 16) that the generally more common 1,2-diacyl phosphoglycerides are accompanied in most tissues by greater or lesser amounts of the corresponding 1-alkyl-2-acyl, 1-alk-1'-enyl-2-acyl, and 1,2-dialkyl analogues. Indeed, the major components of the brain ethanolamine phosphoglycerides are of the 1-alkenyl, 2-acyl type. As has been noted, it is for this reason usually incorrect (for example) to designate ethanolamine-containing phosphoglycerides (EPG) isolated from natural sources as PE unless such analogous substances have been shown to be absent. Physical properties of these analogues tend to be virtually indistinguishable from those of the PEs themselves, although differences in chemical properties (e.g., ease of hydrolysis) permit demonstration and quantification of their presence.

The occurrence of ether bonds in naturally occurring lipids has long been known and such compounds are widely distributed. For example, the neutral lipids of certain shark-liver oils contain large proportions of glyceryl ether diesters, the deacylated products of which are chimyl (16:0), batyl (18:0) and selachyl (18:1) alcohols (Fig. 17.4). Such compounds have also been found in certain tumors. As is often the case with naturally occurring products of unknown function, various medicinal properties have been ascribed to them, such as erythropoietic and leukopoietic functions and the prevention of the toxic effects of benzene and of ionizing radiation on bone marrow.

Small amounts of 1-alkyl-2-acyl phosphoglycerides often accompany the corresponding diacyl phosphoglycerides, but these phosphoglycerides are major components in the lipids of the slug, egg yolk, brain, and the erythrocyte membranes of various species, in which they may amount to 75% of the mixture. The 1,2-dialkyl phosphoglycerides are also found in certain tissues in smaller amounts, e.g., in the brain.

That aldehydogenic lipids exist naturally was known to Feulgen because of the color reactions given by tissues exposed to an acid fuchsin stain. Early work on the structure of such substances was misinterpreted, then regarded as correct

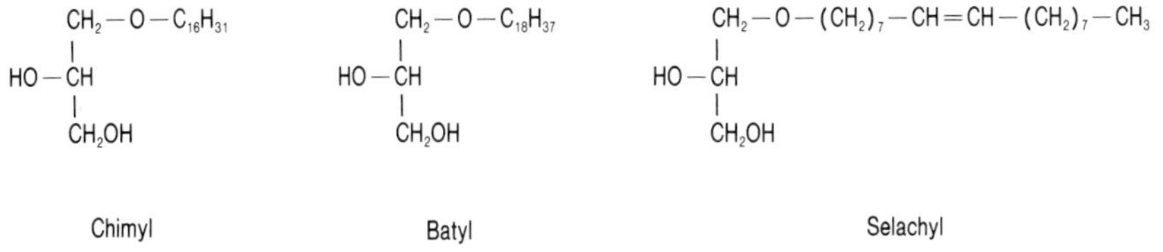

$$\begin{array}{ccc}
CH_2-O-C_{16}H_{31} & CH_2-O-C_{18}H_{37} & CH_2-O-(CH_2)_7-CH=CH-(CH_2)_7-CH_3 \\
| & | & | \\
HO-CH & HO-CH & HO-CH \\
| & | & | \\
CH_2OH & CH_2OH & CH_2OH \\
\\
\text{Chimyl} & \text{Batyl} & \text{Selachyl}
\end{array}$$

FIGURE 17.4. Structures of glyceryl ethers.

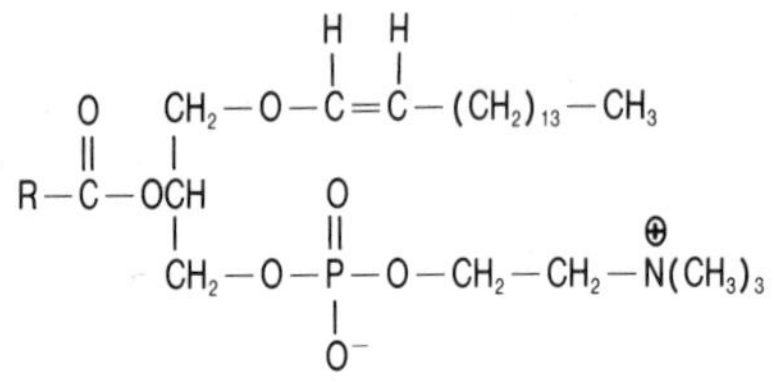

$$\begin{array}{l}\;\;\;\;\;\;\;\;\;\;\;\;\;\;\;\;\;\overset{H}{|}\;\;\overset{H}{|}\\ O\;\;\;CH_2-O-C=C-(CH_2)_{13}-CH_3\\ \|\;\;\;\;|\\ R-C-OCH\;\;\;\;\;\;O\\ |\;\;\;\;\;\;\;\;\;\;\|\\ CH_2-O-P-O-CH_2-CH_2-\overset{\oplus}{N}(CH_3)_3\\ |\\ O^-\end{array}$$

FIGURE 17.5. Structure of an alkenyl ether (phosphatidalcholine or plasmenylcholine).

for many years, and it was not until the work of Klenk and Debuch (1955) and of Rapport and his coworkers (see Rapport and Franzl, 1957) that they were shown, in fact, to be 1-alk-1'-enyl ethers (Fig. 17.5).

Although it was not at first recognized that the ether-containing phosphoglycerides might have any special physiological properties, it has been reported by Demopoulos *et al.* (1979) that IgE-sensitized basophilic leukocytes, and probably other types of mononuclear cells (e.g., mast cells), release in response to antigen stimulation a substance that aggregates platelets and causes the release of their granular constituents, among them serotonin. This substance, known as "platelet-activating factor" (PAF) has been identified as the 1-*O*-alkyl-2-acetyl-*sn*-glyceryl-3-phosphocholine. This is a very potent substance, rivaling the activity of thromboxane A_2; it elicits its effects on platelets (aggregation, release of serotonin) at 3×10^{-11} to 1.1×10^{-10} M concentrations. The structural requirements for the biological activity are very stringent: the 1-*O*-alkyl-2-lyso-*sn*-glyceryl-3-phosphocholine is inactive and, while the 2-propionyl derivative is as active as the 2-acetyl compound, the potency of the 2-butyryl derivative is much less; longer-chain 2-acyl derivatives are inactive. Synthetic 1-acyl-2-acetyl(propionyl)-*sn*-glyceryl-3-phosphocholine (prepared by acylation of 2-lysophosphatidylcholine—made from egg lecithin by the action of phospholipase A_2) also aggregates platelets and causes release of serotonin from the platelets, but is 200 to 300 times less potent than the 1-*O*-alkyl-2-acetyl derivative. The structural specificity needed for biological activity is further emphasized by the fact that the 3-acyl-2-acetyl-*sn*-glyceryl-1-phosphocholine is inert.

The PAF is also a potent antihypertensive agent in "Goldblatt rats" (unilaterally nephrectomized rats with a clamp on the artery to the remaining kidney, restricting the blood flow to the organ. Such animals develop severe hypertension.) Intravenous injection of PAF to such rats causes a drop in mean arterial pressure from 180 to 60–70 mm Hg in less than 2 sec; even 63 ng lowers the blood pressure from 190 to 110 mm Hg. Repeated doses given over 24 hr either intravenously (12–24 µg) or *per os* (20–80 µg) cause prolonged depression of blood pressure lasting 48–72 hr (Blank *et al.,* 1979). PAF at 10^{-6} M to 10^{-5} M concentrations is also a chemotactic agent for human neutrophilic polymorphonuclear leukocytes and monocytes and also causes an increased release of β-galactosidase and lysozyme from human neutrophils. Although neutrophilic polymorphonuclear leukycotes are probably the main source of PAF, basophils and platelets can also release it in response to specific stimuli. Enzymes exist for its synthesis also in spleen, lungs, bone marrow, thymus, lymph glands, liver, and kidney. There is a remarkable similarity between the biological activity of PAF and thromboxane

A_2 and some of the leukotrienes (e.g., leukotriene B_4; cf. Chapters 10 and 11), and even between the stimuli that cause the release of these agents from cells. They may even act synergistically in promoting phenomena of inflammation, although this is a matter of speculation at present.

Even after the structures were known, the biosynthetic pathways of the ether and alkenyl ether linkages generated much more confusion than knowledge and only recently have the mechanisms been clarified. Part of the confusion arose from the similarities in structure of the groups in the 1-position. Acids, aldehydes, and alcohols derived from this position are generally saturated or monounsaturated, containing 16–18 carbon atoms, so it was easy to assume a reductive sequence from the predominant 1-acyl through the 1-alkenyl to the 1-alkyl (see below for a discussion of this point). Indeed, some studies seemed to bear out these ideas, though others indicated a reverse path, and it was not until the extensive investigations of Snyder and his coworkers neared completion that the true mechanism became apparent (Snyder, 1970).

It was first shown that in mouse preputial gland tumors and later in normal glands and other sources such as Ehrlich ascites tumor, glyceryl ethers are formed in the presence of ATP, CoA, Mg^{2+}, NADPH, a long-chain alcohol, and either dihydroxyacetone phosphate or 3-phosphoglyceraldehyde. Three very revealing requirements, in retrospect, gave hints as to the mechanism of biosynthesis. First, it was found that if 3-phosphoglyceraldehyde isomerase were inhibited, only dihydroxyacetone phosphate was effective, thus implicating this compound as the true glyceride precursor. Second, although NADPH was required, it proved to be inhibitory if added before the start of the reaction. Third, coenzyme A was required in a reaction presumably not involving fatty acylation. These curious earlier findings were understood when Hajra (1970) showed that the process is initiated by the 1-acylation of dihydroxyacetone phosphate, a reaction requiring ATP, Mg^{2+}, and CoA for formation of the requisite acyl-CoA (Eq. 13a):

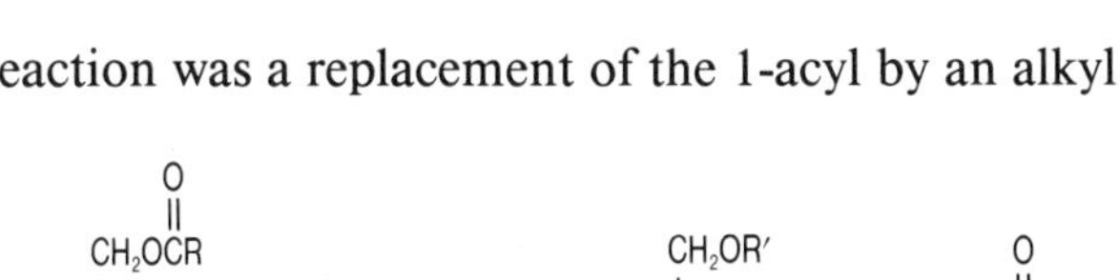

$$\begin{array}{c} CH_2OH \\ | \\ C=O \\ | \\ CH_2OP-O^- \\ | \\ O^- \end{array} + \text{R-COSCoA} \xrightarrow{\text{acyl-CoA:G-3-P acyl transferase}} \begin{array}{c} CH_2OCR \\ \parallel \\ O \\ | \\ C=O \\ | \\ CH_2OP-O^- \\ | \\ O^- \end{array} + \text{CoASH} \quad (17.13a)$$

The next reaction was a replacement of the 1-acyl by an alkyl group:

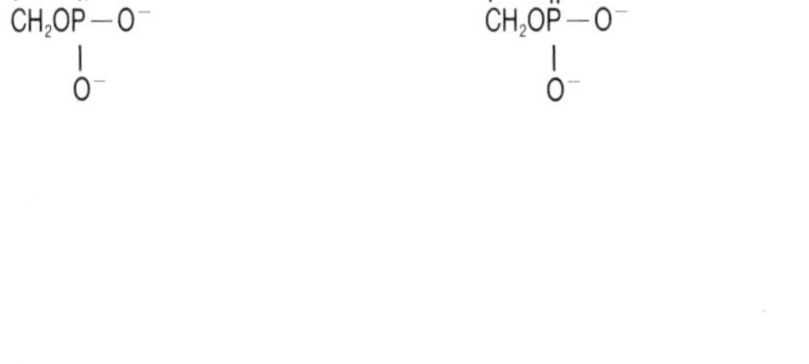

$$\begin{array}{c} O \\ \parallel \\ CH_2OCR \\ | \\ C=O \\ | \\ CH_2OP-O^- \\ | \\ O^- \end{array} + R'OH \xrightarrow{Mg^{2+}} \begin{array}{c} CH_2OR' \\ | \\ C=O \\ | \\ CH_2OP-O^- \\ | \\ O^- \end{array} + R'-C\begin{array}{c}O \\ \parallel \\ \\ O^-\end{array} \quad (17.13b)$$

In accord with this sequence of events, it was shown that with 1-acyldihydroxyacetone phosphate as a starting material, ATP and CoA are no longer required. Premature addition of NADPH thwarted the process by reduction of dihydroxyacetone phosphate to glycerylphosphate.

The formation of the 1-alkyldihydroxyacetone phosphate is followed by reduction to sn-1-alkylglyceryl-3-phosphate:

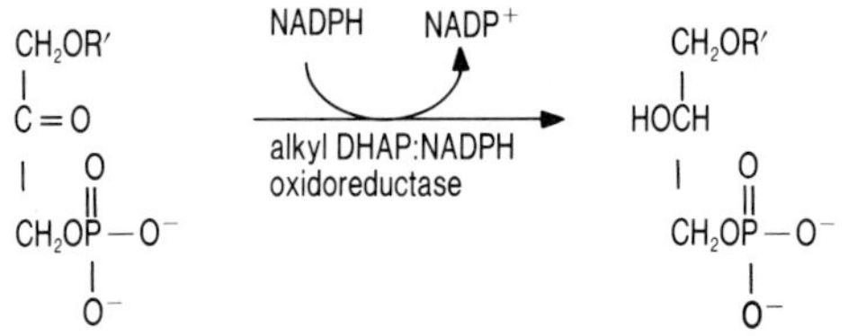

$$\begin{array}{c} CH_2OR' \\ | \\ C=O \\ | \\ CH_2OP(O)-O^- \\ | \\ O^- \end{array} \xrightarrow[\text{oxidoreductase}]{\text{NADPH} \quad \text{NADP}^+ \quad \text{alkyl DHAP:NADPH}} \begin{array}{c} CH_2OR' \\ | \\ HOCH \\ | \\ CH_2OP(O)-O^- \\ | \\ O^- \end{array} \qquad (17.13c)$$

This intermediate is then acylated:

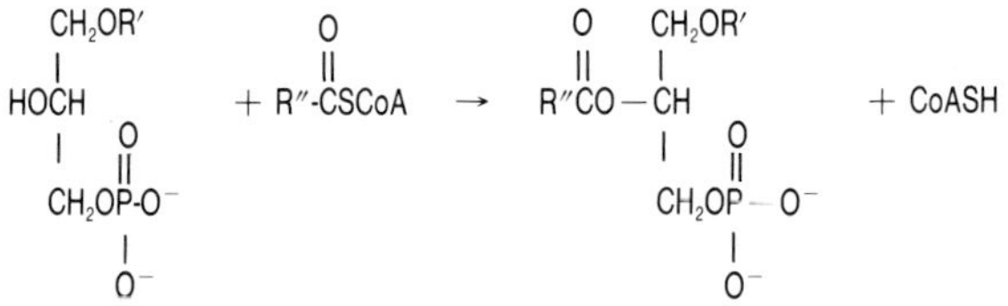

$$\begin{array}{c} CH_2OR' \\ | \\ HOCH \\ | \\ CH_2OP(O)-O^- \\ | \\ O^- \end{array} + R''\text{-CSCoA} \rightarrow \begin{array}{c} O \quad CH_2OR' \\ \| \quad | \\ R''CO-CH \\ | \\ CH_2OP(O)-O^- \\ | \\ O^- \end{array} + CoASH \qquad (17.13d)$$

The resulting phosphatidic acid analogue functions as a precursor of both diacyl glyceryl ethers and the corresponding phosphoglycerides. As a matter of fact, Kiyasu and Kennedy (1960) had already shown that the "plasmalogenic" diglycerides function as precursors of phospholipids about as efficiently as do the normal diglycerides. With these pathways in mind, it became possible to rationalize the fact that glyceryl ethers are particularly prevalent in tumors. Hajra suggested that the typically low activity of 3-phosphoglyceraldehyde dehydrogenase in tumors would result in elevation of concentrations of dihydroxyacetone phosphate, and thus favor formation of glyceryl ethers (Hajra, 1970).

Although the general features of the exchange reaction in the formation of the 1-alkyldihydroxyacetone phosphate were outlined by Hajra, the detailed mechanism by which such an exchange might occur was not immediately apparent. Such a mechanism proposed by Friedberg et al. (1982) has furnished the details and, at the same time, introduced an unusual type of acyl cleavage. The proposed mechanism, as outlined in Fig. 17.6, involves the labilization of the pro-R hydrogen on C-1 of DHAP with formation of an enediol (step I), followed by addition of a proton from the medium to C-2 to give a carbonium ion (step II) which adds the alkoxy portion of the long-chain alcohol with retention of configuration at C-1 (possibly involving a cyclic intermediate) (step III). The acyl group, $RC(O)O^-$, is split off in an unusual reaction in which both carboxyl oxygens are retained (step IV) and a proton is lost from C-2 to give an enediol (step V) which rearranges to the product 1-alkyldihydroxyacetone phosphate.

In the absence of a long-chain alcohol, the acyl group is replaced by the

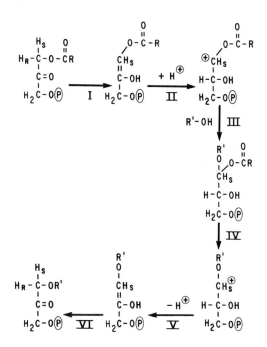

FIGURE 17.6. Proposed mechanism of O-alkyl-dihydroxyacetone phosphate synthesis from acyldihydroxyacetone phosphate. When R' is alkyl, the product is O-alkyldihydroxyacetone phosphate. When R' is hydrogen, the product is dihydroxyacetone phosphate.

hydroxyl from water and the pro-R hydrogen at C-1 exchanges with the hydrogen from water yielding DHAP, again with retention of the configuration at C-1.

Another specialized synthetic pathway involving the ether-containing phosphoglycerides is concerned with the formation of PAF (see p. 420). In general, PAF can be synthesized by either of two reactions catalyzed by microsomal enzyme systems similar to those already considered for the more general case (Fig. 17.7). One that involves the usual transfer of phosphocholine from CDP to sn-1-alkyl-2-acetylglycerol has an absolute requirement for Mg^{2+} (Ca^{2+} ions inhibit the reactions) (path a of Fig. 17.7). Maximum activity is seen at pH 8.0 with sn-1-hexadecyl-2-acetyl- and sn-octadecenoyl-2-acetylglycerol. The rate of transfer of phosphocholine to sn-1-octadecyl-2-acetylglycerol is only about one-half of that observed with the other two substrates. The most abundant source of this enzyme is the spleen, but lungs, liver, kidney and heart contain it also. This phosphocholinetransferase is stabilized by dithiothreitol (DTT) and is thus probably different from the transferase that catalyzes the reaction sn-1,2-diacylglycerol + CDP-choline → 1,2-diacyl-sn-glyceryl-3-phosphocholine + CMP and is inhibited by DTT (Renooij and Snyder, 1981).

In the second reaction, leading to the synthesis of PAF, 1-O-alkyl-2-lyso-sn-glycero 3-phosphocholine is acetylated with acetyl-CoA (Wykle et al., 1980) (path b of Fig. 17.7). This acetyltransferase is also a microsomal enzyme, its best sources being the spleen, lungs, lymph nodes, and thymus. 1-Palmitoyl-2-lyso-sn-glycero 3-phosphocholine can also serve as substrate for the acetyltransferase, but the 3-alkyl-2-lyso-sn-glycero 1-phosphocholine cannot.

The acetylation of a 2-lyso-glycero 3-phosphocholine is probably the more

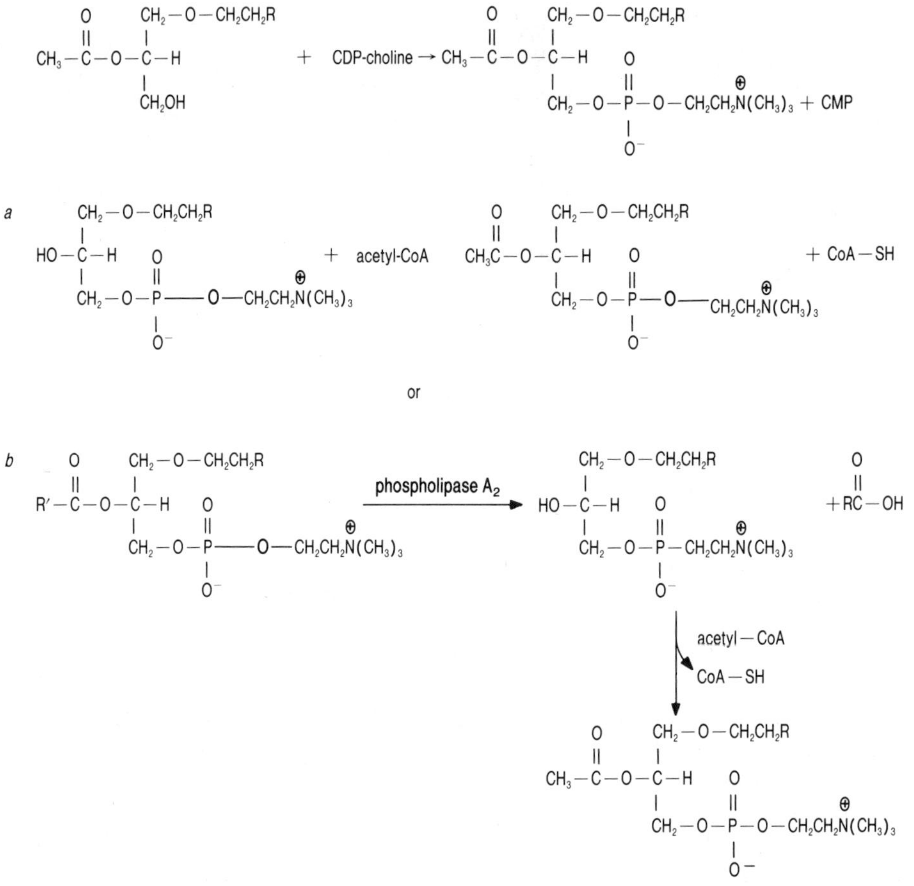

FIGURE 17.7. Reactions leading to the synthesis of the platelet-activating factor, 1-O-alkyl-2-acetyl-sn-glyceryl-3-phosphocholine.

important of the two reactions that can generate PAF. The various stimuli known to cause release of PAF are also those that stimulate the release of polyunsaturated fatty acids (e.g., arachidonic), presumably from C-2 of phospholipids preparatory to prostaglandin or leukotriene biosynthesis (cf. Chapters 10 and 11). Such release of a fatty acid from phospholipids, catalyzed by phospholipase A_2, would generate a 2-lyso-sn-glycero 3-phosphocholine as substrate for the acetyltransferase. The existence of phospholipase A_2 in, e.g., peritoneal mast cells, is known (Martin and Lagunoff, 1982). Such phospholipase A_2 has the interesting property of being activated by the product of the reaction it catalyzes.

The importance of the acetyltransferase reaction in the generation of PAF is emphasized by the observation that during phagocytosis of opsonized zymosan particles by human polymorphonuclear leukocytes—a stimulus that also causes release of PAF—there is a transient tenfold activation of the acetyltransferase reaction without any change in the phosphocholinetransferase reaction (Alonso et al., 1982).

A further report from the laboratory of Hajra served to explain the nature of the alkyl and alkenyl groups commonly found in the ether moiety of these lipids. Early studies of the structures of fatty aldehydes released from plasmalogens suggested simplistically that these substances were formed by more or less direct modification of the sn-1-ester linkage of diacyl phosphoglycerides. Study of the acyl-CoA reductase (from developing rat brain) involved in formation of fatty alcohols (see R'OH, Eq. 17.13b) revealed that it is specific for 16:0, 18:0, and 18:1 acyl groups. In tissue of certain brain tumors the specificity of this enzyme is lost and sn-1-ether groups are therefore more highly unsaturated, and thus more closely analogous in this respect to the 2-acyl moieties (Bishop and Hajra, 1981).

Although these studies clearly delineated the mechanism of formation of the glyceryl ethers and the corresponding ether-group-containing phospholipids, they did not serve to explain the mechanism of formation of the more abundant and widespread alkenyl ether lipids. Actually, evidence had accumulated for some time that, in organisms such as the slug and dogfish, alkyl groups are, in fact, immediate precursors of the 1-alkenyl corresponding groups. This was confirmed in several laboratories by the finding that a reaction occurring in microsomes, with a 1-alkyl-2-acyl phosphoglyceride as precursor and requiring oxygen and NADPH, led to formation of the corresponding alkenyl ether phosphoglyceride. These requirements, in addition to inhibition by EDTA, o-phenanthroline, mercuribenzoate, and CN^-, but not by CO, indicate that the dehydrogenation is effected by a mixed-function oxidase with electron transfer by cytochrome b_5 (Paltauf and Holasek, 1973).

Although it had been reported that, in intestinal cells, fatty alcohols are hydrolyzed from alkyl ether glycerides and then oxidized to acids, the characteristic resistance of saturated ethers to hydrolysis makes this sequence of events appear to be unlikely. It has been subsequently shown that initial oxidation obviates the requirement of hydrolysis in effecting the same overall change. Tietz et al. (1964), using liver microsomes plus the 100,000 × g supernatant of liver homogenate, found that in the presence of oxygen, NADPH, and 6,7-dimethyltetrahydropterin (PtH_4), glyceryl ethers yield long-chain aldehydes, presumably via a hemiacetal that easily and spontaneously dissociates to glycerol and a fatty aldehyde, which, in the presence of NAD, is oxidized to the corresponding fatty acid (the supernatant contains an NADPH dihydropterin reductase) (Eq. 17.14).

$$\begin{array}{c} CH_2OCH_2R \\ | \\ HOCH \\ | \\ CH_2OH \end{array} + O_2 + PtH_4 \rightarrow \left[\begin{array}{c} H \\ | \\ O \\ | \\ CH_2OCHR \\ | \\ HOCH \\ | \\ CH_2OH \end{array} \right] \rightarrow \begin{array}{c} CH_2OH \\ | \\ HOCH \\ | \\ CH_2OH \end{array} + \begin{array}{c} PtH_2 + H_2O \\ + \\ R-CHO \end{array} \xrightarrow[NADH]{NAD} \begin{array}{c} \text{aldehyde} \\ \text{dehydrogenase} \\ RCOOH \end{array}$$

$$NADP^+ \quad NADPH \qquad (17.14)$$

A similar type of oxidative splitting has been reported by Yavin and Gatt (1972) with a brain 250,000 × g supernatant plus microsomes. In this case the reducing agent appears to be ascorbate or an ascorbate derivative in the supernatant, and inhibition by EDTA indicates a metal component, probably iron.

These studies clarifying the mechanism for splitting the alkyl ether bond by an oxidative reaction have not been equally helpful for understanding the formation of alkenyl ethers. Since the two reactions are evidently very similar, it remains to be determined why in the one case the alkenyl ether bond remains intact and in the other case it is broken.

17.4. Control of Phosphoglyceride Biosynthesis

One of the intriguing questions in this field is the means by which the nature of the complex mixture of phosphoglycerides is controlled. Although the determination of the relative proportions of components of such complex mixtures is difficult, it appears at present that the cytidyl transfer reactions are rate-limiting and that these are, at least in part, controlled by concentrations of substrates. Thus, choline or lysophosphatidylcholine stimulates PC synthesis, while ethanolamine and phosphorylethanolamine stimulate PE synthesis. The apparent K_m for chicken brain microsomal 1,2-di(alkyl or acyl)-sn-glycerol:CDP ethanolamine transferase decreases in the presence of either diacyl- or alkylacylglycerols, and unsaturated fatty acids stimulate PE formation, while saturated fatty acids appear to inhibit it. (Analogous characteristics of CDP choline transferase are less clearly understood.)

In general, in the presence of the quite small concentrations of free fatty acids normally present in tissues, their nature and the presence of the diacyl- or dialkylglycerols and phosphorylated bases appear to be controlling. However, as the concentration of exogenous fatty acids increases, the formation of triacylglycerols becomes dominant.

In a study with cultured dissociated brain cells (probably astrocytes), Yavin and Menkes (1974) found that added labeled linolenate (18:3) is first incorporated into di- and triacylglycerols. However, as the conversion of 18:3 into its longer more highly unsaturated analogues progressed, these tended to be transferred to ethanolamine phosphoglycerides from triacylglycerols and choline phosphoglycerides. When the temperature of the culture medium was decreased, fatty acyl elongation and desaturation were interrupted at the 20:5ω3 step, no 22:6ω3 being formed. Moreover, under these conditions PE synthesis was low.

Finally, the composition of the phosphoglyceride mixture may influence the turnover of its individual components. Thus, phospholipase A_2 activity is stimulated by choline-containing phospholipids, while the alkylacyl phosphoglycerides have been reported to inhibit the action of phospholipase A_1. As the dynamics of biomembrane maintenance and modification becomes better understood, it is therefore expected that their extant compositions, in addition to the concentrations and activities of various enzymes involved, may well play significant roles (O'Doherty, 1980).

REFERENCES

Alonso, F., Gil., M. G., Sánchez-Crespo, M., and Mato, J. M., 1982, Activation of 1-alkyl-2-lyso-glycero-3-phosphocholine: acetyl-CoA transferase during phagocytosis in human polymorphonuclear leukocytes, *J. Biol. Chem.* **257**:3376.
Artom, C., and Swanson, M. A., 1948, On the absorption of phospholipides, *J. Biol. Chem.* **175**:871.
Bishop, H. H., and Strickland, K. P., 1970, On the specificity of cytidine diphosphate diglycerides in monophosphoinositide biosynthesis by rat brain preparations, *Can. J. Biochem.* **48**:269.
Bishop, J. E., and Hajra, A. K., 1981, Mechanism and specificity of formation of long-chain alcohols by developing rat brain, *J. Biol. Chem.* **256**:9542.
Blank, M. L., Snyder, F., Byers, L. W., Brooks, B., and Muirhead, E. E., 1979, Antihypertensive activity of an alkyl ether analogue of phosphatidyl choline, *Biochem. Biophys. Res. Commun.* **90**:1194.
Chan, P. H., Yurko, M., and Fishman, R. A., 1982, Phospholipid degradation and cellular edema induced by free radicals in brain cortical slices, *J. Neurochem.* **38**:525.
Demopoulos, C. A., Pinckard, R. N., and Hanahan, D. J., 1979, Platelet-activating factor. Evidence for 1-O-alkyl-2-acetyl-sn-glyceryl-3-phosphorylcholine as the active component (a new class of lipid chemical mediators), *J. Biol. Chem.* **254**:9355.
Friedberg, S. J., Weintraub, S. T., Singer, M. R., and Greene, R. C., 1982, The mechanism of ether bond formation in O-alkyl lipid synthesis in Ehrlich ascites tumor. Unusual cleavage of the fatty acid moiety of acyl dihydroxyacetone phosphate, *J. Biol. Chem.* **258**:136.
Gray, N. C. C., and Strickland, K. P., 1982, On the specificity of a phospholipase A_2 purified from the 106,000 × g pellet of bovine brain, *Lipids* **17**:91.
Hajra, A. K., 1970, Acyldihydroxyacetone phosphate: precursor of alkyl ethers, *Biochem. Biophys. Res. Commun.* **39**:1037.
Hall, M. O., and Nyc, J. F., 1959, Lipids containing mono- and dimethyl-ethanolamine in a mutant strain of *Neurospora crassa*, *J. Am. Chem. Soc.* **81**:2275.
Kennedy, E. P., and Weiss, S. B., 1956, The function of cytidine enzymes in the biosynthesis of phospholipids, *J. Biol. Chem.* **222**:193.
Kiyasu, J. Y., and Kennedy, E. P., 1960, The enzymatic synthesis of plasmalogens, *J. Biol. Chem.* **235**:2590.
Klenk, E., and Debuch, H., 1955, Zur Kenntnis der cholinhaltigen Plasmalogene (Acetalphosphatide) des Rinderherzmuskels, *Hoppe-Seyler's Z. Physiol. Chem.* **299**:66.
Kornberg, A., and Pricer, W. E., Jr., 1953, Enzymatic esterification of α-glycerophosphate by long-chain fatty acids, *J. Biol. Chem.* **204**:345.
Lands, W. E. M., and Merkl, I., 1963, Metabolism of glycerolipids. III. Reactivity of various acyl esters of coenzyme A with α'-acylglycerophosphorylcholine, and positional specificities in lecithin synthesis, *J. Biol. Chem.* **238**:898.
Lie, K. B., and Nyc, J. F., 1962, Effect of lipid composition on niacin acid transport, *Biochim. Biophys. Acta* **57**:341.
Martin, T. W., and Lagunoff, D., 1982, Rat mast cell phospholipase A_2: activity towards exogenous phosphatidyl serine and inhibition by N-(7-nitro-2,1,3-benzoxadiazol-4-yl) phosphatidyl serine, *Biochemistry* **21**:1254.
O'Doherty, P. J. A., 1980, Regulation of ethanolamine and choline phosphatide biosynthesis in isolated rat intestinal villus cells, *Can. J. Biochem.* **58**:527.
Paltauf, F., and Holasek, A., 1973, Enzymatic synthesis of plasmalogens. Characterization of the 1-O-alkyl-2-acyl-sn-glycero-3-phosphorylethanolamine desaturase from mucosa of hamster small intestine, *J. Biol. Chem.* **248**:1609.
Porcellati, G., and di Jeso, F., 1971, Membrane-bound enzymic activity in the base-exchange reactions of phospholipid metabolism, *Adv. Exp. Med. Biol.* **14**:111.
Raetz, C. R. H., and Kennedy, E. P., 1973, Function of cytidine diphosphate-diglyceride and deoxycytidine diphosphate-diglyceride in the biogenesis of membrane lipids in *Escherichia coli*, *J. Biol. Chem.* **248**:1098.
Rapport, M. M., and Franzl, R. E., 1957, Structure of plasmalogens. III. The nature and significance of the aldehydogenic linkage, *J. Neurochem.* **1**:303.

Rehbinder, D., and Greenberg, D. M., 1965, Studies on the methylation of ethanolamine phosphatides by liver preparations, *Arch. Biochem. Biophys.* **109**:110.

Renooij, W., and Snyder, F., 1981, Biosynthesis of 1-alkyl-2-acetyl-*sn*-glycero-3-phosphocholine (platelet activating factor and a hypotensive lipid) by cholinephosphotransferase in various rat tissues, *Biochim. Biophys. Acta* **663**:545.

Scow, R. O., Stein, Y., and Stein, O., 1967, Incorporation of dietary lecithin and lysolecithin into lymph chylomicrons in the rat, *J. Biol. Chem.* **242**:4919.

Shier, W. T., and Du Bourdieu, D. J., 1982, Role of phospholipid hydrolysis in the mechanism of toxic cell death by calcium and ionophore A23187, *Biochem. Biophys. Res. Commun.* **109**:106.

Snyder, F., 1970, The biochemistry of lipids containing ether bonds, in: *Progress in the Chemistry of Fats and Other Lipids,* Vol. 10 (R. T. Holman, ed.), Pergamon, Oxford, pp. 289–329.

Tietz, A., Lindberg, M., and Kennedy, E. P., 1964, A new pteridine-requiring enzyme system for the oxidation of glyceryl ethers, *J. Biol. Chem.* **239**:4081.

Wykle, R. L., Malone, B., and Snyder, F., 1980, Enzymatic synthesis of 1-alkyl-2-acetyl-*sn*-glycero-3-phosphocholine, a hypotensive and platelet-aggregating lipid, *J. Biol. Chem.* **255**:10256.

Yavin, E., and Gatt, S., 1972, Oxygen-dependent cleavage of the vinylether linkage of plasmalogens. 2. Identification of the low-molecular-weight active component and the reaction mechanism, *Eur. J. Biochem.* **25**:437.

Yavin, E., and Menkes, J. H., 1974, Polyenoic acid metabolism in cultured dissociated brain cells, *J. Lipid Res.* **15**:152.

18
SPHINGOLIPID METABOLISM

18.1. Introduction

The complex lipids known as sphingolipids, or at least some representatives of the group, became known at the end of the nineteenth century through the classical work of J. L. W. Thudicum (1884) who also provided their names even though he was unable to determine their structures fully. They all contain a nitrogenous base, sphingosine. The origin of this name is obscure; there are two apocryphal stories of how Thudicum coined it. According to one, the name is derived from the Greek verb sphingein, meaning "to bind tightly," implying that sphingolipids were tightly bound to tissues, particularly in structures of the brain. According to the second legend, Thudicum named these lipids after the Sphinx of Thebes, whose mystery was difficult to solve. The second legend stems probably from Thudicum's own writing. On describing the basic unit of these lipids as of "alkaloidal nature," he wrote: " ... to which, in commemoration of the many enigmas which it presented to the inquirer, I have given the name of *Sphingosin*" (Thudicum, 1884, p. 149). The naming after the Sphinx of Thebes is most apposite as the structural mysteries of sphingosine were not solved for more than 50 years after Thudicum first isolated it from the brain. Thudicum isolated and characterized from human and ox brain not only lecithin (phosphatidylcholine), kephalin (phosphatidylethanolamine), and sphingomyelin, which he correctly recognized to contain two bases, and phosphorus and nitrogen in the ratio of 1:2, but also the cerebrosides—phrenosin, kerasin, nervon, and oxynervon. He showed that the cerebrosides contained sphingosine, a fatty acid, and an optically active hexose, which he named "cerebrose," and which was subsequently identified as galactose. From Thudicum's data we calculate that he must have observed for a freshly prepared aqueous solution of cerebrose a specific rotation of $[\alpha]_D = 76.8°$, which is remarkably close to the present-day value of $[\alpha]_D = 83.3°$ for galactose. He even observed what is undoubtedly the mutarotation of galactose. Thudicum's "neurin," isolated from hydrolysates of lecithin and sphingomyelin, was, in all probability, choline.

18.2. The Structural Unit of Sphingolipids—Sphingosine

Sphingosine, to which Thudicum (1884, p. 150) assigned the elemental formula $C_{17}H_{35}NO_2$, has engaged the attention of many chemists and biochemists since the beginning of this century. Its structure, with all its configurational details, was not established until 1953/54. The early work was reviewed by Carter (1958). According to the Fischer convention sphingosine is D(+)-*erythro*-1,3-dihydroxy-2-amino-4-*trans*-octadecene or, by present-day notation of absolute configuration, the 2*S*-amino-3*R*-hydroxy-4-*trans*(E)-octadecen-1-ol (Fig. 18.1).

Sphingosine is one of a small group of related compounds. Thus, animal sphingolipids (sphingomyelin and cerebrosides) contain, in addition, a small percentage (5–12%) of dihydrosphingosine (Carter, 1958; Sweeley and Moscatelli, 1959). In gangliosides of brain and spinal cord of man and other animals 41–48% of the nitrogenous base is C_{18}-sphingosine and 34–43% is its C_{20} homologue, C_{20}-sphingosine (Sambasivarao and McCluer, 1964). In plant sphingolipids, three nitrogenous bases have been identified: (i) phytosphingosine, found in seeds and yeast, contains a hydroxyl group at C-4 in place of the double bond in sphingosine; its structure is 2*S*-amino-3*R*,4*R*-dihydroxyoctadecan-1-ol (Fig. 18.1) and it arises probably by the *trans*-addition of elements of water to the double bond of sphingosine (cf. Hanahan and Brockerhoff, 1965); (ii) dehydrophytosphingosine,

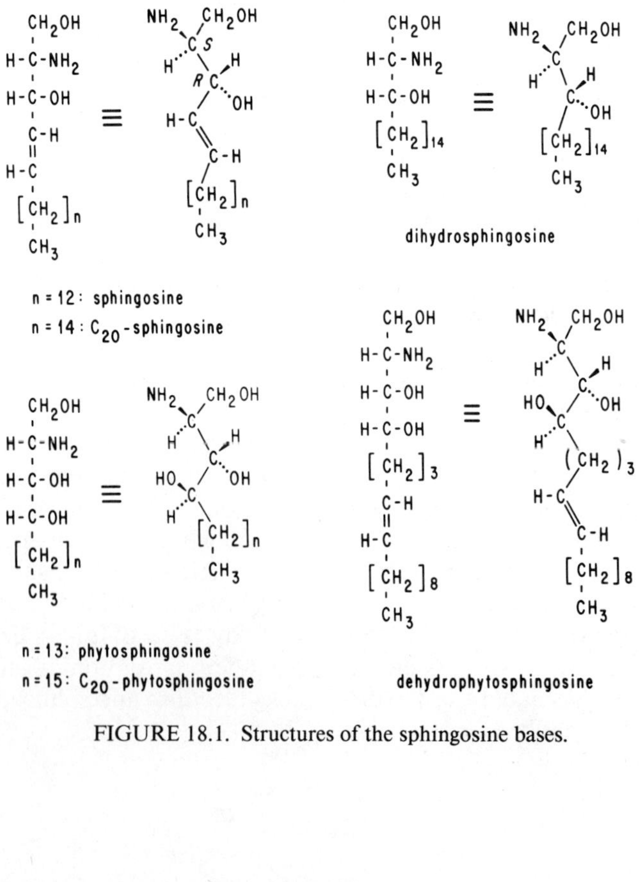

FIGURE 18.1. Structures of the sphingosine bases.

the major nitrogenous base in soybean phosphatides (Sweeley and Moscatelli, 1959), is structurally related to phytosphingosine except that it contains a *trans* double bond at position 8; (iii) C_{20}-phytosphingosine, bis-homo-phytosphingosine, is a major constituent of yeast phosphatides. The structures of the various sphingosines are shown in Fig. 18.1 (for structures of sphingolipids see also Chapter 16).

According to the system of nomenclature first approved by IUPAC–IUB in 1976 (IUPAC–IUB, 1977), sphingosine has been renamed sphingenine and its saturated variant, sphinganine. Both the old and new terms are used in journals and textbooks with some preference for the older system. Because of the subtle difference in the pronunciation of the two new names and because a simple typographical error in their spelling could lead to confusion, we prefer to use sphingosine and dihydrosphingosine instead of sphingenine and sphinganine.

Karlsson (1970) has proposed a shorthand designation for the sphingosine bases, similar to that used for the fatty acids, in which the chain-length, number of double bonds, and number of hydroxy groups are designated. For example, the common C_{18} sphingosine would be d18:1 (d standing for dihydroxy). Positions and configurations could also be shown but would ultimately make the system more cumbersome and they can usually be omitted unless unusual compounds are under discussion.

18.3. Biosynthesis of Sphingolipids

18.3.1. Biosynthesis of Sphingosine

As glycerol is the backbone of the phosphoglycerides, so sphingosine is the basic unit of sphingolipids. Indeed, sphingosine might be compared to a monoglyceride, since the hydrocarbon chains of the sphingosines are comparable to acyl chains of the 1-acylglycerols and confer their hydrophobic properties on the sphingolipids. Moreover, the multiplicity of sphingosine base structures is similar to, though not as extensive as, that of the fatty acids, although this was not appreciated by the earlier workers. Thus, when investigations of the biosynthesis of sphingosine were begun, it was convenient to consider sphingosine as a single compound with a formula $C_{18}H_{34}O_2N$, although it was admitted that the "sphingosines" from many sources might differ markedly in structure.

In vivo studies established that sphingosine is synthesized from a fatty acid (palmitic) and the amino acid serine. *In vitro* experiments with brain homogenates showed that synthesis takes place in the microsomal fraction and requires an acyl(palmitoyl)-CoA, pyridoxal phosphate, Mn^{2+}, and NADPH (Brady *et al.*, 1958). Suggestions that the first step was reduction of the acyl-CoA to an aldehyde proved to be false, but the stage was set for further studies to define the mechanism.

The first step in the synthesis is the pyridoxal phosphate-catalyzed condensation of serine with a fatty acyl-CoA accompanied by the decarboxylation of

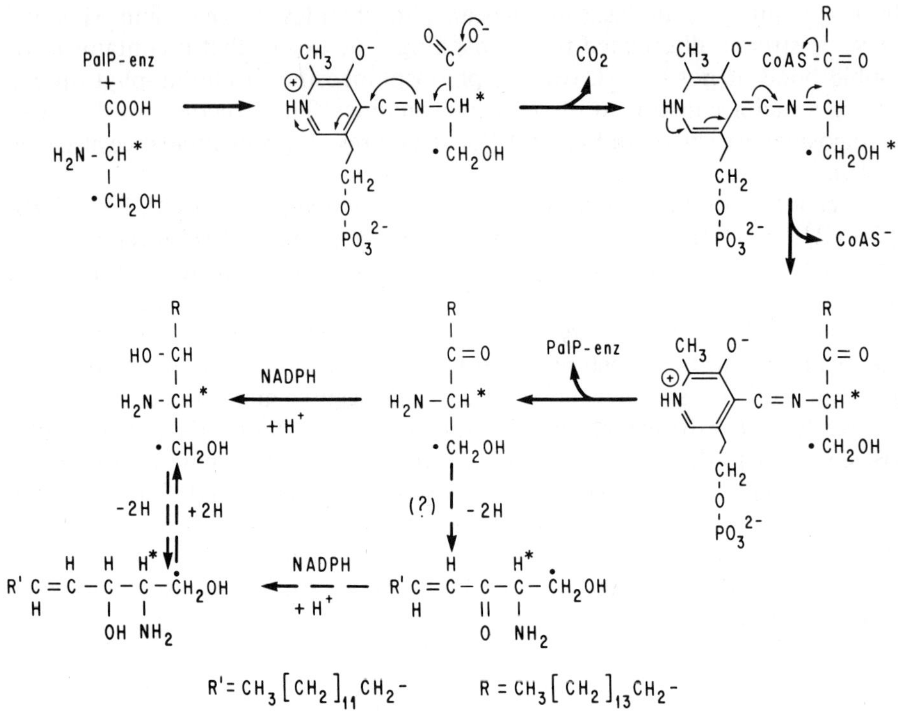

FIGURE 18.2. Possible reaction sequences in the biosynthesis of sphingosine and dihydrosphingosine (after Braun and Snell, 1968).

serine. Presumably the serine is presented to the enzyme as the pyridoxal phosphate Schiff base, but the exact point at which decarboxylation occurs is uncertain. The product of the reaction is 3-ketodihydrosphingosine, reduction of which with NADPH yields dihydrosphingosine. Dehydration of the latter could then yield sphingosine. Braun and Snell (1968) reported that a microsomal fraction of the yeast *Hansenula ciferri* also synthesizes sphingosine from palmitoyl-CoA and serine in a pyridoxal phosphate-dependent reaction. They showed that 3-ketodihydrosphingosine is an intermediate in the process. The mechanism proposed by Braun and Snell (1968), shown in Fig. 18.2, is probably applicable to all biochemical systems synthesizing sphingosine and dihydrosphingosine. The stage of introduction of the double bond into sphingosine is still somewhat uncertain. This could occur either by dehydrogenation of the 3-ketodihydrosphingosine or of dihydrosphingosine itself after reduction of the 3-keto intermediate with NADPH. Braun and Snell's mechanism is compatible with the data of Weiss (1963) who showed that tritium from [α-^{3}H]serine was not lost during synthesis of sphingosine in rats. Hammond and Sweeley (1973) demonstrated the attractive idea that, at least in an oyster microsomal preparation, the dehydrogenation occurs at the keto stage. Nakano and Fujino (1973), on the other hand, have presented evidence for a bizarre reaction in which, in the presence of an isomerase, the ketodihydro compound is converted directly to the final product by an internal oxidation–reduction:

$$\underset{\underset{NH_2}{|}}{CH_3(CH_2)_{14}\overset{\overset{O}{\|}}{C}CH\,CH_2OH} \underset{\longrightarrow}{\overset{\text{oxidoreductase}}{\rightleftharpoons}} CH_3(CH_2)_{12}-CH=CH-\underset{\underset{OH}{|}}{CH}-\underset{\underset{NH_2}{|}}{CH}-CH_2OH \qquad (18.1)$$

At the present time, the exact sequence of reduction and dehydrogenation steps is unclear and, as a matter of fact, several pathways may be followed.

One complicating problem stems from the finding that sphingosine synthesized *in vitro* appears to be rapidly acylated to the ceramide, whereas dihydrosphingosine may react more slowly and can be isolated in the free form. However, Shoyama and Kishimoto (1976) have reported that the initial condensation product may be rapidly incorporated into ceramide and that in this form it can be desaturated and reduced. Thus, free sphingosine may not be an intermediate in the biosynthesis of the sphingolipids. In any event, the compound actually basic to the formation of all complex sphingolipids is ceramide and its biosynthesis is of special interest.

18.3.2. Ceramide (N-Acylsphingosine) Biosynthesis

It is surprising that there is still some uncertainty concerning the mechanism of ceramide formation. The most likely mechanism would seem to be the transfer of the acyl group from CoA to the 2-amino group of sphingosine (Sribney, 1966). Ullman and Radin (1972) have proposed transferases specific for different chain-lengths, with the activity of the individual transferases in each tissue determining to a large extent the acyl composition of its sphingolipids. However, Yavin and Gatt (1969) have reported evidence that the nominally hydrolytic enzyme, ceramidase, has an equilibrium favorable to the synthetic reaction and that a major pathway for ceramide formation is by this mechanism (see below for further discussion of ceramidase).

A third mechanism of ceramide formation has been reported by Kishimoto and Kawamura (1979) to be intimately associated with the alpha hydroxylation system for long-chain fatty acids (see Chapter 7). This system requires a reduced pyridine neucleotide and soluble heat-stable and heat-labile factors, and effects the conversion of lignoceric acid to ceramides and cerebrosides containing both non-hydroxy and hydroxy fatty acyl moieties. Apparently the acyl-CoA is not an intermediate and ceramidase is not involved. Thus, it appears at present that ceramides may be formed in three different ways. It is of interest, in this connection, that despite the high α-hydroxyacyl content of brain cerebrosides, no corresponding ceramides have been found.

18.3.3. Sphingomyelin Biosynthesis

From the close structural analogy of sphingomyelin to phosphatidylcholine, both chemically and functionally (Fig. 18.3, and Chapter 16), it was logical to consider that their biosynthetic pathways might be similar. Indeed, this appeared to be the case when Sribney and Kennedy (1958) found that in the presence of a chicken liver homogenate a ceramide could be converted to sphingomyelin via the CDP-choline reaction:

$$\text{ceramide} + \text{CDP-choline} \xrightarrow{\text{CDP choline:ceramide phosphocholine transferase}} \text{sphingomyelin} + \text{CMP} \quad (18.2)$$

However, difficulties of interpretation were immediately apparent. The ceramide used was the N-acetylsphingosine, a homologue having physical properties quite different from those of the ceramides with much longer fatty acids; moreover, only the derivative of *threo*-sphingosine acted as a sphingomyelin precursor, despite the fact that all naturally occurring sphingolipids are derivatives of the *erythro* isomer. Several other mechanisms have been proposed in attempts to resolve this dilemma.

First, the sequence of steps might be reversed, sphingosylphosphocholine being formed before acylation:

$$\text{sphingosine} + \text{CDP-choline} \rightarrow \text{sphingosylphosphocholine} + \text{CMP} \quad (18.3)$$
$$\text{sphingosylphosphocholine} + \text{acyl-CoA} \rightarrow \text{sphingomyelin} + \text{CoA}$$

These reactions have been shown to occur under *in vitro* conditions, although there remains doubt that this is the normal pathway.

Second, Fujino *et al.* (1968) have reported that most reactions of sphingolipids are nonstereospecific and that, furthermore, in *in vitro* reactions dealing with amphiphilic compounds in which both reactants and products are unlikely to be molecularly dispersed, the presence of detergents can markedly alter their course. Under such conditions, it was found that both the *threo*- and *erythro*-ceramides served efficiently as precursors of sphingomyelin. This subject will be considered in greater detail below.

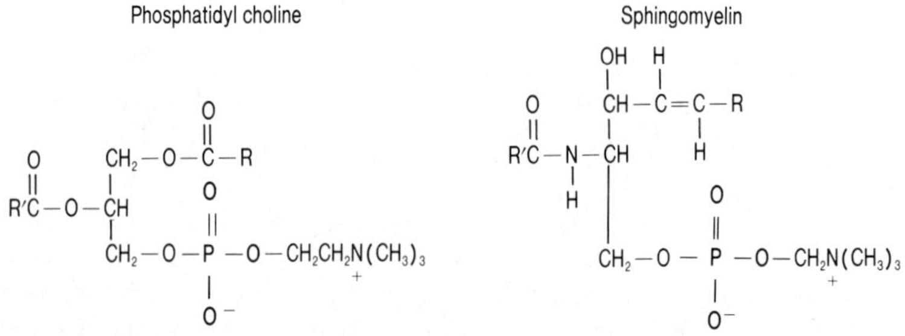

FIGURE 18.3. Structures of (a) phosphatidylcholine and (b) sphingomyelin.

Evidence reported more recently indicates that the major means of formation of sphingomyelin, as demonstrated in mammalian cells in culture, may in fact be by an exchange reaction in which phosphatidylcholine donates its phosphocholine moiety to ceramide (Diringer et al., 1972):

$$\text{PC + ceramide} \rightleftarrows \text{sphingomyelin + diacylglycerol} \qquad (18.4)$$

18.3.4. Cerebroside Biosynthesis

The controversial history of the elucidation of the route of synthesis of cerebrosides resembles that of the sphingomyelins. Part of the problem, in this case, may stem from the occurrence of different cerebrosides that not only have different structures but also, in general, have different functions and, consequently, different locations in the cell. Thus, during brain development, glucosylceramide, the precursor of the gangliosides (see below), is formed at a steady rate, probably in the neurons, in which the gangliosides are an important membrane lipid. Galactosylceramide, on the other hand, is present in only small amounts in the fetal brain, increases rapidly during the synthesis and deposition of myelin (of which it is a major lipid), and remains fairly constant during later life. It must be kept in mind, then, that the two types of cerebrosides might well be expected to show differences in biosynthesis and to respond to different regulatory influences. The high content of very-long-chain, odd-chain, and α-hydroxy fatty acyl groups found in the galacto- (but not gluco-) cerebrosides also suggests different origins. It is readily understandable that such fatty acids might require mechanisms of activation different from those of less unusually constituted acids. Finally, along with the complications of physical state possibly influencing metabolism of all the sphingolipids, there is the possibility of two equally probable routes to the final product. Brady (1962) found that a brain microsomal fraction could use UDP-galactose in elaborating galactosyl cerebroside, presumably by reaction with endogenous ceramide. Morell et al. (1970), however, reported that intracerebrally injected ceramide was generally hydrolyzed rather than incorporated into cerebroside. They reported, however, that with the microsomal fraction from brains of young mice, hydroxy fatty acyl ceramide would react with UDP-galactose (but not UDP-glucose). When several investigators (see Hammarström, 1971) reported that sphingosine reacts with UDP-galactose to give psychosine (galactosylsphingosine) and Brady and coworkers (1965) showed that psychosine could be converted to cerebroside with an acyl coenzyme A, all possible routes of biosynthesis had been shown to be possible:

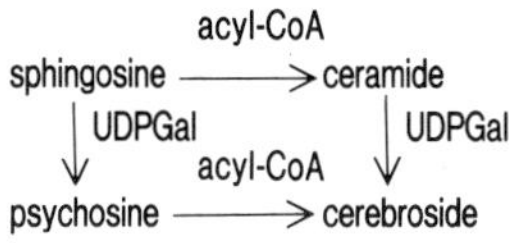

$$\begin{array}{c} \text{acyl-CoA} \\ \text{sphingosine} \longrightarrow \text{ceramide} \\ \Big\downarrow \text{UDPGal} \qquad \Big\downarrow \text{UDPGal} \\ \text{acyl-CoA} \\ \text{psychosine} \longrightarrow \text{cerebroside} \end{array} \qquad (18.5)$$

Again, the confusion may stem from physical state, since inclusion of phosphatidylcholine in the microsomal preparation permitted the formation of cerebroside from both unsubstituted and hydroxyacyl ceramide but not from psychosine. Basu *et al.* (1973) concluded that UDP-glucose transferase is active only in very young brains and declines with development, whereas UDP-galactose transferase is maximal during myelination, and, furthermore, that separate enzymes are responsible for the formation of the unsubstituted and hydroxyacyl cerebrosides. The position of psychosine in the scheme is still unclear.

However, as in many cases of reactions of sphingolipids, the results of studies designed to determine the actual route of biosynthesis may point in opposite directions. For example, specific UDP-galactosyl transferases have been reported for both hydroxyacyl and non-hydroxyacyl ceramides, but careful search has failed to reveal the presence of hydroxyacyl ceramides in brain. Kishimoto and Kawamura (1979) found that the mitochondrial system for hydroxylation of lignoceric acid (24:0) resulted in the production of cerebronylceramide and cerebroside (phrenosine) but no free cerebronic acid (2R-hydroxy-24:0). It therefore seems probable that hydroxylation is the rate-limiting step in hydroxyacyl cerebroside formation and that the ceramide formed from hydroxy acids is rapidly converted to cerebroside without release from the enzyme system.

A similar situation exists with the C_{20} sphingosine. Although it contributes about 70% of the sphingosine base of gangliosides, no C_{20} homologue is found in ceramides. Once again, it is possible that the C_{20} sphingosine, once formed, is converted to ceramide, which is then very rapidly glucosylated and incorporated into the neuronal membrane.

18.3.5. Cerebroside Sulfate Biosynthesis

Cerebroside sulfate, or galactosylceramide-3′-sulfate, is a typical myelin lipid, which is present in only trace amounts before myelination and is then synthesized in the endoplasmic reticulum of the glial cells and transferred to myelin, after which it appears to have a very low turnover rate in the mature brain. The biosynthetic reaction involves the transfer of sulfate from 3′-phosphoadenosine-5′-phosphosulfate (PAPS) to the 3′ position of the galactose moiety of the cerebroside (Cumar *et al.,* 1968) (Fig. 18.4).

It was previously thought that cerebroside sulfate, because of its increased polarity, might be more readily transported than cerebroside from the site of synthesis in the endoplasmic reticulum to myelin. However, it now appears that both cerebroside and cerebroside sulfate are derived from the endoplasmic reticulum or Golgi apparatus of the glial cells and that they are transported from these sites to those actively forming and maturing myelin. Just before myelination (eight days in the rat), cerebroside synthesis increases markedly and the concentration in the cytosol, probably bound to a specific carrier protein, increases more than tenfold. At the onset of myelination (after nine days in the rat) this concentration drops precipitously as myelination proceeds.

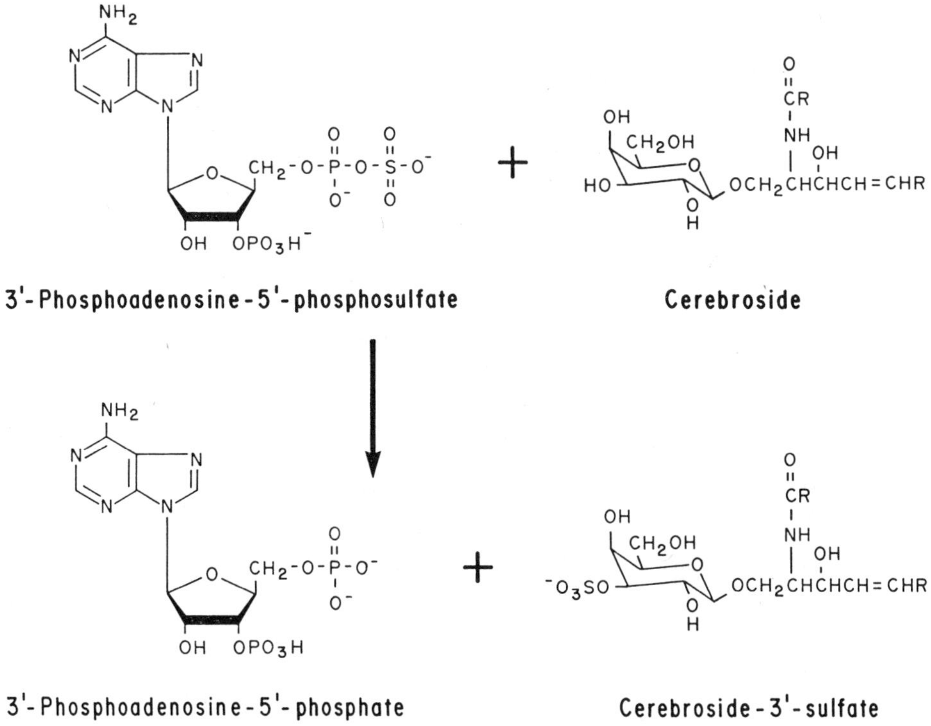

FIGURE 18.4. Biosynthesis of cerebroside sulfate.

18.3.6. Biosynthesis of the Complex Ceramide Hexosides

The routes of biosynthesis of the more complex sphingolipids—the gangliosides, hematosides, and globosides—starting with a ceramide glucoside, or, in rare cases, galactoside, proceed by pathways familiar to the carbohydrate chemist. Unlike the case of the myelin lipids, biosynthesis of the gangliosides starts with the earliest brain development and continues slowly to maturity with an increased rate during myelination. As had been reported by Morell *et al.* (1970), galactosylceramide injected intracerebrally in rats does not serve as a ganglioside precursor but is simply hydrolyzed to its precursors. Newly synthesized glucosylceramide, therefore, which is found in only small amounts in brain, must serve as the precursor of these complex sphingolipids (Nishimura and Yamakawa, 1968). It is of interest also that the brain cerebrosides (galactosides) have a complement of fatty acids quite different from that of the gangliosides, being longer-chain and containing a higher proportion of hydroxy and odd-chain acids. Thus, it appears that the precursor of the cerebrosides is a ceramide with acyl groups about 24 carbons in length and that this precursor is converted largely to the galactoside in the endoplasmic reticulum of the glial cells. The precursor of the gan-

1) Cer + UDPGlc → Glc(β1→1)Cer + UDP
2) UDPGlc ⇌ UDPGal
3) GLc(1→1)Cer + UDPGal → Gal(β1→4)Glc(β1→1)Cer
4) UDPGlcNAc ⇌ UDPGalNAc
5) Gal(β1→4)Glc(β1→1)Cer + UDPGalNAc → UDP + GalNAc(β1→4)Gal(β1→4)Glc(β1→1)Cer
6a) (Rat brain)
   GalNAc(1→4)Gal(1→4)Glc(1→1)Cer + UDPGal →
   UDP + Gal(1→3)GalNAc(1→4)Gal(1→4)Glc(1→1)Cer
         or
6b) (Frog brain)
   GalNAc(1→4)Gal(1→4)Glc(1→1)Cer + CMPNANA →
   CMP + GalNAc(1→4)Gal(1→4)Glc(1→1)Cer
                    3
                    ↑
                    2
                   NANA
7) GalNAc(1→4)Gal(1→4)Glc(1→1)Cer + UDPGal →
                    3
                    ↑
                    2
                   NANA
   UDP + Gal(1→3)GalNAc(1→4)Gal(1→4)Glc(1→1)Cer
                    3
                    ↑
                    2
                   NANA
         or
8) Gal(1→3)GalNAc(1→4)Gal(1→4)Glc(1→1)Cer + CMPNANA →
   Gal(1→3)GalNAc(1→4)Gal(1→4)Glc(1→1)Cer
                    3
                    ↑
                    2
                   NANA

Other sialic acid residues can be added via the CMPNANA derivatives.

FIGURE 18.5. Biosynthesis of gangliosides.

gliosides, on the other hand, is a ceramide with shorter acyl chains (C_{16-18}) that is converted to the glucoside.

The pathway to the higher ceramide hexosides is similar to those leading to other oligosaccharides. Thus the sugars are transferred from their UDP derivatives, and the N-acetylneuraminic acid (NANA) is transferred from the CMP derivative (see Fig. 18.5). At the globoside stage (no NANA), two alternate paths are possible and occur in different tissues. On the one hand, in rat brain (Fig. 18.5, step 6a) an additional galactose can be added to give a more complex globoside. On the other hand, in frog brain (Fig. 18.5, step 6b), NANA may be added to give the ganglioside G_{M2}. Both compounds can undergo alternative additions to give ganglioside G_{M1a}. At this stage additional sialic acid residues can be added to yield the higher sialogangliosides (see Handa and Burton, 1969).

The N-acetylneuraminic acid incorporated into gangliosides of neurones and of cell surfaces in general has a structure (see Chapter 16) indicating its derivation from mannose (C4–C9) and pyruvate (C1–C3). Other sialic acids have almost identical structure but with the *N*-acetyl replaced by *N*-glycolyl.

At this point, it is pertinent to consider a naming system for the gangliosides. A complete chemical name would be entirely too cumbersome and several shorthand methods have been introduced.

The most generally used system is that of Svennerholm (1970) and it will be used in this text. In this system, G refers to gangliosides. The numbers of sialic acid moieties are designated by subscripts M (mono), D (di), T (tri), Q (quadri), and P (penta). The integrity of the polysaccharide chain is indicated by a number—1 for an intact chain (see G_{D1a}, Fig. 18.6), 2 for the loss of galactose, and 3 for the additional loss of galactosamine (in which case, the product is technically no longer a ganglioside, but a hematoside, lactosyl ceramide). Finally, the number of sialic acid residues on the inner galactose is designated by a (1), b (2), or c (3). Thus, the Tay-Sachs ganglioside

$$\begin{array}{c} \text{GalNAc}(1\rightarrow 4)\text{Gal}(1\rightarrow 4)\text{Glc}(1\rightarrow 1)\text{Cer} \\ 3 \\ \uparrow \\ 2 \\ \text{NANA} \end{array}$$

would be G_{M2}. (Since only the inner galactose residue remains, there is no need to designate the location of the NANA.)

The covalent structures and spatial arrangement of the major gangliosides are depicted in Fig. 18.6, illustrating also the naming system for these compounds.

18.4. Degradation of the Sphingolipids

The enzymatic degradation of the sphingolipids appears to be of particular importance to the body, in part because in many cases these lipids cannot be tolerated in high concentration, as is emphasized below, and a number of familial diseases are associated with their excessive accumulation. Even the very long-chain fatty acids associated with certain sphingolipids may be damaging and special means for their disposal have evolved (see Chapter 7). In general the degradation of the complex sphingolipids appears to be a function of the lysosomes, since the enzymes involved have in most cases been shown to be associated with these particles and to have low pH optima. The enzymes have been studied in whole animals, in various tissue preparations (in some cases in purified form), and in cells in culture. The last method is of particular interest, since in many cases the diseases characterized by deficiency of these lysosomal enzymes can be diagnosed in cells cultured from various tissues, such as leukocytes or fetal cells from amniotic fluid.

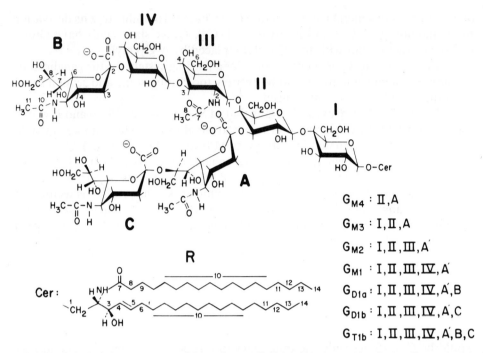

FIGURE 18.6. Covalent structures of the major gangliosides. Shown complete is the oligosaccharide of the most complex member of the series, G_{T1b}, as indicated in the figure. All the gangliosides contain the ceramide portion, residue R. The carbon atoms of the hexopyranoside residues (I, II, III, and IV) are numbered as in residue III. The carbons of the sialic acids (A, B, and C) are numbered as in residue C (from Sillerud et al., 1982).

18.4.1. Ganglioside Degradation

The first steps in the degradation of the higher gangliosides (see Fig. 18.7) involve stepwise removal of all but one of the sialic acid residues, the order of removal determining the particular ganglioside remaining. The sialic acid on the inner galactose moiety resists the action of sialidase until after the removal of the two outermost hexose residues (galactose and galactosamine). Thus, although two monosialogangliosides are theoretically possible, G_{M1a} is the major product resulting from the action of sialidase on the higher sialogangliosides.

Although no disease appears to be associated with lack of a sialidase, it is of interest that certain transformed cells, which have lost contact inhibition, are deficient in sialylated gangliosides, resulting both from a deficiency of the synthetic path and an increase in sialidase activity (Schengrund et al., 1973). It is interesting to speculate that the malignant transformation and the accompanying loss of contact inhibition are in large part properties of the cell surface and that reversal of these changes might be accomplished by replacement of the surface sialic acid residues. The remainder of the degradative steps are accomplished by a β-galactosidase, a galactosaminidase, a sialidase, and a galactosidase, resulting finally in

SPHINGOLIPID METABOLISM

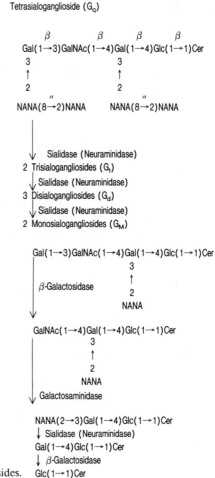

FIGURE 18.7. Enzymatic degradation of the gangliosides.

the formation of a cerebroside. Each of these enzymes, with the exception of the sialidase, has been shown to be missing in a specific genetic disease resulting, in each case, in the accumulation of the lipid normally specifically attacked by the missing enzyme (see below).

18.4.2. Degradation of Sphingolipids of Smaller Molecular Weight

Like the gangliosides, the less complex sphingolipids are also involved in familial diseases and their accumulation has usually been shown to result from deficiency of one or more degradative enzymes; these enzymes have therefore been studied fairly intensively. In the case of cerebroside sulfate, the disease metachromatic leukodystrophy is characterized by low levels of sulfatase and thus by increased proportions of cerebroside sulfate. Myelin deficiency in this disease appears to implicate cerebroside sulfate as an essential factor in the process of myelination.

Cerebroside accumulates in Gaucher's disease, in which specific gluco- or galactosidases are absent from the spleen and other organs.

Sphingomyelin is degraded by a sphingomyelinase that is fairly widespread, occurring in intestine, liver, spleen, etc. Its absence is responsible for Niemann-Pick disease. It may also be missing from the atherosclerotic aorta since sphingomyelin accumulates in the atherosclerotic plaques, although other explanations for this accumulation may be equally possible.

The ceramidase isolated from beef brain by Yavin and Gatt (1969) was claimed by these authors to be active in the synthetic as well as the degradative direction. In a study of this enzyme, Gatt and his coworkers noted that the equilibrium constant for the reaction:

$$\text{ceramide} + H_2O \underset{K_{eq} = 2 \times 10^{-4}}{\overset{K_{eq} = 6 \times 10^{-6}}{\rightleftharpoons}} \text{sphingosine} + \text{fatty acid} \qquad (18.6)$$

was different for the forward and back reactions. This led to a study of the kinetics of reactions involving amphiphilic reactants and products, particularly in the case of compounds such as the simpler sphingolipids, in which the hydrophobic moieties of these molecules are particularly large and hence influential. His conclusions in this regard will be discussed below.

18.4.3. Degradation of Sphingosine

Sphingosine degradation occurs in many tissues in both mitochondria and endoplasmic reticulum. It is preceded by phosphorylation and then appears to resemble a reverse aldolase type of reaction giving ethanolamine phosphate and palmitaldehyde, which is oxidized to palmitic acid by aldehyde dehydrogenase (Stoffel et al., 1969). For dihydrosphingosine, the reactions are given in Eq. 18.7.

An interesting reaction occurs in the case of phytosphingosine for which these same reactions result in the formation of α-hydroxypalmitic acid (Barenholz and Gatt, 1968). Further degradation of this acid by an alpha oxidation system results in the formation of pentadecanoic acid as the major product.

It has been reported that the products of this degradation are rapidly used in synthetic processes. The aldehyde, if not oxidized to the acid, serves efficiently as a source of the alkenyl ether moiety of plasmalogens (perhaps after reduction to the alcohol), while the ethanolamine phosphate is incorporated into the polar portions of PE, PC, and sphingomyelin. (Note that these reactions represent another pathway of conversion of serine to ethanolamine.)

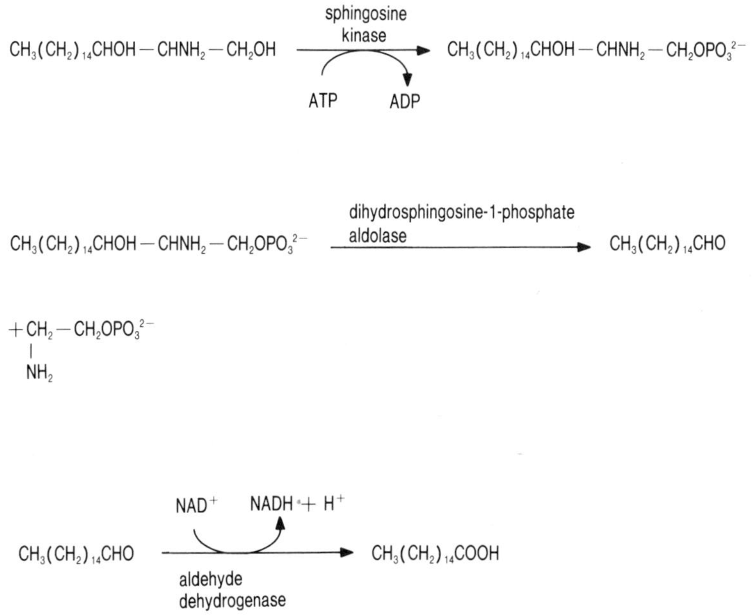

$$CH_3(CH_2)_{14}CHOH-CHNH_2-CH_2OH \xrightarrow[ATP \quad ADP]{\text{sphingosine kinase}} CH_3(CH_2)_{14}CHOH-CHNH_2-CH_2OPO_3^{2-}$$

$$CH_3(CH_2)_{14}CHOH-CHNH_2-CH_2OPO_3^{2-} \xrightarrow{\text{dihydrosphingosine-1-phosphate aldolase}} CH_3(CH_2)_{14}CHO$$

$$+ \begin{array}{c} CH_2-CH_2OPO_3^{2-} \\ | \\ NH_2 \end{array}$$

$$CH_3(CH_2)_{14}CHO \xrightarrow[\text{aldehyde dehydrogenase}]{NAD^+ \quad NADH + H^+} CH_3(CH_2)_{14}COOH \qquad (18.7)$$

18.5. The Enzymes of Lipid Metabolism

The significance of findings in *in vitro* studies of the enzymes of lipid metabolism—particularly, perhaps, in those involved in sphingolipid metabolism—are often clouded by lack of knowledge and concern over possible effects of the *in vivo* physical state of enzymes and/or substrates. For most such enzymes the substrate lipids occur in lipid bilayers of various membranes and the enzymes may have to be incorporated into the membrane for full activity. In other cases, action occurs only after disruption of the membranes and thus separated fragments of bilayers, vesicles, or micelles may be involved. Clearly, classical enzyme kinetics may not be applicable to many of these reactions. Gatt and his coworkers were led by their studies of the enzymes of sphingolipid metabolism to consider the ways in which the kinetics of the reactions of lipid metabolism might differ from those involving soluble enzymes and substrates.

In general, for most first-order enzymatic reactions involving soluble enzymes and substrates, a plot of v_0, initial reaction rate, against [S], substrate concentration, is a rectangular hyperbola, while that of v_0^{-1} against $[S]^{-1}$ is a straight line. Deviations from these characteristics may often be interpreted in terms of invalidity of certain assumptions made in deriving the expressions for first-order enzyme kinetics.

Amphiphiles in aqueous media are typically monodispersed only at low concentrations and tend to aggregate at higher concentrations; beyond some "critical micellar concentration" (CMC), characteristic of the particular amphiphile

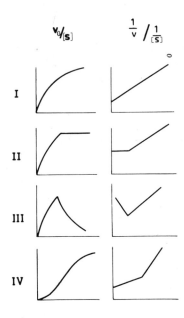

FIGURE 18.8. Classification of the $v_0/[S]$ and $v_0^{-1}/[S]^{-1}$ curves as related to four types of interactions of enzymes with soluble amphiphilic molecules (from Gatt et al., 1972).

involved, the monomer concentration remains constant, while that of the aggregate increases. In such systems the relationship of the concentration of the actual substrate may therefore differ from the formal concentration of added substrate.

Gatt has recognized four idealized types of kinetics which are represented by the plots in Fig. 18.8 and are discussed briefly below. In each case, examples are presented from commonly encountered reactions (Gatt et al., 1972).

Type I represents classical Michaelis–Menten kinetics in which the substrates are monomers and substrate concentrations studied are all less than the CMC, or if the substrates are aggregates of approximately identical structures (e.g., spherical micelles), the CMC is very low and the substrate concentration range does not extend into regions of formation of higher aggregates with different enzyme affinities.

Type II starts as a typical rectangular hyperbola but, at a certain concentration, breaks to become parallel to the abscissa. Such a curve may mean that the enzyme acts on monomers (but not on aggregates), the concentration of which is constant beyond the CMC.

Type III curves start as normal hyperbolas but quickly reach a peak and fall back toward the baseline. This represents the case in which the enzyme uses the monomeric forms but not the aggregates and is inhibited either by high concentration of monodisperse or aggregated substrate. Inhibition of enzymes by amphiphilic compounds (e.g., fatty acids or acyl-CoAs) is a common phenomenon. The inhibition can usually be reversed (as expected) by increasing enzyme protein; however, the fact that other proteins, such as albumin or boiled microsomal protein, produce similar effects suggests that amphiphile–protein interactions of low specificity are involved. In the presence of such protein the curves are converted toward Type I.

Type IV curves result when the enzyme acts preferentially or exclusively on aggregates rather than monomers. Again, addition of albumin alters the v_0 vs [S] curves, which become more like Type I.

This short discussion represents, of course, simplification of complex phenomena. The interested reader is therefore urged to refer to the original discussion of Gatt and his coworkers and to more recent papers on the subject. In these papers, examples of each type are given and deviations from the ideal are rationalized.

18.6. Disorders of Sphingolipid Metabolism

A host of inherited diseases, "inborn errors of metabolism," are associated with the metabolism of sphingolipids. Some of these (Tay-Sachs disease, Gaucher's disease) were recognized as distinct clinical entities a century ago and all were established during the first half of our century as being lipid storage diseases, but it was not until relatively recent times—within the last two decades—that it became clear that, with the exception of a single case of a recently discovered disease, all resulted from the deficiency of a lysosomal hydrolytic enzyme specific for a particular step in the degradation of the sphingolipids. The one exception is a G_{M3} (hematoside) sphingolipodystrophy in which no higher homologues of gangliosides were found and hence it must be assumed that in this case ganglioside synthesis was arrested at the G_{M3} stage.

It is not our intention to describe these abnormalities in detail but only to call attention to their existence and to indicate the specific enzyme deficiency which is their cause. The most detailed description of all these abnormalities, and associated clinical and biochemical defects, may be found in the various editions of the great book by Stanbury and his colleagues, *The Metabolic Basis of Inherited Disease* (Stanbury *et al.,* 1978; 1983). Most of the disorders of sphingolipid metabolism are inherited in an autosomal recessive mode, which means that one pair of abnormal alleles is needed for the disease to manifest itself and also that the parents, heterozygous for a particular trait (e.g., Tay–Sachs disease), usually have no manifestation of the disease, although their heterozygous "carrier" state can be determined by laboratory tests. It follows from the Mendelian law of inheritance that—by probability—one out of four of the offspring of a couple heterozygous for an enzyme deficiency will be homozygous for the abnormality and may have clinical manifestations. One abnormality among the sphingolipidoses, Fabry's disease, is an exception in its mode of inheritance as it is X-linked, and thus the male hemizygote has the full clinical manifestations of the disease; a female heterozygote may be completely free of any manifestations, although heterozygous females with all manifestations of the disease as seen in a male hemizygote have been reported (Desnick *et al.,* 1978).

The ultimate hydrolytic product of sphingolipids is a ceramide, and the substituents are removed from the ceramide one by one by specific hydrolytic

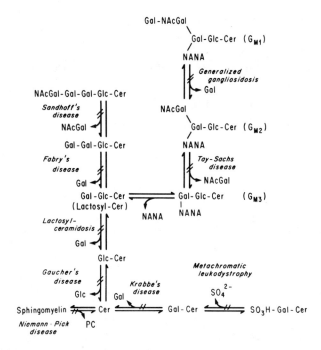

FIGURE 18.9. Relationships of lysosomal enzymes to lipid storage diseases.

enzymes in the reverse order to that in which they were added. This may be best appreciated from the diagram of Fig. 18.9 which summarizes in a flow sheet the known abnormalities of sphingolipid metabolism.

18.6.1. Sphingomyelin Lipidosis, or Niemann-Pick Disease

Sphingomyelin is normally hydrolyzed to ceramide and phosphocholine by sphingomyelinase present in all organs, although the liver is the richest source of this enzyme, but sphingomyelinase can be assayed also in disrupted leukocytes or cultured fibroblasts. The enzyme can be assayed most simply by the use of [^{14}C]sphingomyelin, labeled in the choline moiety, as substrate. The phospho[^{14}C]choline liberated in the enzymatic hydrolysis is water-soluble and thus easily separated from the unhydrolyzed substrate, which is soluble only in organic solvents. A more recently introduced substrate, an analogue of sphingomyelin, 2-hexadecanoylamino-4-nitrophenylphosphorylcholine yields, on hydrolysis by sphingomyelinase, the yellow-colored 4-nitrophenylhexadecanilide and thus allows a spectrophotometric assay of the enzyme.

Absence of sphingomyelinase or its severe deficiency is found in Niemann-Pick disease and leads to accumulation of sphingomyelin in many sites of the body, mainly in the liver, spleen, bone marrow, and lung and in foam-cells in the cerebral cortex. The lipid storage in the liver and spleen results in gross enlargement of these organs, and is referred to as hepatosplenomegaly. The sphingomyelin that accumulates in the various organs, particularly in liver and spleen, need not have originated from those organs. Sphingomyelin is a constituent of all membranes. Therefore turnover of all cells can generate free sphingomyelin and thus, in the absence of sphingomyelinase, the lipid can be transported by the blood and deposited in organs that may have special affinity for it.

Although the disease, inherited in an autosomal recessive mode, occurs in all ethnic groups, there is a prevalence of the most severe and infantile form with involvement of the central nervous system among Ashkenazi Jews. The frequency of the heterozygous state is uncertain, but it has been estimated to be about 1:100 in individuals of Ashkenazic Jewish ancestry. It is now possible to identify heterozygotes by the assay of sphingomyelinase in either fresh disrupted leukocytes or in extracts of cultured fibroblasts by the use of either of the substrates mentioned. In addition, prenatal diagnosis of the disease is possible by the assay for sphingomyelinase in cultured amniotic cells obtained by amniocentesis before the twentieth week of gestation (see Brady, 1978a) (Table 18.1).

18.6.2. Glucosylceramide Lipidosis: Gaucher's Disease

This is one of the commonest abnormalities of sphingolipid metabolism, first described in 1882 by Philip Gaucher after whom the disease is named. It is characterized by hepatosplenomegaly and by the accumulation of glucosylceramide in the reticuloendothelial system, in liver, spleen, lymph nodes, and bone marrow. The blood and aspirates of bone marrow contain highly characteristic "Gaucher cells," 20–100 μm in diameter, with an eccentric nucleus and filamentous inclusion bodies in the cytoplasm, which, when viewed under the phase-contrast microscope, have the appearance of balls of crinkled tissue paper.

In 1966, Brady showed that Gaucher's disease was associated with deficiency of glucocerebrosidase, a β-glucosidase, the enzyme that cleaves glucosylceramide into glucose and ceramide. This hydrolytic enzyme is present in all tissues, but

TABLE 18.1. Sphingomyelinase Activity in Extracts of Fibroblasts from Normal Individuals, Heterozygotes, and Individuals with Type A Niemann-Pick Disease Assayed with [^{14}C]Sphingomyelin ([^{14}C]Sph) or 2-Hexadecanoylamino-4-nitrophenyl phosphocholine (HNP) as Substrates

	Substrate	
	[^{14}C]Sph	HNP
	Substrate hydrolyzed (nmol/hr/mg protein)[a]	
Normal controls ($n = 15$)	72 ± 20	219 ± 70
Type A heterozygotes ($n = 7$)	36 ± 10	93 ± 17
Type A homozygotes ($n = 7$)	1.0 ± 0.4	0

[a]Values and standard deviations calculated from data of Table 35-1 of Brady (1978a).

the spleen is its richest source; it is also present in human placenta. Because several acid β-glucosidases exist that can hydrolyze artificial substrates, such as umbelliferyl-β-D-glucopyranoside or 2-hexadecanoylamino-4-nitrophenyl-β-D-glucopyranoside, but not the natural substrate, the most suitable substrate for the assay of glucocerebrosidase is [^{14}C]glucosylceramide. Homozygotes of Gaucher's disease, and heterozygous carriers of the trait, can be identified by measuring the glucocerebrosidase activity in washed leukocytes obtained from venous blood or in cultured skin fibroblasts. Cultured amniotic cells of normal fetuses contain the glucocerebrosidase at high level, but the level of this enzyme in the cultured amniotic cells of a fetus homozygous for Gaucher's disease was only about 4% of normal (see Brady, 1978b).

With reference to the flow sheet of Fig. 18.9, it may be seen that there are many sources of glucosylceramide. It is believed that most of the glucosylceramide that accumulates in Gaucher's disease is derived from the breakdown of erythrocytes and leukocytes. Erythrocytes contain not only di- and trihexosylceramides, but also gangliosides, the main ganglioside being NANA (2→3) Gal-β(1→4)N-Ac-galactosaminyl-β(1→3) Gal-β(1→4)-Glc-ceramide, identical with a ganglioside found also in spleen. The asialo form of this ganglioside is a major constituent of human polymorphonuclear leukocytes (Wherrett, 1973). Thus the turnover of erythrocytes and leukocytes with the concomitant hydrolysis of the above substances provides the glucosylceramide that accumulates in Gaucher's disease.

18.6.3. Gangliosidoses

There are five main types of gangliosidoses, all inherited in an autosomal recessive mode, and associated with neuronal accumulation in the central nervous system of G_{M2} and G_{M1} gangliosides and related glycolipids. They all result from the absence or severe deficiency of specific lysosomal hydrolytic enzymes. G_{M3} gangliosidosis, referred to at the beginning of Section 18.6 (Max et al., 1974) and described so far only in one case, may be the consequence of the deficiency of N-acetylgalactosamine transferase, as no higher homologues of gangliosides could be identified in the tissues of the infant who died at 3½ months of age. O'Brien (1978) suggests that this abnormality—so far unique—should be called N-acetylgalactosamine transferase deficiency rather than a gangliosidosis, because the term gangliosidosis has generally been used to denote conditions of ganglioside storage that result from an impaired breakdown. The G_{M3} sialidase activity in the liver and brain of the infant described by Max et al. (1974) was normal, whereas the G_{M3}:UDP-N-acetylgalactosaminyl transferase activity was only about 10% of normal.

18.6.4. Tay-Sachs Disease

By far the commonest and most important abnormality among the sphingolipidoses is Tay-Sachs disease, or G_{M2} gangliosidosis. It was first described in

1881 by Waren Tay, a British ophthalmologist who observed the "cherry red spot" as a macular degeneration in the eye of an infant with marked weakness of trunk and limbs. Then in 1887 Sachs, an American neurologist, reported similar cases of infants with blindness and dementia and called the disease "amaurotic familial idiocy." It was the study of the composition of the brain of a child with Tay–Sachs disease that led to the discovery of gangliosides.

The degradation of G_{M2} ganglioside to G_{M3} requires the cleavage of N-acetylgalactosamine from G_{M2}. Robinson and Stirling (1968) first recognized the existence in human spleen of two hexosaminidases A and B, which have both β-N-acetylgalactosaminidase and β-N-acetylglucosaminidase activities. Both enzymes are present in all human tissues and cells—except erythrocytes—in leukocytes, cultured fibroblasts and amniotic cells, blood serum, urine, amniotic fluid, and even in tears. Hexosaminidase A (Hex A) is somewhat more acidic than Hex B and also more labile to heat. Hex A has an M_r of 100,000 and one free SH-group, and is composed of two pairs of identical subunits, $\alpha_2\beta_2$. Each α-subunit also carries one sialic acid residue, whereas the β-subunit carries none. The α-subunits are joined by one disulfide bridge, and the β-subunits contain a total of five inter- and intra-subunit disulfide bridges. Hex B consists of two pairs of β-subunits, identical with the β-subunit of Hex A; thus structurally Hex B is also a tetramer, $\beta_2\beta_2$. The two hexosaminidases can be readily separated from one another by ion-exchange chromatography and by starch–gel electrophoresis, and can also be distinguished from one another by the easy heat inactivation of Hex A. They can be assayed with the artificial substrates p-nitrophenyl- or 4-methylumbelliferyl-β-D-N-acetylglucosamine, or β-D-N-acetylgalactosamine (Geiger and Arnon, 1976). Tay–Sachs disease results from the absence or great deficiency of Hex A. In contrast the levels of Hex B are greatly elevated in the tissues in the disease. As a result of the enzyme deficiency, G_{M2} ganglioside accumulates in greatly swollen neurons which, under the electron microscope, show large lamellar whorls of inclusion bodies.

Tay–Sachs disease is inherited in the autosomal recessive mode. Over 90% of the cases occur among Jews of Ashkenazi descent whose ancestors lived in the nineteenth century in the Lithuanian and Polish provinces of Korno and Grodno. Because of the autosomal recessive mode of inheritance, detection of heterozygotes has become most important. Heterozygotes can be detected by assaying for Hex A in serum, white blood cells, or cultured fibroblasts. In New York City the carrier rate among Jews is estimated to be about 1:30 as compared to 1:300 among non-Jewish Americans. Prenatal diagnosis of the homozygous state is possible with great certainty by assay of Hex A in cultured amniotic cells.

18.6.5. Sandhoff's Disease

Sandhoff's disease is a variant of Tay–Sachs disease. Sandhoff's disease results from the deficiency of both Hex A and Hex B and leads to the accumulation of both G_{M2} and of asialo-G_{M2} (N-AcGal-Gal-Gal-Glc-ceramide).

The genetics of Tay–Sachs and Sandhoff's disease can be readily understood on the assumption of the existence of two alleles, one coding for the α-subunit and the other for the β-subunit of the two hexosaminidases. Since the β-subunit in Hex B is identical with that of Hex A, mutation of the allele for the β-subunit will cause abnormality of both Hex A and Hex B (Sandhoff's disease). Mutation of the allele for the α-subunit affects, of course, only Hex A and is the fundamental cause of Tay-Sachs disease.

18.6.6. G_{M1} Gangliosidosis (Generalized Gangliosidosis)

G_{M1} gangliosidosis, both its infantile and juvenile forms (known also as Norman-Landing disease and Derry's disease, respectively), is a relatively rare panethnic disease inherited in the autosomal recessive manner (see O'Brien, 1978). They are the consequence of severe deficiency of an acid β-galactosidase (β-galactosidase A) activity in brain, liver, or cultured fibroblasts, less than 0.1% of normal when assayed with G_{M1} ganglioside as substrate. The nature of the mutation is an altered structural gene of β-galactosidase A as the liver of afflicted individuals contains normal or more than normal amounts of a protein that cross-reacts with anti-β-galactosidase A antibodies. β-Galactosidase A has an M_r of 72,000, and pH optimum of 4.35. It is a sialoglycoprotein. The pure enzyme can cleave off galactose not only from $\beta(1\rightarrow3)$ linkages, as in G_{M1} ganglioside, but also from other $\beta(1\rightarrow4)$ linkages, as in some glycoproteins (asialofetuin), or in lactose. Oddly enough, the enzyme is inactive with lactosylceramide and galactosylceramide, but hydrolyzes a number of artificial substrates, such as 4-methylumbelliferyl-β-D-galactoside, phenyl-β-D-galactoside, p-nitrophenyl-, o-nitrophenyl-, and m-nitrophenyl-β-D-galactoside, and even 4-methylumbelliferyl-α-L-arabinoside.

A second enzyme, β-galactosidase B, which has a pH optimum of 4.2, is also deficient in G_{M1} gangliosidosis. The molecular weight of this enzyme is about ten times higher than that of β-galactosidase A, but it is also a sialoglycoprotein. It is uncertain whether β-galactosidase B is simply a polymer of enzyme A; however, it must at least share subunits with enzyme A as antibodies raised against the pure enzyme A quantitatively precipitate also enzyme B. The relationship between these two β-galactosidases may be similar to that between hexosaminidase A and B (see Tay-Sachs disease and Sandhoff's disease).

Given the fact that β-galactosidase A has a wide substrate specificity, there is a possibility of accumulation in cells of substances other than G_{M1} in deficiency of β-galactosidase A.

18.6.7. Galactosylceramide Lipidosis: Globoid Cell Leukodystrophy: Krabbe's Disease

The names of this rare disease, inherited in an autosomal recessive manner, describe some of its essential features: the accumulation of galactosylceramide in certain cells in the central nervous system, typically in large, often multinucleated "globoid" cells arranged in clusters in the white matter associated with poor myelination and destruction of myelin. The condition was first fully described by the Danish physician Krabbe (1916) in two pairs of sibs and a fifth infant, all of whom died before two years of age.

The disease, which manifests itself in early infancy, occurs among many nations and is associated with hyperirritability, stiffness of limbs, fevers of unknown origin, regression of psychomotor development, convulsions, optic atrophy, and early death (see Suzuki and Suzuki, 1978).

Among the organs the brain is most severely affected. It is small with shrunken gyri and wide sulci; the white matter in the brain is very deficient, has a grayish color, and is of rubbery consistency. On histological examination the white matter appears largely replaced by extensive gliosis and clusters of globoid cells. On electron microscopic examination the globoid cells are seen to contain hollow tubular inclusion bodies which in some areas appear as twisted tubules. These inclusion bodies consist of bilayers of galactosylceramide. A similar appearance of globoid cells can be produced experimentally by the intracerebral injection of galactosylceramide to rats.

The disease is attributed to severe deficiency of a specific galactosylceramide β-galactosidase in the gray and white matter of brain, and in the liver and spleen (see Suzuki and Suzuki, 1978). The activity of this specific β-galactosidase in Krabbe's disease was found to be only 5–10% of normal. Galactosylceramide β-galactosidase is another of the lysosomal enzymes. It is pertinent to mention that the deficiency can be demonstrated only by the use of galactosylceramide as substrate in assays with tissue homogenates. When an artificial substrate, such as *p*-nitrophenyl β-galactoside is used, no deficiency of β-galactosidase is apparent, indicating that the galactosylceramide β-galactosidase is a highly specific enzyme. Suzuki and Suzuki (1978) point out that this observation " ... is in contrast to G_{M1}-gangliosidosis, in which galactocerebroside β-galactosidase is normal ... while β-galactosidase activity assayed with *p*-nitrophenyl β-galactoside is extremely deficient." However, the enzyme deficiency can be demonstrated also with galactosylsphingosine (galactopsychosine) as substrate. White blood cells can be used for the diagnosis of the disease and detection of heterozygotes.

There is an animal model for globoid cell leukodystrophy. The disease is found among West Highland and Cairn terrier dogs in which the clinical course is said to be similar to that of the human disease and to be inherited in an autosomal recessive manner. It is also associated with deficiency of galactosylceramide β-galactosidase in brain, liver, and kidney, astrocytosis, and appearance of globoid cells in the white matter of brain.

18.6.8. *Metachromatic Leukodystrophy*

Metachromatic leukodystrophy, or sulfatide lipidosis, is another disease inherited in an autosomal recessive manner.

The name of this disease originates in part from a phenomenon noted by histologists and known as metachromasia. Certain cationic dyes, e.g., toluidine blue and cresyl violet, shift their light absorption maxima in the presence of substances containing anionic groups to shorter wavelength and as a consequence transmit light of longer wavelength. Hence sulfatides, heparin granules in mast cells, chondroitin sulfate, gangliosides, and amyloid (in amyloidosis) stain with such dyes not blue but pink or reddish brown.

The disease was first recognized in 1910 by Alzheimer (cf. Alzheimer, 1910, p. 493 ff., and plate XXXII) and many years later was identified as a sulfatide storage disease resulting from a deficiency of a lysosomal enzyme, cerebroside sulfatase (see Dulaney and Moser, 1978).

Cerebroside sulfatase was known for many years as arylsulfatase A, because it was first recognized as an enzyme that hydrolyzes a variety of aromatic sulfate esters and at first no physiological function could be attributed to it. Arylsulfatase A remains even today the favored and commonly used name for the enzyme and either p-nitrocatechol sulfate or 4-methylumbelliferyl sulfate are used for its assay (cf. Fluharty and Edmond, 1978), as the natural substrate, cerebroside sulfate, is not readily available. Cerebroside sulfate (sulfatide) labeled with ^{35}S can be prepared from the brain of animals injected with $^{35}SO_4^{2-}$ and also used for assay of the enzyme.

A number of other sulfatases exist also, the deficiencies of which are also associated with several storage diseases. Among these sulfatases, arylsulfatase A, purified from ox and human liver, is the best characterized, although it probably exists in all tissues and is even excreted in the urine. The liver enzyme from either source is a glycoprotein with a molecular weight of 105,000–107,000 at pH 7.5, but polymerizes into a tetrameric form at pH 5.5. Under reducing conditions and in the presence of sodium dodecyl sulfate (SDS) the monomeric form dissociates into two subunits: the enzyme from ox liver into two identical subunits of about 55,000 molecular weight; the human liver enzyme on the other hand gives two components on SDS/gel electrophoresis with M_r of 49,000 and 59,000 (Fluharty and Edmond, 1978).

The late infantile form of metachromatic leukodystrophy, the symptoms of which appear after the first year of life, leads to progressive neurological deterioration, mental retardation, decerebrate rigidity, and blindness. The tissues in this condition contain no detectable levels of arylsulfatase A activity, but they contain a protein which cross-reacts with antibodies raised against the normal enzyme and which in Ouchterlony double-diffusion plates shows immunological identity with the normal enzyme. Thus it appears that the abnormality results from a structural gene mutation. The enzymatic deficiency leads to abnormal myelination in the central nervous system, the deposition of sulfatides in peripheral nerves and in the central nervous system, where the accumulated sulfatides often appear in granules in Schwann cells around myelinated fibers and in macrophages in spaces around blood vessels. Sulfatides, as metachromatically staining granules, accumulate also in many other tissues and organs.

There is also a juvenile onset between 5 and 21 years and an adult form of the disease; in these forms there are detectable levels of sulfatase A activity in tissues and the appearance of the symptoms have been correlated with the enzyme levels measured in leukocytes or cultured fibroblasts: the higher the activity, the later the onset. All the known cases of the adult form, which appear at 30 to 40 years of age, were diagnosed only post mortem, as the symptoms may appear quite bizarre and can manifest themselves even as schizophrenia with seizures and motor incoordination.

18.6.9. Fabry's Disease

This is one of the oldest well-characterized clinical entities; it was described independently in England by Anderson and by Fabry in Germany in 1898 chiefly on the basis of the characteristic skin lesions of the disease known as *angiokeratoma corporis diffusum universale*. However, it was not until 50 years later that the condition was recognized as a lipid storage disease; the nature of the enzymatic defect responsible for it was established only in 1967 by Brady *et al.* A comprehensive review of the features of the disease has been presented by Desnick *et al.* (1978).

This is the only sphingolipidosis that is transmitted through an X-linked gene, which means that a hemizygous male, with the abnormal X-chromosome, bears the full brunt of the disease, but does not transmit it to his sons, but his daughters will be obligate heterozygous carriers of the trait. It is expected that one-half of the sons from the union of a heterozygous female and normal male will have the disease and one-half of the daughters will be carriers and the other half will be normal.

The enzymatic defect is a deficiency of the lysosomal enzyme α-galactosidase A which causes a progressive accumulation in all organs, connective tissue, blood vessels and body fluids of neutral glycosphingolipids bearing a terminal galactosyl residue in α-linkage, mostly the trihexosylceramide, galactosyl-($\alpha 1 \rightarrow 4$)galactosyl ($\beta 1 \rightarrow 4$)glucosylceramide. Smaller amounts of galactosyl-($\alpha 1 \rightarrow 4$)galactosyl ($\beta 1 \rightarrow 1$)ceramide are also found in tissues in Fabry's disease. In the instances when a patient with Fabry's disease belongs to blood group B or blood group AB, the blood group B glycosphingolipids Gal ($\alpha 1 \rightarrow 3$)Gal($2 \leftarrow 1\alpha$Fuc)($\beta 1 \rightarrow 3$)GlcNAc ($\beta 1 \rightarrow 3$)Gal ($\beta 1 \rightarrow 4$)Glc-($\beta 1 \rightarrow 1$)Cer, and Gal ($\alpha 1 \rightarrow 3$)Gal($2 \leftarrow 1\alpha$Fuc)($\beta 1 \rightarrow 4$)GlcNAc ($\beta 1 \rightarrow 3$)Gal ($\beta 1 \rightarrow 4$)Glc ($\beta 1 \rightarrow 1$)Cer, which also have a terminal galactose unit in α-linkage, can accumulate as well. The structures of these cerebroside polyhexosides are shown in Fig. 18.10.

The Gal-Gal-Glc-Cer, which is the chief substance accumulating in Fabry's disease, can be found in increased amounts in plasma, in urinary sediment, in many tissues, notoriously in the kidneys, and in cultured fibroblasts. The urinary sediment viewed through a polarizing microscope is birefringent and shows a Maltese-cross pattern. Vance *et al.* (1969) found that the concentration of the trihexoside in the plasma in 10 individuals with Fabry's disease ranged from 0.38 to 1.46 μmol/dl (mean 0.76 $\pm$ 0.21) against the normal of 0.21 $\pm$ 0.07 μmol/dl. In two heterozygous females they found values of 0.36 and 0.54 μmol/dl, but found no increases in the trihexoside content of red cells. In the urine (in the sediment) rather large amounts of the trihexoside are found in hemizygotes, 1.57 μmol/24-hr urine; heterozygous females also excrete excessive amounts of the

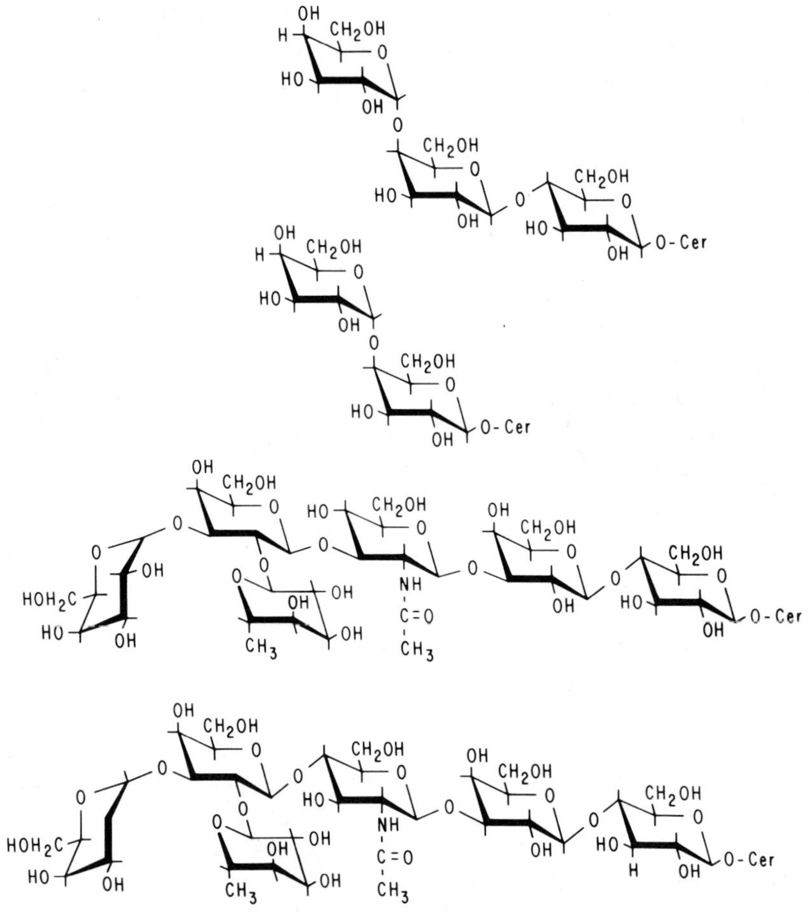

FIGURE 18.10. Cerebroside polyhexosides accumulating in Fabry's disease.

lipid—0.41 μmol/24-hr as compared to the normal of about 0.03 μmol/24-hr. Cultured skin fibroblasts of patients with Fabry's disease contain three to four times more of the Gal-Gal-Glc-Cer (2.15–2.67 μmol/g dry weight) than do cells from normal individuals (0.66 μmol/g dry weight); the fibroblasts from one female Fabry heterozygote were reported to contain high concentration of the trihexoside as well: 2.26 μmol/g dry weight (Dawson et al., 1972).

The α-galactosidase A is present in blood plasma, in circulating leukocytes and even in tears and all three materials can be used for diagnosis with the aid of the artificial substrates 4-methylumbelliferyl- or p-nitrophenyl-α-D-galactoside. The activity of this enzyme in normal plasma ranges between 6 and 14 nmol substrate hydrolyzed per hr per ml; in normal leukocytes the range is between 20 and 40 nmol/hr·mg leukocyte protein; and in tears the values are even higher: 80–160 nmol/hr·ml of tears. In heterozygotes the values are about one-half of the normal, while in afflicted hemizygotes the enzyme activity is undetectable. Prenatal diagnosis of Fabry's disease is possible by the examination of amniotic fluid and of cultured amniotic cells before the 20th day of gestation.

The earliest signs of the disease, which may appear in childhood or adolescence, are the cutaneous lesions, the angiokeratomas. These consist of dilatations of small blood vessels or angiomas in the upper layers of the dermis with hyperkeratosis over larger lesions. They appear as purple or purplish-black spots most commonly around the umbilicus and over the knees. Corneal opacities and tortuosity of conjunctival and retinal blood vessels with aneurysms of small vessels are also among the early signs of the disease and may be found in the eyes of heterozygotes also. Because of the increasing deposition of glycosphingolipids in blood vessels and kidneys, the major symptoms of Fabry's disease with advancing age relate to the cardiovascular system. Thromboses, infarctions, hypertension, cardiac enlargement, angina pectoris, myocardial or cerebral infarctions, renal failure with uremia may be the terminal events. The most debilitating symptom in this disease is the almost constant pain the patients experience. The pain may be nagging, burning sensation or hyperesthesias in palms and soles of feet, with occasional crises of agonizing pain that may last for minutes or days and which may be so severe that the patients contemplate suicide.

Although it is rare among heterozygous carriers of the trait of an inborn error of metabolism to manifest the disease, heterozygotes for Fabry's disease display great variation in this respect, some being completely asymptomatic, others having many of the symptoms of the disease and dying from its sequelae.

REFERENCES

Alzheimer, A., 1910, Beiträge zur Kenntnis der pathologischen Neuroglia and ihrer Beziehung zu den Abbauvorgängen im Nervengewebe, in: *Histologische und Histopathologische Arbeiten,* Vol. 3, No. 3 (F. Nissl and A. Alzheimer, eds.), Fischer, Jena, pp. 401–562.
Anderson, W., 1898, A case of "angeio-keratoma", *Brit. J. Dermatol.* **10**:113.
Barenholz, Y., and Gatt, S., 1968, Degradation of sphingosine, dihydrosphingosine, and phytosphingosine in rats, *Biochemistry* **7**:2603.
Basu, S., Kaufman, B., and Roseman, S., 1973, Enzymatic synthesis of glucocerebroside by a glucosyltransferase from embryonic chicken brain, *J. Biol. Chem.* **248**:1388.
Brady, R. O., 1962, Studies on the total enzymatic synthesis of cerebrosides, *J. Biol. Chem.* **237**:PC2416.
Brady, R. O., 1966, Sphingolipidoses, *New England J. Med.* **275**:312.
Brady, R. O., 1978a, Sphingomyelin lipidosis: Niemann-Pick disease, in: *The Metabolic Basis of Inherited Disease,* 4th ed. (J. B. Stanbury, J. B. Wyngaarden, and D. S. Fredrickson, eds.), McGraw-Hill, New York, pp. 718–730.
Brady, R. O., 1978b, Glucosyl ceramide lipidosis: Gaucher's disease, in: *The Metabolic Basis of Inherited Disease,* 4th ed. (J. B. Stanbury, J. B. Wyngaarden, and D. S. Fredrickson, eds.), McGraw-Hill, New York, pp. 731–746.
Brady, R. O., Formica, J. V., and Koval, G. J., 1958, The enzymatic synthesis of sphingosine. II. Further studies on the mechanism of the reaction, *J. Biol. Chem.* **233**:1072.
Brady, R. O., Bradley, R. M., Young, O. M., and Kaller, H., 1965, An alternative pathway for the enzymatic synthesis of sphingomyelin, *J. Biol. Chem.* **240**:PC3693.
Brady, R. O., Gal, A. E., Bradley, R. M., Martensson, E., Warshaw, A. L., and Laster, L., 1967, Enzy-

matic defect in Fabry's disease: Ceramidetrihexosidase deficiency, *New England J. Med.* **276:**1163.

Braun, P. E., and Snell, E. E., 1968, Biosynthesis of sphingolipid bases. II. Keto intermediates in synthesis of sphingosine and dihydrosphingosine by cell-free extracts of *Hansenula ciferri*, *J. Biol. Chem.* **243:**3775.

Carter, H. E., 1958, Sphingolipides, in: *Chemistry of Lipides as Related to Atherosclerosis: A Symposium* (I. H. Page, ed.), Charles C. Thomas, Springfield, Illinois, pp. 82–94.

Cumar, F. A., Barra, H. S., Maccioni, H. J., and Caputto, R., 1968, Sulfation of glycosphingolipids and related carbohydrates by brain preparations from young rats, *J. Biol. Chem.* **243:**3807.

Dawson, G., Matalon, R., and Dorfman, A., 1972, Glycosphingolipids in cultured human skin fibroblasts. II. Characterization and metabolism in fibroblasts from patients with inborn errors of glycosphingolipid and mucopolysaccharide metabolism, *J. Biol. Chem.* **247:**5951.

Desnick, R. J., Klionsky, B., and Sweeley, C. C., 1978, Fabry's disease (α-galactosidase A deficiency), in: *The Metabolic Basis of Inherited Disease*, 4th ed. (J. B. Stanbury, J. B. Wyngaarden, and D. S. Fredrickson, eds.), McGraw-Hill, New York, pp. 810–840.

Diringer, H., Marggraf, W. D., Koch, M. A., and Anderer, F. A., 1972, Evidence for a new biosynthetic pathway of sphingomyelin in SV 40 transformed mouse cells, *Biochem. Biophys. Res. Commun.* **47:**1345.

Dulaney, J. T., and Moser, H. W., 1978, Sulfatide lipidosis: Metachromatic leukodystrophy, in: *The Metabolic Basis of Inherited Disease*, 4th ed. (J. B. Stanbury, J. B. Wyngaarden, and D. S. Fredrickson, eds.), McGraw-Hill, New York, pp. 770–809.

Fabry, J., 1898, Ein Beitrag zur Kenntniss der Purpura haemorrhagica nodularis (Purpura papulosa haemorrhagica Hebrae), *Arch. Dermatol. Syphilis* **43:**187.

Fluharty, A. L., and Edmond, J., 1978, Arylsulfatases A and B from human liver, *Methods Enzymol.* **50:**537.

Fujino, Y., Nakano, Mo., Negishi, T., and Ito, S., 1968, Substrate specificity for ceramide in the enzymatic formation of sphingomyelin, *J. Biol. Chem.* **243:**4650.

Gatt, S., Barenholz, Y., Borkovski-Kubiler, I., and Leibovitz-Ben Gershon, Z., 1972, Interaction of enzymes with lipid substrates, in: *Advances in Experimental Medicine and Biology*, Vol. 19 (B. W. Volk, and S. M. Aronson, eds.), Plenum Press, New York, pp. 237–256.

Gaucher, P. C. E., 1882, De l'epithelioma primitif de la rate. Hypertrophie idiopathique de la rate sans leucémie (Thèse pour le doctorat en médecine), Octave Doin, Paris.

Geiger, B., and Arnon, R., 1976, Chemical characterization and subunit structure of human N-acetylhexosaminidases A and B, *Biochemistry* **15:**3484.

Hammarström, S., 1971, On the biosynthesis of cerebrosides containing non-hydroxy acids. 1. Mass spectrometric evidence for the psychosine pathway, *Biochem. Biophys. Res. Commun.* **45:**459.

Hammond, R. K., and Sweeley, C. C., 1973, Biosynthesis of unsaturated sphingolipid bases by microsomal preparations from oysters, *J. Biol. Chem.* **248:**632.

Hanahan, D. J., and Brockerhoff, H., 1965, Phospholipids and glycolipids, in: *Comprehensive Biochemistry*, Vol. 6 (M. Florkin and E. H. Stotz, eds.), Elsevier, Amsterdam, pp. 83–140.

Handa, S., and Burton, R. M., 1969, Biosynthesis of glycolipids: Incorporation of N-acetyl galactosamine by a rat brain particulate preparation, *Lipids* **4:**589.

IUPAC–IUB, 1977, The nomenclature of lipids. Recommendations (1976), *Lipids* **12:**455.

Karlsson, K.-A., 1970, Sphingolipid long chain bases, *Lipids* **5:**878.

Kishimoto, Y., and Kawamura, N., 1979, Ceramide metabolism in brain, *Mol. Cell. Biochem.* **23:**17.

Krabbe, K., 1916, A new familial, infantile form of diffuse brain-sclerosis, *Brain* **39:**74.

Max, S. R., Maclaren, N. K., Brady, R. O., Bradley, R. M., Rennels, M. B., Tanaka, J., Garcia, J. H., and Cornblath, M., 1974, GM_3 (hematoside) sphingolipodystrophy, *New England J. Med.* **291:**929.

Morell, P., Costantino-Ceccarini, E., and Radin, N. S., 1970, The biosynthesis by brain microsomes of cerebrosides containing nonhydroxy fatty acids, *Arch. Biochem. Biophys.* **141:**738.

Nakano, M., and Fujino, Y., 1973, Enzymatic conversion of labeled ketodihydrosphingosine to sphingosine, *Biochim. Biophys. Acta* **296:**457.

Nishimura, K., and Yamakawa, T., 1968, Isolation of cerebroside containing glucose (glucosyl ceramide) and its possible significance in ganglioside synthesis, *Lipids* **3:**262.

O'Brien, J. S., 1978, The gangliosidoses, in: *The Metabolic Basis of Inherited Disease,* 4th ed. (J. B. Stanbury, J. B. Wyngaarden, and D. S. Fredrickson, eds.), McGraw-Hill, New York, pp. 841–865.
Robinson, D., and Stirling, J. L., 1968, N-Acetyl-β-glucosaminidases in human spleen, *Biochem. J.* **107**:321.
Sachs, B., 1887, On arrested cerebral development, with special reference to its cortical pathology, *J. Nerv. Ment. Dis.* **14**:541.
Sambasivarao, K., and McCluer, R. H., 1964, Lipid components of gangliosides, *J. Lipid Res.* **5**:103.
Schengrund, C.-L., Lausch, R. N., and Rosenberg, A., 1973, Sialidase activity in transformed cells, *J. Biol. Chem.* **248**:4424.
Shoyama, Y., and Kishimoto, Y., 1976, *In vivo* conversion of 3-ketoceramide to ceramide in rat liver, *Biochem. Biophys. Res. Commun.* **70**:1035.
Sillerud, L. O., Yu, R. K., and Schafer, D. E., 1982, Assignment of the carbon-13 nuclear magnetic resonance spectra of gangliosides G_{M4}, G_{M3}, G_{M2}, G_{M1}, G_{D1a}, G_{D1b}, and G_{T1b}, *Biochemistry* **21**:1260.
Sribney, M., 1966, Enzymatic synthesis of ceramide, *Biochim. Biophys. Acta* **125**:542.
Sribney, M., and Kennedy, E. P., 1958, The enzymatic synthesis of sphingomyelin, *J. Biol. Chem.* **233**:1315.
Stanbury, J. B., Wyngaarden, J. B., and Fredrickson, D. S. (eds.), 1978, *The Metabolic Basis of Inherited Disease,* 4th ed., McGraw-Hill, New York.
Stanbury, J. B., Wyngaarden, J. B., Fredrickson, D. S., Goldstein, J. L., and Brown, M. S. (eds.), 1983, *The Metabolic Basis of Inherited Disease,* 5th ed., McGraw-Hill, New York.
Stoffel, W., LeKim, D., and Sticht, G., 1969, Distribution and properties of dihydrosphingosine-1-phosphate aldolase (sphinganine-1-phosphate alkanal-lyase), *Hoppe-Seyler's Z. Physiol. Chem.* **350**:1233.
Suzuki, K., and Suzuki, Y., 1978, Galactosylceramide lipidosis: globoid cell leukodystrophy (Krabbe's disease), in: *The Metabolic Basis of Inherited Disease,* 4th ed. (J. B. Stanbury, J. B. Wyngaarden, and D. S. Fredrickson, eds.), McGraw-Hill, New York, pp. 747–769.
Svennerholm, L., 1970, Ganglioside metabolism, in: *Comprehensive Biochemistry* (M. Florkin and E. H. Stotz, eds.), Elsevier, Amsterdam, pp. 201–227.
Sweeley, C. C., and Moscatelli, E. A., 1959, Qualitative microanalysis and estimation of sphingolipid bases, *J. Lipid Res.* **1**:40.
Tay, W., 1881, Symmetrical changes in the region of the yellow spot in each eye of an infant, *Trans. Ophth. Soc. U. K.* **1**:55.
Thudicum, J. L. W., 1884, *A Treatise on the Chemical Constitution of the Brain.* (A facsimile edition of the original published by Bailliere, Tindall and Cox, London, with a new historical introduction: "Reflections upon a Classic," by D. L. Drabkin.) 1962, Archon Books, Hamden, Connecticut.
Ullman, M. D., and Radin, N. S., 1972, Enzymatic formation of hydroxy ceramides and comparison with enzymes forming nonhydroxy ceramides, *Arch. Biochem. Biophys.* **152**:767.
Vance, D. E., Krivit, W., and Sweeley, C. C., 1969, Concentrations of glycosyl ceramides in plasma and red cells in Fabry's disease, a glycolipid lipidosis, *J. Lipid Res.* **10**:188.
Weiss, B., 1963, The biosynthesis of sphingosine. I. A study of the reaction with tritium-labeled serine, *J. Biol. Chem.* **238**:1953.
Wherrett, J. R., 1973, Characterization of the major ganglioside in human red cells and of a related tetrahexosyl ceramide in white cells, *Biochim. Biophys. Acta* **326**:63.
Yavin, E., and Gatt, S., 1969, Enzymatic hydrolysis of sphingolipids. VIII. Further purification and properties of rat brain ceramidase, *Biochemistry* **8**:1692.

19
NUTRITIONAL VALUE OF LIPIDS

Dietary fat is the most concentrated source of energy of all the nutrients, supplying 9 kcal/g, about double that contributed by either carbohydrate or protein. In an adequately fed adult, these fat calories are used to a limited extent for immediate energy but primarily for storage in a compact form. In the body of "normal" weight adults, approximately 141,000 kcal are stored as fat as compared to 1,000 kcal as carbohydrate and 24,000 kcal as protein (Masoro, 1977). Because of the ability of the body to store fat in adipose cells, all calories ingested in excess of need are converted to triacylglycerols (triglycerides).

Fat contributes to the palatability and flavor of food, since most food flavors are fat-soluble, and to the satiety value, since fatty foods remain in the stomach for longer periods of time than do foods containing protein and carbohydrate. Fat is a carrier of the fat-soluble vitamins, A, D, E, and K, and the essential fatty acids, which have been shown to be important in growth and in the maintenance of many body functions. Because essential fatty acids are not synthesized by animal species, these or their precursors must be supplied in the diet (see Chapter 9).

The amount of fat in the diet in the United States has increased in the last 60 years by about 25%, and the kind of fat in foods and, thereby, in the diet, has changed as well. Technological advances in the food industry, including efficient and safe hydrogenation procedures, have led to the availability of margarines and shortenings as substitutes for butter and lard. Both margarines and shortenings are produced from a variety of vegetable oils expressed from grains, legumes, and seeds. Although hydrogenation techniques decrease the content of unsaturated and essential fatty acids in vegetable oils, the solid, more stable, margarines produced still contain greater amounts of linoleic acid than do the dairy products they are designed to replace.

In the United States, dietary fat is derived from several sources. At present, the average diet of adults in the United States contains about 160 g of fat per day, which corresponds to approximately 42% of the caloric needs. Of this amount, approximately 14% are as butter and margarine and other spreads, 12% as shortenings, and 6% as cooking and salad oils. The rest of the fat is consumed as "invisible" fat which includes the fat in marbled meats, that present in the flesh and skin of poultry and fish, in nuts, in olives, in avocados, in corn, and in other fat-containing foods. Some fat associated with certain animal products is often lost as a result of cooking. In the present diet in the United States, a larger proportion of fat calories is supplied by vegetable fat than by animal fat.

Most dietary fat is ingested in the form of triacylglycerols, with additional small amounts of fatty acids present in cholesteryl esters and phospholipids. Free fatty acids occur to a limited extent in some vegetable oils.

Fats and oils are arbitrarily differentiated by their melting points since their other properties are quite similar. The melting point of a fatty acid depends on its chain-length, degree of unsaturation, and spatial configuration, and the melting point of a triacylglycerol is a function of its fatty acid composition. The more saturated and the longer the fatty acid chain, the higher the melting point (see Chapter 3). Dairy fats and some vegetable oils (e.g., palm and coconut oils) contain a relatively high concentration of saturated short- and medium-chain fatty acids (6–12 carbons), whereas in most other edible vegetable and animal fats, the triacylglycerol fatty acids are C_{16} and C_{18}, either saturated or unsaturated to varying degrees. The polyunsaturated fatty acid (PUFA), *cis, cis*-9, 12-octadecadienoic acid, linoleic, is one of the "essential" fatty acids (EFA) required by animal species. These are discussed in Chapter 9.

High-fat diets supply primarily triacylglycerols whereas diets low in fat provide a larger proportion of phospholipids. The most common phospholipid present in foods is lecithin (phosphatidylcholine). The phospholipids are components of all cells, primarily located in membranes; they aid in the emulsification of the triacylglycerols prior to their digestion and absorption (see Chapter 12).

19.1. Cholesterol

Cholesterol occurs only in foods of animal origin and, therefore, the amount ingested depends on the content of animal products in the diet. At present, the usual diet in the United States contains 400–600 mg of cholesterol per day. However, the amount that is absorbed varies with the other components of the diet. Cholesterol is poorly absorbed; approximately 50% of that ingested is absorbed (Grundy, 1978). In addition, plant sterols, some fibrous material, and possibly other constituents of the diet, may interfere with cholesterol absorption (see Chapter 12). Cholesterol ingestion starts at birth; human milk contains from 26 to 52 mg of cholesterol per 8 ounces and cow's milk about 34 mg/8 ounces (Jensen *et al.*, 1978).

Cholesterol plays a vital role in the body; it gives stability to membranes and plays a role in the transport of substances in and out of cells. Its catabolic products include bile acids, adrenal hormones, sex hormones, and, in the skin, 7-dehydrocholesterol—the precursor of vitamin D_3. However, since cholesterol is readily biosynthesized in the body, no dietary requirement has been established. In human beings, as well as in some laboratory animals, its endogenous synthesis may be decreased by dietary cholesterol. Hepatic cholesterogenesis is increased when the caloric intake is elevated, and is reduced by caloric restriction. The degree of saturation of dietary fat does not affect the rate of synthesis (Grundy, 1978).

Although it has been generally accepted that hypercholesterolemia is only one of many risk factors for coronary heart disease (McGill, 1979), several long-term epidemiologic studies have implicated an elevated serum cholesterol level as an important risk factor for the development of myocardial infarction (Heyden *et al.*, 1971; Hill and Wynder, 1976). For men with serum cholesterol levels of 260 mg/dl and above, the risk of developing ischemic heart disease has been reported to be twice as high as that for men with serum cholesterol levels below 220 mg/dl. In general, epidemiologic studies indicate that serum cholesterol levels over 200 mg/dl imply an increased risk of coronary heart disease (Kannel *et al.*, 1964).

Over several decades, and in various parts of the world, numerous epidemiologic studies have been performed which attempted to correlate high levels of dietary cholesterol with the morbidity due to atherosclerosis (Eskin, 1971; Armstrong *et al.*, 1975; Connor, 1961). While some of the investigations showed direct associations, other studies found no difference in serum cholesterol concentrations between men who consumed diets high in cholesterol and those whose diets were low in cholesterol (Keys *et al.*, 1950). The review of many such studies leads to the conclusion that even though there may be an association between dietary and serum cholesterol concentration in some individuals, there is no proof of a cause and effect relationship in the population at large (McGill, 1979). In fact, in studies based on individual values, little or no association of dietary cholesterol with serum cholesterol, coronary atherosclerosis, or coronary heart disease has been found (Yano *et al.*, 1978).

Since many variables, including age, length of dietary exposure, previous diet, and fat content of the diet, contribute to and confound the issues, experimental studies in human beings are difficult to compare and evaluate. Some of these investigations suggest a possible relationship between dietary cholesterol and serum cholesterol, but the relationship is weak and varies greatly among individuals. Most often, when dietary factors are considered, the correlation of serum cholesterol levels with dietary saturated fatty acids is much stronger than with the dietary cholesterol content. It appears that most animal species are more sensitive to dietary cholesterol than is man, and, on the other hand, most animals are less responsive to saturated fat than is man (Mahley, 1979).

Several recent investigations show evidence of marked variations in plasma cholesterol response in man to food cholesterol challenges (Slater *et al.*, 1976; Porter *et al.*, 1977; Kummerow, 1975; Flynn *et al.*, 1979). In each of these studies, increasing dietary cholesterol ingestion by feeding eggs or meat did not significantly increase the plasma cholesterol level in the great majority of men.

In a study from the Netherlands (Bronsgeest-Schoute *et al.*, 1979b) no correlation was found between changes in serum cholesterol and the number of eggs eaten. A very variable response to dietary cholesterol was confirmed in the human population. In addition, it was concluded (Bronsgeest-Schoute *et al.*, 1979a) that the presence of a high content of linoleic acid in the diet counteracted the possible effect of dietary cholesterol on serum cholesterol, if the cholesterol was provided as egg yolk. The investigators emphasize the need for more research to facilitate the identification of "hyperresponders" who are more sensitive to changes in dietary cholesterol.

The possibility remains, of course, that dietary cholesterol could affect atherogenesis without changing serum cholesterol concentrations. However, it has been reported (Mahley et al., 1978) that the addition of 4–6 eggs per day to the unrestricted diets of eleven men and women increased a high-density lipoprotein subfraction (the lipoprotein protective against heart disease) regardless of the effect on plasma cholesterol.

Some years ago a hypothesis was presented to the effect that high cholesterol intake early in life helps to establish a permanent mechanism for maintaining low serum cholesterol concentration in adulthood (Reiser and Sidelman, 1972; Reiser et al., 1979). However, this presumed stabilization of the anabolic and catabolic processes of cholesterol in adults by preconditioning young rats with high-cholesterol diets was not confirmed (Kris-Etherton et al., 1979). But even though early exposure to exogenous cholesterol did not protect the rat against subsequent dietary-induced hypercholesterolemia, cholesterol metabolism was affected in early life.

There have been attempts to test this hypothesis in human beings. In an abnormal (hyperlipoproteinemic) population of infants, it was reported (Glueck and Tsang, 1972) that a commercial formula low in cholesterol but high in PUFA led to lower plasma cholesterol levels than those seen in infants fed a regular bovine milk formula. However, no follow-up studies of plasma cholesterol values in adults were reported. In another study (Friedman and Goldberg, 1975), comparing breast-fed with formula-fed infants, no differences in plasma cholesterol levels were found in the children after one year. Similarly, a later investigation with slightly older children (7–12 years of age) also proved to be inconclusive (Hodgson et al., 1976). However, here again the theory was not tested by follow-up studies when the children became adults.

In a recent study with rats (Naseem et al., 1980) prenatal exposure to a high-cholesterol, high-fat diet influenced both the synthetic and degradative enzymes of cholesterol metabolism. The inhibition of HMG-CoA reductase (an enzyme important in cholesterol biosynthesis) was over 75% in suckling and mature animals whose mothers had been fed the experimental diet (1% cholesterol).

Some oxidation products of cholesterol, when given at extremely low dosages, have a toxic effect on arterial smooth muscle cells, whereas pure cholesterol has but a minimal angiotoxic effect (Imai et al., 1976; Peng et al., 1978). The cholesterol in dehydrated foods, such as powdered milk and powdered egg yolk, is highly susceptible to autoxidation. Evidence suggests that endogenously synthesized cholesterol is protected from autoxidation by antioxidants in the animal organism. On the other hand, traces of oxidation products of cholesterol have been isolated from human atheromata (Smith and Van Lier, 1970). There is an indication that pure cholesterol (either endogenously synthesized or chemically isolated) is not atherogenic (Taylor et al., 1979). Present methods used in isolation and storage of some cholesterol-containing foods seem to promote the formation of angiotoxic cholesterol derivatives. It may be advisable to protect cholesterol-containing foods with antioxidants to minimize the formation of these toxic derivatives.

Ecologic associations of coronary heart disease with cancer have led to suggestions that cholesterol may play a role in both of these diseases (Rose et al., 1974; Lea, 1966). Serum cholesterol levels were found to be lower in persons subsequently developing colon cancer, which suggests that low serum cholesterol levels may pose their own noncardiovascular risks.

When more than 3,000 individuals in Evans County, Georgia, were followed for over a decade, it was observed that in 127 cancer cases the mean serum cholesterol level was significantly lower than that of the population without cancer. This finding was consistent for the various cancer sites, with particular significance in males (Kark et al., 1980).

By virtue of its having a regulatory role in biological membranes, cholesterol may function as a bioregulator in the development and inhibition of leukemia (Inbar and Shinitzky, 1974).

Since the various hypotheses as to the cause of atherosclerosis are still being investigated, it is premature to advise drastic dietary changes. A number of risk factors for cardiovascular disease have been identified from epidemiologic studies. These include male sex, genetic trait, hypercholesterolemia, hypertension, obesity, diabetes, smoking, and physical inactivity. In addition, also implicated are high concentrations of low-density lipoproteins, enhanced platelet aggregation, transformed smooth muscle cells, altered prostaglandin metabolism, and steroid hormone effects on arterial metabolism.

19.2. Polyunsaturated Fatty Acids

Since reducing hypercholesterolemia appears to decrease one of the risk factors for heart disease, a decrease in dietary saturated fatty acids and an increase in the proportion of PUFA have been advocated (Anderson et al., 1973; Kummerow, 1975). A positive role of PUFA in human nutrition and metabolism has long been proposed, particularly since it has been recognized that some of the essential fatty acid deficiency symptoms (e.g., dermatitis), long observed in animals, occur in human beings under certain conditions (Söderhjelm et al., 1970; Carroll, 1977), e.g., in children and adults receiving prolonged parenteral nutrition with fat-deficient preparations. An intake of EFA as 1–2% of total calories is recommended to prevent deficiency (see Chapter 9).

In middle-aged people with hypercholesterolemia, as much as a 20% reduction in serum cholesterol levels has been achieved through a relatively high intake of polyunsaturated fat (up to 20% of total calories as linoleic acid) (Vergroesen, 1972). Nevertheless, clinical trials still have not definitely established the advantage of extremely high dietary PUFA levels in human nutrition (Carroll, 1977; Kaunitz, 1978).

Experiments on rats, when extrapolated to human beings, led to the recommendation that PUFA in human diets should not exceed half of the total fat calories in the average diet (Narayan et al., 1974). Later, however, it was advocated that 10% of total fat calories as EFA should be the upper limit (Carroll, 1977), although the American Heart Association recommends a total fat intake of 30%

with a P/S (polyunsaturates to saturates) ratio of 1:1 (AHA Committee Report, 1978).

Whether dietary modifications resulting in lowering plasma lipids will prove to be of benefit in decreasing the risk for heart disease still has not been definitely resolved.

However, there are potentially undesirable effects of dietary polyunsaturated fats. Excessive levels of dietary fats lead not only to obesity but apparently have a positive relationship to certain types of carcinogenesis (Carroll, 1977). Animals on a high-fat diet develop mammary tumors more readily than do similarly treated controls on low-fat diets (Carroll, 1975). Epidemiologic data on human populations reveal a correlation between breast cancer and dietary fat intake, but not exclusively polyunsaturated fat.

Even though most vegetable oils with a high linoleic acid content usually contain considerable amounts of vitamin E, some of it may be lost during storage, processing, or cooking. Since PUFAs are easily oxidized to form toxic and possibly carcinogenic peroxides and other oxidized products, large amounts of dietary PUFA will increase vitamin E, or other antioxidant, requirements in order to prevent the formation of these lipid peroxides in tissues (see Chapter 6).

A higher incidence of gallstones has been observed in animals fed PUFAs (Lofland, 1975) as well as in patients on an experimental high polyunsaturated fat diet (Sturdevant *et al.,* 1973).

EFAs are known to be precursors of prostaglandins, prostacyclins, and thromboxanes, which are potent vasoactive agents with a variety of other actions depending on the target organ and the type of prostaglandins formed (see Chapter 10).

Toxic and potentially carcinogenic substances such as polycyclic aromatic hydrocarbons are produced by heating oils containing high levels of PUFAs (Michael *et al.,* 1966). Various symptoms of toxicity, including irritation of the digestive tract, organ enlargement, growth depression, and even death, have been observed when highly abused (oxidized and heated) fats were fed to laboratory animals (Andia and Street, 1975). It has also been demonstrated that laboratory-heated oils may act as cocarcinogens (Sugai *et al.,* 1962). However, it has been shown that commercially used frying fats contain only small amounts of potentially toxic substances and produce no appreciable ill effects in animals consuming them (Nolen *et al.,* 1967).

19.3. Trans Fatty Acids

The hydrogenation of vegetable oils, in the manufacture of margarine, results in the reduction of the levels of the highly unsaturated fatty acids. In addition, the hydrogenation process produces shifts in the position of the double bonds and also isomerization of the *cis* to the *trans* forms of the fatty acids. All these effects of hydrogenation produce fats (triacylglycerols) with a higher melting point and convert the liquid oils into semi-solid fats which can be packaged and stored for longer periods of time. Of the 10 billion (10×10^9) lb of fats and oils produced in the U.S.A., over 6 billion (6×10^9) lb are partially hydrogenated (Dutton, 1979). The fatty acid isomers occur to varying extents in the commercial products. Margarines may contain up to 30–45% of *trans* fatty acids. Most natural food fats and oils contain only *cis* isomers of fatty acids. Small amounts of *trans* isomers occur in fats from ruminants.

The presence of fats containing *trans* fatty acids in the diet is of recent concern to nutritionists. Since properties of cell membranes such as permeability, enzyme activity, cell membrane fragility, and even cell division may be influenced by the character of their lipid components, it is of great importance to recognize that potential modifications of these parameters may be due to changes in fatty acid composition.

The *trans* fatty acid content of a variety of human tissues has been determined (Johnston *et al.*, 1957). As much as 12.2% and 14.4% of *trans* isomers have been found in adipose tissue and liver respectively; human atheromata contain up to 8.8%. Approximately 4% of the *trans* isomer of oleic acid 18:1 (elaidic acid) was found in human milk fat (Picciano and Perkins, 1977).

The digestibility of hydrogenated fats in man depends on the melting point and ranges from 79% to 98% (Emken, 1979). In the rat, the absorption coefficient for elaidic acid (9-*trans*-18:1) is 95.6%, indicating that its absorption by the mucosal cells of the intestine is not affected by the *trans* configuration of the fatty acid.

When fats containing approximately 45% *trans* monounsaturated fatty acids and 6% *trans*-linoleic acid (9,12-*trans*-18:2) were fed to rats for long periods of time, it was found that the phospholipids in tissues preferentially accumulated the *trans* fatty acids (Alfin-Slater *et al.*, 1976). Since phospholipids play a significant role in cell membranes, the incorporation of *trans* fatty acids into membrane phospholipids may affect metabolic processes (Alfin-Slater *et al.*, 1976; Alfin-Slater and Aftergood, 1979).

In well-controlled feeding studies with hydrogenated fats, a higher degree of saturation of dietary fats was positively correlated with increases in chylomicron aggregation in blood samples (Swank, 1951). Plasma cholesterol increases due to the intake of hydrogenated fats were also noted and further investigations implicated isomers of the fatty acids as the causative agents in elevating the blood cholesterol levels (Horlick, 1960). However, in another study, fats containing up to 21% total *trans* fatty acids and 8% 9,12-*cis*-18:2 acid fed to men did not elevate serum cholesterol levels (McOsker *et al.*, 1962). Similarly, when a diet containing no *trans* fat was compared with one that contained 44% total *trans* fatty acids, no significant differences in plasma cholesterol and triacylglycerols were observed (Mattson *et al.*, 1975). On the other hand, Vergroesen (1972) concluded that in the presence (but not in the absence) of dietary cholesterol, feeding elaidic acid resulted in higher serum cholesterol levels than were obtained with oleic and palmitic acids. In general, it appears that when adequate amounts of oleic and linoleic acids are available, the effect of hydrogenated fat on serum cholesterol levels is negligible. It is possible, however, that elaidic acid is utilized for energy less readily than its *cis* isomer since it accumulates in the plasma free fatty acid fraction bound to albumin (FFA—see Chapter 13).

So far, no short-term toxic effects have been reported in human beings as a result of the ingestion of hydrogenated fats. Similarly, long-term studies on rats fed a margarine fat containing approximately 35% *trans* fatty acids for 75 generations have not shown any apparent deleterious effects (Alfin-Slater *et al.,* 1973). Nevertheless, it is important to ensure an adequate supply of EFAs in diets that contain fatty acid isomers (Alfin-Slater and Aftergood, 1979). It has been shown that EFA deficiency is accentuated by dietary *trans* fatty acids as well as by completely saturated fatty acids.

Moore and coworkers found decreased serum lecithin:cholesterol acyltransferase (LCAT) activity when *trans* fatty acids were fed to rats for nine months. Moreover, a negative correlation was found between serum free cholesterol and LCAT activity (Moore *et al.,* 1980a).

Furthermore, investigations seem to indicate that certain enzymes are able to discriminate among fatty acids of various structures and, therefore, some fatty acids may be incorporated selectively into different positions of different cellular lipids. If *trans* fatty acids are incorporated into lipids of cell membranes, they may be able to exert changes in membrane permeability.

Trans fatty acids have also been implicated in certain pathological conditions. Although their effect on elevating plasma cholesterol levels is controversial, nevertheless they are suspect as far as atherogenesis is concerned and it has also been suggested that they may be carcinogens (Enig *et al.,* 1978). However, in a study on the incorporation of *trans* isomers into tissues of rats that had been fed these fatty acids for three months, and on their disappearance after the diet was discontinued (Moore *et al.,* 1980b), in all tissues examined, i.e., plasma, liver, kidney, heart, adipose tissue, and red blood cells, there was incorporation of the *trans* acids into phospholipids and triglycerides. After eight weeks of feeding the *trans* fatty acid-free diet, only negligible amounts of the *trans* acids were present in tissues, indicating that the tissues metabolized these acids.

19.4. Interesterified Fats

In addition to the chain-length and unsaturation of their fatty acids, the physical nature of fats is determined by the distribution of the constituent fatty acids among the three positions of the glycerol molecule. The process of interesterification of fats (Sreenivasan, 1978) has resulted in better melting qualities, spreadability and edible qualities of margarines, without the formation of fatty acid isomers.

Under certain conditions interesterification of certain fats has been shown to be of benefit. For example, peanut oil has been shown to be atherogenic when fed to rabbits (Kritchevsky *et al.,* 1976). In natural peanut oil the arachidic (20:0) and behenic (22:0) acids are present only in the *sn*-3 position of the triacylglycerol. When peanut oil is randomized (e.g., by heating with small amounts of sodium methoxide) or autointeresterified, these fatty acids become evenly distributed throughout the molecule and the peculiar atherogenicity of the peanut oil disappears (Kritchevsky *et al.,* 1973).

19.5. Rapeseed Oil

Rapeseed oil, which is derived from seeds of *Brassica* species, is used in the diet in many countries. A characteristic fatty acid present in this oil is erucic acid, 22:1ω9, which accounts for 20–55% of its total fatty acids; in addition, it may contain as much as 10% of the 20:1ω9 acid. Other vegetable oils contain only traces of these fatty acids.

It has long been known that the consumption of large amounts of erucic acid by laboratory animals leads to retarded growth and the accumulation of lipids in heart, adrenals, and liver. It has been suggested that erucic acid is less efficiently oxidized than are other fatty acids and, in fact, that it affects the oxidation of other fatty acids. The *in vitro* oxidation of palmitoyl carnitine was reduced by more than 50% in tissue slices or homogenates of liver from animals fed even low levels of erucic acid. In addition, erucoyl carnitine was found to inhibit the oxidation of other acyl carnitines (Christiansen *et al.*, 1977). With high levels of erucic acid in the diet, the cardiac triacylglycerols were found to contain more than 50% of this acid. It also is incorporated into the cardiolipin fraction of rat liver mitochondria, thereby influencing energy production. When erucic acid is present in the diet, the accumulated triacylglycerols in rat heart reflect the dietary fatty acids, suggesting an inhibited oxidation of all dietary fatty acids in this organ. Lipid droplets were observed in the myocardial cells as early as three hours after the administration of rapeseed oil to rats (Ziemlanski *et al.*, 1973). The inability of cardiac tissue to metabolize these fatty acids readily leads to their incorporation into triacylglycerols which are deposited *in situ*.

Young animals are particularly susceptible to the effects of erucic acid. Therefore, rapeseed oil is no longer incorporated into infant formulas in countries such as Canada and Sweden where this oil has been extensively used. However, in general, there have been no cases reported of toxicity or changes in man similar to those which have been observed in experimental animals (Vles, 1975).

In recent years *Brassica* cultivars, low in erucic acid, have been developed. These low erucic acid-containing rapeseed oils have fatty acids predominantly of C_{18} chain-length. The use of such oils permitted growth of rats comparable to that achieved with other fats and oils; no deleterious effects were observed (Craig and Beare, 1968). Rapeseed oil that is low in erucic acid is preferred over the regular rapeseed oil for partial hydrogenation and eventual incorporation into margarines. However, for human use, the *Brassica* oils are usually blended or mixed with other oils to ensure adequate availability of EFAs.

19.6. Marine Oils

Marine oils, many of which contain long-chain fatty acids, are also used as edible oils. Some of these oils are particularly high (10–30%) in docosenoic acids (cis-22:1ω11 (cetoleic)). Partial hydrogenation of this fatty acid does not completely saturate the double bond of docosenoic acid but rather promotes the formation of isomers. A cardiac lipidosis in laboratory animals has been shown to occur as a result of long-term feeding experiments. In human populations consuming high-fat diets (some of which contained a partially hydrogenated marine oil) a mild lipidosis was occasionally found in the heart. But, in general, no harmful effect in human beings has been attributed to the intake of unprocessed marine lipids (Vles, 1975). Again, the presence of an adequate level of EFA counteracts any ill effects observed when partially hydrogenated marine oils are used in the diet (see Chapter 5, Table 5.5).

On the other hand, when the effect of a diet containing large amounts of mackerel, supplying approximately 8 g/day of polyunsaturated $\omega 3$ fatty acids, was studied for a period of three weeks in human volunteers, a lowering of serum lipids accompanied by an increase in HDL-cholesterol resulted (von Lossonczy et al., 1978). A replacement of $\omega 9$ and $\omega 6$ acids by $\omega 3$ acids was observed. Even though the mackerel fat contained at least 30% of 20:1 and 22:1 acids, these were not found in blood serum lipids. It had been concluded earlier that the hypocholesterolemic effect of fish oils was not predictable on the basis of its degree of unsaturation (Peifer et al., 1965).

19.7. Cyclopropenoid Acids

Cyclopropenoid acids, such as sterculic and malvalic acids, normally occur in small amounts in unprocessed cottonseed oil, and in large amounts in *Sterculia foetida* oil. These acids inhibit the fatty acid desaturase system, resulting in increased levels of stearic acid and decreased levels of oleic acid (Reiser and Raju, 1964). Marked alterations of membrane permeability take place which can be manifested in laying hens by the "pink egg white" syndrome. When fed in conjunction with the carcinogen, aflatoxin, sterculic acid proves to be a cocarcinogen for certain species such as trout.

The implications for man of the ingestion of cyclopropenoid fatty acids do not appear to be serious. Commercial cottonseed oils contain no more than 0.04–0.5% of these acids and this amount is constantly being decreased by improved technology (Mattson, 1973). In addition, cooking destroys the biological activity of these compounds.

19.8. Thrombogenic Properties of Fats

A factor in coronary heart disease, not necessarily related to atherosclerosis, is the formation of thrombi, which result from increased platelet aggregation and clotting activity. Such conditions have been observed in animals fed saturated fats (Renaud, 1977). *In vitro* studies have also shown that the long-chain saturated fatty acids, particularly 18:0, induce platelet aggregation and accelerate coagulation. On the other hand, 18:2 and 18:3 prevent this action. The thrombogenic fatty acids apparently act by modifying the composition of the platelet phospholipids (Renaud and Gautheron, 1975).

In patients with cardiovascular disease, an increase in 18:0 and 18:1 at the expense of the PUFAs was found in platelet lipids (Renaud et al., 1970). The susceptibility of platelets to aggregation in normal subjects was reduced by increasing the polyunsaturated/saturated fatty acid (P/S) ratio of dietary fat and by markedly reducing the total fat intake (Iacono et al., 1974).

Arachidonic acid, a precursor of prostaglandins, also seems to be involved both in the susceptibility of platelets and the resistance of platelets to aggregation (see Chapter 10).

Arachidonic acid in the vascular epithelium is a precursor of prostacyclin (PGI_2), which is antiaggregatory. However, through a separate though related pathway, arachidonic acid is metabolized in platelets to thromboxane A_2 which, together with its precursor PGG_2 endoperoxide, induces platelet aggregation (see Chapter 10). Thus it appears that the excess of PUFAs may be involved in aggregation as well as in its prevention. Thromboxane A_2 is very unstable and breaks down spontaneously into thromboxane B_2, which is biologically inactive (Moncada and Vane, 1979).

Eicosapentaenoic acid ($20:5\omega3$) derived from linolenic acid is probably converted to an antiaggregatory prostacyclin but not to an active thromboxane. This may explain the observation that Eskimos are protected from thromboembolic disorders (Gryglewski and Moncada, 1979) and have prolonged clotting times, since their high-fish diets are rich in linolenic acid and its products.

The understanding of how thromboxane formation might be inhibited without preventing prostacyclin production is the subject of ongoing research. Aspirin, for example, can inhibit cyclo-oxygenase activity, thereby preventing the formation of both thromboxane and prostacyclin. However, it appears that the PG cyclo-oxygenase in the platelets is more sensitive to aspirin than are similar enzymes in the vascular epithelium. In addition, inactivation of the cyclo-oxygenase by aspirin is irreversible and the non-nucleated platelets cannot regenerate the enzyme. Therefore, only newly derived platelets, formed after the disappearance of the aspirin, can produce thromboxane (TXA_2). In the vascular epithelium, on the other hand, in addition to being less sensitive to aspirin, the enzyme continues to be synthesized and, as a result, continues to form prostacyclins (see Chapter 10).

19.9. Medium-Chain Triacylglycerols

Fatty acids with chain-lengths of 6–10 carbon atoms are known as medium-chain, and medium-chain triacylglycerols (MCT) arc important in the diet under conditions in which longer-chain fatty acids are not absorbed. They usually originate from vegetable oils such as coconut oil and palm oil and their water solubility is significantly higher than that of the longer-chain fatty acid triacylglycerols because of their smaller molecular size. As a result, the digestion, absorption, and metabolism of MCT differ from those of the longer-chain fatty acid triacylglycerols. They are absorbed directly into the portal circulation. As pointed out in Chapter 12 on lipid absorption, MCT are of particular value for patients with a variety of fat malabsorption problems. Some milk fats (e.g., rat, rabbit) are rich in MCT; in such animal species, these are of special importance during the developmental period of the young.

19.10. Summary

Epidemiologic evidence linking high-fat diets, principally diets rich in saturated fat and cholesterol, with a higher incidence of cardiovascular disease and cancer, has prompted many health care professionals to recommend for populations with a high incidence of atherosclerosis, obesity, and maturity-onset diabetes, a diet adequate to maintain ideal body weight, i.e., 10–15 energy percent protein and 30–35 energy percent fat. Increase in the ratio of polyunsaturated to saturated fat and decrease in dietary cholesterol has also been recommended (FAO and WHO Report, 1978). Since a cause and effect relationship between fat and cardiovascular disease or between fat and cancer has not been demonstrated experimentally, many investigators do not believe that drastic changes in the diet should be made until further documentation is available (Ahrens, 1979). Recently the results of a ten-year study to test the effect of lowering cholesterol levels in hypercholesterolemic men showed that reducing serum cholesterol levels resulted in a decreased risk of coronary heart disease. However, the reduction in serum cholesterol levels was achieved through the addition of cholestyramine to the diet rather than through a cholesterol-lowering diet (Lipid Research Clinics Program, 1984). Cholestyramine, a resin, was found to interfere with the absorption of substances other than cholesterol and bile acids, e.g., fat-soluble vitamins, as well as producing undesirable side effects, including gastrointestinal disturbances and a higher incidence of gallstones.

REFERENCES

Ahrens, E. H., [Jr.], 1979, Dietary fats and coronary heart disease: unfinished business, *Lancet* **2**:1345.
Alfin-Slater, R. B., and Aftergood, L., 1979, Nutritional role of hydrogenated fats, in: *Geometrical and Positional Fatty Acid Isomers* (E. A. Emken and H. J. Dutton, eds.), American Oil Chemists' Society, Champaign, Ill., pp. 53–74.
Alfin-Slater, R. B., Wells, P., Aftergood, L., and Melnick, D., 1973, Dietary fat composition and tocopherol requirement: IV. Safety of polyunsaturated fats, *J. Am. Oil Chem. Soc.* **50**:479.
Alfin-Slater, R. B., Aftergood, L., and Whitten, T., 1976, Nutritional evaluation of *trans* fatty acids, *J. Am. Oil Chem. Soc.* **53**:468A.
American Heart Association Committee Report, 1978, Value and safety of diet modification to control hyperlipidemia in childhood and adolescence. A statement for physicians, *Circulation* **58**:381A.
Anderson, J. T., Grande, F., and Keys, A., 1973, Cholesterol-lowering diets, *J. Am. Diet. Assoc.* **62**:133.
Andia, A. M. G., and Street, J. C., 1975, Dietary induction of hepatic microsomal enzymes by thermally oxidized fats, *J. Agr. Food Chem.* **23**:173.

Armstrong, B. K., Mann, J. I., Adelstein, A. M., and Eskin, F., 1975, Commodity consumption and ischemic heart disease mortality, with special reference to dietary practices, *J. Chron. Dis.* **28**:455.

Bronsgeest-Schoute, D. C., Hautvast, J. G. A. J., and Hermus, R. J. J., 1979a, Dependence of the effects of dietary cholesterol and experimental conditions on serum lipids in man, I. Effects of dietary cholesterol in a linoleic acid-rich diet, *Am. J. Clin. Nutr.* **32**:2183.

Bronsgeest-Schoute, D. C., Hermus, R. J. J., Dallinga-Thie, G. M., and Hautvast, J. G. A. J., 1979b, Dependence of the effects of dietary cholesterol and experimental conditions on serum lipids in man, III. The effect on serum cholesterol of removal of eggs from the diet of free-living habitually egg-eating people, *Am. J. Clin. Nutr.* **32**:2193.

Carroll, K. K., 1975, Experimental evidence of dietary factors and hormone-dependent cancers, *Cancer Res.* **35**:3374.

Carroll, K. K., 1977, Essential fatty acids: what level in the diet is most desirable? in: *Advances in Experimental Medicine and Biology*, Vol. 83, *Function and Biosynthesis of Lipids* (N. G. Bazán, R. R. Brenner, and N. M. Giusto, eds.), Plenum Press, New York, pp. 535–546.

Christiansen, R. Z., Christophersen, B. O., and Bremer, J., 1977, Monoethylenic C_{20} and C_{22} fatty acids in marine oil and rapeseed oil. Studies on their oxidation and on their relative ability to inhibit palmitate oxidation in heart and liver mitochondria, *Biochim. Biophys. Acta*, **487**:28.

Connor, W. E., 1961, Dietary cholesterol and the pathogenesis of atherosclerosis, *Geriatrics* **16**:407.

Craig, B. M., and Beare, J. L., 1968, Nutritional properties of Canadian Canbra oil, *Can. Inst. Food Sci. Technol. J.* **1**:64.

Dutton, H. J., 1979, Hydrogenation of fats and its significance, in: *Geometrical and Positional Fatty Acid Isomers* (E. A. Emken and H. J. Dutton, eds.), American Oil Chemists' Society, Champaign, Ill., pp. 1–16.

Emken, E. A., 1979, Utilization and effects of isomeric fatty acids in humans, in: *Geometrical and Positional Fatty Acid Isomers* (E. A. Emken and H. J.Dutton, eds.), American Oil Chemists' Society, Champaign, Ill., pp. 99–129.

Enig, M. G., Munn, R. J., and Keeney, M., 1978, Dietary fat and cancer trends—a critique, *Fed. Proc.* **37**:2215.

Eskin, F., 1971, The role of the egg as a factor in the aetiology of coronary heart disease, *Commun. Health* (Bristol) **2**:179.

FAO and WHO, 1978, Dietary Fats and Oils in Human Nutrition, Report of an Expert Consultation (21–30 September 1977), Food and Agriculture Organization of the United Nations, Rome.

Flynn, M. A., Nolph, G. B., Flynn, T. C., Kahrs, R., and Krause, G., 1979, Effect of dietary egg on human serum cholesterol and triglycerides, *Am. J. Clin. Nutr.* **32**:1051.

Friedman, G., and Goldberg, S. J., 1975, Concurrent and subsequent serum cholesterols of breast- and formula-fed infants, *Am. J. Clin. Nutr.* **28**:42.

Glueck, C. J., and Tsang, R. C., 1972, Pediatric familial type II hyperlipoproteinemia: effects of diet on plasma cholesterol in the first year of life, *Am. J. Clin. Nutr.* **25**:224.

Grundy, S. M., 1978, Cholesterol metabolism in man, *West. J. Med.* **128**:13.

Gryglewski, R. J., and Moncada, S., 1979, Polyunsaturated fatty acids and thrombosis, *Eur. J. Clin. Invest.* **9**:1.

Heyden, S., Walker, L., Hames, C. G., and Tyroler, H. A., 1971, Decrease of serum cholesterol level and blood pressure in the community, *Arch. Intern. Med.* **128**:982.

Hill, P., and Wynder, E. L., 1976, Dietary regulation of serum lipids in healthy, young adults, *J. Am. Diet. Assoc.* **68**:25.

Hodgson, P. A., Ellefson, R. D., Elveback, L. R., Harris, L. E., Nelson, R. A., and Weidman, W. H., 1976, Comparison of serum cholesterol in children fed high, moderate, or low cholesterol milk diets during neonatal period, *Metabolism* **25**:739.

Horlick, L., 1960, The effect of artificial modification of food on the serum cholesterol level, *Can. Med. Assoc. J.* **83**:1186.

Iacono, J. M., Binder, R. A., Marshall, M. W., Schoene, N. W., Jencks, J. A., and Mackin, J. F., 1974, Decreased susceptibility to thrombin and collagen platelet aggregation in man fed a low fat diet, *Haemostasis* **3**:306.

Imai, H., Werthessen, N. T., Taylor, C. B., and Lee, K. T., 1976, Angiotoxicity and arteriosclerosis due to contaminants of USP-grade cholesterol, *Arch. Pathol. Lab. Med.* **100**:565.

Inbar, M., and Shinitzky, M., 1974, Cholesterol as a bioregulator in the development and inhibition of leukemia, *Proc. Nat. Acad. U.S.A.* **71**:4229.
Jensen, R. G., Hagerty, M. M., and McMahon, K. E., 1978, Lipids of human milk and infant formulas: a review, *Am. J. Clin. Nutr.* **31**:990.
Johnston, P. V., Johnson, O. C., and Kummerow, F. A., 1957, Occurrence of *trans* fatty acids in human tissue, *Science* **126**:698.
Kannel, W. B., Dawber, T. R., Friedman, G. D., Glennon, W. E., and McNamara, P. M., 1964, Risk factors in coronary heart disease. An evaluation of several serum lipids as predictors of coronary heart disease. The Framingham study, *Ann. Intern. Med.* **61**:888.
Kark, J. D., Smith, A. H., and Hames, C. G., 1980, The relationship of serum cholesterol to the incidence of cancer in Evans County, Georgia, *J. Chron. Dis.* **33**:311.
Kaunitz, H., 1978, Toxic effects of polyunsaturated vegetable oils, in: *Symposium on the Pharmacological Effect of Lipids* (J. J. Kabara, ed.), American Oil Chemists' Society, Champaign, Ill., pp. 203–210.
Keys, A., Mickelsen, O., Miller, E. v. O., and Chapman, C. B., 1950, The relation in man between cholesterol levels in the diet and in the blood, *Science* **112**:79.
Kris-Etherton, P. M., Layman, D. K., York, P. V., and Frantz, I. D., Jr., 1979, The influence of early nutrition on the serum cholesterol of the adult rat, *J. Nutr.* **109**:1244.
Kritchevsky, D., Tepper, S. A., Vesselinovitch, D., and Wissler, R. W., 1973, Cholesterol vehicle in experimental atherosclerosis. Part 13. Randomized peanut oil, *Atherosclerosis,* **17**:225.
Kritchevsky, D., Tepper, S. A., Kim, H. K., Story, J. A., Vesselinovitch, D., and Wissler, R. W., 1976, Experimental atherosclerosis in rabbits fed cholesterol-free diets. 5. Comparison of peanut, corn, butter, and coconut oils, *Exp. Mol. Pathol.* **24**:375.
Kummerow, F. A., 1975, Symposium: Nutritional perspectives and atherosclerosis. Lipids in atherosclerosis, *J. Food Sci.* **40**:12.
Lea, A. J., 1966, Dietary factors associated with death-rates from certain neoplasms in man, *Lancet* **2**:332.
Lipid Research Clinics Program, 1984, The Lipid Research Clinics coronary primary prevention trial results. I. Reduction in incidence of coronary heart disease, *J. Am. Med. Assoc.* **251**:351.
Lofland, H. B., 1975, Animal model of human disease: cholelithiasis, *Am. J. Pathol.* **79**:619.
Mahley, R. W., 1979, Dietary fat, cholesterol, and accelerated atherosclerosis, in: *Atherosclerosis Reviews,* Vol. 5 (R. Paoletti and A. M. Gotto, Jr., eds.), Raven Press, New York, pp. 1–34.
Mahley, R. W., Bersot, T. P., Innerarity, T. L., Lipson, A., and Margolis, S., 1978, Alterations in human high-density lipoproteins, with or without increased plasma-cholesterol, induced by diets high in cholesterol, *Lancet* **2**:807.
Masoro, E. J., 1977, Fat metabolism in normal and abnormal states, *Am. J. Clin. Nutr.* **30**:1311.
Mattson, F. H., 1973, Potential toxicity of food lipids, in: *Toxicants Occurring Naturally in Foods,* 2nd ed., National Academy of Sciences, Washington, D.C., pp. 189–209.
Mattson, F. H., Hollenbach, E. J., and Kligman, A. M., 1975, Effect of hydrogenated fat on the plasma cholesterol and triglyceride levels of man, *Am. J. Clin. Nutr.* **28**:726.
McGill, H. C., Jr., 1979, The relationship of dietary cholesterol to serum cholesterol concentration and to atherosclerosis in man, *Am. J. Clin. Nutr.* **32**:2664.
McOsker, D. E., Mattson, F. H., Sweringen, H. B., and Kligman, A. M., 1962, The influence of partially hydrogenated dietary fats on serum cholesterol levels, *J. Am. Med. Assoc.* **180**:380.
Michael, W. R., Alexander, J. C., and Artman, N. R., 1966, Thermal reactions of methyl linoleate. I. Heating conditions, isolation techniques, biological studies and chemical changes, *Lipids* **1**:353.
Moncada, S., and Vane, J. R., 1979, Arachidonic acid metabolites and the interactions between platelets and blood-vessel walls, *New England J. Med.* **300**:1142.
Moore, C. E., Alfin-Slater, R. B., and Aftergood, L., 1980a, Effect of *trans* fatty acids on serum lecithin: cholesterol acyltransferase in rats, *J. Nutr.* **110**:2284.
Moore, C. E., Alfin-Slater, R. B., and Aftergood, L., 1980b, Incorporation and disappearance of *trans* fatty acids in rat tissues, *Am. J. Clin. Nutr.* **33**:2318.
Narayan, K. A., McMullen, J. J., Butler, D. P., Wakefield, T., and Calhoun, W. K., 1974, The influence of a high level of dietary corn oil on rat serum and liver lipids, *Nutr. Rep. Int.* **10**:25.
Naseem, S. M., Khan, M. A., Heald, F. P., and Nair, P. P., 1980, The influence of cholesterol and fat

in maternal diet of rats on the development of hepatic cholesterol metabolism in the offspring, *Atherosclerosis* **36**:1.
Nolen, G. A., Alexander, J. C., and Artman, N. R., 1967, Long-term rat feeding study with used frying fats, *J. Nutr.* **93**:337.
Peifer, J. J., Lundberg, W. O., Ishio, S., and Warmanen, E., 1965, Studies of the distributions of lipids in hypercholesteremic rats, *Arch. Biochem. Biophys.* **110**:270.
Peng, S.-K., Taylor, C. B., Tham, P., Werthessen, N. T., and Mikkelson, B., 1978, Effect of autooxidation products from cholesterol on aortic smooth muscle cells, *Arch. Pathol. Lab. Med.* **102**:57.
Picciano, M. F., and Perkins, E. G., 1977, Identification of the *trans* isomers of octadecenoic acid in human milk, *Lipids* **12**:407.
Porter, M. W., Yamanaka, W., Carlson, S. D., and Flynn, M. A., 1977, Effect of dietary egg on serum cholesterol and triglyceride of human males, *Am. J. Clin. Nutr.* **30**:490.
Reiser, R., and Raju, P. K., 1964, The inhibition of saturated fatty acid dehydrogenation by dietary fat containing sterculic and malvalic acids, *Biochem. Biophys. Res. Commun.* **17**:8.
Reiser, R., and Sidelman, Z., 1972, Control of serum cholesterol homeostasis by cholesterol in the milk of the suckling rat, *J. Nutr.* **102**:1009.
Reiser, R., O'Brien, B. C., Henderson, G. R., and Moore, R. W., 1979, Studies on a possible function for cholesterol in milk, *Nutr. Rep. Int.* **19**:835.
Renaud, S., 1977, Dietary fats and thrombosis, *Bibl. Nutr. Dieta* **25**:92.
Renaud, S., and Gautheron, P., 1975, Influence of dietary fats on atherosclerosis, coagulation and platelet phospholipids in rabbits, *Atherosclerosis* **21**:115.
Renaud, S., Kuba, K., Goulet, C., Lemire, Y., and Allard, C., 1970, Relationship between fatty-acid composition of platelets and platelet aggregation in rat and man: Relation to thrombosis, *Circulation Res.* **26**:553.
Rose, G., Blackburn, H., Keys, A., Taylor, H. L., Kannel, W. B., Paul, O., Reid, D. D., and Stamler, J., 1974, Colon cancer and blood-cholesterol, *Lancet* **1**:181.
Slater, G., Mead, J., Dhopeshwarkar, G., Robinson, S., and Alfin-Slater, R. B., 1976, Plasma cholesterol and triglycerides in men with added eggs in the diet, *Nutr. Rep. Int.* **14**:249.
Smith, L. L., and Van Lier, J. E., 1970, Sterol metabolism. Part 9. 26-Hydroxycholesterol levels in the human aorta, *Atherosclerosis* **12**:1.
Söderhjelm, L., Wiese, H. F., and Holman, R. T., 1970, The role of polyunsaturated acids in human nutrition and metabolism, in: *Progress in the Chemistry of Fats and Other Lipids* (R. T. Holman, ed.), Vol. 9, Part 4, Pergamon Press, Oxford, pp. 555–585.
Sreenivasan, B., 1978, Interesterification of fats, *J. Am. Oil Chem. Soc.* **55**:796.
Sturdevant, R. A. L., Pearce, M. L., and Dayton, S., 1973, Increased prevalence of cholelithiasis in men ingesting a serum-cholesterol-lowering diet, *New England J. Med.* **288**:24.
Sugai, M., Witting, L. A., Tsuchiyama, H., and Kummerow, F. A., 1962, The effect of heated fat on the carcinogenic activity of 2-acetylamino-fluorene, *Cancer Res.* **22**:510.
Swank, R. L., 1951, Changes in blood produced by a fat meal and by intravenous heparin, *Am. J. Physiol.* **164**:798.
Taylor, C. B., Peng, S-K., Werthessen, N. T., Tham, P., and Lee, K. T., 1979, Spontaneously occurring angiotoxic derivatives of cholesterol, *Am. J. Clin. Nutr.* **32**:40.
Vergroesen, A. J., 1972, Dietary fat and cardiovascular disease: possible modes of action of linoleic acid, *Proc. Nutr. Soc.* **31**:323.
Vles, R. O., 1975, Nutritional aspects of rapeseed oil, in: *The role of fats in human nutrition* (A. J. Vergroesen, ed.), pp. 433–477, Academic Press, New York.
von Lossonczy, T. O., Ruiter, A., Bronsgeest-Schoute, H. C., van Gent, C. M., and Hermus, R. J. J., 1978, The effect of a fish diet on serum lipids in healthy human subjects, *Am. J. Clin. Nutr.* **31**:1340.
Yano, K., Rhoads, G. G., Kagan, A., and Tillotson, J., 1978, Dietary intake and the risk of coronary heart disease in Japanese men living in Hawaii, *Am. J. Clin. Nutr.* **31**:1270.
Ziemlanski, S., Rosnowski, A., and Opuszyńska-Freyer, T., 1973, Ultrastructure of early fatty infiltration in the myocardium after administration of rapeseed oil in the diet, *Acta Med. Pol.* **14**:279.

Index

A 23187
  arachidonic acid derivatives and, 223
  mechanism of action of, 223
  SRS-A release and, 233
  structure of, 223
Abetalipoproteinemia, 268
Acetic acid, crystal structure of, 30
Acetone, stearic acid solubility in, 19
Acetonitrile, stearic acid solubility in, 19
Acetyl-CoA
  carboxylase, 127
  in fatty acid synthesis, 117, 119, 126
N-Acetylneuraminic acid (NANA): see Sialic acid
N-Acetylsphingosine, 434
ACTH
  fat mobilization and, 289
  prostaglandin antagonism of, 186
Adenylate cyclase, potentiation by prostaglandins, 290
Adipose tissue
  brown, 292
  function of, 292
  glycerophosphate in, 412
  triglycerides (triacylglycerols) in, 70
  white, 292
Adrenal cortex, cholesterol in, 296
Alcohol, fat absorption and, 265
Amphiphilic lipid(s), 369–403
  conformation in biomembranes, 395–401
  definition of, 369–370
  formation of monomolecular films of, 394–395
  interactions with water, 392–394
  isolation from tissue, 370
  physical properties of biomembrane, 390–401
  structures of, 371–390
Antioxidants, 92–93
  therapeutic uses of, 97
Apolipoproteins (apoproteins), 280
  metabolism of, 280
  physical properties of major, 280

Apoprotein A-I, as an activator of lecithin:cholesterol acyltransferase, 280
Apoprotein A-II, 280
Apoprotein B, 280
Apoprotein C-I, 280
Apoprotein C-II, 280
Apoprotein C-III, 280
Apoprotein D, 280
Apoprotein E, 280
Arachidonic acid, 140
  analogue of, 186
  autoxidation of, 96
  15-HPETE from, 230
  hydroperoxidation of, 230
  leukotrienes derived from, 240
  metabolism of, 192
  oxygenated derivatives of, 231
  platelet aggregation and, 193
  prostanoids and, 149, 160, 192–200
  SRS-A as derivative of, 233–242
  structure of, 80
2-Arachidonoyldipalmitoyl glycerol, crystal structure of, 47
Argentation chromatography, 69
Ascorbic acid, 95
Aspirin, prostaglandin endoperoxide synthetase inhibition by, 186
Autoxidation
  of arachidonic acid, 96
  of a diene, 85
  hydroperoxides formed during, 84
  of linoleic acid, 46, 84
  of oleic acid, 86
  rate of, 83–84
Avocado, fats in, 70, 460

Beef fat, fatty acids of, 72
Benzene, stearic acid solubility in, 19
BHA, 93
BHT, 93
Bile acids
  biosynthesis of, 354, 356

Bile acids (*cont.*)
  from cholesterol, 359
  hydroxylation reactions in biosynthesis of, 357
  intestinal modification of, 358
  primary, 356
  reabsorption of, 354
  role in fat digestion, 257
  secondary, 356
  structural correlation with cholesterol, 355
  triglyceride transport and, 266
Bile salts, function of, 259
Biotin, in fatty acid synthesis, 120
Bloor lipids, 5
Boltzmann equation, 36, 42
Brain
  free fatty acid uptake by, 276
  lipoprotein lipase in, 276
  prostaglandins in, 191
Breath test, for lipid peroxidation, 89
Butane
  conformational analysis of, 35–37
  infrared spectrum of, 39
Butane arrays, 35–37
  gauche, 39, 40, 41
  in stearic acid, 38
Butanol, stearic acid solubility in, 19
Butanone, stearic acid solubility in, 19
Butter
  fatty acids in, 72
  triglycerides in, 255

Calcium-mobilizing ionophore, *see* A 23187
Cancer
  antioxidants in, 97
  coronary heart disease and, 463
  dietary fat intake and breast, 464
Carbohydrates
  dietary, 290
  fatty acid biosynthesis and, 108
Carbon tetrachloride
  carboxylic acid association in, 29–30
  peroxidation and, 90
  peroxidation induced by, 96
  stearic acid solubility in, 19
Carboxyl group, dissociation of, 27
Carboxylic acids, 23–33; *see also* Fatty acids
  association, 27–32
    in carbon tetrachloride, 29–30
    in water, 31
  conformation of C1–C2 bond, 32
  dimer, 28
  hydrogen bonds in, 29
  dissociation of, 23–27
  hydrogen-bond donors/acceptors and, 30
  hydrogen exchange between molecules of, 28

Carboxylic acids (*cont.*)
  melting points, alternation of, 54
  molecular weights of, 27
  monomolecular films of, 55–57
  resonance hybridization and, 28
  solubility in water, 17
Carcinogenesis, polyunsaturated fatty acids and, 464
Cardiolipins, 383–384, 395
Carnitine, in mitochondrial fatty acid oxidation, 103
Carotenoids, 300
  acyclic, 301
  bicyclic, 301
Catecholamine, prostaglandin antagonism, 186
Cephalin: *see* Phosphatidylethanolamine
Ceramidase, 442
Ceramide(s), 386
  biosynthesis of, 433
  1-deoxy-1-phosphonic acid, 386
  1-deoxy-1-sulfonic acid, 377
  hexosides, 437–439
Cerasin, 371
Cerebron: *see* Phrenosin
Cerebroside, 389
  biosynthesis of, 435–436
  crystal structure of, 395–398
  solubility of, 20
  sphingosine in, 429
Cerebroside sulfate, 389
  biosynthesis of, 436, 437
Chelators, metal, 93
Chemiluminescence, prostaglandin endoperoxide synthetase and, 174
Chemotaxis
  leukotriene B_4 and, 244
  f-Met-Leu-Phe and, 244
  prostaglandins and, 217
Chicken, fatty acids in, 72
Chloroform, stearic acid solubility in, 19
Cholesterol
  absorption, 266-267
  acetate carbons in prenyl units of, 314, 315
  acetate as precursor of, 313–315
  autoxidation of, 326, 462
  in bile, 266
  in biomembranes, 394–395
  biosynthesis of, 295–367
    rate-limiting enzyme of, 324
    sequences of, 318
  chemical structure, 303
  in myelin sheaths, 296
  in daily diet, 460
  in dehydrated foods, 462
  dietary, 266

INDEX

Cholesterol (*cont.*)
  discovery of, 295
  5(6)-double bond in, 311
  function of, 394–395, 460
  high-density lipoprotein transport of, 282
  intestinal, 266
  lanosterol as intermediate in biosynthesis of, 315–316
  micelle uptake of, 267
  oxidation products of, 462
  pectin and absorption of, 267
  plant sterols and, 267
  squalene as intermediate in biosynthesis of, 315–316
  stereochemistry, 311
  structural correlation with bile acids, 355
  structure of, 296
  tissue distribution of, 296
  transformation to bile acids, 359
  in vertebrates, 296
Choline, from hydrolysates of sphingolipids, 429
  biosynthesis of, 415
  requirement by rat, 415
Chromatography, 69
    gas–liquid, 154
    reversed phase, 69
    silicic acid, 21
    SRS-A analysis via, 248
Chylomicrons
  composition of, 265, 273
  discharge from absorption cells, 265
  flotation rate of, 277
  formation of, 257
  function of, 276
  Golgi complex and formation of, 264
  particle size of, 261, 265, 273
  physicochemical properties of, 279
  in thoracic duct lymph, 257
  zinc deficiency and formation of, 265
Ciliatine, 381
Cirrhosis, biliary, 278
Citrate, as substrate for fatty acid biosynthesis, 118
Coconut oil, 71, 469
  triglycerides (triacylglycerols) in, 255
Colipase, 259
Compactin
  discovery of, 327
  effects on cultured cells, 328, 329
  as inhibitor of HMG-CoA reductase, 327
  structure of, 327
Configuration, activation energy required for changes in, 41
Corn, fat in, 460
Corn oil, fatty acids in, 71

Coronary heart disease
  cancer and, 463
  hypercholesterolemia and, 461
Cottonseed oil, 71
Critical micellar temperature, 58, 392
Crystallization, fatty acids and, 49–53
Cyclic AMP, 149
  prostaglandins and, 187
Cycloartenol, 307
  methyl groups in, 308
  stereochemistry of, 312
Cyclohexane
  fatty acid solubility in, 27
  stearic acid solubility in, 19
Cyclopropenoid fatty acids, 468
Cysteine, hydroperoxide reduction by, 86
Cytolipins, 389

Dahle hypothesis, 88
Desaturation, 133–147
  aerobic, 134–137
  anaerobic, 133–134
  in higher plants, 136–137
  mechanism of control of, 145
  rates, 142, 143
  temperature control in, 137–139
Diabetes mellitus, plasma-free fatty acids and, 108
Diacylglycerols, formation of, 287
Dihydrosphingosine, 430
Dipole-dipole interactions, 15
Diprenols, 298
Dispersion forces: *see* London forces
Dissociation constants
  of carboxylic acids, 24
  determination of, 25
Dithiothreitol, 423
DTPA, as antioxidant, 93

EDTA, as antioxidant, 93
Egg, fatty acids in, 72
Elaidic acid
  conformation of, 45
  crystal structure of, 52
Elaidinization, 42
Elaidoyl group, conformation of, 44
Electronegativity, 9
Electrostatic interactions, 9–11
Enthalpy, solubility and, 7
Entropy, solubility and, 7, 8
Epinephrine, fat mobilization and, 289
Ergosterol
  discovery of, 292
  structure of, 305
  trivial name for, 309

Essential fatty acids, 139
  deficiency disease, 144
  mechanism of action of, 144
Ethane, conformation of, 35
Ethanol
  peroxidation induced by, 96
  stearic acid, solubility in, 19
Ethyl acetate, stearic acid solubility in, 19
Euphol, 307

Fabry's disease, 453–455
Farnesyl diphosphate
  biosynthesis of, 324
  from isopentenyl diphosphate, 333–338
  as precursor of presqualene pyrophosphate, 341
  as precursor of squalene, 338
Farnesyl diphosphate synthetase, 334, 337
Fasting
  FFA levels and, 289
  lipoprotein lipase and, 275
  triacylglycerol (triglyceride) levels and, 277
Fat
  absorption, 257, 260
    alcohol and, 265
    in jejunum, 263
  calories stored as, 459
  cancer and dietary, 464
  dairy, 460
  deposition of, 288
  depots in mammals, 75
  dietary, 108, 459
    composition of, 255
  digestion of, 255–272
  in fish, 74
  in fruit, 70
  interesterified, 466
  intestinal absorption of, 263
  malabsorption of, 268
  polyunsaturated, 87
  rancidity of, 86
  saturated, 71, 72
  synthesis, inhibition of, 291
  thrombogenic properties of, 468–469
  unsaturated, 71, 72
Fat digestion
  bile acids in, 257
  pancreatic lipase and, 258
  partition hypothesis of, 256–257
Fat malabsorption, 268
Fatty acid(s), *see also* Carboxylic acids
  acidity of, 26
  α,β-unsaturated, dissociation of, 25–26
  autoxidation of, 83
  biohydrogenation of, 145

Fatty acid(s) (*cont.*)
  biosynthesis of, 108, 115–132
    control of, 126–128
    electron donor in, 116
    enzyme system in, 116
  catabolism of, 101–113
  composition in triglycerides, 264
  conformation of hydrocarbon chains in, 32
  conversion to coenzyme A derivatives, 103
  crystallizing power of, 49
  crystals, X-ray diffraction analysis of, 50
  cyclopropenoid, 468
  desaturation of, 133–147
  dimeric, crystallization of, 50
  eicosapolyenoic, 162, 218, 232
  essential, 139, 144
  in fish body fats, 74
  hydrocarbon moiety of, conformation of, 33–47
  hydroxy, dissociation of, 26
  in mammalian depot fat, 75
  methyl groups in crystals of, 55
  in microorganisms, 70
  mobilization of, 288, 289, 290
  monounsaturated, 83
  naming of, 77–81
  oxidation of, 104–112
    in liver, 291
  peroxidation, 83–99, 221, 222
  plasma-free, increases in, 108
  polyenoic, 139–145, 221, 222
  polymorphism in crystallization of, 53–54
  polyunsaturated, formation of, 139–145
    toxic substances from heated, 94
  properties of, 23
  reaction with oxygen, 83
  salts of, 57–68
  saturated, 83
  solubility in cyclohexane, 27
  structure of, 23, 77–81
  in tissue lipids, 69–81
  trans forms of, 464–466
  translocation of, 102–104
  unsaturated, 9, 83, 133
Fatty acid synthetase, structure of, 121–123
Fatty acyl chain, saturated, length of, 40
Fish, fat in, 460
Folch-Bligh-Dyer procedure, 20, 371
Follicle-stimulating hormone, prostaglandins and, 189
Formic acid, crystal structure of, 30
FPL 55712, as leukotriene antagonist, 234
  structure of, 234
Fractional crystallization, 69

Free fatty acids (FFA)
  fasting levels of, 289
  half-life of, 276
  in plasma, 108
  production of, 287
  starvation and levels of, 277

Gangliosides, 389–390, 440–441
  biosynthesis of, 437–439
  degradation of, 440–441
  in lipid storage diseases, 445–455
Gangliosidoses, 448, 450
Gaucher's disease, 389, 442, 445, 447–448
Geraniol, 298
Glanzmann's thrombasthenia, 202
Globoside, 389
Glucagon, prostaglandin antagonism of, 186
Glucocorticoids, fat mobilization and, 290
Glutathione, hydroperoxide reduction by, 86
Glyceroglycolipids, 384–386
  sulfate esters of, 385
Glycerol
  as backbone of phosphoglycerides, 431
  structural features of, 372
  tissue utilization of, 275
Glycerophospholipids, 377–384
Glyceryl ethers, structure of, 419
Growth hormone, fat mobilization and, 290

Halobacterium cutirubrum, lipids of, 373, 383, 385
Heart, free fatty acid uptake by, 276
Hematin, prostaglandin endoperoxide synthetase and, 175
n-Heptane, stearic acid solubility in, 19
Hermansky-Pudlak syndrome, 202
5-HETE, precursor, 225
High-density lipoproteins
  cholesterol transport and, 282
  electrophoretic analysis, 280
5-HPETE
  formation of, 224
    by lipoxygenase, 223
  reduction of, 224
15-HPETE, 230
Hydrocarbon moiety
  conformation in fatty acids, 33–47
    unsaturation and, 41–47
  of crystalline fatty acids, 54
  flexibility of saturated, 59
  hydrophobicity of, 56
  London forces, interaction between, 51
Hydrogen bonding, competitive characteristics of, 30
Hydrogen bonds, 12–13

Hydroperoxides
  dietary, 90
  formation of, 84
  of linoleate, 85
  as oxidizing agents, 86
  structure of, 84
Hydroperoxyeicosatetraenoic acid, *see* 5-HPETE and 15-HPETE
Hydrophilic groups, variability in structure of, 376–377
Hydrophobic bonding, 13
  in soap solutions, 59
Hydrophobic effect, 7
  dissociation and, 25
Hydrophobic groups
  micelle formation and, 64
  variability in structure of, 374–375
Hydroxyeicosatetraenoic acid, *see* 5-HETE and 15-HETE
3-Hydroxy-3-methylglutarate (HMG)
  chemical structure of, 317
  as a source of prenyl units, 317
HMG-CoA
  from leucine metabolism, 320
  synthesis of, 319
HMG-CoA lyase, 321–322
HMG-CoA reductase, 322
  in endoplasmic reticulum, 322
  inactivation of, 325
  inhibitors of, 326–330
  in liver, 324
  mechanism of action, 319
  molecular size, 329
  regulation of, 324
  stereospecificity, 319
  substrate for cytosolic, 322
Hydrolecithins, solubility of, 20
Hydroxydecanoyl-ACP desaturase, 133
Hypercholesterolemia, 278
  heart disease and, 461
Hyperchylomicronemia, lipoprotein lipase and, 276
Hyperglyceridemia, 278
Hypoglycin, 107

Indomethacin
  prostaglandin biosynthesis and, 184
  prostaglandin endoperoxide synthetase inhibition and, 186
"Inositol effect," 161
Intermediate-density lipoprotein, physicochemical properties, 279
Intermolecular forces, 9–16
  relative strengths of, 14–16

Iron
  in vitro peroxidation and, 95
  peroxidation and excess, 90
Isopentenyl diphosphate
  from mevalonate, 330–333
  structure of, 333
Isopentenyl diphosphate isomerase, 334

Krabbe's disease, 450–451
Krafft point, 58, 61, 67

Lamellae, 67–68
Lanosterol, 307
  biosynthesis of, 345–348
  conversion to cholesterol, 348–354
  cyclization of squalene to, 316
  as intermediate in cholesterol biosynthesis, 315–316
    methyl groups in, 308
    stereochemistry of, 311–312
    systemic name for, 308
Lard, fatty acids in, 72
Lecithin(s), 378, *see also* Phosphatidylcholines
  hydrolysates of, 429
Lecithin:cholesterol acyltransferase (LCAT)
  activation of, 280
  deficiency, 282
  trans fatty acids and activity of, 466
Leucine, HMG-CoA from metabolism of, 320
Leukodystrophy, metachromatic, 442, 451–453
Leukotriene(s)
  from arachidonic acid, 240
  discovery of, 217
  enzymatic formation of, 226, 227
  metabolism of, 243–244
  from polyenoic fatty acids, 222–232
  structure–function relations among, 249
Leukotriene A_4 (LTA_4), 235
Leukotriene B_4 (LTB_4),
  monohydroxyeicosatetraenoic acids and, 244
Leukotriene C_3 (LTC_3), 240
  structure of, 240
  synthesis of, 241
  trans-, 240
  UV spectrum, 241
Leukotriene C_4 (LTC_4)
  A 23187 inhibition of, 241
  cardiovascular effects of, 246
  LTD_4 from, 238
  vs. LTD_4, 246
  pulmonary pressure and, 247
  synthesis of, 237
  UV spectrum, 235

Leukotriene C_5 (LTC_5)
  vs. LTC_4, 240
  structure of, 240
  synthesis of, 241
Leukotriene D_4 (LTD_4)
  cardiovascular effects, 246
  vs. LTC_4, 246
  synthesis of, 236, 237
Leukotriene E_4 (LTE_4)
  chromatographic analysis, 239
  synthesis of, 237
Lewis bases, 12
Linoleate
  autoxidation of, 84
  hydroperoxides, 85
Linoleic acid
  autoxidation of, 91
  conversion by potato lipoxygenase, 220
  polyunsaturated fatty acids from, 149
  transformation of, 141
Linoleoyl group conformation, 45
Lipamino acids, 380
Lipase
  gastric, 257
  hormone-sensitive, 290
  lingual, 257
  lipoprotein, 275
  pancreatic, 257, 258
Lipid(s)
  age and clearance of, 277
  amphiphilic, 369–403
  Bloor, 5
  chromatographic methods for extraction of, 23
  classification of, 2–3
  defined, 5
  digestion and, 255–272
  metabolism, enzymes of, 443–445
  in muscle, 70
  in nervous tissue, 70
  oleoyl groups in, 43
  serum levels, 277
  solubility of, 5–22
  pH and, 6
Lipid peroxidation, 83–99
  breath test for, 89
  chemistry of, 83–87
  in living organisms, 93–99
  measurement of, 87–90
  in model membranes, 90–91
  NADPH-dependent, 95
  termination of, 92
  in tissues, 88

INDEX

Lipoprotein(s)
  high-density, 279, 280, 282
  lipase, 275
  metabolism of blood, 280
  synthesis of, 264–265
  triacylglycerol-rich, 280
  very-low-density, 277
Lipoprotein lipase (LPL)
  in brain, 276
  fasting and, 275
  hyperchylomicronemia and activity of, 276
  production of, 287
Liposome, unilamellar, 91
Lipoxygenase(s)
  conversion of linoleic acid by potato, 220
  generation of 5-HPETE by, 223
  inhibition of, 230
  mechanism of action, 221, 222
  molecular weights of, 219
  of platelets, 193
  in polymorphonuclear leukocytes, 222
  potato tuber, 218, 219, 220, 221
  soybean, 218, 219, 220, 221
  substrates for, 218
Liver
  cholesterol in, 296
  fatty acid oxidation in, 291
  fatty acid uptake, 276
London forces, 11–12
  qualitative comparison with electrostatic forces, 12
  between hydrocarbon moieties, 51
  between methylene groups and water, 56
Long-chain bases (LCBs), *see* Sphingoids
Low-density lipoproteins (LDL)
  physicochemical properties of, 279
  receptor pathway, regulation of cholesterol biosynthesis via, 352–354
Lowry-Brønsted acids, 12
Luteinizing hormone (LH), prostaglandins and, 189
Lyso amphiphilic lipids, 386

Malabsorption syndrome, 268
Malondialdehyde
  formation of, 88–89
  "Schiff base" reaction, 89
Margaric acid, melting point of, 54
Marine oil, 467–468
Mass spectrometry
  in analysis of squalene biosynthesis, 340
  prostaglandins, 153

Melting point
  alternating, 55
  defined, 54
Metachromatic leukodystrophy, 442, 451–453
Methane, dissolution in water, 8
Methanol, stearic acid solubility in, 19
Methyl groups, in crystals of fatty acids, 55
Methyl linoleate, reaction with oxygen, 84
Methyl oleate, oxygen absorption by, 83
Methylene chloride, stearic acid solubility in, 19
Methylene groups, London forces between water and, 56
Mevaldate reductase, 324
  reaction catalyzed by, 323
Mevaldic acid, reduction to mevalonate, 323
Mevalonate 5-diphosphate, conversion to squalene, 332
Mevalonate 5-diphosphate decarboxylase, 333
Mevalonate kinase
  activation of, 331
  specificity of, 331
Mevalonic acid, *see also* Mevalonolactone
  conversion to isopentenyl diphosphate, 330–333
  discovery of, 318
  formation of, 319–324
  structure of, 317
Mevalonolactone, *see also* Mevalonic acid
  chemical structure of, 317
  compared with mevinolin, 327–328
Mevinolin
  chemical structure of, 327
  discovery of, 327
  effects on cultured cells, 328, 329
  effects on HMG-CoA reductase, 327–328
Micelle(s), 60–66
  carboxyl groups of, 62
  composition of, 260
  formation of, 63–64
  model of, 65–66
  of potassium laurate, 62
  size and shape of, 62
  spherical, 66
  stearate, 66
Microsomes, prostaglandin biosynthesis and, 168
Microvilli, lipid-protein membrane of, 263
Milk
  fatty acids in, 72, 73, 76, 77
  triacylglycerols in, 255
Monoacylglycerol, 288
  sphingosine and, 431
Muscle, lipids of, 70
Myelin figures, 393

Myocardial infarction, hypercholesterolemia and, 461
Myristoyl group, gauche shortening of, 41

Nerol, 298
Nerolidol, 298
Nervone, 388
Nervous tissue, lipids in, 70
Neurotransmitters, prostaglandins and, 192
Niemann-Pick disease, 442, 446–447
Nitrogen dioxide, peroxidation induced by, 96
Nordihydroguaiaretic acid, 93
Norepinephrine, fat mobilization and, 289
Nuclear magnetic resonance(NMR), in analysis of squalene biosynthesis, 340
Nuts, fats in, 459

Oleates, in fruit fats, 70
Olefins, cyclization of, 346
Oleic acid, 30
  autoxidation of, 86
  conformational analysis of, 45
  crystal structure of, 51–52
  from hydrolysis products of natural lipids, 41
  hydroperoxides, 84
  melting point of crystalline, 51
  prevalence of, 73
  trans isomer of, 52
Oleoyl group, 43
  anticoplanar conformation of, 46
Olive oil, 71
  fats in, 70
Oxidation, 92
  alpha, 110–112
  beta, 104–108
  omega, 109–110
2,3-Oxidosqualene lanosterol cyclase, 348
Oxygen
  absorption by methyl oleate, 83
  fatty acid reaction with, 83
  hyperbaric, 97
  in vitro peroxidation and, 95
Oxynervone, 388
Ozone
  peroxidation and exposure to, 90
  peroxidation induced by, 96

Palm oil, 469
Palmitates, 124
  in fruit fats, 70
  production of, 129
  synthesis of, 76
Palmitic acid
  melting point, 54

Palmitic acid (*cont.*)
  sphingosine biosynthesis and, 431
  stearic acid and, 53
Palmitoyl group, gauche shortening of, 41
Pancreatic lipase, 257, 258
  reaction catalyzed by, 258
  specificity, 259
Peanut oil, 71
Pectin, cholesterol absorption and, 267
Peroxidation, 83–99
  carbon tetrachloride induced by, 96
  enzymatically induced, 95
  ethanol-induced, 96
  in vitro, 94
  nitrogen dioxide-induced, 96
  ozone-induced, 96
Phagocytosis, prostaglandins and, 217
Phosphatases, reactivation of HMG-CoA reductase by, 325–326
Phosphatidalethanolamines, *see* Plasmenylethanolamines
Phosphatidic acids, 375, 378
  esters of, 378
  formation of, 408
Phosphatidylcarnitine, 379
Phosphatidylcholine, 282, 378, *see also* Lecithin(s)
  biosynthesis of, 409, 410
  dipalmitoyl, 379, 394
  in omega oxidation, 110
  phosphorolysis of, 408
  sonification of an unsaturated, 91
  as source of prostaglandin precursors, 161
  stearoyl moiety of molecule of, 40
  structural analogy to sphingomyelin, 434
  triglyceride (triacylglycerol) transport and, 266
Phosphatidylethanolamine, 379
  N-α-carboxyethyl, 380
  crystal structure of, 395–398
  N-fatty-acylated, 380
  N-mono- and N,N-dimethyl, 380
  phosphono analogs of, 380
  source of prostaglandin precursors, 161
Phosphatidylglycerol(s)
  α-aminoacyl, 382
  biosynthesis of, 383
  glucosaminides of, 382
  hydrophilicity of, 382
  3′-phosphates (PGPs), 383
  structures of derivatives of, 383
Phosphatidylinositols, 381
  biosynthesis of, 415
  hydrophilicity of, 382

Phosphatidylinositols (*cont.*)
  mannosides of, 382
  phospho (4'-mono- and 4',5'-di-), 382
Phosphatidyl-β-methylcholine, 379
Phosphatidyl-L-serine, solubility of, 20
  structure of, 381
Phosphatidyl-L-threonine, 381
Phosphocholinetransferase, 423
Phosphoglycerides
  biosynthesis of, 408–419
  glycerol as backbone of, 431
  metabolism of, 405–428
Phospholipases, 405–408
  mechanism of action, 405–406
  phospholipase A_1, 405–406
  phospholipase A_2, 405–407
  phospholipase B, 407
  phospholipase C, 407
  phospholipase D, 407
  prostaglandin release and, 161
Phospholipase: acyltransferase, 418
Phospholipids
  absorption of, 266
  biosynthesis of, 418–426
    control of, 426
  in chylomicrons, 273
  solubility of, 20
  structure of, 70
  synthesis of, 266
Phospholipid exchange proteins, 418
Phosphomevalonate kinase, 332
Phrenosin, 371
Phytanic acid, 112
Phytoene, 300
Phytosphingoids, 376
Plant sterols, 308
  cholesterol and, 267
Plasmalogens, hydrolysis of, 374
Plasmanic acids, 375, 378
Plasmenic acids, 375, 378
Plasmenylethanolamines, 375, 380
Plasminogen activator, 189
Platelet-activating factor, 420
  synthesis of, 424
Platelet aggregation
  arachidonic acid and, 193
  collagen-induced, 201
  prostacyclin and, 203
  prostaglandins and, 185
  and thromboxane A_2, 191, 200
Plexaura homomalla as source of prostaglandins, 158

Polarity, solubility and, 16–19
  of solvents and solutes, 6
Polarized covalent bonds, 11
Polycyclic aromatic hydrocarbons, 464
Polyisoprenoid compounds, 297–302
  acyclic, 301, 302
  chemical structure of, 297–298
  in naturally occurring substances, 301–302
Polymorphism, in fatty acid crystallization, 53–54
Polyunsaturated fatty acids, 87, 463–464
  determinations of, 140
  formation, in animals, 139–145
  polycyclic aromatic hydrocarbons and, 464
  for prostaglandin synthesis, 160–162
  role of, 149
  in tissue lipids, 143
  toxic substances from, 94
  transformations of, 141
Poriferasterol, systemic name for, 309
Prenyltransferase, 334
  mechanism of reactions of, 337
Propane, 33, 34
2-Propanol, stearic acid solubility in, 19
Propionic acid, vapor density of, 29
Prostacyclin, 150, 196–200
  analogue, 199
  biosynthesis of, 198, 199
  body distribution of, 200
  discovery of, 196–197, 200
  free acid, 199
  interaction with thromboxane A_2, 200–203
  metabolism, 205, 206
  platelet aggregation and, 203
  structure of, 197
  synthetase, 207
  therapeutic use of, 203
Prostaglandin(s), 96, 149
  ACTH and, 186
  biological effects, 189–192
  biosynthesis of, 150, 160–180
    degradation products of, 163, 164
    enzymes of, 167–180
    indomethacin and, 184
    mechanism of, 162–167
    microsomes and, 168
  in brain, 191
  catecholamines and, 186
  in cerebrospinal fluid, 191
  chemotaxis and, 217
  cyclic AMP and, 187
  "cytoprotective" effect, 191
  dehydration and isomerization products of, 185

Prostaglandin(s) (cont.)
  fat mobilization and, 289
  FSH and, 189
  in gastrointestinal tract, 190–191
  glucagon and, 186
  as local hormones, 149
  luteinizing hormone and, 189
  mechanism of action of, 186–188
  metabolism, 180–186
    products formed during, 184
    species differences in, 183
  neurotransmitters and, 192
  occurrence of, 159–160
  phagocytosis and, 217
  phospholipase and the release of, 161
  platelet aggregation and, 185
  potentiation of adenyl cyclase by, 290
  precursors of, 150, 160, 218
  quantitative analysis of, 160
  receptors, 186
  in semen, 189–190
  structural analogues in marine life, 158
  structural identification of, 153–158
  structure and chemistry of, 151–153
  synthetase system, 167, 168
  as thermoregulators, 191
  thromboxane A_2 and, 196
  TSH and, 186
  vascular effects of, 189
  vasopressin and, 187
Prostaglandin dehydrogenase, 181
Prostaglandin endoperoxide, analogues of, 186
Prostaglandin endoperoxide:E isomerases, 177–179
Prostaglandin endoperoxide:F_2 reductase, 179–180
Prostaglandin endoperoxide synthetase
  assays, 177
  chemiluminescence and, 174
  functions of, 168
  hematin activation of, 175
  inhibitors of, 186
  mechanism of action, 169–171
  purification of, 168
  self-catalyzed inactivation of, 175, 176
Prostaglandin isomerases, 177–179
Prostaglandin peroxidase, role of tryptophan in activation of, 174
Prostaglandin reductases, 179–180
Prostanoic acid, 151, 152
Prostanoid(s), 149, 207
Protons, apparent mobility of, in ice, 29
Psychosines, 387
Pyruvate, as a source of carbons for fatty acid biosynthesis, 126

Radioimmunoassay, 160
  for SRS-A determination, 248
Rapeseed oil, 467
Refsum's disease, 112
Resonance hybridization, carboxylic acid and, 28
Robinson's hypothesis, 315
Rod-type micelles, in soap solutions, 66

Sandhoff's disease, 449–450
Serine, sphingosine biosynthesis and, 431
Sialic acid, 389–390
Silica gel, adsorption of unsaturated fatty acids on, 91
Skeletal muscle, free fatty acid uptake, 276
Skin, cholesterol in, 296
Slow-reacting substances of anaphylaxis, see SRS-A
Soaps
  in aqueous solution, 57–68
  crystal structure of, 67
  "middle," 66
  solutions, micellar, 60–68
  surface concentration of, 58
Solanesol, 299
Solubility
  enthalpy and, 7
  entropy and, 7, 8
  polarity and, 16–19
Soybean oil, 71
Spadicol, 299
Spectrophotometric analysis, 87
Sphinganine, see Dihydrosphingosine
Sphingenine, see Sphingosine
Sphingoids
  phyto, 376
  structure of, 372, 375–376
Sphingolipids
  degradation of, 439–443
  hydrolysis of, 373
  less complex, 441–442
  metabolism, 429–457
    disorders of, 445–455
  structure of, 70, 386–389
Sphingomyelin
  of biomembranes, 387
  biosynthesis of, 434–435
  ethanolamine analogues of, 387
  hydrolysates of, 429
  solubility of, 20
  structural analogy to phosphatidylcholine, 434
  structure of, 387
Sphingomyelinase, 442
  deficiency of, 446
Sphingophospholipids, 386–388

Sphingophosphonolipids, 381, 386–388
Sphingosine
  biosynthesis of, 431–433
    in vitro, 433
  in cerebrosides, 429
  chemical structure of, 430
  degradation of, 442
  in gangliosides, 430
  monoglycerides and, 431
Squalene, 300
  biosynthesis of, 318, 324, 338–345
    intermediary stages of, 330
    in a liver enzyme system, 339
    mass spectrometry in, 340
    NMR in, 340
  cyclization of, 318
  cyclization to lanosterol, 345–348
  cyclization products of oxide of, 307
  as intermediate in cholesterol biosynthesis, 315–316
Squalene epoxidase, 348
Squalene synthetase, 338, 344
SRS-A, 217
  A 23187 and release of, 233
  antagonist, 234
  as arachidonic acid derivative, 233–242
  assays, 232
  biological activity of, 246–248
  chemical structure of, 233
  discovery of, 218, 232
  in vitro effects, 247
  mechanism of action of, 247
  metabolism of, 243–244
  precursors of, 218
  solubility of, 232
  synthesis of, 239
  UV absorption spectrum of, 234
Stearic acid
  conformation of, 35, 38, 45
  melting point of, 54
  palmitic acid and, 53
  solubility in common solvents, 19
Stearoyl moiety, gauche shortening of, 41
Steatorrhea, 257, 268, 269
Steroids, planar representation of structures of, 309
Sterols, 295–367
  absolute configuration at C-20 of natural, 306
  nomenclature of, 303–312
  oxygenated, 326, 327
  planar representation of structure of, 309
  plant, 308
  with saturated side-chain, 308
  stereochemistry of, 303–312
  trimethyl, 307

Sulfatides: see Cerebroside sulfate
Sulfolecithin, 379
Sulfonolipids, 385
Superoxide dismutase, 95
Synthetases, mitochondrial as compared with cytosolic, 320

Tay-Sachs disease, 445, 448–449
Terpenoids, see Polyisoprenoid compounds
"Theorell" enzyme, 219
Thermogenesis, 292, 293
Thiokinase, 104
Thrombanoic acid, 194, 204
Thromboxane A_2, 150
  formation of, 193–194
  half-life of, 194
  hemorrhage and, 191
  hydrolysis of, 203
  interaction with prostacyclin, 200–203
  metabolism of, 203
  prostaglandins and, 196
  synthetase, 203
Thromboxane B_2, 203
  discovery of, 194, 198
  metabolites of, 204, 205
Thyroid-stimulating hormone (TSH)
  fat mobilization and, 289
  prostaglandin antagonism of, 186
Tirucallol, 307
Toluene, stearic acid solubility in, 19
Triacylglycerols, see also Triglycerides
  biosynthesis of, 285–287
  chylomicron, 274
  deposition and mobilization of, 287
  fasting levels of, 277
  hydrolysis, 275, 287–290
  medium-chain, 469–470
  metabolism, 285
Trichloroethylene, stearic acid solubility in, 19
Triglycerides, see also Triacylglycerols
  in adipose tissue, 70
  bile acids and transport of, 266
  in butter, 255
  in coconut oil, 255
  fatty acid composition of, 264
  formation of, 263
  as major constituent of dietary fat, 255
  medium-chain, 268
  in milk, 255
  solubility of, 255
Triparanol, 352
Triprenol, 298
Tryptophan, in activation of prostaglandin peroxidase, 174

Unsaturated fatty acids, 255
　adsorption on silica gel, 91
　occurrence of, 133
Unsaturated fatty acyl groups, conformation of hydrocarbon moiety of, 41–47

Vegetable oils, 70, 469
　fatty acids in, 71
Very-low-density lipoproteins (VLDL), 277
　lipid composition of, 280
　lipoprotein physicochemical properties of, 279
Vitamin A, precursor of, 301
Vitamin E, peroxidation and deficiency of, 90

Water
　hydrogen-bonding characteristics of, 8
　stearic acid, solubility in, 19

X-ray diffraction analysis
　of a cerebroside, 395–398
　of elaidic acid, 52
　fatty acid crystals, 50
　of hydrocarbon, 39
　of oleic acid, 51–52
　of a phosphatidylethanolamine, 395–398

Zinc deficiency, chylomicron formation and, 265